An Introduction to Behavioural Ecology

COMPANION WEBSITE

This book is accompanied by a companion website:

www.wiley.com/go/davies/behaviouralecology

With figures and tables from the book for downloading

An Introduction to Behavioural Ecology

Nicholas B. Davies, John R. Krebs and Stuart A. West

FOURTH EDITION

WILEY-BLACKWELL
A John Wiley & Sons, Ltd., Publication

This edition first published 2012

Blackwell Publishing was acquired by John Wiley & Sons in February 2007. Blackwell's publishing program has been merged with Wiley's global Scientific, Technical and Medical business to form Wiley-Blackwell.

Registered Office
John Wiley & Sons, Ltd, The Atrium, Southern Gate, Chichester, West Sussex, PO19 8SQ, UK

Editorial Offices
9600 Garsington Road, Oxford, OX4 2DQ, UK
The Atrium, Southern Gate, Chichester, West Sussex, PO19 8SQ, UK
111 River Street, Hoboken, NJ 07030-5774, USA

For details of our global editorial offices, for customer services and for information about how to apply for permission to reuse the copyright material in this book please see our website at www.wiley.com/wiley-blackwell.

Library of Congress Cataloging-in-Publication data has been applied for

ISBN 978-1-4443-3949-9 (hardback) – 978-1-4051-1416-5 (paperback)

A catalogue record for this book is available from the British Library.

Wiley also publishes its books in a variety of electronic formats. Some content that appears in print may not be available in electronic books.

Set in 10/12pt Photina by SPi Publisher Services, Pondicherry, India

1 2012

Contents

Photo © Oliver Krüger

COMPANION WEBSITE

This book is accompanied by a companion website:

www.wiley.com/go/davies/behaviouralecology

With figures and tables from the book for downloading

Photo © osf.co.uk.

Preface

In this Preface, we summarise the history and organisation of this book. The previous editions, by John Krebs and Nick Davies (1981, 1987, 1993) celebrated the early years of behavioural ecology. Our aim was to understand how behaviour evolves in the natural world. This requires links between studies of behaviour, evolution and ecology. The link with evolution is central because we expect natural selection to favour those behaviour patterns which maximise an individual's chances of surviving and passing copies of its genes on to future generations. The link with ecology comes in because ecology sets the stage on which individuals play their behaviour, so the best way to behave depends on ecological selection pressures, such as the distribution in space and time of food, enemies and places to live. The social environment will be important too, because individuals will often have to compete for scarce resources. So we need to consider how behaviour evolves when there are social interactions, with the potential for both conflict and cooperation.

This new edition celebrates a maturing and flourishing field, with exciting new links being forged with other disciplines. We now have three co-authors. John was an adviser for Nick's doctorate research in Oxford and Nick, in turn, lectured to Stu in undergraduate courses in Cambridge. So we span three (short!) academic generations and we have all enjoyed learning from each other during the preparation of this book. All the chapters have been heavily revised or completely reorganised and re-written to incorporate the many new ideas and examples which have emerged since the last edition - in some cases overturning what used to be the conventional wisdom. The central themes remain: a reductionist approach to consider the costs and benefits of decision making and how trade-offs are resolved by selection; a "gene's eye" view of behaviour; and a game theoretic approach to analyse the resolution of conflicts of interest. John and Nick remember the early days when Bill Hamilton, Robert Trivers and John Maynard Smith began to explore the ideas of kin selection, family conflicts and game theory for the analysis of conflict resolution, and when Richard Dawkins was testing drafts of *The Selfish Gene* as undergraduate lectures in Oxford. It is exciting to see how these ideas have evolved to stimulate the new research we discuss here. Throughout the book, we emphasise the theoretical background, but we prefer to develop the theory with examples rather than with abstract arguments. Some of the more complicated arguments are presented in boxes.

Chapter 1 begins with "watching and wondering" as we introduce how to frame different kinds of questions about behaviour. We then describe field experiments on clutch size, which show that individuals tend to maximise their lifetime reproductive success. Because an individual's survival and reproductive success depends critically on its behaviour, selection is expected to design individuals to be efficient at feeding, avoiding predators, finding mates, and so on.

In Chapter 2 we discuss how to test hypotheses for the adaptive advantage of behaviour. One method is comparison among species, in effect analysing the results of evolutionary "experiments", correlating differences between species in behaviour with differences in ecological and social selection pressures. There have been recent improvements in methodology, using phylogenies to identify independent evolutionary transitions and the order in which traits change. The second method, pioneered by Niko Tinbergen, is to perform experiments, for example to change behaviour and measure the consequences for an individual's survival and reproductive success.

In Chapter 3, we focus on individual "decision making" between alternative courses of action. We show how optimality models can be used to predict decision rules, and how the same basic models can often be applied to what at first sight seem very different problems, such as feeding and searching for mates. We discuss the roles of social learning and teaching in the development of individual decision making, and use examples from food storing to explore the links between behavioural ecology, cognition and neuroscience. Chapter 4 considers decisions over evolutionary time, and how these change during arms races between predators and prey, and brood parasites and hosts.

The next two chapters consider how individuals should behave when they have to compete with others for scarce resources. Chapter 5 introduces a game theoretic approach to contest behaviour and shows that the outcome is often variability in the population, as individuals distribute across different habitats in space and time, or choose alternative strategies or tactics as they compete for food and mates. We also discuss the concept of animal personalities, a current growing field of research. In Chapter 6, we review the costs and benefits of group living, particularly in relation to foraging and avoiding predation. Recent studies have shown how local decision rules made by individuals can have remarkable consequences for group dynamics, leading to spectacular coordinated movements in bird flocks, fish shoals and ant trails.

The next four chapters are concerned with sexual reproduction. Chapter 7 shows that fundamental differences between the sexes in gamete size and parental investment often leads to males competing for females, either by force or by charm (Darwin's theory of sexual selection). Females may chose males based on the resources they provide or the genetic benefits for their offspring. There is often sexual conflict and this continues after mating (sperm competition and female choice of sperm). Chapter 8 reviews parental care across the animal kingdom, and three inter-related conflicts: between male and female parent over who should care, and how much care to provide; between siblings; and between parents and their offspring. We consider the theory and evidence for each of these conflicts and distinguish between "battleground" models (which define the conflict) and "resolution" models (which explore the outcomes).

In Chapter 9, we show how different mating systems emerge depending on the economics of parental care and mate defence. The use of DNA profiles to measure parentage has now become routine and has revolutionised our view of family life, revealing (for example) that social monogamy does not necessarily mean genetic monogamy. Chapter 10 examines sex allocation: the problem of how a parent should divide its investment between male and female offspring. Sex allocation in hymenopteran insects provides some of the most convincing quantitative tests of evolutionary theory.

In the final four chapters, we turn our attention to social behaviour. Under what circumstances would we expect the evolution of altruism, namely helping others to reproduce at the expense of one's own reproductive output? (Chapter 11). When would it pay individuals to cooperate with either related or unrelated individuals? (Chapter 12). We show how social theory can be tested in a variety of animals from microbes to meerkats. Chapter 12 is devoted to the social insects, where altruism reaches its most sophisticated development in the form of sterile worker castes. We discuss new theory for the genetic predispositions and ecological factors promoting this remarkable behaviour, and show that there are often conflicts of interest even within the most cooperative societies. In Chapter 14, we discuss how natural selection shapes signals, focussing on the evolution of honesty and deception.

Finally, in the last chapter (15) we return for a critical re-assessment of some of our main premises: the "gene's eye" view of behaviour, optimality models and evolutionarily stable strategies. We also point to the flourishing interactions with other fields of research.

The literature has become vast since the earlier editions of this book, so in this edition we have had to be even more selective. We hope that lecturers using this book will add their own favourite examples to those we have mentioned. Throughout, we have tried to point to gaps in current theory and evidence. We hope that readers will be inspired not only to fill these, but also to discover new problems to solve.

Acknowledgements

Photo © Elizabeth Tibbetts.

We thank the editorial staff at Wiley-Blackwell for all their help and encouragement, especially Ward Cooper, Kelvin Matthews and Delia Sandford. Robert Campbell was instrumental in encouraging us in the earlier editions. For help with various chapters, we thank: Joao Alpedrinha, Staffan Andersson, Tim Birkhead, Koos Boomsma, Lucy Browning, Max Burton, Tim Clutton-Brock, Bernie Crespi, Emmett Duffy, Claire El Mouden, Andy Gardner, Ashleigh Griffin, James Higham, Camilla Hinde, Rebecca Kilner, Loeske Kruuk, Carita Lindstedt, Robert Magrath, Allen Moore, Nick Mundy, Hazel Nichols, David Reby, Thom Scott-Phillips, Ben Sheldon, Martin Stevens, Claire Spottiswoode, Mary Caswell (Cassie) Stoddard, Joan Strassmann, Alex Thornton, Rose Thorogood.

And a special thank you to Ann Jeffrey for her truly heroic help with the preparation of the manuscript.

Finally, we thank all our colleagues who so generously provided figures and photographs for the book, and particularly Oliver Krüger for his magnificent cover photograph of a parent Adelie Penguin, introducing its chick to the delights of behavioural ecology.

CHAPTER 1

Natural Selection, Ecology and Behaviour

Photo © Craig Packer

Watching and wondering

Imagine you are watching a bird searching in the grass for food (Fig. 1.1). At first your curiosity may be satisfied simply by knowing what species it is, in this case a starling *Sturnus vulgaris*. You then watch more closely; the starling walks along and pauses every now and then to probe into the ground. Sometimes it finds a prey item, such as a beetle larva, and eventually, when it has collected several prey items, it flies back to the nest to feed its hungry brood.

For students of behavioural ecology, a whole host of questions comes to mind as this behaviour is observed. The first set of questions concerns how the bird feeds. Why has it chosen this particular place to forage? Why is it alone rather than in a flock? Does it collect every item of food it encounters or is it selective for prey type or size? What influences its decision to stop collecting and fly back to feed its chicks?

Asking questions

Another set of questions emerges when we follow the starling back to its nest. Why has it chosen this site? Why this number of chicks in the nest? How do the two adults decide on how much food each should bring? Are these two adults the mother and father of all the chicks? Why are the chicks begging so noisily and jostling to be fed? Surely this would attract predators to the nest. If we could follow our starlings over a longer period, we may then begin to ask about what determines how much effort the adults put into reproduction versus their own maintenance, about the factors influencing the timing of their seasonal activities, their choice of mate, the dispersal of their offspring and so on.

Behavioural ecology provides a framework for answering these kinds of questions. In this chapter we will show how it combines thinking about behaviour, ecology (the 'stage' on which individuals play their behavioural strategies) and evolution

An Introduction to Behavioural Ecology, Fourth Edition. Nicholas B. Davies, John R. Krebs and Stuart A. West.

Fig. 1.1 A foraging starling. Photo © iStockphoto.com/ Dmitry Maslov

(how behaviour evolves by natural selection). But first, we need to be clear about exactly what we mean when we ask the question 'why?'

Tinbergen's four 'why' questions

Niko Tinbergen (1963), one of the founders of scientific studies of animal behaviour in the wild, emphasized that there are four different ways of answering 'why' questions about behaviour. For example, if we asked why male starlings sing in the spring, we could answer as follows:

(1) In terms of *causation*. Starlings sing because the increasing length of day triggers changes in their hormones, or because of the way air flows through the vocal apparatus and sets up membrane vibrations. These are answers about the mechanisms that cause starlings to sing, including sensory and nervous systems, hormonal mechanisms and skeletal–muscular control.

(2) In terms of *development or ontogeny*. For example, starlings sing because they have learned the songs from their parents and neighbours, and have a genetic disposition to learn the song of their own species. This answer is concerned with genetic and developmental mechanisms.

(3) In terms of *adaptive advantage or function*. Starlings sing to attract mates for breeding, and so singing increases the reproductive success of males.

(4) In terms of *evolutionary history or phylogeny*. This answer would be about how song had evolved in starlings from their avian ancestors. The most primitive living birds make very simple sounds, so it is reasonable to assume that the complex songs of starlings and other song birds have evolved from simpler ancestral calls.

Proximate versus ultimate explanations

Causal and developmental factors are referred to as *proximate* because they explain how a given individual comes to behave in a particular way during its lifetime. Factors influencing adaptive advantage and evolution are called *ultimate* because they explain why and how the individual has evolved the behaviour. To make the distinction clearer, an example is discussed in detail.

Reproductive behaviour in lions

In the Serengeti National Park, Tanzania, lions (*Panthera leo*) live in prides consisting of between three and twelve adult females, from one to six adult males and several cubs (Fig. 1.2a). The group defends a territory in which it hunts for prey, especially gazelle and zebra. Within a pride all the females are related; they are sisters, mothers and daughters, cousins and so on. All were born and reared in the pride and all stay there to breed. Females reproduce from the age of four to eighteen years and so enjoy a long reproductive life.

For the males, life is very different. When they are three years old, young related males (sometimes brothers) leave their natal pride. After a couple of years as nomads they attempt to take over another pride from old and weak males. After a successful takeover

Fig. 1.2 (a) A lion pride. (i) The females are returning to the middle of their territory after chasing away a neighbouring pride. (ii) Females and cub. (iii) Males patrolling the territory and (iv) relaxing. (v) Male with cub. Photos © Craig Packer (b) Infanticide: a male that has just taken over ownership of a pride, with a cub in his jaws that he has killed. Photo © Tim Caro

they stay in the pride for two to three years before they, in turn, are driven out by new males. A male's reproductive life is therefore short.

The lion pride thus consists of a permanent group of closely related females and a smaller group of separately interrelated males present for a shorter time. Brian Bertram (1975) considered two interesting observations about reproductive behaviour in a pride.

Female lions show synchronous oestrus

(1) Lions may breed throughout the year but, although different prides may breed at different times, within a pride all the females tend to come into oestrus at about the same time. The mechanism, or causal explanation, is likely to be the influence of pheromones on oestrus cycles (Stern & McClintock, 1998). But why are lionesses designed to respond in this way? One adaptive advantage of oestrus synchrony is that different litters in the pride are born at the same time and cubs born synchronously survive better. This is because there is communal suckling and, with all the females lactating together, a cub may suckle from another female if its mother is out hunting. In addition, with synchronous births there is a greater chance that a young male will have a similar-aged companion when it reaches the age at which it leaves the pride. With a companion a male is more likely to achieve a successful take-over of another pride (Bygott *et al.* 1979; Packer *et al.* 1991).

Males kill cubs after take-over

(2) When a new male, or group of males, takes over a pride they sometimes kill the cubs already present (Fig. 1.2b). The causal explanation is not known but it may be the unfamiliar odour of the cubs that induces the male to attack them. But, whatever the mechanism, why are male lions designed to respond in this way?

The benefit of infanticide for the male that takes over the pride is that killing the cubs fathered by a previous male brings the female into reproductive condition again much more quickly. This hastens the day that he can father his own offspring. If the cubs were left intact then the female would not come into oestrus again for 25 months. By killing the cubs the male makes her ready for mating after only nine months. Remember that a male's reproductive life in the pride is short, so any individual that practises infanticide when he takes over a pride will father more of his own offspring and, therefore, the tendency to commit infanticide will spread by natural selection.

The take-over of a pride by a new coalition of adult males also contributes to the reproductive synchrony of the females; because all the dependent offspring are either killed or evicted during the take-over, the females will all tend to come into oestrus again at about the same time (Packer & Pusey, 1983b). Interestingly, the sexual activity of the females is most intense during the first few months after a take-over. The females play an active role in soliciting copulations from several males and this appears to elicit competition between different male coalitions for the control of the pride, with the result that larger coalitions eventually become resident. This is of adaptive advantage to the female because she needs protection from male harassment of her cubs for over two years in order to rear her cubs successfully (3.5 months gestation plus 1.5–2 years with dependent young) and only large male coalitions are likely to remain in the pride for more than two years. High sexual activity in females at around the time of take-overs may therefore incite male–male competition and so result in the best protectors taking over the pride (Packer and Pusey, 1983a).

Observation	Causal explanations	Functional explanations
1 Females are synchronous in oestrus	Chemical cues? Take-overs by males	Better cub survival Young males survive better and have greater reproductive success when they leave pride if in a group
2 Young die when new males take over pride	Abortion Take-over males kill or evict young	Females come into oestrus more quickly Male removes older cubs which would compete with his young

Table 1.1 Summary of causal and functional explanations for two aspects of reproductive behaviour in lions (Bertram, 1975; Packer and Pusey, 1983a, 1983b).

Causal and functional explanations of lion behaviour

The differences between the causal and functional explanations of these two aspects of reproductive behaviour in the lions are summarized in Table 1.1. The key point is that causal explanations are concerned with mechanisms, while functional explanations are concerned with why these particular mechanisms (rather than others) have been favoured by natural selection.

Natural selection

The aim of behavioural ecology is to try and understand how an animal's behaviour is adapted to the environment in which it lives. When we discuss adaptations we are referring to changes brought about during evolution by the process of natural selection. For Charles Darwin, adaptation was an obvious fact. It was obvious to him that eyes were well designed for vision, legs for running, wings for flying and so on. What he attempted to explain was how adaptation could have arisen without a creator or, put another way, how you could get the appearance of design without a designer. His theory of natural selection, published in the *Origin of Species* (Darwin, 1859), can be summarized as follows:

(1) Individuals within a species differ in their morphology, physiology and behaviour (*variation*).

Heritable variation with competition for survival and reproduction

(2) Some of this variation is *heritable*; on average offspring tend to resemble their parents more than other individuals in the population.

(3) Organisms have a huge capacity for increase in numbers; they produce far more offspring than give rise to breeding individuals. This capacity is not realized because the number of individuals within a population tends to remain more or less constant over time. Therefore, there must be *competition* between individuals for scarce resources, such as food, mates and places to live.

(4) As a result of this competition, some variants will leave more offspring than others. These will be those that are best at competing for the scarce resources. Their offspring

will inherit the characteristics of their successful parents and so, through *natural selection* over the generations, organisms will come to be *adapted* to their environment. The individuals that are selected, naturally, will be those best able to find food and mates, avoid predators and so on.

(5) If the environment changes, then new variants may do best and so natural selection can lead to *evolutionary change*.

When Darwin formulated his idea he had no knowledge of the mechanism of heredity. The modern statement of the theory of natural selection is in terms of genes. Although selection acts on differences in survival and reproductive success between individual organisms, or phenotypes, what changes during evolution is the relative frequency of genes. We can restate Darwin's theory in modern genetic terms as follows:

(1) All organisms have genes which code for proteins. These proteins regulate the development of the nervous system, muscles and structure of the individual, and so influence its behaviour.

(2) Within a population many genes are present in two or more forms, or alleles, which code for slightly different forms of the same protein or determine when, where and how much of the protein is expressed. These will cause differences in development and function, and so there will be variation within a population.

Selection causes changes in gene frequency

(3) Any allele that results in more surviving copies of itself than its alternative will eventually replace the alternative form in the population. Natural selection is the differential survival of alternative alleles through their effects on replication success.

The individual can be regarded as a temporary vehicle or survival machine by which genes survive and replicate (Dawkins, 1976). Because selection of genes is mediated through phenotypes, the most successful genes will usually be those that are most effective in enhancing an individual's survival and reproductive success (or that of relatives, as we shall show later in the book).

Genes and behaviour

Natural selection can only work on genetic differences, so for behaviour to evolve: (a) there must be, or must have been in the past, behavioural alternatives in the population; (b) the differences must be, or must have been, heritable; in other words a proportion of the variation must be genetic in origin; and (c) some behavioural alternatives must confer greater reproductive success than others.

Behavioural differences may have a genetic basis

Some examples to show how *genetic differences* between individuals can lead to *differences in behaviour* are now discussed. Note the emphasis on the word *difference*. When we talk about 'genes for' a particular structure or behaviour, we do not imply that one gene alone codes for the trait. Genes work in concert and many genes together will influence an individual's mating preference, foraging, migration and so on. However, a *difference* in behaviour between two individuals may be due to a *difference* in one (or more) genes. A useful analogy is the baking of a cake. A difference in one word of a recipe (one versus two spoonfuls) may mean that the taste of the whole cake is different, but this does not mean that the one word is responsible for the entire cake (Dawkins, 1978). Whenever we talk about 'genes for' certain traits, this is shorthand for gene differences bringing about differences in behaviour.

Three other important points should be borne in mind when reading these examples. Firstly, the molecular path linking genes and behaviour is complicated (transcription, translation, influence on sensory systems, neural activity, brain metabolism and so on). Secondly, the arrow linking genes and behaviour goes in both directions (Robinson *et al.*, 2008). Not only do genes influence behaviour, through effects on brain development and physiology, but behaviour can also influence gene expression. Thirdly, just because it can be shown that genes influence behaviour does not imply that genes alone produce the behaviour. Behavioural development is an outcome of a complex interaction between genes and environment. The examples now discussed help to make these general points clearer.

Drosophila and honeybees: foraging, learning and singing

Rovers and sitters in *Drosophila*

Larvae of the fruit fly *Drosophila melanogaster* feed in one of two distinct ways. 'Rovers' wander around in search of food while 'sitters' tend to remain in one small area to feed. These differences persist into the adult stage, with rover flies also searching more widely when foraging. In the absence of food, rovers and sitters (larvae or adults) do not differ in general activity. This difference in foraging strategies is caused by a difference in just one gene (the *foraging* gene, *for*) which codes for an enzyme which is rather snappily called cyclic guanosine monophosphate (cGMP) dependent protein kinase (PKG). This enzyme is produced in the brain and influences behaviour. Flies with the 'rover' allele (for^R) show higher PKG activity than those homozygous for the 'sitter' allele (for^s). When the for^R allele is inserted into the genome of sitter larvae, they become rovers (Osborne *et al.*, 1997).

Individuals with the for^R allele also have better short-term memory for olfactory stimuli, while those with the for^s allele perform better at long-term memory tasks involving odour cues. These differences may be coadapted with the differences in foraging behaviour: rovers may benefit from fast learning as they move between food patches, while sitters, with a sedentary feeding style, may benefit from long-term memory (Mery *et al.*, 2007).

In one orchard population in Toronto, 70% of larvae was rovers while 30% was sitters. Why do the two feeding types persist? Laboratory experiments reveal that rovers do best under patchy food and high larval densities (rovers are better at finding new food patches) while sitters do best with more uniformly distributed food and at low larval density (when roving is unnecessary as local food is abundant; Sokolowski *et al.*, 1997). Therefore, each morph does best under different ecological conditions. However, a further factor is involved in maintaining the polymorphism. When food is scarce, competition is most intense between individuals of the same morph: sitters compete most with sitters within local food patches, while rovers compete most with other rovers over the discovery of new food patches. This leads to the situation where the rarer type has an advantage, which is termed negative frequency-dependent selection; in a population of rovers a sitter does especially well, while in a population of sitters a rover does especially well. Because each type does better when rare, this will tend to maintain the behavioural polymorphism (Fitzpatrick *et al.*, 2007). This topic is discussed further in Chapter 5.

Gene expression and behaviour changes with age in honeybees

The same *foraging* gene, *for*, regulates age changes in foraging worker honeybees, *Apis mellifera*. When they are young, adult worker bees perform various tasks inside the hive, such as storing food and caring for the brood. Then, when they are about three weeks old, they begin to go off on long foraging flights to collect pollen and nectar for the colony. This marked change from 'sitting at home' to 'roving for food' involves changes in the expression of *for*, with foragers having increased production of the enzyme PKG. When young workers were induced to switch to foraging earlier (one week of age) by removal of older workers, these precocious foragers also had increased *for* expression. Therefore, expression of *for* was related to social information (presence or absence of older workers), which then influenced foraging activities; it was not just a response to age. Finally, experimental elevation of PKG activity in young workers also led to a switch to foraging behaviour (Ben-Shahar *et al.*, 2002).

Thus, in *Drosophila* different individual foraging behaviours are caused by differences in alleles of the *for* gene, while in honeybees the switch in behaviour within individuals is caused by changes in *for* gene expression.

Drosophila courtship song

Single gene differences can also cause differences in *Drosophila* courtship song. Males produce a courtship song by vibrating their wings and the temporal pattern of the song varies between species. Breeding experiments and molecular genetic analysis reveal that these differences in song structure are caused by differences in the *period* gene. Transfer of a small piece of the *period* gene from *D. simulans* to *D. melanogaster* causes *melanogaster* males to produce the *simulans* song rather than *melanogaster* song (Wheeler *et al.*, 1991).

MC1R: mate choice and camouflage

A gene influencing melanin

The lightness or darkness of skin, hair or feathers depends primarily on the amount of a pigment, melanin, produced by specialised skin cells (melanocytes). The MC1R gene (melanocortin-1 receptor) encodes a receptor that is expressed in melanocytes. The activity of this receptor regulates the amount and type of melanin synthesis. Point mutations in this gene are associated with colour variation in fish, reptiles, birds and mammals, so this gene has been conserved through a long evolutionary history.

In lesser snow geese (*Anser chen caerulescens*) there are two colour morphs, white and blue. Individuals that are homozygous for one variant allele at MC1R are white, while those that are heterozygous or homozygous for the other allele are blue. Curiously, there is no evidence for any selective advantage in being either white or blue. However, colour influences the choice of mate. There is assortative mating by colour (white with white, blue with blue) and young goslings imprint on their parents' colour and then favour a mate of the same colour (Mundy *et al.*, 2004).

Variation in the same gene controls colour in the rock pocket mouse (*Chaetodipus intermedius*). In the Pinacate desert of Arizona, the mouse occurs in two colour forms. Dark, melanic mice live on black lava flows while sandy-coloured mice live in sandy, desert habitat. There is selective predation by owls against mice which do not match their background (Nachman *et al.*, 2003).

Blackcaps: migratory behaviour

The cases discussed so far involve single gene differences causing marked differences in phenotype. Often, however, phenotype differences reflect the effects of many genes acting in concert. Migration behaviour provides an excellent example.

Most species of warblers are summer visitors to Europe. If individuals are kept in a cage, they show a period of 'restlessness' in the autumn at the time they would migrate south to the Mediterranean or beyond to Africa. Quantitative comparisons between populations breeding at different latitudes have shown that the duration of restlessness correlates with migration distance, while the direction of fluttering in the cage correlates with migration direction. Therefore, migration behaviour can be studied experimentally in caged birds.

Selection experiments for migration behaviour

Peter Berthold and colleagues have investigated the genetic basis for migration distance and direction in blackcaps, *Sylvia atricapilla* (Fig. 1.3a). Populations in southern Germany are highly migratory while those in the Canary Islands are sedentary. When birds from these two populations were cross-bred in aviaries, their offspring showed intermediate migratory restlessness, suggesting genetic control (Fig. 1.3b). Selection experiments confirmed that there was a genetic basis to differences in migration behaviour. Among 267 hand-raised blackcaps from a population in the Rhone Valley of southern France, three-quarters showed migratory restlessness while one quarter did not. By selectively breeding from either migratory or non-migratory parents, lines of blackcaps were produced that were either 100% migratory (in three generations) or 100% resident (in six generations). Furthermore, among the migrant individuals migratory activity had also responded to selection (Fig. 1.3c). Not only does this experiment reveal a genetic basis to migratory behaviour, it also shows how rapidly migration may evolve.

A new migration habit – evolution in action

Finally, and thrillingly, Berthold and coworkers have discovered an example of evolution in action. Central European populations of blackcaps traditionally winter to the southwest of their breeding grounds in the western Mediterranean (Fig. 1.3d). During the past 40 years, however, the number of blackcaps wintering in Britain and Ireland (1500km to the north of the traditional wintering grounds) has steadily increased. At first, it was assumed that these must be British breeding birds, remaining in response to milder winters. However, ringing recoveries indicated that they were breeders from central Europe with an entirely new migration habit. Blackcaps wintering in Britain were caught and kept in aviaries. When their migration behaviour was tested in cages, they exhibited a westerly autumn migration direction, shifted c70° from the traditional south-westerly route. Furthermore, their offspring inherited this new autumnal orientation (Fig. 1.3d).

The new migration direction is probably being favoured because of milder winters and more winter food in Britain, both from garden feeders and winter fruit bushes planted in recent decades. This new population of migrants enjoys a shorter distance to winter quarters and an earlier arrival back in the central European breeding grounds in spring. This enables them to gain the best breeding territories and to produce more offspring (Bearhop *et al.*, 2005). The different arrival times on the breeding grounds also lead to assortative mating by wintering area (males wintering in Britain tend to pair with females wintering in Britain) and hence restricted gene flow, which has likely contributed to the rapid evolution of the new migration behaviour (Bearhop *et al.*, 2005).

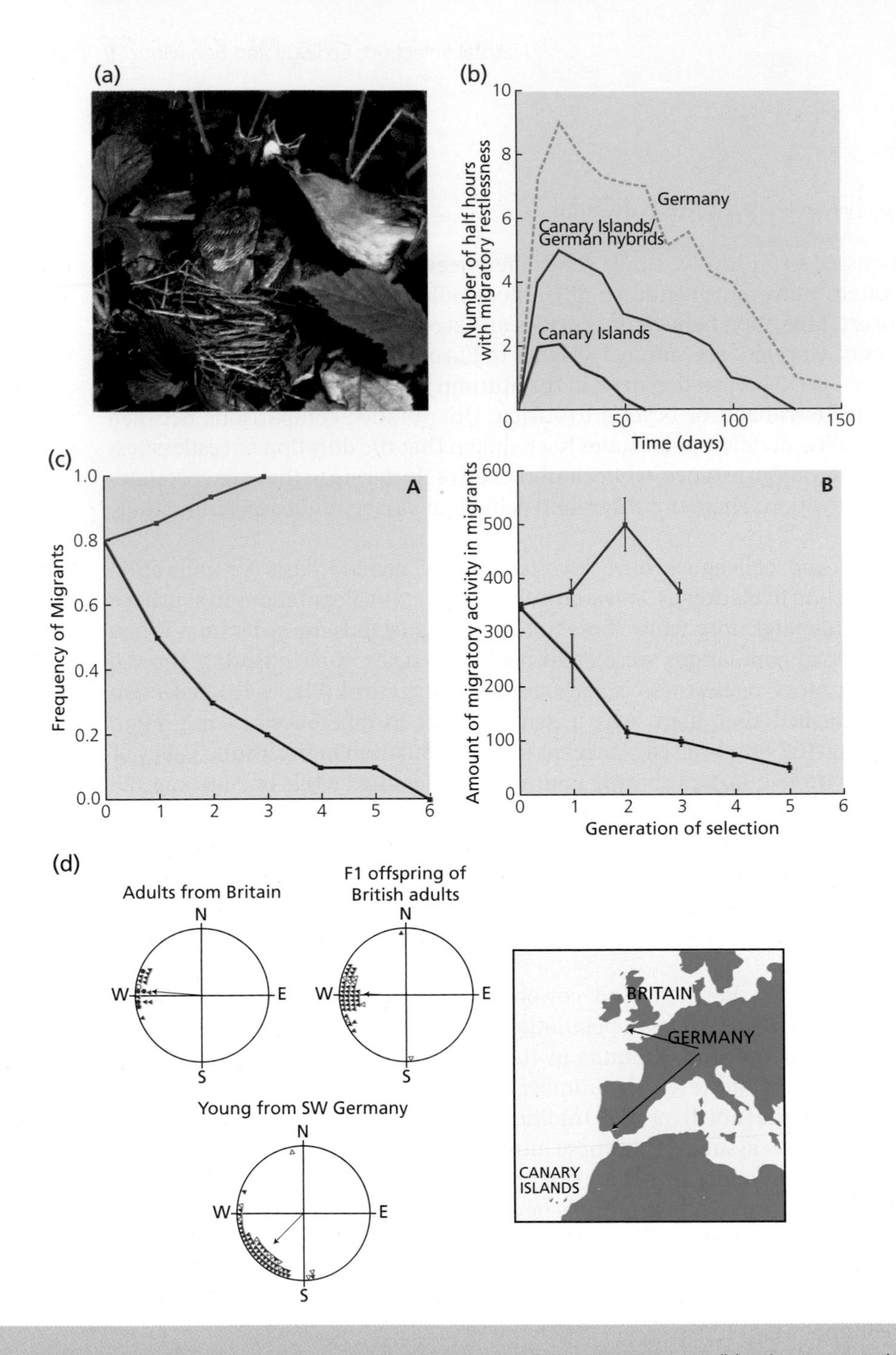

Fig. 1.3 Blackcap migration. (a) Male blackcap with nestlings. Photo © W. B. Carr (b) Migratory restlessness (measured in cages) during the time of autumn migration in blackcaps from Germany, the Canary Islands and F_1 hybrids of these two populations. From Berthold and Querner (1981). Reprinted with permission from AAAS. (c) Artificial selection in aviary populations for higher (red) and lower (blue) migratory behaviour in a partially migratory blackcap population from southern France; A: frequency of migrant individuals; B: migratory activity in migrants. From Berthold *et al.* (1990) and Pulido *et al.* (1996). (d) Traditionally, in autumn blackcaps from southern Germany migrate in a south-west direction to winter in the western Mediterranean region. During the past 40 years a new migration habit has evolved, with some blackcaps migrating west to Britain; F_1 offspring from these adults inherit the new direction. Each point in the circles to the left refer to the direction of migration of one caged individual and the arrows indicate the mean direction). From Berthold *et al.* (1992). Reprinted with permission from the Nature Publishing Group.

Selfish individuals or group advantage?

Behaviour of advantage to individuals may be disadvantageous to the group

We now return to our theme of studying the adaptive significance of behaviour, how it contributes to an individual's chances of survival and its reproductive success. We interpreted the behaviour of the lions in relation to individual advantage, reflecting Darwin's emphasis on evolution as a struggle between individuals to out-compete others in the population. Many traits evolve because of their advantage to the individual even though they are disadvantageous to others in the population. For example, it is not to the species' advantage to have a cub killed when a new male takes over a lion pride. It is not to the lionesses' advantage either! However, she is smaller than the male and often there is probably not much that she can do about it. Infanticide has evolved simply because the advantage to the male that practises it outweighs the cost to the female in resisting.

Not so long ago, however, many people thought that animals behaved for the good of the group, or of the species. It was common to read (and sometimes still is) explanations like, 'lions rarely fight to the death because, if they did so, this would endanger survival of the species' or, 'salmon migrate thousands of miles from the open ocean into a small stream where they spawn and die, killing themselves with exhaustion to ensure survival of the species'. Because 'group thinking' is so easy to adopt, it is worth going into a little detail to examine why it is the wrong way to think about the evolution of behaviour.

The most famous proponent of the idea that animals behave for the good of the group was V.C. Wynne-Edwards (1962, 1986). He suggested that if a population over-exploited its food resources it would go extinct, and so adaptations have evolved to ensure that each group or species controls its rate of consumption. Wynne-Edwards proposed that individuals restrict their birth rate to prevent over-population, by producing fewer young, not breeding every year, delaying the onset of breeding and so on. This is an attractive idea because it is what humans ought to do to control their own populations. However, there are two reasons for thinking that it is unlikely to work for animal populations.

Theoretical considerations

Imagine a species of bird in which a female lays two eggs and there is no over-exploitation of the food resources. Suppose the tendency to lay two eggs is inherited. Now consider a mutant that lays three eggs. Since the population is not over-exploiting its food supplies, there will be plenty of food for the young and because the three-egg genotype produces 50% more offspring it will rapidly increase at the expense of the two-egg genotype.

Will the three-egg type be replaced by birds that lay four eggs? The answer is yes, as long as individuals laying more eggs produce more surviving young. Eventually a point will be reached where the brood is so large that the parents cannot look after it as efficiently as a smaller one. The clutch size we would expect to see in nature will be the one that results in the most surviving young because natural selection will favour individuals that do the best. A system of voluntary birth control for the good of the group will not evolve because it is unstable; there is nothing to stop individuals behaving in their own selfish interests.

Group selection

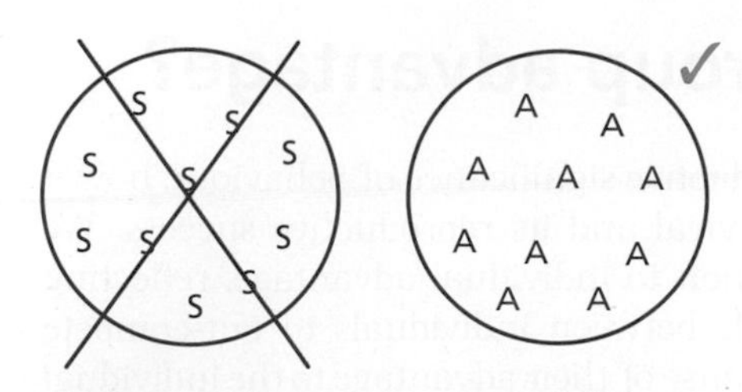

Fig. 1.4 Wynne-Edwards' model of group selection. Groups of selfish individuals (S) over-exploit their resources and so die out. Groups of altruistic individuals (A), who do not over-exploit resources (e.g. by having fewer offspring than they could potentially raise) survive.

Wynne-Edwards realized this and so proposed the idea of 'group selection' to explain the evolution of behaviour that was for the good of the group. He suggested that groups consisting of selfish individuals died out because they over-exploited their food resources. Groups that had individuals who restricted their birth rate did not over-exploit their resources and so survived. By a process of differential survival of groups, behaviour evolved that was for the good of the group (Fig. 1.4).

In theory this can work, but it would require that groups are selected during evolution, with some groups dying out faster than others. In practice, however, groups usually do not go extinct fast enough for group selection to be an important force in evolution. Individuals will nearly always die at a faster rate than groups, so individual selection will be more powerful. In addition, for group selection to work populations must be isolated, such that individuals cannot successfully migrate between them. Otherwise there would be nothing to stop the migration of selfish individuals into a population of individuals all practising reproductive restraint. Once selfish individuals arrive, their genotype would soon spread. In nature, groups are rarely isolated sufficiently to prevent such immigration. So group selection as proposed by Wynne-Edwards is usually going to be a weak force and probably rarely very important (Williams, 1966a; Maynard Smith, 1976a). We revisit this topic in the final chapter.

Individual selection more powerful

Empirical studies: optimal clutch size

Apart from these theoretical objections, there is good field evidence that individuals do not restrict their birth rate for the good of the group but rather maximize their individual reproductive success. A classic example is the long-term study of the great tit (*Parus major*) in Wytham Woods, near Oxford, UK, started in 1947 by David Lack (Lack, 1966).

Clutch size in great tits ...

In this population the great tits nest in boxes (Fig. 1.5a) and lay a single clutch of eggs in the spring. All the adults and young are marked individually with small numbered metal rings round their legs. The eggs of each pair are counted, the young are weighed and their survival after they leave the nest is measured by re-trapping ringed birds. This intensive field study involves several people working full-time throughout the year, and it has been going on for over 60 years! Most pairs lay 8–9 eggs (Fig. 1.5b, bars). The limit is not set by an incubation constraint because when more eggs are added the pair can still incubate them successfully. However, the parents cannot feed larger broods so well. Chicks in larger broods get fed less often, are given smaller caterpillars and, consequently, weigh less when they leave the nest (Fig. 1.6a). It is not surprising that feeding the young produces a limit for the parents because they have to be out searching for food from dawn to dusk and may deliver over 1000 items per day to the brood at the peak of nestling growth. In a survey of the sustainable

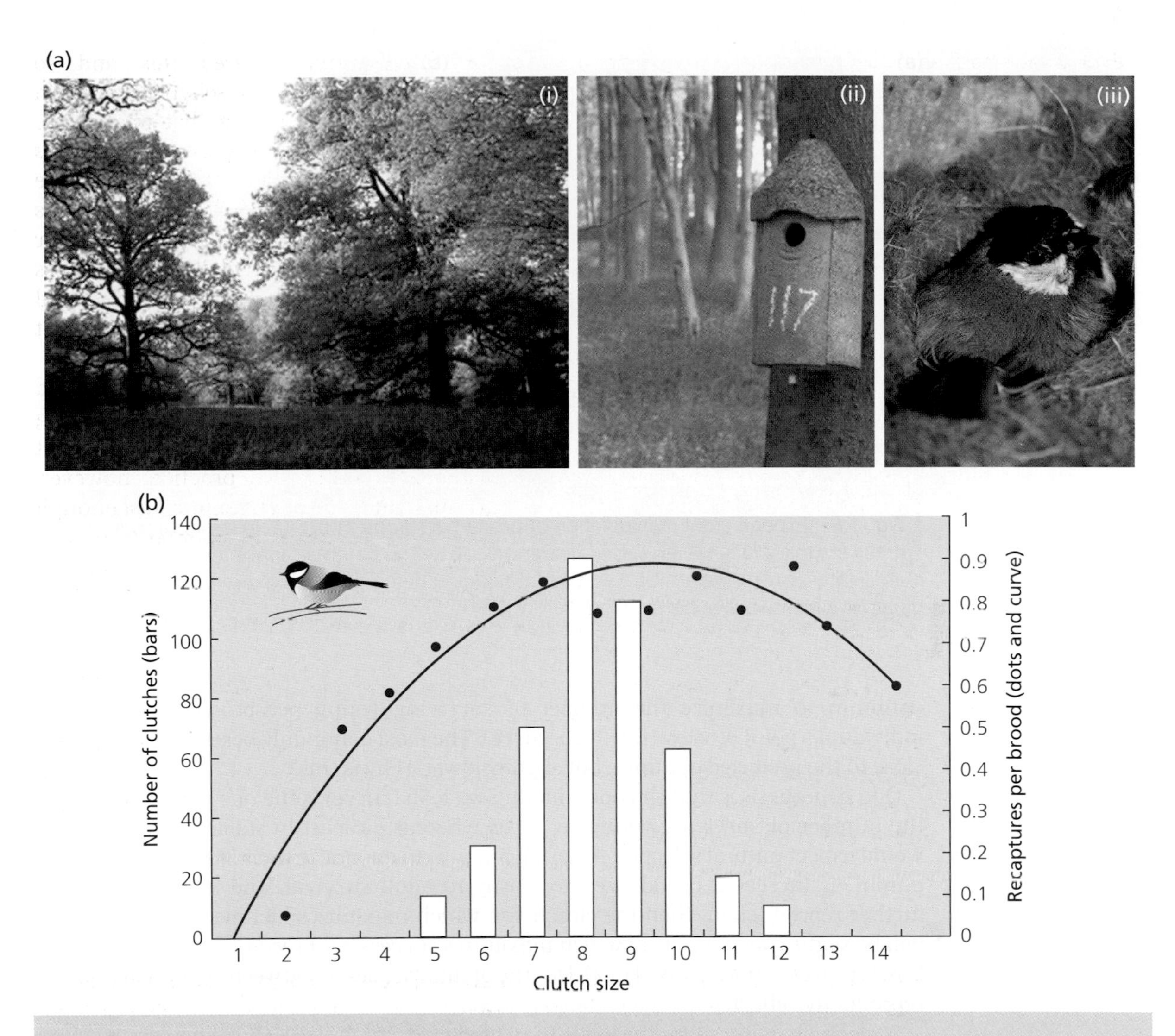

Fig. 1.5 (a) (i) Wytham Woods, Oxford, the site of a long-term study of great tit reproductive behaviour. Photo © Jane Carpenter (ii) A nest box. Photo © Ben Sheldon. (iii) Female great tit incubating a clutch. Photo © Sandra Bouwhuis (b) Bars: The frequency distribution of the clutch size of great tits in Wytham Woods. Most pairs lay 8–9 eggs. Curve and blue dots: Experimental manipulation of brood size shows that the clutch size that maximizes the number of surviving young per brood is slightly larger than the average observed clutch size. From Perrins (1965).

metabolic rates of animals, only two examples were found of animals working at more than seven times their resting metabolic rate: breeding birds and cyclists on the Tour de France cycle race (Peterson *et al.*, 1990).

... is less than that predicted to maximize the number of surviving young per brood

The significance of nestling weight is that heavier chicks survive better (Fig. 1.6b). Therefore, an over-ambitious parent will leave fewer surviving young because it cannot feed its nestlings adequately. By creating broods of different sizes experimentally and allocating them at random to different nests, it was demonstrated that there is an

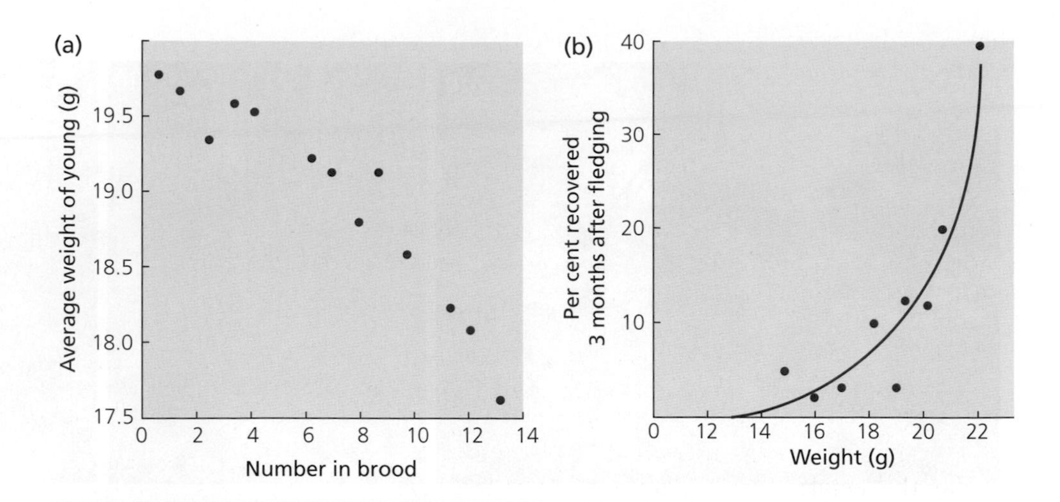

Fig. 1.6 Experimental manipulation of brood sizes in great tits. (a) In larger broods of great tits the young weigh less at fledging because the parents cannot feed them so efficiently. (b) The weight of a nestling at fledging determines its chances of survival; heavier chicks survive better. From Perrins (1965).

optimum to maximize the number of surviving young per brood from a selfish individual's point of view (Fig. 1.5b, curve). The most commonly observed clutch size is close to the predicted optimum but slightly lower. Why is this?

Two hypotheses for the mismatch between observed and predicted ...

One hypothesis is that the optimum in Fig. 1.5b (curve) is the one which maximizes the number of surviving young *per brood* whereas, at least in stable populations, we would expect natural selection to design animals to maximize their *lifetime* reproductive output. If increased brood sizes are costly to adult survival, and hence chances of further reproduction, then the clutch size which maximizes lifetime breeding success will be slightly less than that which maximizes success per breeding attempt (Fig. 1.7). Box 1.1 gives a more general model for the optimal trade-off between current and future reproductive effort.

A second hypothesis for the lower than predicted clutch size is that when great tits are experimentally given extra eggs or chicks they may well be able to rear some extra young efficiently, but we have ignored the costs of egg production and incubation (Monaghan & Nager, 1997). A fairer test would be to somehow manipulate birds into laying extra eggs, rather than giving them extra eggs or chicks for free. If females were forced to pay the 'full cost' of laying and incubating the extra eggs, then this may reduce the predicted optimal brood size to maximize the number of surviving chicks per brood.

.... involve considering further trade-offs

Note that both hypotheses involve measuring further trade-offs. David Lack's predicted optimum (Fig. 1.5b, curve) involved the trade-off between offspring number and quality. Our first hypothesis for the mis-match between his prediction and the observed clutch size is that we need to consider, in addition, the trade-off between adult reproductive effort and adult mortality. The second hypothesis concerns another trade-off, that between investment in egg production and incubation versus chick care. As we shall see throughout this book, resources are limited and one of the main themes of behavioural ecology is investigating how various trade-offs are solved by natural selection.

Marcel Visser and Kate Lessells (2001) measured the effects of these two extra trade-offs on great tit optimal clutch size by a clever experimental design (first used by Heany & Monaghan (1995) for studying clutch size in a seabird). In a nest-box population of great tits in the Hoge Veluwe, a large national park in The Netherlands, they had three experimental groups of females, each raising two extra chicks:

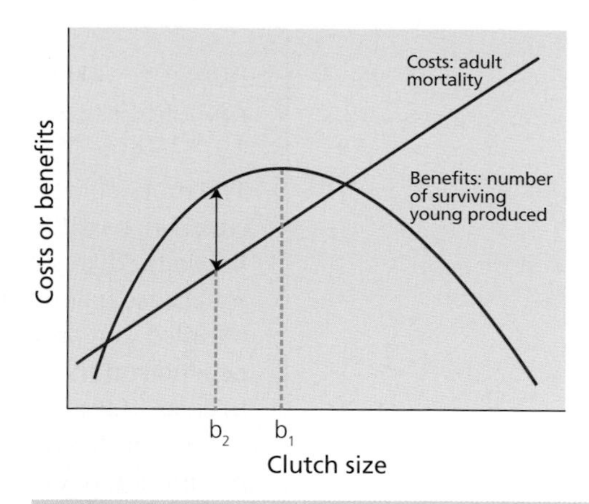

Fig. 1.7 The influence of adult mortality on the optimal clutch size. The number of young produced versus clutch size follows a curve, as in Fig. 1.5, with b_1 being the clutch size which maximizes the number of young produced per brood. Increased clutch size, however, has the cost of increased adult mortality, shown here for simplicity as a straight line. The clutch size which maximizes lifetime reproductive success is b_2, where the distance between the benefit and cost curves is a maximum. This is less than the clutch size b_1, which maximizes reproductive success per brood. From Charnov and Krebs (1974).

(i) *Free chicks*. Two extra nestlings were added to the nest, soon after the female's own brood hatched. These females, therefore, only had to raise two extra chicks.
(ii) *Free eggs*. Two extra eggs were added to the clutch on the day the female began to incubate her own clutch. These females, therefore, had to incubate two extra eggs as well as raise the two extra chicks.
(iii) *Full costs*. The female was induced to lay two extra eggs by removing the first four eggs of the clutch on the day they were laid (previous experiments had shown that removal of four led to two extra eggs being laid). These four removed eggs were kept in a bed of moss and were returned to the clutch before incubation began. So this third group had to lay the two extra eggs, as well as incubate them, and raise the two extra chicks, thus paying the full cost of an increased clutch size.

A trade-off between reproductive effort and adult survival to maximize lifetime success

The results showed that the number of young produced who survived to breeding age (recruits) did not differ between the three treatments. Therefore, there was no support for the second hypothesis; *full costs* females produced just as many surviving young as those given free eggs or chicks. However, female survival *was* affected; *full costs* females had the lowest survival to the next breeding season, while *free chicks* females survived the best, with *free eggs* females having intermediate survival. These results, therefore, support the first hypothesis; there is a trade-off between increased reproductive effort and adult survival. When female fitness was calculated, *full costs* females had lower fitness than control females (who were left to raise the clutch size they initially chose; Fig. 1.8). Therefore, when the costs of both egg production and incubation are taken into account, the observed clutch size is optimal (at least in comparison with an increase in clutch size of two eggs).

Brood size manipulations are most easily done with birds, but similar studies with mice (König *et al*., 1988) and insects (Wilson, 1994) also suggest that reproductive rate tends to maximize individual success, though the trade-offs involved vary from case to case, and they are often tricky to measure.

Individuals may have different optima

Clutch size may vary from year to year and during the season depending on food supplies, so individuals do show some variation. However, the variations are in relation to their own selfish optima, not for the good of the group. A good example of individual optimization is provided by Goran Högstedt's study (1980) of magpies, *Pica pica*,

BOX 1.1 THE OPTIMAL TRADE-OFF BETWEEN SURVIVAL AND REPRODUCTIVE EFFORT (PIANKA AND PARKER, 1975; BELL, 1980)

The more effort an individual puts into reproduction, the lower its chances of survival, so the lower its expectation of future reproductive success. Reproductive costs include allocation of resources to reproduction which would otherwise have been spent on own growth and survival and the increased risks entailed in reproduction, such as exposure to predators. The optimal life history depends on the shape of the curve relating profits in terms of present offspring to costs in terms of future offspring.

The families of straight lines represent fitness isoclines, that is equal lifetime production of offspring (Fig. B1.1.1). In a stable population, present and future offspring will be of equal value and these lines will have slopes of -1. In an expanding population, current offspring are worth more than future offspring (current offspring gain a greater contribution to the gene pool) and the slopes are steeper. In a declining population, future offspring are worth more and slopes will be less than -1.

The point of intersection of the curves relating the trade-off between current and future reproductive success, with the fitness isocline furthest from the origin, gives the optimal reproductive tactic (indicated by a solid dot). When the trade-off curve is convex (a), fitness is maximized by allocating part of the resources to current reproduction and part to survival (i.e. iteroparity, or repeated breeding). When the curve is concave (b), it is best to allocate all resources to current reproduction, even at the expense of own survival (semelparity, or 'big bang' suicidal reproduction). If maximal future reproductive success is greater than maximal current reproductive success in case (b), then the optimal tactic is to not breed and save all resources for the future.

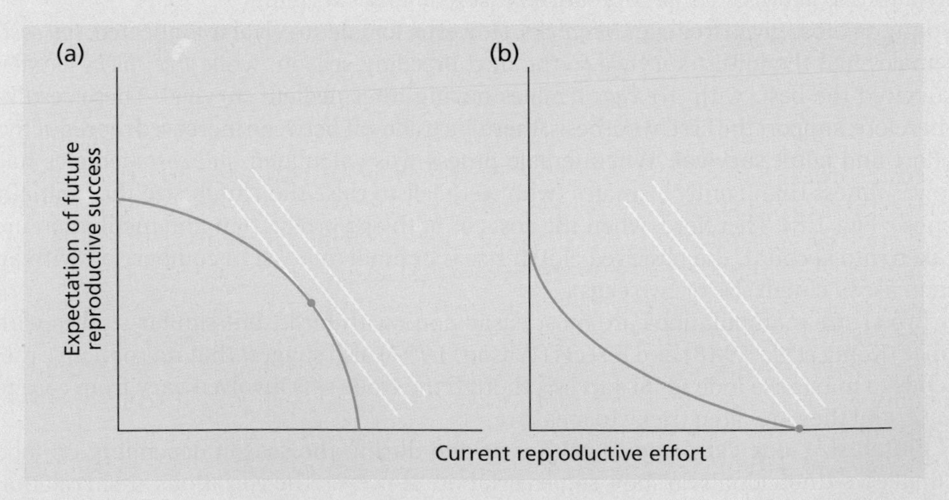

Fig. B1.1.1

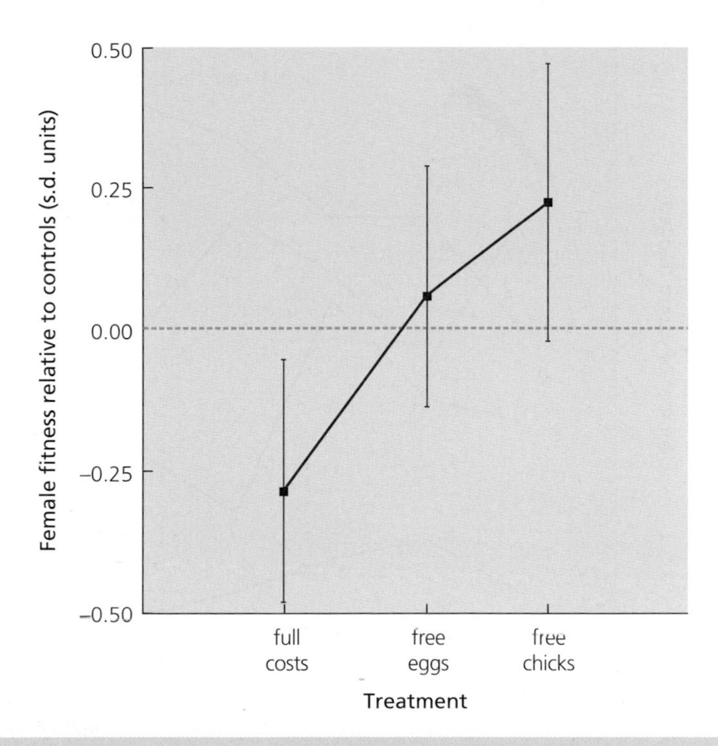

Fig. 1.8 Fitness of female great tits in the experiment of Visser and Lessells (2001). Fitness was measured as (female survival to next breeding season) + (0.5 × number of offspring surviving to next breeding season). The logic behind this measure is that each offspring has only half the female's genes, so has half the 'genetic value' of the female herself. Female fitness is measured relative to controls (who raised the clutch size they initially laid) for three experimental groups who each had two extra chicks to raise, but with varying extra costs (see text). Females given *free chicks* or *free eggs* did better than controls but females forced to pay the *full costs* of laying and incubation had lower fitness than controls. From Visser and Lessells (2001).

breeding in southern Sweden. Observed clutch sizes varied from five to eight depending on feeding conditions in different territories. To test the hypothesis that some females laid only five eggs because this was the maximum number of young they could raise efficiently on their particular territories, Högstedt manipulated clutch sizes experimentally. He found that pairs that had produced large clutches did best with large broods, while those which had laid small clutches did best with smaller broods (Fig. 1.9). Variation in clutch size occurred because there was a range of territory quality and each pair raised a brood size appropriate for its own particular territory. Experiments have shown similar individual optimization of clutch size in great tits (Pettifor *et al.*, 1988; Tinbergen & Daan, 1990) and collared flycatchers (Gustafsson & Sutherland, 1988).

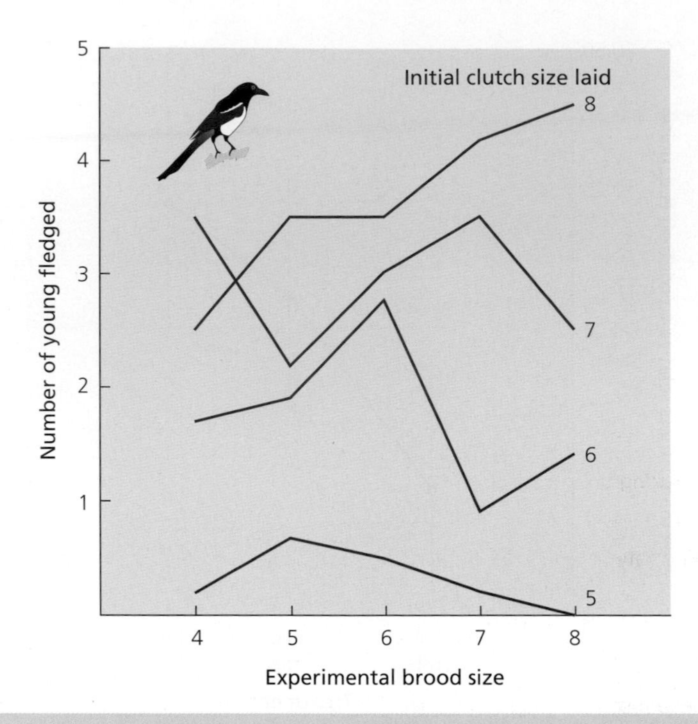

Fig. 1.9 Experiments on clutch size in magpies. Pairs that had initially laid 5, 6, 7 or 8 eggs were given experimentally reduced or enlarged broods. Pairs that had naturally laid large clutches did better with large broods and those naturally laying small clutches did better with small broods. From Högstedt (1980). Reprinted with permission from AAAS.

Phenotypic plasticity: climate change and breeding times

Reaction norms

The ability of a single genotype to alter its phenotype in response to environmental conditions is termed *phenotypic plasticity*. For example, we have just seen that clutch size is a phenotypically plastic trait which varies with season and food availability. When the phenotypic variation is continuous, the relationship between phenotype and the environment for each genotype is called a *reaction norm* (Fig. 1.10). There may be genetic variation in both the elevation of the line (the trait value) and its slope (the way the trait value changes in response to the environment). Recent studies of the earlier breeding of songbirds in response to climate change provide a good example of phenotypic plasticity. They also show how useful it is to study both proximate and ultimate explanations of behaviour together, hand in hand.

Warmer springs and earlier breeding in great tits

Over 47 years (1961–2007), the mean egg laying date of female great tits in the Wytham Wood population (near Oxford, UK) has advanced by about 14 days (Fig. 1.11a). The main changes have been from the mid 1970s, since when there has been a marked increase in spring temperatures (Fig. 1.11b). This has led to the earlier

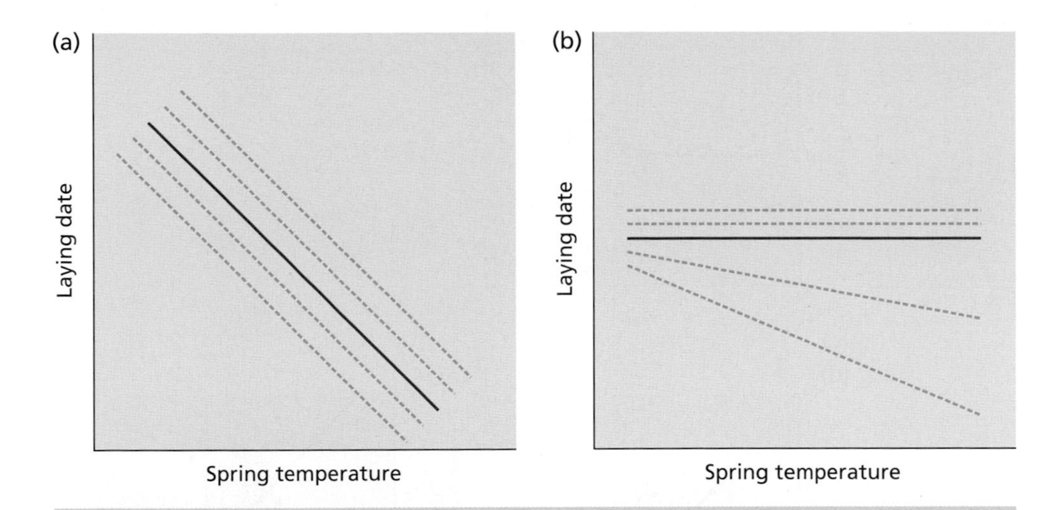

Fig. 1.10 Phenotypic plasticity in laying dates in response to spring temperatures. Dashed lines represent examples of reaction norms for different individual females, who may differ in their average laying date (elevation) or in their plasticity in response to spring temperatures (slope). In Wytham Woods, UK, the great tits respond as in (a), with no significant variation between females in plasticity and a strong average population response to temperature (solid line). In the Hoge Veluwe, The Netherlands, the great tits respond as in (b), with no significant average population response (solid line) but significant variation in individual female plasticity. After Charmantier *et al.* (2008). Reprinted with permission from AAAS.

emergence of oak leaves (*Quercus robur*) and of winter moth caterpillars (*Operophtera brumata*), which feed on the oak leaves and are a key food for nestling tits. The rates of change of egg laying date with temperature (Fig. 1.11c) and of caterpillar emergence with temperature (Fig. 1.11d) are similar, so the tits have closely tracked the temporal changes in food availability over almost five decades.

Genetic change or phenotypic plasticity?

How have the tits managed to do this? For temperate breeding birds, an increasing photoperiod in the spring is the primary proximate cue that initiates gonadal growth and the hormonal changes involved in breeding. However, the response can be fine-tuned by other cues, such as temperature, food availability and social stimulation (Dawson, 2008). One possible explanation for the earlier breeding is that there has been micro-evolutionary change in the tit population, with selection favouring new genotypes with different thresholds of response to these proximate cues (e.g. breeding at shorter lengths of day). The other possibility is that earlier breeding has simply arisen through phenotypic plasticity, with no need for any genetic change.

Anne Charmantier, Ben Sheldon and colleagues have shown that this second hypothesis explains the response to climate change by the Wytham great tits. They analysed the laying dates of 644 individual females who had bred in three or more years. They found no significant variation among these individual responses to spring temperature, so all females had similar reaction norms (Fig. 1.10a). Furthermore, the slope of these individual responses was similar to that for the population as a whole

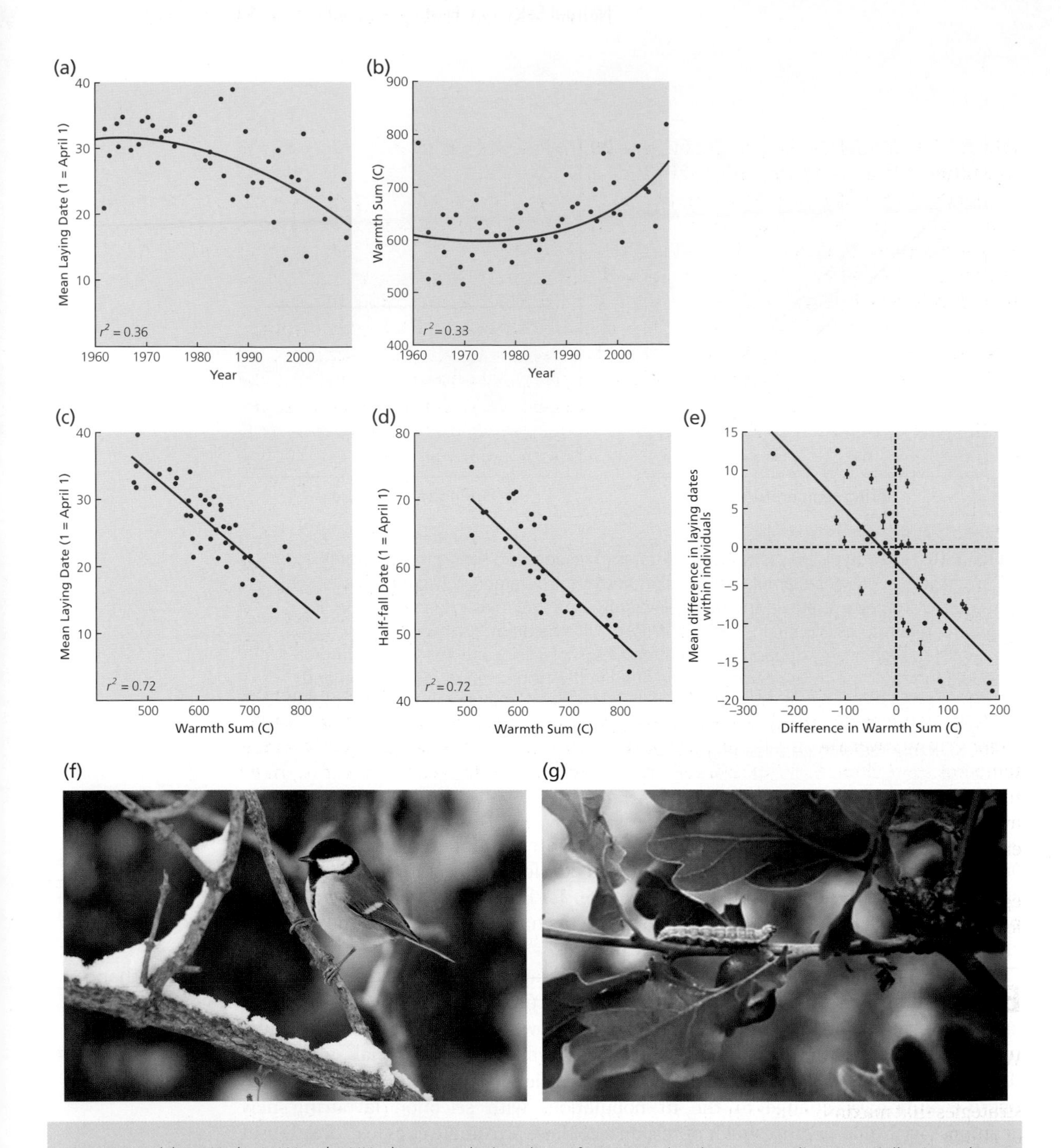

Fig. 1.11 (a) In Wytham Woods, UK, the mean laying date of great tits has become earlier, especially since the mid 1970s. (b) Spring temperatures have also increased, as measured by 'Warmth Sum', which is the sum of daily maximum temperatures between 1 March and 25 April (the pre-laying period). The rates of change in mean egg laying date with temperature (c) and caterpillar emergence with temperature (d) are similar. (e) Phenotypic plasticity in response of individual female great tits, measured as their difference in laying date in successive years plotted against the difference in spring warmth in the same pair of years. Figures a-e from Charmantier *et al* (2008). Reprinted with permission from AAAS. (f) Female great tit. Photo © Thor Veen. (g) Winter moth caterpillar on oak. Photo © Jane Carpenter.

(Fig. 1.11e). Therefore the population level change can be explained entirely by the magnitude of the plastic responses of individual females.

Differences between populations

Studies of a great tit population in the Hoge Veluwe, The Netherlands, paint a very different picture (Visser *et al.*, 1998; Nussey *et al.*, 2005). Here, there has also been a similar environmental change during the last three decades (1973–2004), with warmer late spring temperatures and earlier emergence of the tits' caterpillar food supply. However, there has been no change in the tits' egg laying date, with the result that many of the Dutch birds are now breeding too late to catch the caterpillar peak for their hungry offspring. As a result, female lifetime reproductive success has declined over the study period (in contrast to the Wytham population, which is flourishing). Analysis of the variation shown by individual females over successive years, showed that (unlike the Wytham population) females varied in their phenotypic plasticity. Some responded little to annual variation in temperature whereas others showed a marked response (Fig. 1.10b). Furthermore, the variation in plasticity is heritable. In theory, then, the more plastic genotypes should now be favoured by natural selection.

Why do the Dutch and British tits differ? One possibility is that females in the two populations use different proximate cues to time their egg laying (Lyon *et al.*, 2008). For example, if only photoperiod was used as a cue, then individuals would not breed earlier in warmer springs. By contrast, if both birds and caterpillars responded to temperature, or some other common environmental cue, then individual tits would automatically track any yearly variation in caterpillar emergence.

Another possibility is that British and Dutch tits use the same cues but in Britain the cues are better predictors of the food supply that will be available to nestlings. In the Hoge Veluwe, over the last three decades there has been little change in early spring temperatures (when the adult tits are forming food reserves to breed) in contrast to the markedly warmer late spring temperatures, which influence the caterpillar food available to nestlings. The adult tits may, therefore, not have been able to predict the earlier food availability for their offspring (Visser *et al.*, 1998).

The conclusion is that we need to understand the proximate mechanisms used to time egg laying in order to predict how populations will evolve to cope with changing food supplies.

Behaviour, ecology and evolution

We can now summarize the main themes of this book.

Firstly, during evolution natural selection will favour individuals who adopt life history strategies that maximize their gene contribution to future generations. The optimization of clutch size in great tits provides a convincing quantitative test of this, but we shall see later in the book that having offspring is only one of the ways of passing genes on to the future. Another pathway is by helping close relatives to reproduce. One of the questions we shall ask is what factors influence which pathway individuals choose.

Four main themes

Secondly, because an individual's success at survival and reproduction depends critically on its behaviour, selection will tend to design individuals to be efficient at foraging, avoiding predators, finding mates, parental care and so on. Resources are limited, so there will always be trade-offs involved, both within and between these various activities. For example, will an individual avoid predation best by seeking the safety of a group or by hiding away alone? The best place to feed may have the highest predation risk. How are these trade-offs solved by natural selection?

Thirdly, individuals are likely to have to compete with others for scarce resources. As we shall discover, conflict occurs not only between rivals for mates or territories, but also between members of a breeding pair and even between parents and their own offspring. How are such conflicts resolved? Can the outcome sometimes be cooperation rather than overt conflict?

Fourthly, individuals play their behaviour on an ecological stage. Different species live in different habitats and exploit different resources. This, too, is expected to influence an individual's best options. So we will also be exploring how ecological conditions influence how individuals behave.

We will show how the same basic theories can be applied to a wide range of organisms, from microbes to meerkats, and we will see the ingenuity required to design careful experiments to test the theories, both in the field and the laboratory. Most of all, we hope to show how ideas from behavioural ecology can help us to understand and appreciate the marvels of the natural world.

Summary

Behavioural ecology aims to understand how behaviour evolves in relation to ecological conditions, including both the physical environment and the social environment (competitors, predators and parasites). It is important to distinguish proximate factors, which explain how individuals come to behave in a particular way during their lifetime, from ultimate factors, which concern adaptive advantage in evolution. Natural selection works on genetic differences. Examples were discussed to illustrate how genetic differences cause differences in phenotype and behaviour: foraging, learning and courtship in *Drosophila*; foraging in honeybees; colour and mate/habitat choice in geese and mice; and migration strategies in the blackcap, which provide an example of a recent evolutionary change in behaviour.

Individuals are not generally expected to behave for the good of the group but rather to maximize their own gene contribution to future generations. Field experiments reveal that clutch size in great tits maximizes individual lifetime reproductive success. Life history trade-offs include those between quantity and quality of offspring within a brood, and between current and future reproduction.

Recent studies of how great tits have advanced their time of breeding in relation to climate warming provide a good example of phenotypic plasticity (the ability of a single genotype to produce different phenotypes in response to environmental conditions). They also show that a full understanding of evolutionary responses requires studies of proximate and ultimate factors to go hand in hand.

Further reading

The classic books by Niko Tinbergen (1974) and Bert Hölldobler and Edward O. Wilson (1994) convey the delight of watching and wondering in the field. The books by Richard Dawkins (1982, 1989) explain why evolution favours behaviour that benefits individuals and genes, rather than species and groups. Reeve and Sherman (1993) provide a lucid discussion of the distinctions between Tinbergen's four questions and the inter-relationships between them. Scott-Phillips *et al.* (2011) discuss the distinction between proximate and ultimate questions about human behaviour. Robinson *et al* (2008) review genes and social behaviour. Pulido (2007) reviews the genetics and evolution of bird migration. Godfray *et* al. (1991) review clutch size. Both and Visser (2001) show how migrant birds, which breed in northern Europe, may be constrained in their responses to advanced springs on the breeding grounds due to climate change, because their migration from African winter quarters is triggered by day-length variation on the wintering grounds.

TOPICS FOR DISCUSSION

1. Is it possible to investigate the function of a behaviour pattern without understanding also its causation, development and evolution?
2. In this chapter it was concluded that infanticide had evolved because of its advantage to male lions when they take-over a pride. An alternative hypothesis is that it is simply the non-adaptive outcome of the mayhem involved when a new group of males takes over. How would you distinguish between these hypotheses?
3. Discuss how the production of a larger clutch size in one year could lead to decreased reproductive success of the parents in future years. How would you test your hypotheses?
4. Discuss the problems of investigating whether clutch sizes are 'optimal' from field studies. What would you conclude if your study showed that clutch size was sometimes apparently not optimal? (Read: Tinbergen & Both, 1999). Could rapid climate change lead to sub-optimal clutch sizes?

Photo © Susana Carvalho

CHAPTER 2
Testing Hypotheses in Behavioural Ecology

A rigorous scientific approach to the function of behaviour involves four stages: observations, hypotheses, predictions and tests. The first two, observation and hypotheses, often go hand in hand. It may take some time getting to know a particular species before it is possible to ask good questions about its behaviour and ecology. Niko Tinbergen's work on gulls was the result of many years' painstaking observations of their behaviour in the wild. Having observed some aspect of behaviour that we do not understand, how should we proceed?

Let us assume, for example, that we want to discover why our animal lives in a group as opposed to on its own. We may get a strong hint about the function of this simply from observation. For example, if the animal only lived in a group in the breeding season we might suspect that it gained some advantage in terms of increased reproductive success, whereas if it only lived in a group in winter we may suspect some advantage concerned with improved adult survival through feeding efficiency or avoiding predation. We can test our ideas in three main ways:

Three methods of hypothesis testing

(1) *Comparison between individuals within a species.* Individuals in groups may have greater success at feeding or avoiding predators than solitary individuals. Furthermore, success may vary with group size. The problem, however, is that there may be confounding variables: solitary individuals may be poorer competitors and this, rather than their solitary existence per se, may explain their lower success, or individuals in groups may live in better quality habitats, and so on.

(2) *Experiments.* It is often better, therefore, to perform an experiment. With an experiment we can vary one factor at a time; for example, we could change group size and see how this influenced success under a particular set of conditions. Niko Tinbergen pioneered the method of elegant field experimentation to answer

An Introduction to Behavioural Ecology, Fourth Edition. Nicholas B. Davies, John R. Krebs and Stuart A. West.

functional questions. For example, to test the hypothesis that spacing of gull nests reduced predation he put out experimental plots of eggs with different spacing patterns and found that those with a clumped distribution suffered greater predation than those that were spaced out as in nature (Tinbergen *et al.*, 1967).

(3) *Comparison among species.* Different species have evolved in relation to different ecological conditions and so comparison among species may help us to understand how differences in feeding ecology or predation pressure, for example, influence the tendency to live in groups or to be solitary. Using the comparative method is rather like looking at the result of experiments done by natural selection over evolutionary time. The results of these 'experiments' are the designs of the various species' behaviour which we now observe. For example, group living may occur most often in species which experience particular conditions of food or predation.

In this chapter we will focus on these last two methods for investigating adaptation, beginning with comparison among species.

The comparative approach

Correlating species differences in behaviour with differences in ecology

The idea of comparison lies at the heart of most hypotheses about adaptation. It is the comparative study of different species which gives us a feel for the range of strategies that animals adopt in nature. When we ask functional questions about the behaviour of a particular species we are usually asking why it is different from other species. Why does species A live in groups, compared with species B which is solitary? Why do males of species B mate monogamously, compared with males of species A which mate with multiple females (polygyny), and so on? A powerful method for studying adaptation is to compare groups of related species and attempt to find out exactly how differences in their behaviour reflect differences in ecology.

Darwin himself often used comparisons to test ideas about adaptation. In *The Origin of Species* (1859; Chapter 6), he wondered if sutures in the skulls of young mammals had evolved specifically to facilitate birth. However, he noted that sutures also occurred in the skulls of young birds and reptiles 'which only have to escape from a broken egg', so he concluded that sutures have simply risen from 'the laws of growth' and had then 'been taken advantage of in the parturition of the higher animals'. In *The Descent of Man* (1871; Chapter 8), he frequently used comparisons to show the effects of sexual selection, noting that in monogamous species of seals males and females were similar in size, whereas in species where males defended harems the males were 'vastly larger' than the female, and that it was polygynous birds where males had the most elaborate plumage ornaments.

We will first describe some examples which pioneered the comparative approach to behavioural adaptations and inspired workers to use the method with other animal groups. Then we will point out some of the methodological difficulties in formulating and testing hypotheses based on comparison. Finally, we will describe some recent examples of the comparative method which have attempted to overcome these problems.

Breeding behaviour of gulls in relation to predation risk

A suite of adaptations to reduce predation

Most species of gulls nest on the ground, where their eggs and chicks are vulnerable to predation by mammals (such as foxes and stoats) and by birds (such as crows and other gulls). Many of the breeding traits of these ground nesters, such as the black-headed gull (Fig. 2.1a), seem to make good sense as adaptations to reduce predation. For example, the adults take flight whenever a predator approaches and they give alarms and attack it. They maintain the camouflage of the nest by refraining from defecation nearby and, soon after hatching, they remove the empty eggshells, which have white interiors likely to attract predators. The chicks, like the eggs, are cryptically coloured; they leave the nest soon after hatching and hide in the vegetation. This leads to some mixing of neighbouring broods, so it makes sense that adults learn to recognize the calls of their own chicks early on to ensure they direct their parental care to their own young. Parents also give food calls to signal to their hidden young that food is available.

How might we test our hypothesis that these traits have evolved in response to predation? Without testing, of course, our idea is no more than a plausible story. We will see later in the chapter that the function of one of these traits, eggshell removal, can easily be tested by experiment. However, other traits, such as chick behaviour or parental food calling, cannot be so easily manipulated. It would be hard, for example, to manipulate chicks so that they did not hide, or to manipulate parents so they did not call.

Comparing a ground-nester with a cliff-nester

Esther Cullen, one of Niko Tinbergen's research students, showed that a comparison with the breeding traits of a cliff-nesting gull, the kittiwake (Fig. 2.1b), provided comparative support for the anti-predator hypothesis. Kittiwake nests are safer from mammalian predators, who cannot so easily climb down steep cliffs, and they are also safer from avian predators because the gusty winds around the cliffs make attacks from the air more difficult. The different breeding traits of the kittiwake, compared to the black-headed gull, make

(a)

(b)

Fig. 2.1 (a) The black-headed gull nests on the ground. Photo © osf.co.uk. All rights reserved. (b) The kittiwake nests on tiny ledges on steep cliffs. Photo © iStockphoto.com/Liz Leyden.

Traits	Black-headed gull	Kittiwake
Nest site	On ground	On ledge on steep cliffs
Predation risk to nest	High	Low
Adult response to predators	Take flight early; alarm calls, attack predator	Remain on nest until predator close; rarely alarm, weak attack
Nest construction	Loosely built, shallow cup	Elaborate, deep cup
Nest concealment	Adults do not defecate near nest	Adults defecate near nest
	Adults remove eggshells	Adults do not remove eggshells
Chick behaviour	Cryptic colouration (brown with black markings) and behaviour (crouch or hides in vegetation)	Not cryptic (white and grey) and ignores disturbance.
	Weaker claws	Strong claws and muscles for clinging
	Leaves nest after a few days	Remains in nest until can fly (about six weeks)
	Runs off when attacked	Does not run off
	Vigorous wing flapping and jumping during development	Less vigorous movements
Chick recognition by parents	Within a few days	Not until about five weeks, just before young fledge
Chick feeding	Parents give food calls to attract hidden young	No parent food calls
	Adults often regurgitate food onto ground	Adults pass food directly to young's bill

Table 2.1 Comparison of breeding traits of two gulls: the ground-nesting black-headed gull *Larus ridibundus* and the cliff-nesting kittiwake *Rissa tridactyla* (Cullen, 1957).

good sense as a response to this reduced predation pressure (Table 2.1). Thus, adults rarely give alarms and remain on their nests if a predator flies past. The nest cup is more elaborate (to retain the eggs on the tiny cliff ledges). There is no need for a safe nest to be camouflaged, so adults often defecate by the nest (the cliff ledges become splashed with white) and they do not remove empty eggshells. The chicks are not cryptic, they ignore predators and remain in the nest until they can fly (around six weeks of age). With no chances that their own chicks will wander off and become mixed up with other chicks, there is no need for parents to recognize their young soon after hatching. Cullen showed experimentally that kittiwakes accepted foreign young on their nests and did not recognize their own young until much later on, just before fledging. Finally there was no need for parental food calls to announce the arrival of food because the young were not hidden.

The different traits of black-headed gulls and kittiwakes, therefore, provide strong support for our hypothesis that these various suites of behaviour have evolved as adaptations in response to predation differences between the two sites.

Social organization of weaver birds

The gull comparison involved just two species. Including a larger number of species will obviously improve the power of the analysis. The first person to attempt a systematic comparative analysis of social organization was John Crook (1964), who studied about 90 species of weaver birds (Ploceinae). These are small sparrow-like birds which live throughout Africa and Asia, and although many look rather alike there are some striking differences in their social organization. Some are solitary, some go around in large flocks. Some build cryptic nests in large defended territories while others cluster their nests together in colonies. Some are monogamous, with a male and a female forming a permanent pair bond; others are polygamous, the males mating with several females and contributing little to care of the offspring. How can we explain the evolution of this great diversity in behaviour (Fig. 2.2)?

Social behaviour correlated with diet

Crook's approach was to search for correlations between these aspects of social organization and the species' ecology. The ecological variables he considered were the type of food, its distribution and abundance, predators and nest sites. His analysis showed that the weaver birds fell into two broad categories (Table 2.2):

Fig. 2.2 Differences in social organization in weaver birds. (a) Red-headed weaver *Anaplectes melanotis*; a woodland insectivore which often breeds in monogamous pairs on dispersed territories. (b) Southern masked weaver *Ploceus vellatus*; a savannah seed-eater which nests in colonies and is polygynous. (c) Village weaver *Ploceus cucullatus*, another colonial, polygynous savannah species. All photos © Warwick Tarboton.

		Number of species in each category				
		Pair bond		Sociality		
Habitat	**Main food**	**Monogamous**	**Polygynous**	**Solitary**	**Grouped territories**	**Colonial**
Forest	Insects	17	0	17	0	1
Savannah	Insects	5	1	4	0	2
Forest	Insects + seeds	3	0	2	0	1
Savannah	Insects + seeds	1	7	1	0	7
Grassland	Insects + seeds	1	1	1	0	1
Savannah	Seeds	2	11	0	1	16
Grassland	Seeds	0	15	0	13	3

Table 2.2 Social organization of weaver bird species (Ploceinae) in relation to habitat and diet (Crook, 1964; Lack, 1968).

(1) Species living in the forest tended to be insectivorous, solitary feeders, defend large territories and build cryptic solitary nests. They are monogamous and males and females have similar plumage.

(2) Species living in the savannah tended to eat seeds, feed in flocks and nest colonially in bulky conspicuous nests. They are polygamous and there is sexual dimorphism in plumage, the males being brightly coloured and the females rather dull.

Predation and food dispersion are key selective forces

Why is the behaviour and morphology of the weaver birds linked to their ecology in such a striking way? Crook invoked predation and food as the main selective pressures that have influenced the evolution of social organization. His argument was as follows:

(1) In the forest, insect food is dispersed. Therefore, it is best for the birds to feed solitarily and defend their scattered food resources as a territory. Because the food is difficult to find, both parents have to feed the young and, therefore, they stay together as a pair throughout the breeding season. With the male and female visiting the nest, both must be dull coloured to avoid attracting predators. Cryptic nests spaced out from those of neighbours decrease their vulnerability to predation.

(2) In the savannah, seeds are patchy in distribution and locally superabundant. It is more efficient to find patches of seeds by being in a group because groups are able to cover a wider area in their search. Furthermore, the patches contain so much food

> that there is little competition within the flock while the birds are feeding. In the savannah, birds cannot hide their nests and so they seek safety in protected sites, such as spiny acacia trees. Nests are sometimes bulky to provide thermal insulation against the heat of the sun. Because good breeding sites are few and scattered, many birds nest together in the same tree. Within a colony, males compete for nest sites and those that defend the best sites attract several females while males in the poorer parts of the colony fail to breed. In addition, because food is abundant, the female can feed the young by herself, so the male is emancipated from parental care and can spend most of his time trying to attract more females. This has favoured brighter plumage coloration in males and the evolution of polygamy.

Supporting evidence for this interpretation comes from species with intermediate ecology (Table 2.2). The grassland seed eaters have patchy food supplies, so group living is favoured for efficient food finding. However, in grassland the nests are vulnerable, so predation favours spacing out. The result is a compromise; these species have an intermediate social organization, nesting in loose colonies and feeding in flocks.

These results suggest that food and predation are important in determining social organization. They also reveal how several different traits, such as nests, feeding behaviour, plumage colour and mating system, can all be considered together as a result of the same ecological variables. Crook's work with the weaver birds inspired several people to use the comparative method to study social organization in other groups. David Lack (1968) extended the argument to include all bird species and Peter Jarman (1974) used the same approach for the African ungulates.

Social organization in African ungulates

Jarman (1974) considered 74 species of African ungulates; all eat plant material but differences in the precise type of food eaten are correlated with differences in movements, mating systems and anti-predator behaviour (Fig. 2.3). The species were grouped into five ecological categories (Table 2.3). Just as in the weaver birds, several adaptations seem to go together.

Body size, diet and social organization

The major correlate of diet and social organization is body size. Small species have a higher metabolic requirement per unit weight and need to select high quality food, such as berries and shoots. These tend to occur in the forest and are scattered in distribution, so the small species are forced to live a solitary existence. The best way to avoid predators in the forest is to hide. Because the females are dispersed, the males are also dispersed and the commonest mating system is for a pair to occupy a territory together.

At the other extreme, the largest species eat poor quality food in bulk and graze less selectively on the plains. It is not economical to defend such food supplies and these species wander in herds, following the rains and fresh grazing. In these large herds there is potential for the strongest males to monopolize several females by defence of a harem or a dominance hierarchy of mating rights. When predators come along these species cannot hide on the open plains, so either flee or rely on safety in numbers in the herd. Ungulates of intermediate size show aspects of ecology and social organization in between these two extremes (Table 2.3).

(a)

(b)

(c)

Fig. 2.3 Differences in social organization of African ungulates. (a) Kirk's dik-dik *Madoqua kirki* live in pairs in woodland. Photo © Oliver Krüger. (b) Impala live in small groups in open woodland and grassland. Photo © Bruce Lyon. (c) Wildebeest graze in huge herds out on the open plains. Photo © iStockphoto.com/William Davies.

Limitations of early comparative studies

These early studies revealed the promise of the comparative approach in behavioural ecology. However, there were limitations in the methodology. Many of these are not unique to comparative studies and it is worth bearing them in mind throughout the book.

(a) *Alternative hypotheses*

Problems in interpreting comparative data

The explanations for the differences in behaviour are certainly plausible, but alternative hypothesis have not been considered in a rigorous manner. For example, the nest site difference between the black-headed gull and kittiwake is also likely to be correlated with differences in shelter, competition for nest sites and proximity to the feeding grounds. How can we be sure that predation is the key selective pressure, rather than one of these other variables?

Table 2.3 The social organization of African ungulates in relation to their ecology (Jarman, 1974).

	Exemplary groups	Body weight (kg)	Habitat	Diet	Group size	Reproductive unit	Anti-predator behaviour
Grade I	Dik-dik Duiker	3–60	Forest	Selective browsing; fruit, buds	1 or 2	Pair	Hide
Grade II	Reedbuck Gerenuk	20–80	Brush, riverine grassland	Selective browsing or grazing	2–12	Male with harem	Hide, flee
Grade III	Impala Gazelle Kob	20–250	Riverine woodland, dry grassland	Graze or browse	2–100	Males territorial in breeding season	Flee, hide in herd
Grade IV	Wildebeest Hartebeest	90–270	Grassland	Graze	Up to 150 (thousands on migration)	Defence of females within herd	Hide in herd, flee
Grade V	Eland Buffalo	300–900	Grassland	Graze unselectively	Up to 1000	Male dominance hierarchy in herd	Mass defence against predators

(b) *Quantification of ecological variables*
The ecological variables have not been quantified. For weaver birds, for example, are insects 'dispersed' and are seeds 'patchy'? How exactly will these differences influence the economics of exploitation by individuals?

(c) *Cause and effect*
Consider the observation that weaver birds with a diet of seeds go about in flocks. Our explanation was that seed eating selects for flocking because this is the best way to find a patchy food supply. However, we could equally well have suggested that predation selects for flocking and, as a consequence, the birds are forced to select locally abundant food so all the flock can get enough to eat. In this case a diet of seeds is a consequence, or effect, of flocking, not a cause. Maybe predation also selects for flocking in the forest insectivores but because their diet is incompatible with flocking they have to forage singly.

(d) *Alternative adaptive peaks or non-adaptive differences*
It is tempting when comparing between species to assume that differences are always adaptive but some differences may simply be alternative solutions to the same ecological pressures. An ecologist from Mars who visited the Earth would observe that in the United States people drive their cars on the right hand side of the road while in the United Kingdom they drive on the left. He would then perhaps make lots of measurements in an attempt to find ecological correlates to explain the adaptive significance of the difference. In fact, driving on the right and driving on the left may just be equally good alternatives for preventing accidents (Dawkins, 1980).

Some differences between animals may be like this. Sheep use horns for fighting and deer use antlers. Horns are derived from skin while antlers are derived from bone. The differences between horns and antlers need not necessarily reflect ecological differences; it may simply be a case of evolution working with different raw materials to produce the same functional end. The problem with non-adaptive explanations is that they are hypotheses of the last resort. Further scientific enquiry is stifled. Maybe there is an adaptive explanation for the difference but we just haven't discovered it yet. For example, antlers are dropped and then renewed each year whereas horns are not. Perhaps this difference is related to the extent of seasonal variation in mating competition and food supply?

Some differences between species may reflect different solutions to the same problem

(e) *Statistical analysis and independent data*

We need statistical analysis to tell us how confident we can be in our conclusions. To do this, we need to think carefully about what constitutes independent data points. For example, in Crook's analysis of the weaver birds (Table 2.2), 14 of the 16 grassland species belong to one genus, *Euplectes*. Can we consider all these as the outcomes of independent 'evolutionary experiments'? Congeneric species may often have similar behaviour simply through common ancestry. In this case, analysis of species data will be statistically biased by those genera containing large numbers of species.

What are the independent data?

Summary

These criticisms are important, but they certainly do not mean that the comparative method is a failure. On the contrary, the approach is impressive in the way it brings together such a wide diversity of behavioural and morphological traits within the same ecological framework. Crook's study of the weaver birds and Jarman's work on the antelopes have served as models for ecological work on other groups of species. However, the most recent comparative studies have attempted to control for these various problems, and we will now discuss other examples, bearing the criticisms in mind, to illustrate how changes in methodology have made comparison between species a more rigorous exercise.

Comparative approach to primate ecology and behaviour

As with the weaver birds and antelopes, primates vary in their social organization (Fig. 2.4). There are solitary insectivores, like tarsiers, which live in forests and are nocturnal. There are diurnal forest monkeys, like colobus monkeys, which go around in small groups, feeding on leaves or fruit. Other monkeys, like baboons, are terrestrial and live in large troupes of 50 or several hundred individuals. Among the apes, the orang-utan is solitary, the gibbon lives in pairs and small family units, while the chimpanzee may live in bands of 50.

In the 1970s and 1980s, Tim Clutton-Brock and Paul Harvey used the comparative approach to analyse the evolution of this diversity. Their analyses introduced three marked improvements in methodology over the earlier comparative studies. Firstly, they measured the various aspects of behaviour and morphology on a continuous scale (rather than categorizing primates into groups with different traits). Secondly, they considered alternative hypotheses, and used multivariate statistics to tease out the effects of different

Three improvements in comparative studies

Fig. 2.4 Differences in social organization in primates. (a) A solitary insectivorous tarsier. Photo © iStockphoto.com/ Holger Mette. (b) A small group of black and white colobus monkeys, which eat leaves in the forest. Photo © iStockphoto.com/ Henk Bentlage. (c) A large group of gelada baboons, which feed on the ground on grass leaves and roots. Photo © iStockphoto.com/ Guenter Guni.

ecological variables on each trait. Thirdly, they used different genera as independent data points for analysis, rather than species, to reduce the problem of similarity through common ancestry. Some examples are now discussed to illustrate their approach.

Home range size

Larger animals need to eat more food and so, in general, we would expect them to have larger home ranges. Therefore, if we want to examine the influence of an ecological variable, such as diet, on home range size, we have to control for body weight as a confounding variable. When home range size is plotted against the total weight of the group that inhabits it, as expected the larger the group weight the larger the home range (Fig. 2.5).

Variation with diet

The influence of diet on home range size can be seen when the specialist feeders (insectivores, frugivores) are separated from the leaf eaters (folivores); the specialist feeders have larger home ranges for a given group weight. The probable explanation is

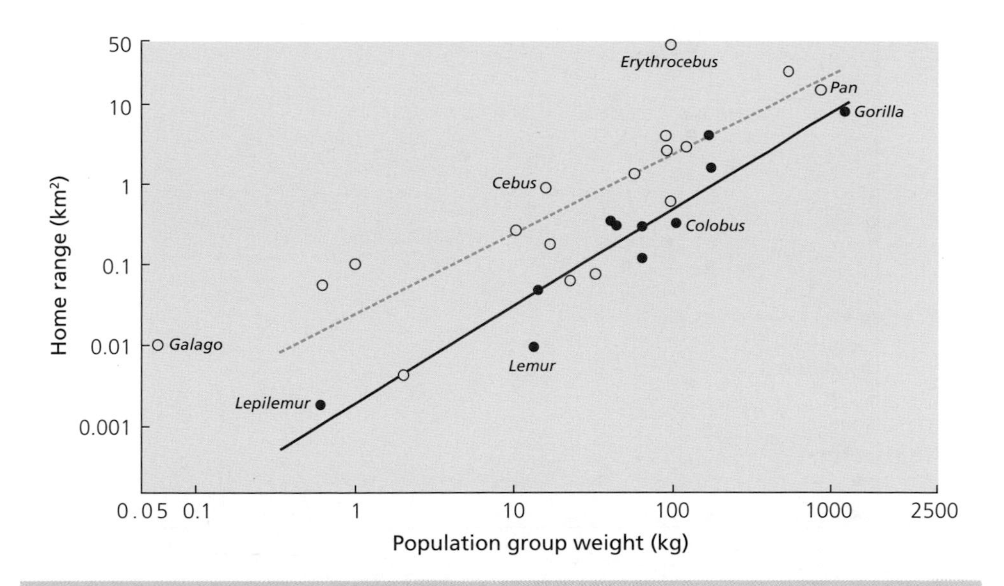

Fig. 2.5 Home range size plotted against the weight of the group that inhabits the home range for different genera of primates. The solid circles are folivores, through which there is a solid regression line. The open circles are specialist feeders (insectivores or frugivores) and the regression line through these points is dashed. Some of the genera are indicated by name. From Clutton-Brock and Harvey (1977).

that fruit and insects are more widely dispersed than leaves, so specialist feeders need a larger foraging area in which to find enough food.

Sexual dimorphism in body weight

In primates, males are often larger than females. Two hypotheses could explain this observation. Sexual dimorphism could enable males and females to exploit different food niches, and thus avoid competition (Selander, 1972). If this was true, then we might predict that dimorphism would be greatest in monogamous species where males and females usually associate together and feed in the same areas. Alternatively, it could have evolved through sexual selection, large body size in males being favoured because this increases success when competing for females (Darwin, 1871). If sexual competition is important then we would predict that dimorphism should be greater in polygamous species, where large male size would be especially advantageous because a male could potentially monopolize several females.

Sexual dimorphism evolves from sexual competition

The comparative data show no sign of the trend predicted by the niche separation hypothesis but do support the sexual competition hypothesis; the more females per male in the breeding group, the larger the male is in relation to the female (Fig. 2.6).

Sexual dimorphism in tooth size

Males often have larger teeth than females. Again, two hypotheses can be suggested (Harvey *et al.*, 1978). Large teeth may have evolved in males for defence of the group

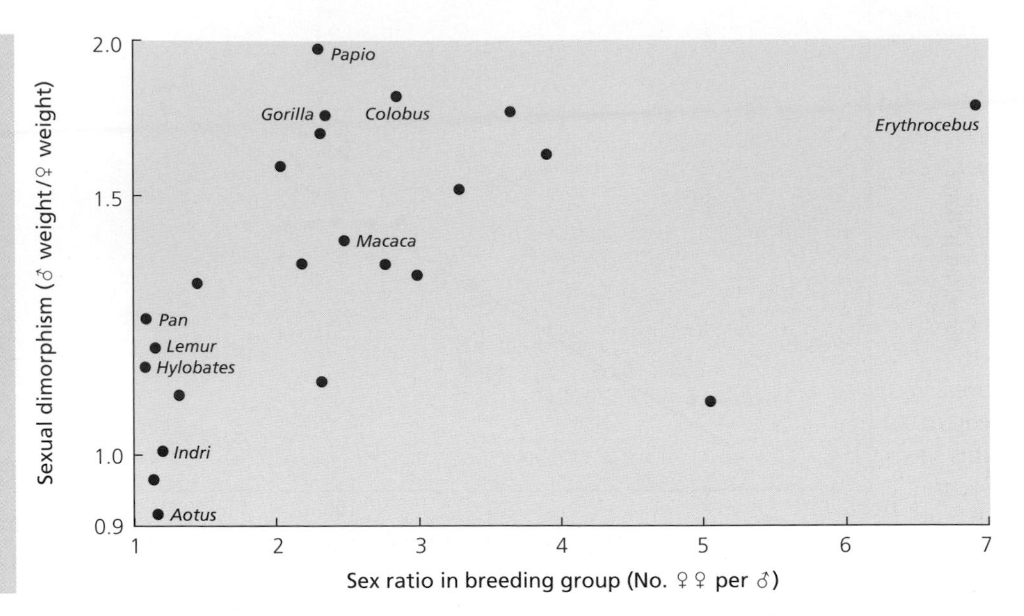

Fig. 2.6 The degree of sexual dimorphism increases with the number of females per male in the breeding group. Each point is a different genus, some of which are indicated by name. From Clutton-Brock and Harvey (1977).

against predators. Alternatively, males may have larger teeth for competition with other males over access to females. There is the problem here of body weight as a confounding variable; males are larger than females, so a difference between the sexes in tooth size could just reflect a difference in body size.

This can be controlled for by calculating the line of best fit when female tooth size is plotted against body weight. If the tooth size of a male is now plotted on the same graph, it can be seen whether its size is greater than expected for a female of the same body weight. The results show that in monogamous species male tooth size is as expected for a female of equivalent body weight. However, it is larger than expected in harem-forming species. These data support the sexual competition hypothesis for the evolution of larger teeth in males. Nevertheless, we cannot exclude the predator defence hypothesis because maybe the harem-forming species are the ones most vulnerable to predation.

The analysis can be taken a step further by considering species where several males live together in a group (multimale troops). It is found that, within this type of social organization, the males of terrestrial species have larger teeth for their body size than arboreal species. Therefore, even within the same mating system there is a difference in tooth size in different habitats. The terrestrial environment is usually thought to present greater risks of predation, so predation pressure may have been responsible for the evolution of larger teeth in terrestrial species.

Sexual competition and defence against predators may both be important

Our conclusion is that both sexual competition and predation may have influenced the evolution of sexual dimorphism in tooth size. There is also the further possibility that differences in tooth size are important in reducing diet overlap between the sexes, so preventing competition for food. This example shows that, even with careful analysis, it may be difficult to tease out the effect of several variables on the evolution of a trait.

Testis size and breeding system

The heaviest primates, the gorilla (*Gorilla gorilla*) and orang-utan (*Pongo pygmaeus*) have breeding systems that involve one male monopolizing mating with several females, and have testes that weigh 30 and 35 g, respectively (average weight of both testes). The smaller chimpanzee (*Pan troglodytes*), by contrast, has a breeding system where several males copulate with each oestrus female and this species has testes weighing 120 g! It seems likely that the marked differences in testes weights are related to differences in breeding system. In single-male breeding systems (gorilla and orang-utan) each male needs ejaculate only enough sperm to ensure fertilization. In multimale systems (chimpanzee), however, a male's sperm has to compete with sperm from other males. Selection should, therefore, favour increased sperm production and, hence, larger testes.

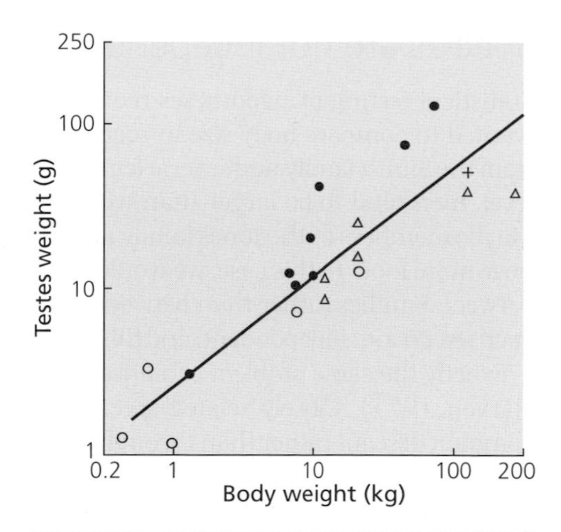

Fig. 2.7 Log combined testes weight (g) versus body weight (kg) for different primate genera. Solid circles are multimale breeding systems. Open circles are monogamous. Open triangles are single-male systems (one male with several females). The cross is our own species, *Homo*, for comparison. From Harcourt *et al.* (1981). Reprinted with permission from the Nature Publishing Group.

Larger testes in multimale groups

Harcourt *et al.* (1981) tested this hypothesis by comparing 20 genera of primates, varying in body size from the 320 g marmoset (*Callithrix*) to the 170 kg gorilla. Figure 2.7 shows that, as expected, testes weight increases with body weight. For a given body weight, however, it is clear that genera with multimale breeding systems have heavier testes than genera with single-male or monogamous breeding systems. The data points for the former group lie above the line, and those for the latter lie below ('single-male' indicates that there is only one breeding male although, as in the gorilla, there may be more than one male in the social group; 'monogamous' indicates that there is just one male and one female in a group). These data therefore support the sperm competition hypothesis.

Using phylogenies in comparative analysis

Since the mid 1980s, there has been a further major advance in comparative analysis. This is to use phylogenies, firstly to identify independent evolutionary transitions and, secondly, to elucidate the order in which traits have evolved (Felsenstein, 1985; Grafen, 1989; Harvey & Pagel, 1991). Before we describe *how* this is done, we must first expand on *why* it needs to be done.

Species are not independent

Statistical testing of hypotheses requires that data points are independent. Imagine we wanted to compare body size in men and women, and that we measured several males from the Smith family and several females from the Jones family. Although, at a population level, men tend to be larger than women, we might not obtain this result in our study. Maybe members of the Jones family are particularly tall, or maybe they are wealthier and have more food. In this case, we would obtain a spurious result that is driven by differences between families rather than between the sexes. Put another way, the data points within each sex are not independent, and this can increase the likelihood of incorrect conclusions.

Species may be similar through common descent ...

Exactly the same problem can arise when comparing between species (Clutton-Brock & Harvey, 1977). Closely related species tend to be similar because they share traits by common descent rather than through independent evolution. To give an extreme example, the Australian mammals are mostly marsupials that carry infants in pouches, whereas British mammals are all placental, using a placenta to nourishing their young in the mother's uterus. If we compared British and Australian mammals, then we would find that having a pouch would correlate with any environmental variable that differed between Britain and Australia, such as mean temperature, rainfall, proportion of land that is desert and so on. However, to infer that any of these variables has selected for this difference would be foolish. The difference is much more easily explained by the historical fact that the placental mammals evolved after Australia became isolated from the other continents.

... which may bias comparative analyses

The problem of non-independence does not rely on such extreme historical patterns or a lack of evolutionary flexibility in a trait (Ridley, 1989). For example, considering the patterns across primates, the gibbons (*Hylobates* spp.) are all monogamous, eat fruit and hold territories. Consequently, every time you add a new species of gibbon to a comparative study, you increase the extent to which these traits are correlated across species. Whilst this could plausibly be because natural selection links these traits (e.g. large territories are required to obtain enough fruit, and this spreads individuals out, favouring monogamy), it could also be explained by an almost infinite number of possibilities (e.g. gibbons are fruit eating specialists, but it is something else that they all do that favours monogamy).

Phylogenies

To solve the problem of species non-independence, it is necessary to take phylogenies into account. A phylogeny is a tree which shows the evolutionary relationships among species (Fig. 2.8a; ignore the details for the moment). Initially, morphological traits were used to construct these trees, but nowadays they are usually based on similarity in DNA sequences in nuclear or mitochondrial genes; the more similar two species are, the more recently they must have shared a common ancestor. If the mutation rate is known (by calibration from fossils or geological events of known dates), then the magnitude of the difference in DNA sequence becomes a 'molecular clock' which estimates the time since two species last shared a common ancestor. Branch lengths in the tree can then indicate the time that has elapsed since divergence. Various statistical methods are used for reconstructing the most likely phylogeny, given the DNA sequences of extant species.

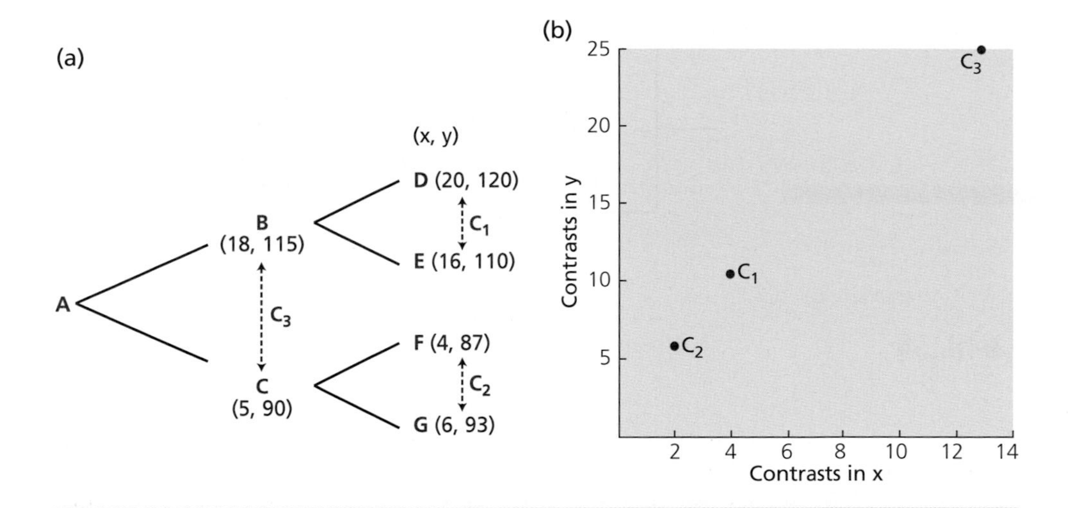

Fig. 2.8 (a) A simple phylogeny. In this tree, A gave rise to two descendants, B and C, each of which gave rise to two more descendants: D and E, and F and G. The character states (x, y) of the four living species (D, E, F, G) are measured. Those of their ancestors (B and C) have to be estimated. In this case they are assumed to have values intermediate between those of their descendants. There are three independent contrasts in this tree: C_1, C_2 and C_3. (b) Plotting contrasts in x against contrasts in y shows that there has been correlated evolution in these two traits.

Estimating ancestral states

The character states (diet, brain size, mating system and so on) of extant species can be measured. How can we know what their extinct ancestors were like? For morphological traits, fossils can be useful but behavioural traits rarely leave a fossil record. We have, therefore, to make an educated guess about ancestral states. Again, various statistical methods are available. The simplest method is parsimony: assign ancestral states to minimize the number of evolutionary changes in the tree from ancestral to extant species. More complex methods are maximum likelihood and Bayesian statistics, which consider which are the most likely ancestral states among various possibilities. In general, if there are frequent changes in the tree then ancestor reconstructions become more uncertain, especially for more distant ancestors (Schluter *et al.*, 1997).

Independent contrasts

Using phylogenies to identify independent evolutionary changes ...

Joe Felsenstein (1985) introduced the method of independent contrasts to solve the problem that species are not independent. Fig. 2.8 is a simple example to explain the method. The key point is that we can assume that two species have evolved independently since their divergence. Therefore, their degree of divergence is independent (statistically) from other changes in the tree. These divergences between related taxa provide independent changes, or contrasts, for our analysis (D versus E and F versus G in Fig. 2.8a). In addition to comparing pairs of species at the tips of the tree, we can also work backwards and compare at higher levels, effectively comparing groups of species or ancestors (B versus C). This is often done by assuming that ancestral values of continuous traits (e.g. brain size)

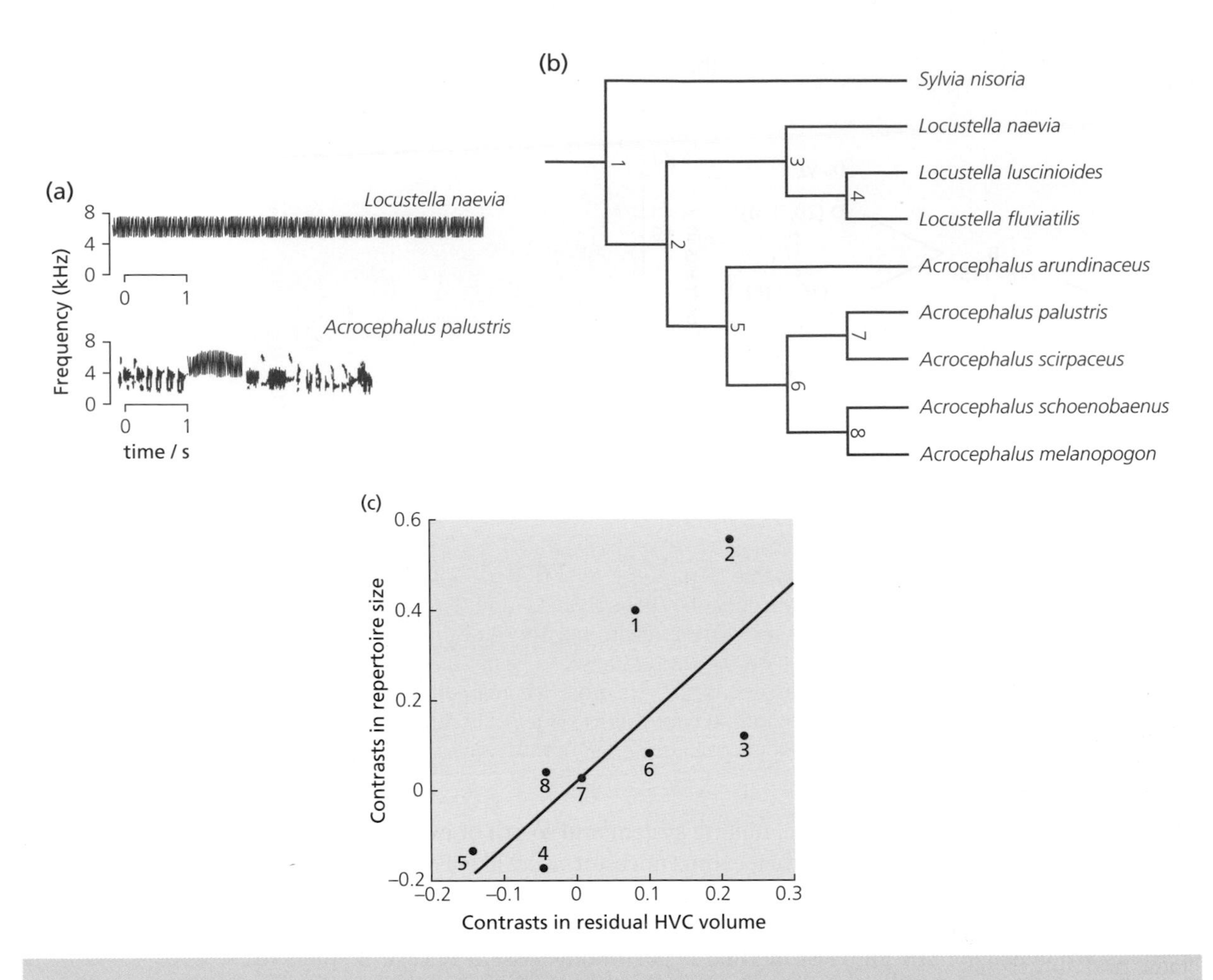

Fig. 2.9 Song complexity and brain anatomy in European warblers (family Sylviidae) from the genus *Acrocephalus* and *Locustella*. (a) Some species, like the grasshopper warbler *L. naevia*, have very simple songs (in this species, one syllable is repeated). Others, like the marsh warbler *A. palustris*, have a complex song with up to a hundred different syllable types in their repertoire. (b) Phylogeny of the *Acrocephalus* and *Locustella* warblers. The numbers refer to the eight independent contrasts used in the analysis. (c) Correlation between contrasts in syllable repertoire size and contrasts in volume of the higher vocal centre (HVC) of the brain (corrected for body size). The eight independent contrasts are labelled. From Szekely *et al.* (1996).

are intermediate between those of two descendant species (e.g. B is the mean of D and E). Fig. 2.9 is an example using this method, which shows that during evolution an increase in song complexity in warblers is correlated with an increase in the volume of a brain nucleus (the higher vocal centre) involved in song learning. This relationship is evident, too, in a wider comparison across 45 species of passerine birds (De Voogd *et al.*, 1993).

This phylogenetic method encouraged a re-analysis of the data on primate testis size (Fig. 2.7). A phylogeny of 58 primate species revealed seven independent pairwise comparisons between multimale and single-male taxa. In all seven cases, the multimale

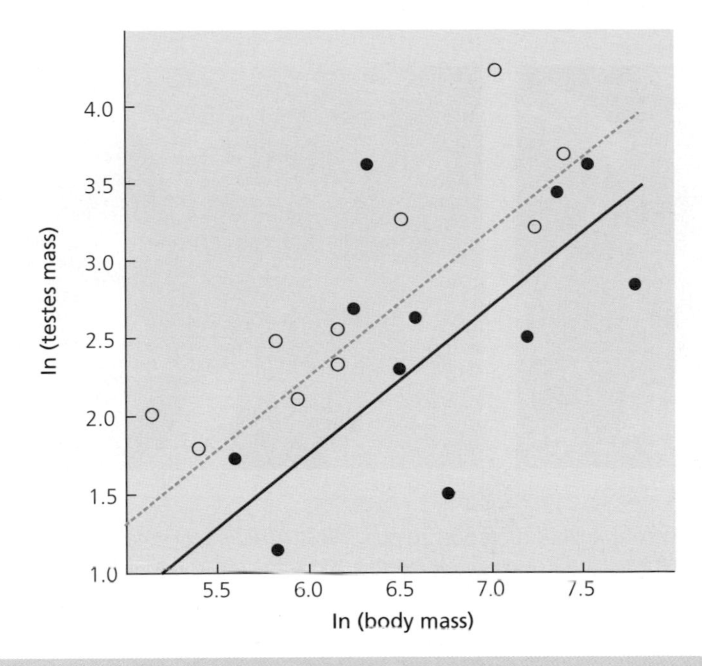

Fig. 2.10 Relationship between testis mass and body mass in bushcricket species (Tettigoniidae) with low (filled circles) and high (open circles) degrees of polyandry. Phylogenetic information was incorporated into the statistical analysis by weighting current species values by the distance separating them in the phylogeny. The lines, fitted from a phylogenetic model, are for low (solid line) and high (dashed line) degrees of polyandry. From Vahed *et al.* (2010).

clade had larger testes, relative to body weight, confirming the original study's conclusion (Harcourt *et al.*, 1995). A comparative analysis has revealed that increased sperm competition also leads to larger testes in bushcrickets (Fig. 2.10).

Discrete variables and the order of change during evolution

... and the order of changes

Characters such as brain size and testis size are continuous variables. Others we might want to consider in a comparative analysis are discrete variables, for example diet (insect versus seeds) or mating system (single versus multimale). Statistical methods have been developed to determine likely ancestor states and which order of character change is most likely in a tree (Pagel, 1994; Pagel & Meade, 2006). For example, in Crook's weaverbirds, did an evolutionary change to a seed diet lead to an increase in flocking, or was it a change to flocking that favoured the evolution of a seed diet? If we could determine this, then we could answer the cause–effect question we raised earlier.

We now turn to another primate example to show how the analysis of discrete traits in a phylogeny can help to identify both independent evolutionary transitions and the order of change.

Fig. 2.11 Sexual swellings in female chimpanzees (Bossou, Guinea, West Africa). (a) Female on the left with male retreating on the right. Photo © Kathelijne Koops. (b) A 42-year old female carrying her five-year old daughter on her back, being inspected by an adult male. She became pregnant soon after this photo was taken. Photo © Susana Carvalho.

Sexual swellings in female primates

Darwin was puzzled

In some species of old world monkeys and apes, females advertise their sexual receptivity with visually conspicuous sexual swellings (Fig. 2.11). 'In my *Descent of Man*', wrote Charles Darwin (1876), 'no case interested or perplexed me so much as the brightly-coloured hinder ends and adjoining parts of certain monkeys'. Species with swellings, for example baboons and macaques, tend to live in groups with several sexually active males ('multimale' groups; Clutton-Brock & Harvey, 1976). Similarly, among the apes, pronounced swellings occur in female chimpanzees, which live in large multimale groups, but not in gibbons, orang-utans or gorillas, which live in smaller, single-male groups involving either monogamy or a male with a harem. In total, across 70 species, none of 29 species living in single-male groups has sexual swellings compared to 29 out of 41 species (71%) living in multimale groups (Nunn, 1999).

Female sexual swellings occur in multimale groups

A phylogenetic analysis supports the association between swellings and multimale groups (Fig. 2.12). Swellings have evolved three times independently in the old world monkeys and apes, and in all three cases this transition is associated with the evolution of multimale groups from a single-male ancestral state. A further phylogenetic analysis, using statistical methods to reconstruct probable ancestral states, suggests that multimale mating systems most likely evolved before sexual swellings (Fig. 2.13). Therefore, it was a new selection pressure in multimale groups that led to the evolution of swellings, rather than swellings favouring the evolution of multimale groups.

There are also two losses of swellings in Fig. 2.12. In one of these (number 2) this was associated with a change from multimale back to single-male grouping, as predicted. In the other case, however, loss of swellings occurred within a multimale system (loss number 1 in Fig. 2.12). A likely reason for this is discussed later on.

The graded signal hypothesis

Why are swellings associated with multimale grouping? When there are several males in her group, a female is faced with a delicate balancing act. On the one hand, she

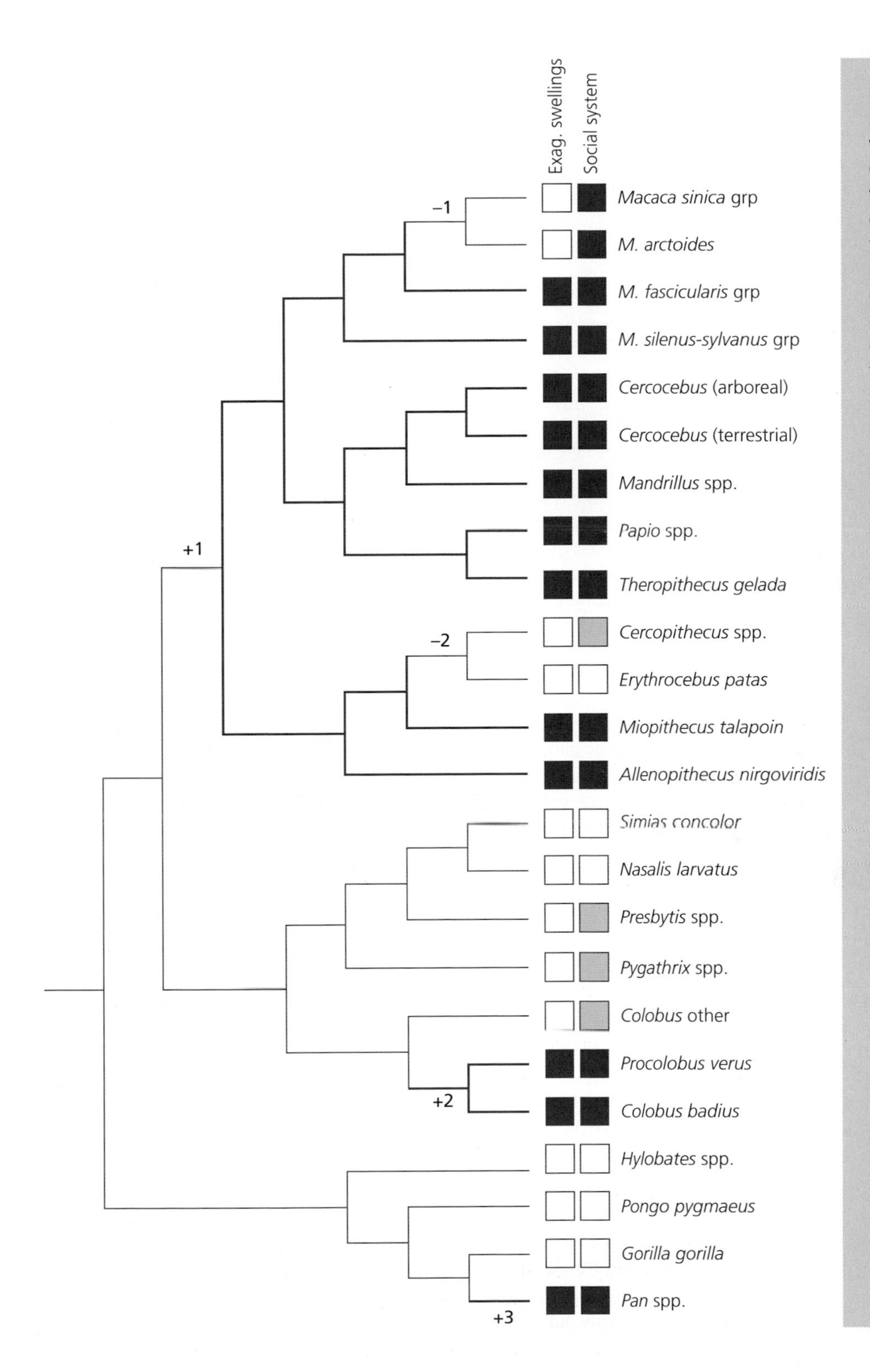

Fig. 2.12 Phylogeny of old world monkeys (Purvis, 1995), with some clades collapsed to facilitate presentation of data. The terminal branches are extant species and their traits are indicated. In the left hand column is presence (purple box) or absence (white box) of sexual swellings. In the right hand column is multimale (purple box) or single-male (white box) mating system. The pale shaded boxes indicate taxa with both single-male and multimale mating systems. The ancestral state is most likely that of no swellings. There have been three gains of sexual swellings in this tree (indicated by +1, +2, +3) and two losses (−1, −2). From Nunn (1999). With permission from Elsevier.

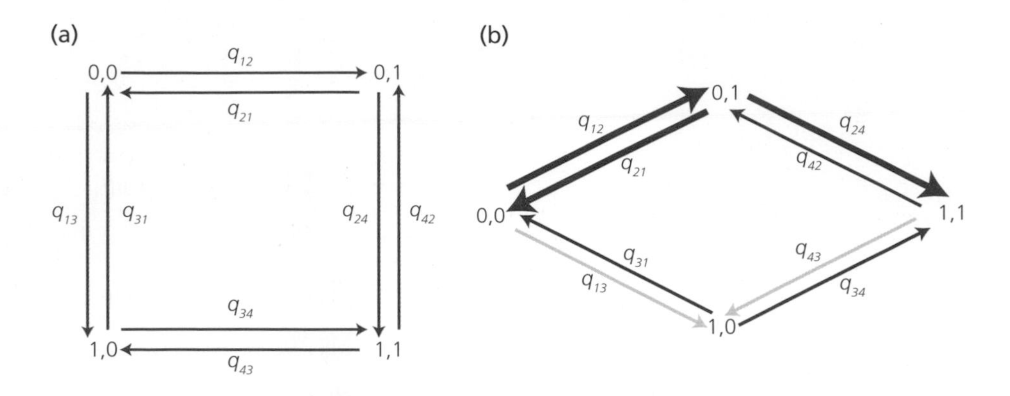

Fig. 2.13 (a) The correlated evolution of discrete traits. Consider two discrete traits in primates: the first (x) is absence (0) or presence (1) of oestrus advertisement by sexual swellings; the second (y) is a single-male (0) or multimale (1) mating system. The eight arrows indicate the possible transitions between the four states. Statistical methods are used to quantify the evolutionary rates of these transitions in a primate phylogeny and to find the model which best describes the evolutionary transitions. For example, if $q_{1,2} = q_{3,4}$, then this implies that the transition to a multimale mating system was independent of the absence or presence of oestrus signals. (b) Flow diagram, showing the statistically most probable evolutionary routes in the primate phylogeny, from an ancestral state of no sexual swellings and a single-male/monogamous mating system (0,0) to a derived state of sexual swellings and a multimale mating system (1,1). Thinnest arrows correspond to transition rates with a high (>94%) posterior probability of being zero. Thickest arrows are the most frequent transitions, with thinner arrows less frequent. The combination of arrows indicates that the mating system changes firstly in evolution (state 0,1) and this then selects for a change in female display to sexual swellings (1,1). The alternative hypothesis, that swellings evolve first (1,0) and this then selects for a multimale system (1,1) is not supported. From Pagel & Meade (2006). With permission of the University of Chicago Press.

may gain from copulating with the dominant male because he is likely to be the best genetic sire and the one best able to protect her and her offspring from predators, and from harassment by other males in the group. On the other hand, if it was clear to all males that the dominant male had certain paternity, then subordinate males might harm or even kill her offspring (Hrdy, 1979). Charles Nunn (1999) suggested that sexual swellings might provide a graded signal, which both enables the female to bias paternity chances to the dominant male while at the same time enhancing opportunities of mating with subordinate males, too, at times when she is still potentially fertile. This would give each male sufficient paternity chances that they would all protect her and her offspring, or at least not be a threat.

In support of the graded signal hypothesis, ovulation is most likely during a female's period of maximal swelling, which is when the dominant male guards her and copulates most frequently (Fig. 2.14). During this period, he is likely to use olfactory cues, too, to fine-tune his assessment of female fertility (Higham *et al.*, 2009). However, individual females may ovulate over several days around maximal swelling. Furthermore, females of species with swellings are sexually active, and hence attractive to males, for about twice the length of time per ovulatory cycle (11 days) compared to species without swellings (five days;

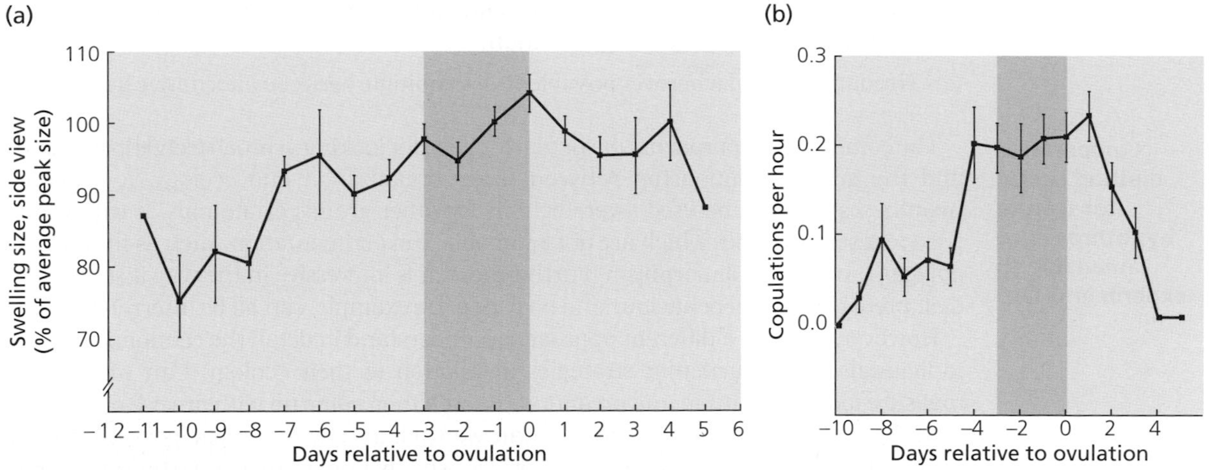

Fig. 2.14 Sexual swellings in a group of wild chimpanzees studied in an evergreen forest in the Tai National Park, Côte d'Ivorie. (a) Swelling size in 12 females (mean ± SE) aligned to the day of ovulation (day 0). Swellings measured from photographs and ovulation determined by enzyme immunoassays from urine samples. The shaded area indicates the fertile phase, when fertilization is most likely. (b) Alpha male copulation rate (mean ± 1SD) during the phase of maximum swellings. Data from 10 females, aligned to the day of ovulation (day 0). Fertile period shaded. From Deschner *et al.* (2004). With permission from Elsevier.

Nunn, 1999). This makes it less likely that the dominant male can monopolise a female for her entire fertile period and so increases her chances of copulating with several males.

Breeding synchrony may also be important

A further factor influencing subordinate male access will be the synchrony of fertile females in the group. Subordinate males will have more chances to mate if the dominant male shifts his attention to other attractive females. This is the likely explanation for why swellings tend to occur most in multimale species that are non-seasonal breeders (e.g. baboons) and not in species which are strongly seasonal in their reproduction (e.g. vervet monkeys). In seasonal breeders, all the females come into oestrus within a few weeks, so females are more likely to overlap in their receptivity. This means that a dominant male is less likely to attempt to monopolise a female, so subordinate males can more readily get access. Hence, sexual swellings are not such an advantage to females. In this context, it is interesting to note that the evolutionary transition from swellings to no swellings within a multimale system involved an increase in breeding seasonality (the *sinica* group of macaques; Fig. 2.12; Nunn, 1999).

The comparative approach reviewed

The statistical approach we have described is certainly a major improvement on the first applications of the comparative method. To summarize, the main improvements are:

Main improvements in modern studies

(1) Different aspects of social organization are treated independently.

(2) Confounding variables are dealt with in a rigorous manner.

(3) Phylogenies are used to identify independent evolutionary transitions and the likely order of trait changes during evolution.

(4) The data are used wherever possible to discriminate between alternative hypotheses.

Comparative method useful for testing hypotheses not amenable to experimentation

The comparative approach is particularly useful for looking at broad trends in evolution and the general relationship between social organization and ecology. It generates hypotheses that can be used as predictions for other groups of animals. It can also be used to test hypotheses which are not amenable to experimentation, such as the effect of polygamy on sexual dimorphism. Furthermore, it is impressive in the way it shows how diet, predation, social behaviour and body size, for example, can all be interrelated.

However, we need a different approach to understand in detail the economics of why individuals adopt particular strategies in relation to their ecology. Can we actually measure food distribution and predation risk and then come up with precise predictions as to how an individual will behave? Can we explain why a monkey goes round in a group of 20 rather than in one of 15 or 25, why its home range is 10 ha rather than 8 or 12 ha, and why it spends one hour in a patch of fruit before moving on? Indeed we can attempt to answer precise questions like these using optimality theory and an experimental approach.

Experimental studies of adaptation

We now turn to a different, and complementary, way of looking at how selection moulds behaviour. Instead of broad scale comparisons between species, the emphasis will be on the behaviour of individuals of the same species and analysing their behaviour in terms of *costs* and *benefits*.

Costs and benefits of eggshell removal in gulls

The idea of trying to measure costs and benefits grew out of Niko Tinbergen's experimental approach to studying the adaptive advantage of behaviour. For example, Tinbergen observed that in a colony of black-headed gulls nesting on sand dunes in north-western England, incubating parents always pick up the broken eggshell after a chick has hatched and carry it away from the nest (Fig. 2.15a). Although carrying the shell takes only a few minutes each year it is crucial for the survival of the young. The eggs and young of the black-headed gull are well camouflaged against the grass, sand and twigs around the nest. The inside of the broken shell, however, is white and highly conspicuous. Tinbergen carried out an experiment to test the hypothesis that the conspicuous white broken shell reduces the camouflage of the nest. He painted hens' eggs to resemble cryptic gull eggs and laid them out at regular intervals in the gull colony. Next to some he placed a broken shell. The results confirmed his prediction that the cryptic eggs were much more likely to be discovered and eaten by predators, such as crows, if they were close to a broken shell (Fig. 2.15b). So it is easy to see why the parent benefits by removing the conspicuous empty shell soon after the chick has hatched: the camouflage of the brood is preserved and the likelihood of the parent perpetuating its genes is increased.

A trade-off involving timing

But there is more to the story than this. The parent does not remove the eggshell immediately; it stays with the newly hatched chick for an hour or more and then goes off with the shell. To explain the delay in removing the shell we have to introduce the idea of a trade-off between costs and benefits. If the parent flies off with the shell at

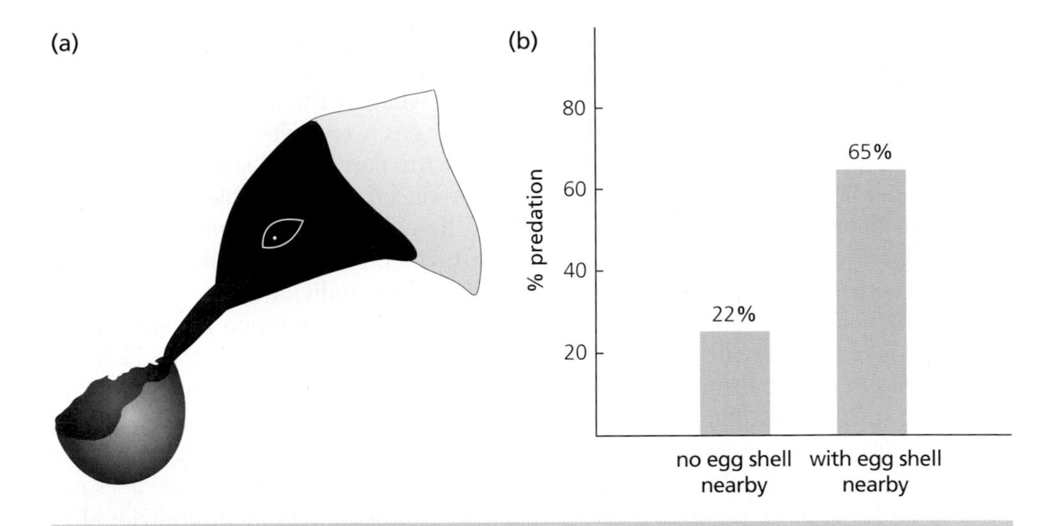

Fig. 2.15 (a) Black-headed gull removing an eggshell. (b) Results of an experiment in which single hen's eggs, painted to resemble black-headed gull eggs, were placed in the dunes, near a nesting colony. Those with an empty eggshell next to them (5 cm away) were more likely to be taken by predators (n = 60 in each treatment). From Tinbergen *et al.* (1963).

once, it has to leave the newly hatched chick unattended (the second parent is away at the feeding grounds fuelling up for its next stint at the nest). Tinbergen observed that the new chick, with its plumage still wet and matted, is easily swallowed and therefore makes a tempting meal for a cannibalistic neighbouring adult. However, when the chick's down has dried out and become fluffy it is much harder to swallow, and is therefore less vulnerable to attacks from neighbours. The parent's delay before removing the shell, therefore, probably reflects a balance between the benefits of maintaining the camouflage of the brood and the costs associated with leaving a newly hatched chick at its most vulnerable moment.

When the balance between costs and benefits is changed, the length of the parent's delay might also be expected to change. This is borne out by observations of the oystercatcher (*Haematopus ostralegus*), another ground nesting bird with camouflaged eggs and young. The oystercatcher is a solitary nester and cannibalism by neighbours is therefore not a risk associated with leaving the newly hatched chicks. The parents benefit by restoring camouflage of the nest as soon as possible after hatching and, as expected, the parent removes broken eggshells more or less as soon as a chick has hatched and before its down is dry.

Optimality models

Quantitative models of costs and benefits

Tinbergen's study of eggshell removal illustrates how experimental studies of costs and benefits can be used to unravel behavioural adaptations, but it has an important limitation. The hypothesis about the trade-off between camouflage and chick vulnerability made only a qualitative prediction. The idea would be consistent with observations of gulls removing eggshells one, two, three or perhaps even four hours after the chick has hatched, so that it is hard to test whether the hypothesis is right or wrong. One way of trying to make a hypothesis more easily testable is to try to generate quantitative predictions. If one could predict that the parent gull should remove its eggshell after 73.5 minutes then one would have produced a very testable model indeed. This is an approach which has been developed by using *optimality models* to study adaptations. An optimality model seeks to predict which particular trade-off between costs and benefits will give the maximum net benefit to the individual.

Thinking back to the gulls, if one could measure exactly how much the survival of the brood is reduced by the conspicuous broken eggshell next to the nest, and exactly how the risk of cannibalism by neighbours changes with time since the chick hatched, one could start to calculate the optimum time for the parent to delay removal of the shell. In this case the optimum might well be defined as the time that maximizes total reproductive success for the season. But the currency of an optimality model does not have to be survival or production of young. The overall success of an individual at passing on its genes may depend on finding enough food, choosing a good place to nest, attracting many mates and so on. In solving any of these problems an animal makes decisions, and the decisions can be analysed in terms of an optimal trade-off between appropriate costs and benefits. For a foraging animal, for example, currencies might be energy and time. This will form the main theme of the next chapter. We close this chapter with a simple example.

Crows and whelks

On the west coast of Canada, as in many coastal areas, crows feed on molluscs. They hunt for whelks at low tide, and having found one they carry it to a nearby rock, hover and drop it from the air to smash the shell on the rock and expose the meat inside. Reto Zach (1979) observed the behaviour of north-western crows in detail and noted that they take only the largest whelks and on average drop the shell from a height of about 5 m. Zach carried out experiments in which he dropped whelks of different sizes from various heights. This, together with data on the energetic costs of flying and searching, gave him the information to carry out calculations of the costs and benefits associated with foraging. The benefit obtained by the crow and the cost paid could both be measured in calories, and Zach's calculations revealed that only the largest whelks (which contain the most calories and break open most readily) give enough energy for the crow to make a net profit while foraging. As predicted from these calculations, the crows ignored all but the very largest whelks even when different sizes were laid out in a dish on the beach.

Crows minimize ascending flight to break the whelk

Usually the crow has to drop each whelk twice or more in order to break it open. Since ascending flight is very costly, Zach thought that the crow might have chosen the dropping height which would minimize the total expenditure of energy in upward flight. If each drop is made from close to the ground, a very large number of drops is required to break open the shell, while at greater and greater heights the shell becomes more and more likely to break open on the first drop (Fig. 2.16a). The experiment of

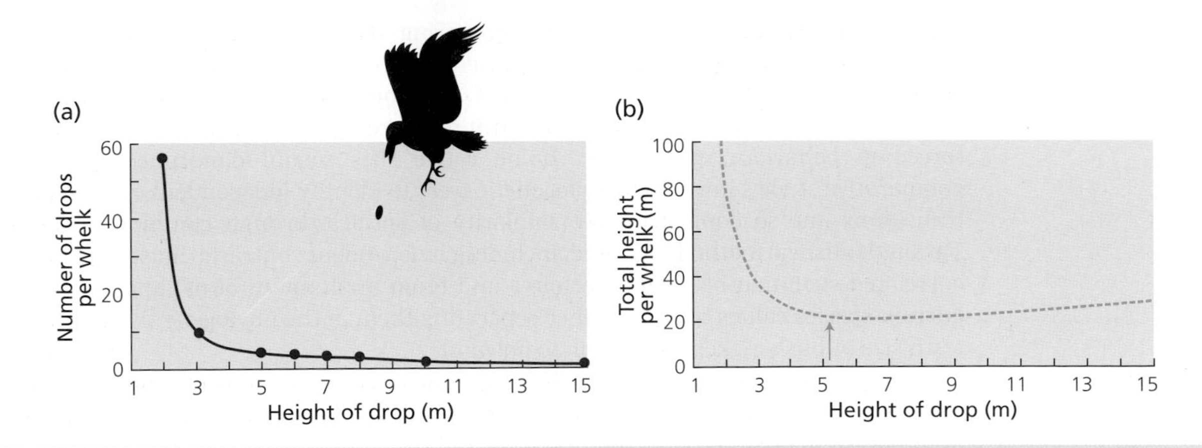

Fig. 2.16 Dropping of whelks by crows. (a) When whelks are dropped, experimentally, from different heights it is found that fewer drops are needed to break the shell when it is dropped from a greater height. (b) Calculation of the total ascending flight needed to break a shell (number of drops × height of each drop). This is minimized at the height most commonly used by the crows (arrow). From Zach (1979).

dropping shells from different heights allowed Zach to calculate the total vertical flight needed to break an average shell from different dropping heights (Fig. 2.16b). The observed average dropping height chosen by the crows (5.2 m) is indeed close to that predicted to minimize the total vertical flight per whelk. However, the crow would have to undertake almost the same total upward flight even if each drop was made from a height somewhat greater than 5.2 m (this is indicated by the very shallow U-shaped curve of Fig. 2.16b) because slightly fewer drops would be needed. Zach suggests that there may be an additional penalty for dropping from too great a height: the whelk may bounce away and be lost from view or may break into so many fragments that the pieces are too small to retrieve.

The story of crows and whelks shows how calculations of costs and benefits can be used to produce a quantitative prediction. The crow seems to be programmed to choose a dropping height that comes close to minimizing the total vertical flight per whelk. Other currencies, such as maximizing net rate of energy gain, predict much greater drop heights (Plowright *et al.*, 1989).

Summary

Throughout this book we will be using three methods to test hypotheses about behavioural adaptations: comparison between individuals within a species, experimental analysis of costs and benefits to individuals and comparison among species (which, in effect, examines the outcome of 'experiments' done by natural selection over evolutionary time).

The comparative method involves comparing different species to see whether differences in behaviour or morphology are correlated with differences in ecology. Early comparative studies of gulls, weaverbirds, antelopes and primates identified food availability and predation as major selective pressures influencing social organization (breeding behaviour, group size, home range size, sexual dimorphism). Recent comparative studies have used phylogenetic trees to identify independent evolutionary transitions and so control for the similarity of species through common ancestry. Various statistical methods are used, including independent contrasts (illustrated by the correlated evolution of song repertoires and brain anatomy in birds) and weighting current species values by the distance separating them in the phylogeny (illustrated by testis size of bushcrickets in relation to polyandry).

Phylogenetic trees can also help illuminate the order in which traits change during evolution. For example, in primates, the transition from a single-male to a multimale mating system preceded the evolution of sexual swellings. Sexual swellings are likely to be advantageous to females in multimale groups because they provide a graded signal of fertility which enables the female to bias paternity to the dominant male while, at the same time, enhancing opportunities of mating with subordinate males too, to reduce the chance that they will harm her offspring.

Comparative studies are especially useful for studying broad trends in evolution and for testing hypotheses which are not amenable to experiments. The experimental approach involves a detailed analysis of the costs and benefits of a behaviour pattern to an individual of a particular species. Behaviour can be viewed as having costs and benefits and animals should be designed by natural selection to maximize net benefit. Ultimately, the net benefit must be measured in terms of gene contribution to future generations. This will depend on shorter-term goals, such as foraging efficiency, mating success and efficiency of avoiding predation. Optimality models can be used to predict which particular trade-offs between costs and benefits give maximum net benefit.

Further reading

The comparative method has been reviewed by Harvey & Purvis (1991). Ridley (1989) provides a lucid summary of why species should not be used as independent data points, even for evolutionarily labile traits. The statistical methods for modern comparative analyses are beyond the scope of this book; for recent introductions into the literature see Freckleton and Harvey (2006), Pagel and Meade (2006), Felsenstein (2008) and Hadfield and Nakagawa (2010). Freckleton (2009) reviews the seven deadly sins of comparative analysis.

Further examples of comparative studies can be found in Fitzpatrick *et al.* (2009) (female promiscuity promotes the evolution of faster sperm in cichlid fish) and Kazancioglu and Alonzo (2010) (evolution of sex change in fish). Höglund's (1989) study of size and plumage dimorphism in birds with different mating systems provides an example of how taking account of phylogeny can change conclusions.

Huchard *et al.* (2009) suggest that once sexual swellings in primates have evolved as fertility signals, they might then be further selected to signal female quality, and they present supporting evidence from a study of wild chacma baboons in Namibia.

TOPICS FOR DISCUSSION

1. What are the relative merits of the comparative method and the experimental approach for studying behavioural adaptations?
2. How can we decide what are the independent units of observation in comparative studies?
3. Read Esther Cullen's (1957) classic study of the kittiwake, in which she compares its cliff-nesting adaptations with those of ground-nesting gulls. How would modern studies improve: (a) her experimental methodology, and (b) extend her comparative analysis?

Photo © Leigh Simmons

CHAPTER 3

Economic Decisions and the Individual

In this chapter we will discuss in more detail how the idea of economic analysis of costs and benefits can be used to understand behaviour. Most of our examples will refer to foraging, but our aim is to illustrate general principles that can be applied to all aspects of behaviour.

The economics of carrying a load

Starlings

Starlings feed their young mainly on leatherjackets (*Tipula* fly larvae) and other soil invertebrates. A busy parent at the height of the breeding season makes up to 400 round trips from its nest to feeding sites every day, ferrying loads of food to its nestlings (Fig. 3.1). In this section we are going to focus our economic analysis on one aspect of the parent starling's behaviour and ask: How many leatherjackets should the parent bring home on each trip? This may seem like an inconsequential question, but the size of load has a critical effect on the parent's overall rate of delivering food to the nest, which in turn determines whether or not the chicks survive to become healthy fledglings. As we saw in Chapter 1, the reproductive success of small birds is often limited by their ability to feed their young. There is, therefore, strong selective pressure on the parents to perform as effective food deliverers.

The starling's problem of load size can be summarized as a graph (Fig. 3.2a). The graph shows time along the horizontal axis and load (measured in leatherjackets) on the vertical axis. Consider a starling at the nest about to embark on a round trip. It has to fly to (and eventually from) the feeding site; the times of these two trips are added together and plotted on the graph as 'travelling time'. When it arrives at the part of the meadow where the leatherjackets are abundant it starts to load up with food. The first couple of leatherjackets are found quickly and easily but, because of the encumbrance of the prey in its beak, the bird takes longer and longer to find each successive prey. The result is a 'loading curve' (or 'gain curve' as it is sometimes called) that rises steeply at first but then flattens off. This is a curve of diminishing returns and the starling's problem is when to give up. If it gives up too early it spends a lot of time travelling for a small load; if it struggles on too long it spends time in ineffective search which could be better spent by going home to dump its load and starting again at the beginning of the loading curve. Somewhere in between these extremes is the starling's 'best' option. A reasonable hypothesis (but at the moment it is no more than this), is that for the starling 'best' means 'providing the maximum net rate of delivery of food to the chicks'. Any starling that is slightly better at producing chicks than its rivals will be at a selective advantage, so selection should in the long run favour behaviour that maximizes chick production.

Fig. 3.1 Starlings fly from their nest to a feeding site, search for a beak-full of leatherjackets by probing in the grass, then take them home to the nestlings. The question examined in the first part of this chapter is how many items the parent should bring on each trip in order to maximize the rate of delivery of food to the nestlings.

Optimal load size in starlings: diminishing returns

The best load can be found by drawing the tangent AB in Fig. 3.2a. The slope of this line is (load/[travel time + foraging time]) or in other words rate of delivery of food; this can be seen by the fact that it forms the hypotenuse of a right angled triangle with a base measured in 'time' and a vertical corresponding to 'load'. Travel time and the loading curve are constraints – fixed properties of the environment (or, more precisely, of the interaction between the starling and its environment), and the line AB gives the maximum slope, hence the maximum rate of delivery of food. Any other line you could draw from A to the loading curve will have a shallower slope (that is, give a lower rate of delivery) than the line AB. A couple of examples are shown in Fig. 3.2a.

Figure 3.2b develops the argument a little further. Suppose that the starling now switches to feeding at a closer site with a short travel time; how should its load per trip change? Using the same method as before, we can now draw two lines (Fig. 3.2b): when the travel time is shorter, the load that maximizes rate is smaller. One way to think of this is to imagine the starling at its moment of decision to go home. If it goes it loses the opportunity to continue foraging, if it stays it loses the opportunity to go home and start afresh. When it is far away the expected returns from going home are relatively low,

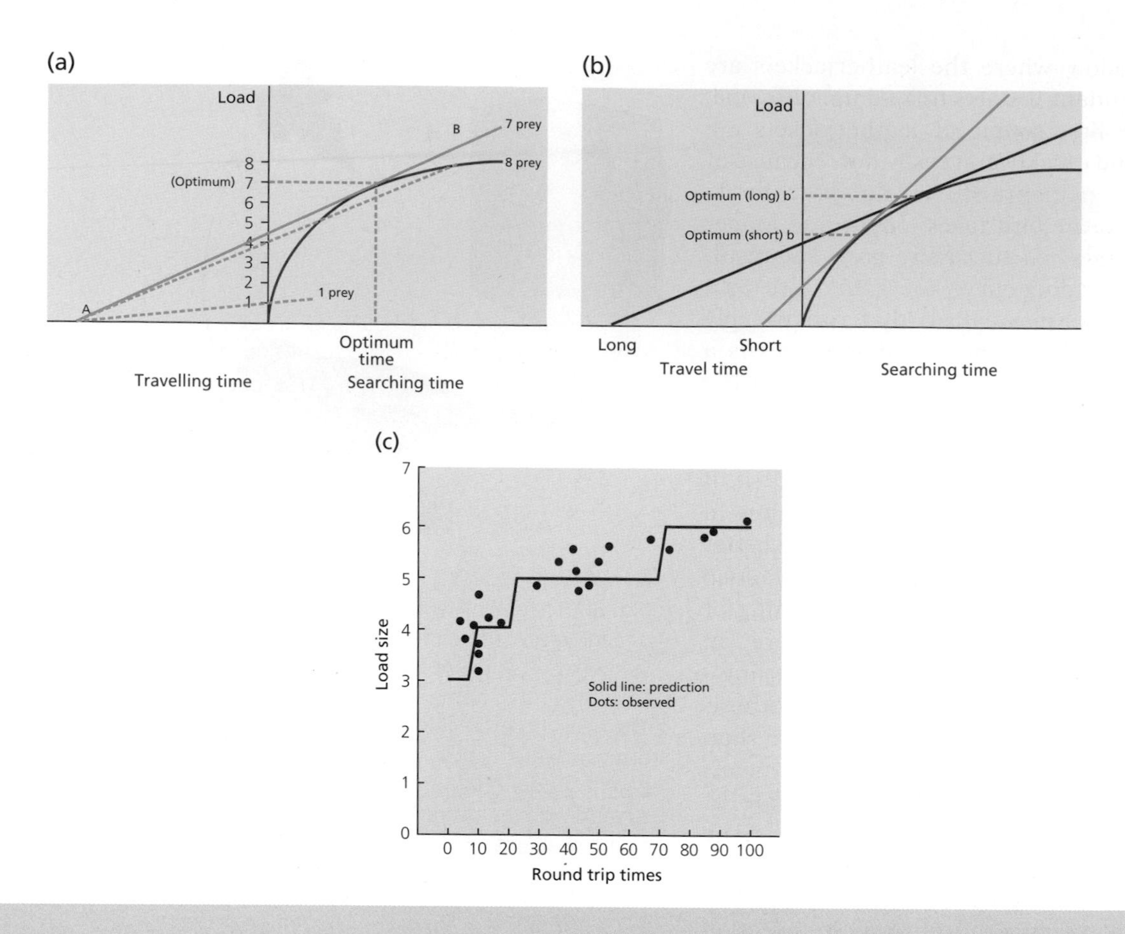

Fig. 3.2 (a) The starling's problem of load size. The horizontal axis shows 'time' and the vertical axis shows 'load'. The curve represents the cumulative number of leatherjackets found as a function of time spent searching. The line AB represents the starling's maximum rate of delivery of food to the nestlings. This rate is achieved by taking a load of seven leatherjackets on each trip. Two other lines, corresponding to loads bigger (eight) and smaller (one) than seven, are shown to make the point that these loads result in lower rates of delivery (shallower slopes). Note that although the cumulative load is shown here as a smooth curve, in reality it is a stepped line since each food item is a discrete package. (b) When the round trip travel time is increased from short to long the load size that maximizes delivery rate increases from b to b′. (c) When starlings were trained to collect mealworms from a feeder, they brought bigger loads from greater distances. Each dot is the mean of a large number of observations of loads brought from a particular distance. The predicted line goes up in steps because the bird is predicted to change its load size in steps of one worm (of course the mean loads do not have to be integers). The prediction shown here is one based on the model of Fig. 3.2b, but it also includes the refinement of taking into account the energetic costs to the parent of foraging and to the chicks of begging. From Kacelnik (1984).

since there is a long way to fly before the next chance to forage. It therefore pays to persist a little longer on the present trip, until current gains drop to a slightly lower level.

The model predicts smaller loads with shorter travel times

Alex Kacelnik (1984) tested this prediction of the model of load size in the following way. He trained parent starlings in the field to collect mealworms for their young from a wooden tray onto which he could drop mealworms through a long piece of plastic pipe. Rather than letting the birds generate their own loading curve by diminishing search efficiency, Kacelnik generated the curve for them by dropping mealworms at successively longer and longer intervals. The trained bird would simply wait on the wooden tray for the next worm to arrive, until eventually it flew home with a beak-load for its chicks. The beauty of this experimental method is that Kacelnik knew the shape of the loading curve precisely, and was hence able to present exactly the same loading curve at randomly varying distances (ranging from 8 to 600 m) from the nest on different days. The results were striking (Fig. 3.2c): not only did the load size increase with distance from the feeder to the nest, but there was a close quantitative correspondence between the observed load sizes and those predicted by the model of maximizing delivery rate.

A field test with starlings

Let us briefly summarize what the results of the starling study show. We started off by considering load size from the point of view of costs and benefits. We formulated a specific hypothesis about how costs and benefits might influence load size in the form of a model (Fig. 3.2a), and then used the model to generate a quantitative prediction (Fig. 3.2b). In making the model we did three important things. Firstly, we expressed a general conviction that starlings are designed by natural selection to be good at their job of parenting. This is not something that we aimed to test, but it is our general background assumption to justify thinking in terms of maximizing pay-off in relation to costs and benefits. Secondly, we made a guess about the *currency* of costs and benefits; we suggested that for a parent starling the crucial feature of doing a good job is maximizing net rate of delivery of food to the nestlings. Thirdly, we specified certain *constraints* on the starling's behaviour. Some of these constraints are to do with features of the environment (the time required to travel, the shape of the loading curve). Another important assumption about constraints is that the starling is assumed to 'know', or at least to behave as if it knows, the travel time and the shape of the loading curve. When we worked out the optimum load size we assumed that these were known. The experimental results supported the predictions of the model, and in so doing they supported the hypotheses about the currency and the constraints that were used to construct the model. Kacelnik compared the predictions of models based on several different currencies and he found, for example, that one based on energetic efficiency (energy gained/energy spent) as opposed to rate gave a rather poor fit to the data.

Optimality models include assumptions about currencies and constraints

Box 3.1 shows that the same economic model we have used for the starlings can be applied to other situations where individuals experience diminishing returns from a patch.

Bees

A similar problem is faced by a worker honeybee as it flies from flower to flower filling its honey crop with nectar to take back to the hive. Bees also often return to the hive with less than the maximum load they could carry and their behaviour can be explained by a model similar to that used for the starling. There is, however, an important difference: the bee experiences a curve of diminishing returns neither because the nectar in its crop makes it less able to suck more flowers nor because of resource depression (Box 3.1) but because the weight of nectar in the crop adds an appreciable energetic cost to flight. The more the bee loads up its crop the more of its load it will burn up as fuel before it gets home. As a consequence, while the gross quantity of nectar harvested increases at a constant rate, the *net* yield of energy for the hive increases at a diminishing rate as the crop fills (producing, in effect, a loading curve like that of the starling).

BOX 3.1 THE MARGINAL VALUE THEOREM AND REPRODUCTIVE DECISIONS

The model of load carrying for starlings is applicable to many other situations in which animals experience diminishing returns within a patch and is known as the 'marginal value theorem' (Charnov, 1976a). It has been used to predict how much time an animal foraging for itself (as opposed to carrying loads) will spend in each site before moving on (Cowie, 1977). Diminishing returns in each patch (generally referred to as 'resource depression') might arise, for example, simply because of depletion, or because prey in the patch take evasive action and become harder to catch, or because the predator becomes less likely to search new areas in the patch (it crosses its own path more) as time goes by, or because the predator starts with the easy prey and then goes on to hunt for those that are more difficult to catch or are less rewarding. An example of the last of these is when bumblebees or other nectar feeders visit the biggest and most rewarding flowers on an inflorescence first, and then go on to the smaller flowers which hold less nectar (Hodges & Wolf, 1981).

Reproductive decisions can be analysed with the same model. An example is Geoff Parker's (1970a) analysis of how male dung flies search for mates (see also Parker & Stuart, 1976). Males compete with one another for the chance to mate with females arriving at cowpats to lay their eggs. Often one male will succeed in kicking another male off a female during copulation and take her over. When two males mate with the same female the second one is the individual whose sperm fertilizes most of the eggs. Parker (1970a) showed this by the clever technique of irradiating males with a synthetic isotope of cobalt (^{60}Co). The sperm of irradiated males can still fertilize an egg but the egg does not develop. If a normal male is allowed to mate after a sterile one about 80% of the eggs hatch, whereas if the sterile male mates second only 20% of them hatch. The conclusion from these 'sperm competition' experiments is clear: the second male's sperm fertilizes about 80% of the eggs. It is not surprising, therefore, that after a male has copulated he sits on top of the female and guards her until the eggs are laid, only relinquishing his position to a rival male after a severe struggle.

When a second male takes over (or when a male encounters a virgin) how long should he spend copulating? Parker carried out sperm competition experiments in which he interrupted the second male's copulation after different times; this showed that the longer the second male mates the more eggs he fertilizes, but the returns for extra copulation time diminish rapidly (see Fig.B3.1.1). There is a cost associated with a long copulation: the male misses the chance to go and search for a new female. After the male has copulated for long enough to fertilize about 80% of the eggs, the returns for further copulating are rather small and the male might do better by searching elsewhere for a new mate.

The analogue of travel time in the starling model is the time the male dung fly must spend guarding the present female until she has laid her eggs plus the time he spends searching for a new female. This total is 156 minutes on average. As shown below, this estimate of travel time can be used to predict with reasonable accuracy how long the male spends copulating with a female.

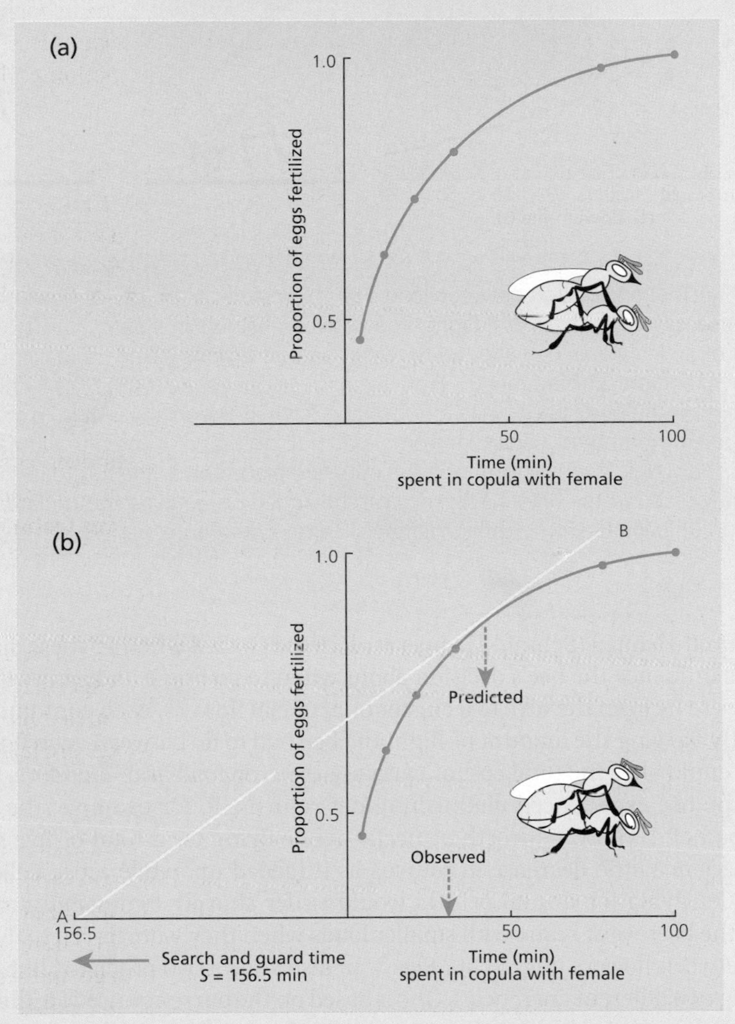

Fig. B3.1.1 (a) The proportion of eggs fertilized by a male dung fly (*Scatophaga stercoraria*) as a function of copulation time: results from sperm competition experiments. (b) The predicted optimal copulation time (that which maximizes the proportion of eggs fertilized per minute), given the shape of the fertilization curve and the fact that it takes 156 min to search for and guard a female, is 41 min. The optimal time is found by drawing the line AB. The observed average copulation time, 36 min, is close to the predicted value (Parker, 1970a; Parker & Stuart, 1976) (photo of a pair of dung flies © Leigh Simmons).

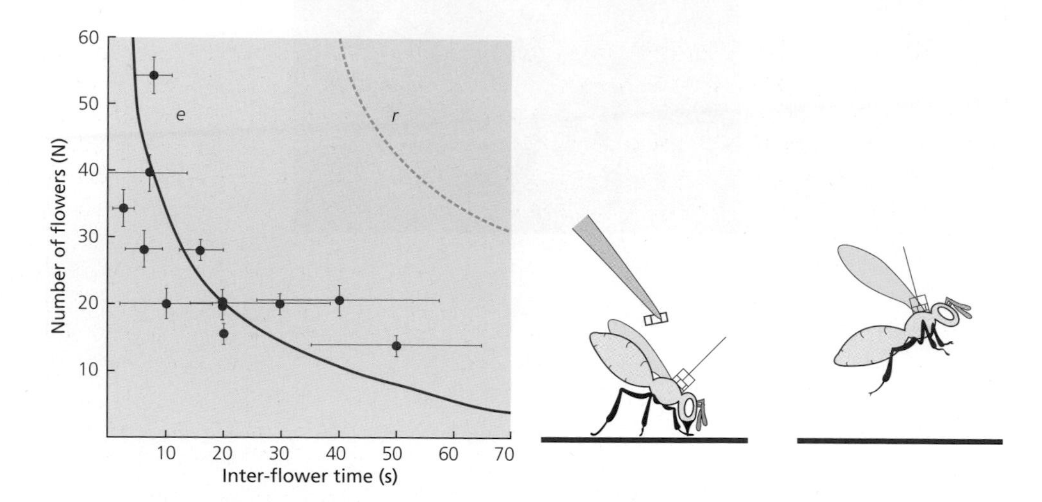

Fig. 3.3 (a) The relationship between load size (expressed as number of flowers visited) carried home by worker bees and flight time between flowers in a patch. Each dot is the mean of an individual bee and the two lines are predictions based on maximizing efficiency (*e*) and maximizing rate (*r*). From Schmid-Hempel *et al.* (1985). (b) By placing tiny weights on the bee's back while it is foraging Schmid-Hempel was able to study the bee's rule of thumb for departure from a patch to go home to the hive with a load of nectar. The weights, in the form of brass nuts, are placed on a fine rod that is permanently glued to the bee's back. They can be added or removed to simulate loading and unloading. From Schmid-Hempel (1986). With permission from Elsevier.

In bees, diminishing returns arise from the cost of carrying nectar

How much nectar should a bee carry home?

Paul Schmid-Hempel (Schmid-Hempel *et al.*, 1985) tested whether these diminishing net returns influence the bee's decision about when to go home and empty its crop. He trained bees to fly from the hive to a cluster of artificial flowers, each containing 0.6 mg of nectar. By varying the amount of flight the bee had to do between each flower in the cluster, he could alter the total cost of carrying the crop load and, therefore, the extent to which the bee experienced diminishing net returns. If, for example, the bee could collect a load of 10 flower's-worth of nectar while flying for a total of five seconds, it would experience little decrease in returns as it loaded up, while a bee collecting the same load by flying for a total of 50 s would suffer sharply diminishing returns. As predicted, the bees went home with smaller loads when they were forced to fly a greater distance between flowers (Fig. 3.3a). Fig. 3.3a also shows two predicted lines based on maximizing two different currencies. One is based on the currency used for the starlings, net rate of energy delivery, while the other is based on a currency that did not work for the starlings, energetic efficiency. In contrast to the starlings, the second currency but not the first accounts for the bees' behaviour.

Bees maximise efficiency, not rate of energy gain

Why should there be this difference between bees and starlings? A simple example shows why the 'starling currency' is normally a sensible one to consider. Compare a starling that forages for one hour, spending 1 kJ and gaining 9 kJ with one that spends 10 kJ and gains 90 kJ. Both have the same efficiency (nine) but the former has 8 kJ to spend on its chicks and the latter has 80. In other words net rate ((gain – cost)/time) tells us how much the animal has left at the end of the day to spend on reproduction or survival while efficiency does not. On the other hand, efficiency may be a sensible currency when the crucial variable for the animal is not just the amount of gain, but also the amount spent. If, for example you have to drive from A to B on a fixed amount of fuel, efficiency might be very important indeed. It turns out that bees may be in this position. The equivalent for a foraging bee to having a fixed amount of fuel would be if it had a more or less fixed total lifetime capacity for expenditure of energy.

Wolf and Schmid-Hempel tested this idea by manipulating the rate of energy expenditure of individuals, either by varying the time they were allowed to forage each day (Schmid-Hempel & Wolf, 1988) or by fixing different sized permanent weights onto their backs (Wolf & Schmid-Hempel, 1989). Both experiments showed that the bees that worked hardest survived for a shorter time than controls. For example, when workers carried a permanent weight of >20 mg, their survival was reduced from 10.8 to 7.5 days. These experiments lend some support to the hypothesis that workers, by maximizing efficiency, might extend their lifespan and thus contribute more nectar overall to the colony than they would by maximizing net rate.

Life expectancy of bees depends on work load

The contrast between bees and starlings serves to underline the point that one of the aims of economic cost–benefit analyses is to compare alternative currencies and to try to understand why a particular currency is appropriate in each case. In each study one of the major advantages of the quantitative analysis was that it allowed us to see when there was a *discrepancy* between observed and predicted results. Without this potential for discrepancy it would have been impossible, for example, to tell whether bees were maximizing rate or efficiency, or nothing at all.

The bee example also illustrates another important point. We have been thinking of animals as well-designed problem solvers making decisions that maximize an appropriate currency, but of course we do not believe that bees and other animals calculate their solutions in the same way as the behavioural ecologist. Instead the animals are programmed to follow rules of thumb which give more or less the right answer. The bees, for example, might use a rule that involved a threshold body weight ('if weight greater than x then go home'). Schmid-Hempel (1986) investigated this by adding tiny (7 mg) weights to the bee's back while it was foraging (Fig. 3.3b). He found that when he added five weights at intervals during a foraging bout the bees went home with a smaller load, as predicted if they were using a threshold weight rule. However, another experiment showed that the rule is not this simple. Instead of adding five weights gradually, Schmid-Hempel added five weights at the start of a foraging bout and then took them off gradually as the bee filled its crop. These bees also went home with smaller loads than unmanipulated bees (or than controls where the weights were placed on the bee's back for a brief moment). The most reasonable interpretation of these results is that the bee in some way integrates the total weight it has carried since arriving at the foraging site.

Adding weight to the bee's back causes it to fly home with a smaller load

The economics of prey choice

The same kind of economic approach that we have used for bees and starlings can also be used to account for the kinds of prey items that predators decide to eat.

When shore crabs are given a choice of different sized mussels they prefer the size which gives them the highest rate of energy return (Fig. 3.4). Very large mussels

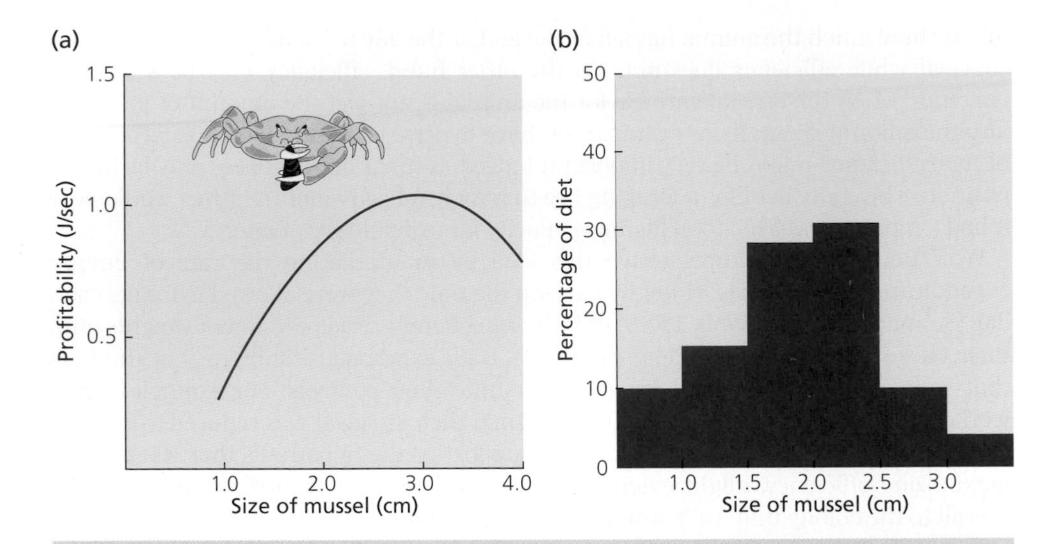

Fig. 3.4 Shore crabs (*Carcinus maenas*) prefer to eat the size of mussel which gives the highest rate of energy return. (a) The curve shows the energy yield per second of handling time used by the crab in breaking open the shell and eating the flesh; (b) the histogram shows the sizes eaten by crabs when offered a choice of equal numbers of each size in an aquarium. From Elner and Hughes (1978).

take so long for the crab to crack open in its chelae that they are less profitable in terms of energy yield per unit breaking time (E/h) than the preferred, intermediate sized, shells. Very small mussels are easy to crack open, but contain so little flesh that they are hardly worth the trouble. However, the crabs eat a range of sizes centred around the most profitable ones. Why should they sometimes eat smaller and larger mussels? One hypothesis is that the time taken to search for the most profitable sizes influences the choice. If it takes a long time to find a profitable mussel, the crab might be able to obtain a higher overall rate of energy intake by eating some of the less profitable sizes.

Optimal prey choice depends on energy values, handling time ...

... and search time: three predictions

To calculate exactly how many different sizes should be eaten we need to develop a more precise argument based on handling time, searching time and the energy values of the various prey (Box 3.2). The equations in Box 3.2 make the following three predictions for the simple example of a predator faced with a choice of two sizes of prey. Firstly, when the more profitable type (higher E/h) is very abundant the predator should specialize on this alone. This is intuitively obvious: if something giving a high rate of return is readily available, an efficient predator should not bother with less profitable items. Secondly, the availability of the less profitable prey should have no effect on the decision to specialize on the better prey. This also makes sense: if good prey are encountered sufficiently often to make it worthwhile to ignore the bad ones, it is never worth taking time out to handle bad prey regardless of how common they are. Thirdly, as the availability of the best prey increases, there should be a sudden change from no preference (the predator eats both types when encountered) to complete preference (the predator eats only the best prey and always ignores the less profitable ones).

BOX 3.2 A MODEL OF CHOICE BETWEEN BIG AND SMALL PREY (CHARNOV, 1976B; KREBS *ET AL.*, 1977)

Consider a predator which encounters two prey types, big prey$_1$ with energy value E_1 and handling time h_1, and small prey$_2$ with energy value E_2 and handling time h_2. The profitability of each prey (energy gain per unit handling time) is E/h. Imagine that the big prey are more profitable, so:

$$\frac{E_1}{h_1} > \frac{E_2}{h_2}$$

How should the predator choose prey so as to maximize its overall rate of gain? Let us assume that the predator has encountered a prey – should it eat it or ignore it?

(a) If it encounters prey$_1$, it should obviously always eat it. Therefore, choice of the more profitable prey$_1$ does not depend on the abundance of prey$_2$.

(b) If it encounters prey$_2$, it should eat it provided that:

Gain from eating > gain from rejection and searching for a more profitable prey$_1$ that is, if:

$$\frac{E_2}{h_2} > \frac{E_1}{S_1 + h_1} \tag{B3.2.1}$$

where S_1 is the search time for prey$_1$.

Re-arranging, the predator should eat prey$_2$ if:

$$S_1 > \frac{E_1 h_2}{E_2} - h_1 \tag{B3.2.2}$$

Thus the choice of the less profitable prey, prey$_2$, does depend on the abundance of the more profitable prey, prey$_1$.

This model makes three predictions. Firstly, the predator should either just eat prey$_1$ (specialize) or eat both prey$_1$ and prey$_2$ (generalize). Secondly, the decision to specialize depends on S_1, not S_2. Thirdly, the switch from specializing on prey$_1$ to eating both prey should be sudden and should occur when S_1 increases such that Equation B3.2.2 is true. Only when the two sides of the equation are exactly equal will it make no difference to the predator whether it eats one or both types of prey.

An experiment which tested these predictions is illustrated in Fig. 3.5. The predators were great tits and the prey were large and small pieces of mealworm. In order to control precisely the predator's encounter rate with the large and small worms the experiment involved the unusual step of making the prey move past the predator rather than vice versa (Fig. 3.5a). The big worms in the experiment were twice as large as the small ones ($E_1/E_2 = 2$) and h_1 and h_2 could be accurately measured as the time needed for the bird to pick up a worm and eat it. During the experiment the bird's encounter rate with large

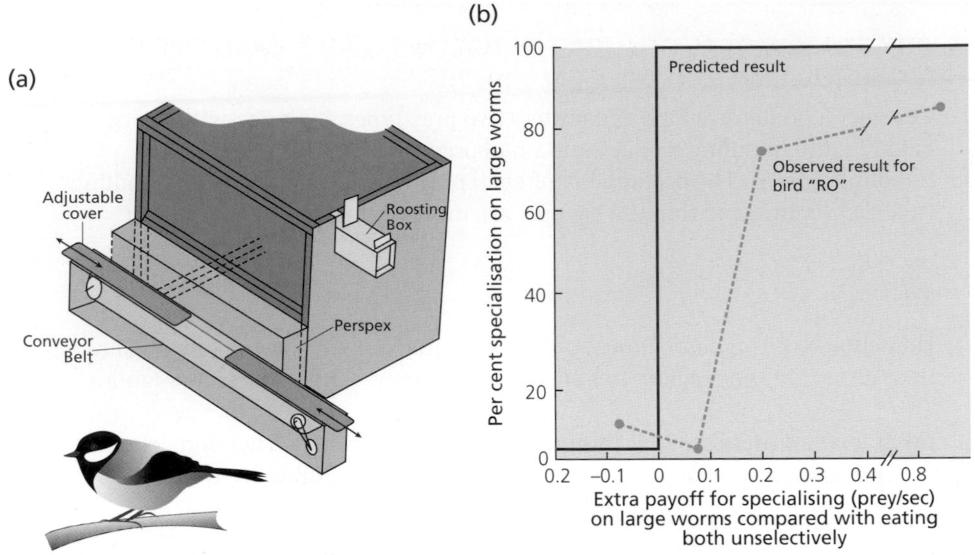

Fig. 3.5 (a) The apparatus used to test a model of choice between big and small worms in great tits (*Parus major*). The bird sits in a cage by a long conveyor belt on which the worms pass by. The worms are visible for half a second as they pass a gap in the cover over the top of the belt and the bird makes its choice in this brief period. If it picks up a worm it misses the opportunity to choose ones that go by while it is eating. (b) An example of the results obtained. As the rate of encounter with large worms increases the birds become more selective. The x-axis of the graph is the extra benefit obtained from selective predation. As shown in Box 3.2, the benefit becomes positive at a critical value of S_1, the search time for large worms. The bird becomes more selective about the predicted point, but in contrast with the model's prediction this change is not a step function. From Krebs *et al.* (1977). With permission from Elsevier.

A test of the optimal diet model

worms was varied so as to cross the predicted threshold from non-selective to selective foraging (Equation B3.2.2 in Box 3.2). The results were qualitatively but not quantitatively as predicted, the main difference between observed and expected results being that the switch was not a step but a gradual change (Fig. 3.5b). When big worms were abundant the birds, as predicted, were selective even if small worms were extremely common.

Sampling and information

The discussion so far has referred to animals that know their environment. Sometimes this may be a reasonable assumption, but at other times it may be more realistic to assume that the animal learns as it goes along. Steve Lima (1984) studied this problem in downy woodpeckers. He trained woodpeckers in the field to hunt for seeds hidden in holes drilled in hanging logs. Each log had 24 holes and in each experiment some logs were quite empty and others had seeds hidden in some or all of the holes. The woodpeckers could not tell in advance which were the empty logs, so they had to use information gathered at the start of foraging on each log to decide whether or not it was likely to be empty and therefore should be abandoned. When the logs contained 0 or 24 seeds the task was easy: looking in a single hole in theory gave sufficient information to decide and the woodpeckers, in fact, took an average of 1.7 looks in an empty log before moving on. The task was more complicated when the two kinds of log contained 0 and 6 or 0 and 12 seeds: finding one empty hole is no longer enough to reject a log, but there must be some point at which the information gained from seeing a succession of empty holes makes it worthwhile giving up. Lima calculated how many empty holes the woodpeckers ought to check before giving up on a log in order to maximize their rate of food intake. The calculated values were 6 and 3 while the observed means were 6.3 and 3.5; thus, the woodpeckers use information gleaned while foraging in a way that comes close to maximizing their overall rate of intake.

The risk of starvation

Two kinds of currency for foraging animals – rate of food intake (starlings, great tits) and efficiency (bees) – have come up so far. Another currency that may be important for foraging animals is the risk of starvation. This is especially likely to be important when the animal lives in an environment that is unpredictable; the exact amount of food the animal will obtain is uncertain.

For example, imagine you are offered the choice of two daily food rations: one is fixed at 10 sausages per day, the other is uncertain; on half the days you get five sausages and on the other half, 20 sausages. Although the *average* of the second diet is higher than that of the first, it is a riskier option. Which is the better option? The answer depends on the benefit (or 'utility' in economic jargon) of eating different numbers of sausages per day. If a diet of 10 is enough to survive on while five is not, then nothing is to be gained by choosing the risky option. If, on the other hand, 10 is not quite enough to survive on, the only viable option may be to take the risk and hope for 20 sausages. This option offers a 50% chance of survival while the certain option offers no chance.

In short, animals should be sensitive not only to the mean rate of return from a particular foraging option but also the variability. Whether or not animals prefer high variability should depend on the relationship between the animal's needs (usually called its *state*) and the expected rewards. If energy requirements are less than the average expected reward, it pays to choose the less variable option (*risk-averse behaviour*) whilst, if requirements are above average, it usually pays to choose the more variable option (*risk-prone behaviour*).

Risk-averse versus risk-prone behaviour

This idea has been tested in an experiment by Caraco *et al.* (1990). They offered yellow-eyed juncos (*Junco phaeonotus*) (small birds) in an aviary a sequence of choices between two feeding options: one variable and one with a fixed pay-off. For example, the variable option in one treatment was either 0 or 6 seeds with a probability of 0.5 each, whilst the corresponding fixed option was always three seeds. The experiment was carried out at two temperatures: 1 and 19°C. At the low temperature the rewards from the fixed option were inadequate to meet daily energy needs, whilst at 19°C they were sufficient. As predicted by the theoretical argument, the birds switched from risk-averse behaviour at 19°C to risk-prone behaviour at 1°C. An equivalent result was obtained by

BOX 3.3 RISK AND OPTIMAL SEQUENCES OF BEHAVIOUR (HOUSTON & MCNAMARA, 1982, 1985)

Consider a small bird in winter, which has to put on energy reserves by dusk in order to survive the night. In the example below, the bird needs to have eight energy units at dusk to survive.

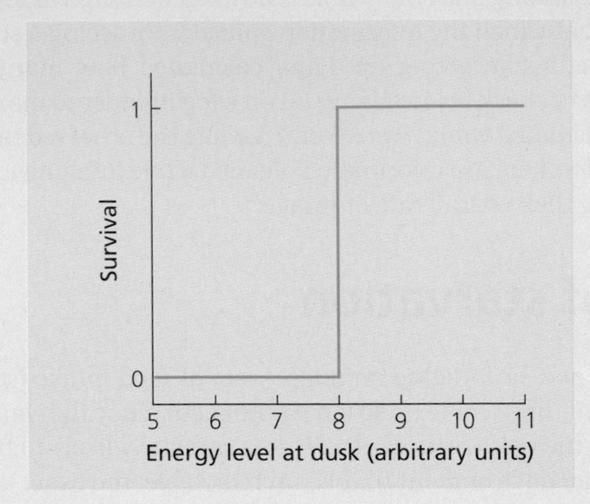

Now, imagine the bird has to choose between two foraging options:

(i) Provides one unit with probability 1.

(ii) Provides two units with probability ½, and zero units with probability ½.

So the average pay-offs are the same, but option (ii) has greater variance. If the bird has time to make just one choice before dusk, then it is easy to see that its best decision should vary depending on its state (current energy reserves) as shown below.

	Energy level if choosing:		
State	**(i)**	**(ii)**	**Best to choose**
6	7	8 or 6	(ii) i.e. take risk
7	8	9 or 7	(i) i.e. play safe

Now imagine there's time for two choices before dusk. If the current state is six units, the bird should play safe (choose (i)) both times, so it goes to seven units, then to eight units, and so survives the night.

This simple theoretical argument shows that the optimal choice depends on both current state and the time available to forage. In general, hungry individuals tend to take more risks.

Cartar and Dill (1990) in a study of bumblebee foraging. They augmented or depleted the energy reserves of the nest, and found that workers switched to risk-prone behaviour when the reserves were low. In this case the reserves of the colony as a whole were treated as equivalent to the reserves of an individual.

These experiments suggest that foragers are able to respond to variability in the amount of reward obtained, and that preference depends on state. However, they do not investigate the question of whether preference changes with time. Two examples of how time of day could be important are suggested by Houston and McNamara (1982, 1985). Firstly, if the animal starts off the day risk-prone, but has good luck in its first few choices, it might be expected to become risk-averse later on. Secondly, as dusk approaches, for a diurnal forager the long period of enforced overnight fasting might sometimes favour a switch to risk-prone behaviour to increase the likelihood of overnight survival (Box 3.3).

Environmental variability, body reserves and food storing

Small birds in winter often experience large daily fluctuations in body mass: the 20-g great tit, for example, typically loses 10–15% of its body mass overnight in winter and regains the mass during the following day (Owen, 1954). The daily gain and overnight loss is almost entirely made up of fat, which acts as fuel for overnight survival: thus each day in winter a small bird faces an uphill struggle to build up sufficient reserves for surviving the next night. Given this observation, should we expect small birds to carry as much fat as possible at all times, as an insurance against starvation? Both empirical observation and optimality models suggest that, in fact, birds usually carry less than the maximum reserves. In winter, birds are usually heaviest on the coldest/harshest days, suggesting that on other days they are carrying fewer reserves than the maximum. Furthermore, if the trajectory of weight gain through the day is examined it is found that birds increase their weight rapidly in the afternoon (Owen, 1954; Bednekoff & Krebs, 1995), implying that earlier in the day they do not carry as much fat as they could. Lima (1986) and McNamara and Houston (1990) explained these observations by hypothesizing that the reserves carried by a bird reflect an optimal trade-off between costs and benefits. The benefit of carrying extra reserves is reduced risk of overnight starvation, whilst the cost is increased danger of death from predation. The danger might arise simply because heavier birds are less agile at escaping or, more subtly, because birds with more reserves spend more time foraging rather than hiding from predators. This hypothesis predicts that the optimal level of reserves will increase (i.e. birds will be heavier) when the energy cost of overnight survival is higher, or more unpredictable, or when the danger of predation is lower (Fig. 3.6).

Optimal fat reserves: trade-off between starvation and predation

Andy Gosler and colleagues found evidence for these predictions from the long-term study of great tits in Wytham Woods, Oxford, UK. When sparrowhawks *Accipiter nisus* re-colonized the wood in the 1980s (after a decline due to pesticides), the winter mass of great tits trapped in the wood declined by about 0.5 g. Mass declines in other parts of the United Kingdom also coincided with the local re-colonization by hawks

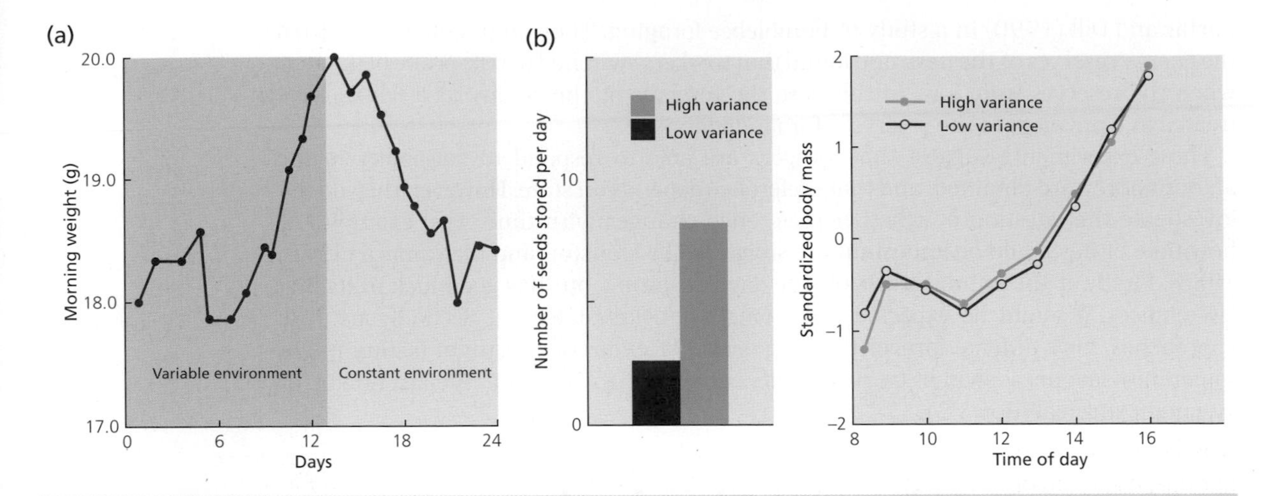

Fig. 3.6 (a) Body reserves and environmental variability. The graph shows the body mass of a captive great tit (one of eight in the experiment) which was transferred from a constant to a variable environment for 12 days before returning to the constant environment. Variability in this experiment was produced by randomly altering the length of the night-time period of no foraging. From Bednekoff & Krebs (1995). (b) Food storing and variability. In this experiment, captive marsh tits (one example is shown) stored more food (left), but did not put on more body reserves (right), in a more variable environment. These results suggest that food storing, like fat storage, is a method of coping with environmental variability: whilst great tits, which do not store food, cope with environmental variability by putting on extra fat reserves, marsh tits store extra food in the environment. The right-hand graph also shows the daily weight trajectory of a marsh tit. In the afternoon, the bird transfers food from its hoards to its body, so reserves rise steeply towards the end of the day. From Hurly (1992).

(Gosler *et al.*, 1995). Experimental presentation of model hawks at feeders led to a similar mass decline in individual birds, so it is likely that the mass change reflects a strategic choice by individuals to carry less fat reserves in the face of increased predation risk. Furthermore, dominant birds (which had priority of access to food) tended to be lighter than subordinates. These results suggest that fat reserves are an insurance and the amount carried is tempered by both starvation risk and predation costs (Gentle & Gosler, 2001).

Food storing birds: from behavioural ecology to neuroscience

The Clark's nutcracker (*Nucifraga columbiana*) of Western North America collects seeds from a variety of pine (*Pinus*) tree species in the autumn, carrying them in a special throat pouch, and stores them in scattered places on a steep hill side, hidden a few centimetres below the soil surface. The seeds are then retrieved as a source of food during the winter and spring, and even in the following breeding season for feeding young. Each nutcracker, it is estimated, stores around 30 000 seeds in 2500–4000 separate places (VanderWall, 1990). Many members of the chickadee and titmice family (Paridae) also use stored food to survive during the winter (Pravosudov & Smulders, 2010). The scale of storing is prodigious. In some titmice an individual bird may store between 100 000 and 500 000 tiny seeds during a winter, each in a separate place (Pravosudov, 1985; Brodin, 1994).

Food hoarding for short-term or long-term stores

Stored food can be thought of as analogous to body fat, stored up in times of plenty and used in times of scarcity (Hitchcock & Houston, 1994; Pravosudov & Lucas, 2001). Some species, like the nutcrackers, use their stores over a whole season, whilst others, such as some of the Paridae, use them on a shorter time scale of days or weeks, perhaps as a strategy for overnight survival in cold weather.

Ornithologists used to assume that stored food was communal property that improved the survival of the group, partly because it seemed inconceivable that birds could actually remember the huge number of places in which they had stored food. After all, many of us have difficulty remembering where we have left one bunch of keys! However, if hoarding has a cost, then free-loaders that did not pay the cost of hoarding but reaped the benefits would replace hoarders in a population (Andersson & Krebs, 1978). Hoarding is advantageous only if the hoarding individual gains from its hoard more than do others in the area: one way of gaining this advantage would be for hoarders to remember the locations of their stores.

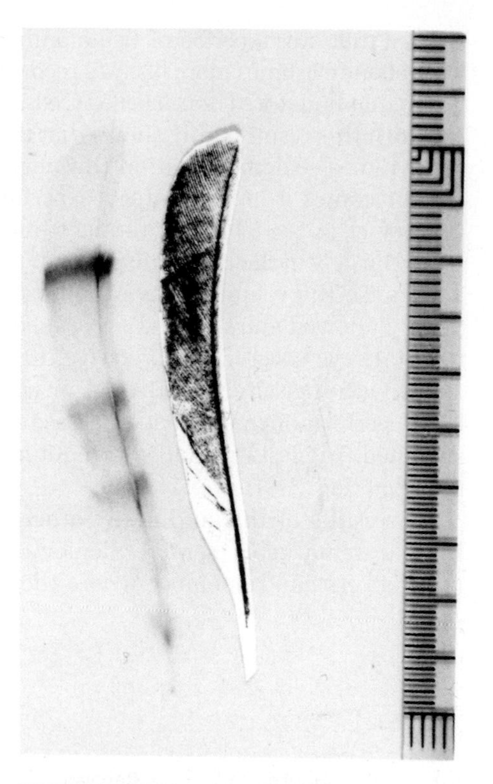

Fig. 3.7 This pair of pictures shows an autoradiograph (left) and a photostat (right) of a willow tit tail feather. The upper edge of the dark bands on the autoradiographs indicate that the owner of the feather ate a radioactive labelled seed on that day, the sulfur having been incorporated into a growing feather. Feathers were induced to grow by pulling out the original, and a replacement grows over the next 40 days. The right hand, photostat, images show faint horizontal lines that are daily growth bars. (Brodin & Ekman, 1994). Reprinted with permission from the Nature Publishing Group.

This evolutionary argument has stimulated many studies of the memory of scatter hoarding birds (Brodin, 2010) that have revealed a remarkable story linking ecology, behaviour and neuroanatomy. In an ingenious winter field experiment, Anders Brodin and Jan Ekman (1994) offered individual willow tits in Sweden 20 sunflower seeds labelled with a radioactive isotope of sulfur (^{35}S). The birds stored the seeds in their home range. The radioactive sulfur was incorporated into growing feathers when the individuals retrieved and ate the labelled seeds, so Brodin and Ekman could work out which birds in the flock recovered the seeds, and when they did so, by autoradiography of growing feathers (Figure 3.7). The results showed that over a period of two months, the individuals that stored the labelled seeds were about five times more likely to recover their items than were other birds in the same group that had stored non-labelled seeds.

Individuals recover their own stores

Whilst this result neatly shows that the storing bird benefits from its own hoarding behaviour, the demonstration that memory is, at least in part, responsible for this benefit comes from laboratory experiments. In a pioneering study, David Sherry (Sherry *et al.*, 1981) used the fact that information from the right eye in birds is stored largely in the left hemisphere of the brain, and vice versa for the right eye. This is because the visual pathways almost completely cross over at the optic chiasm. Sherry allowed marsh tits (*Parus palustris*) in an aviary to store seeds in moss trays with one eye covered, and retrieve their stores after an interval of up to 24 hours, either with the same eye covered, or with the eye cap switched to the other eye. The birds' performance at retrieving seed dropped dramatically when the eye cap was switched, suggesting that information stored in the brain is crucial for successful retrieval (Fig. 3.8).

Spatial memory in food hoarding birds

The results of this and many other experiments showed that food storing birds have a remarkable spatial memory, including the demonstration that Clark's nutcrackers can remember where they have hidden their food after 9–10 months

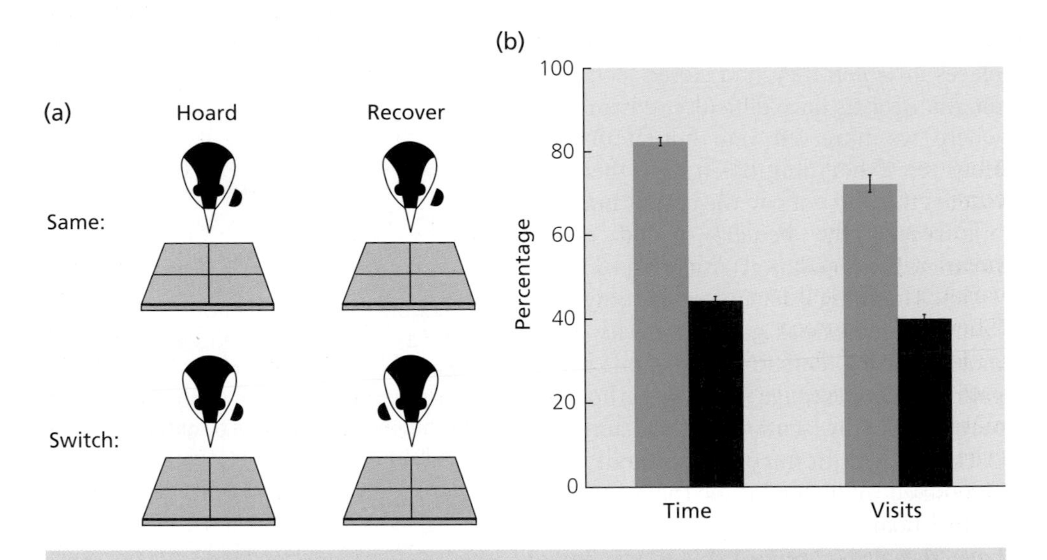

Fig. 3.8 (a) A schematic of the experimental design used by Sherry *et al.* (1981). The birds stored and retrieve food with one eye covered by a little plastic eye cap. In the treatment called 'same', the cap was on the same eye for storing and retrieval, but in the 'switch' condition the eye cap was changed between storage and retrieval of food. During retrieval, the birds searched for seeds that they had stored 24 hours earlier in moss-filled trays on the floor of an aviary. (b) The percentage of time and of visits made to quadrants of the moss trays with stored seeds during recovery, when searching with the same eye (pale blue) and the other eye (dark blue).

(Balda & Kamil, 1992). But is it any more remarkable than the memory of related species that do not store food, or do food storers simply deploy their memory in this particular way?

Alan Kamil, Russ Balda and colleagues have investigated this question by studying four species of North American corvids that rely on food storing to different extents: the Clark's nutcracker, pinyon jay, Mexican jay and Western scrub jay (Balda & Kamil, 2006). They used two laboratory tests of spatial memory and found that the Clark's nutcracker, which is the most reliant on stored food, performs better than the other species. On the other hand, when the task involved remembering colour instead of location, the nutcrackers were no better than the other species. These results, along with others (Shettleworth, 2010a) seem to show that food storing species do, indeed, have an especially good spatial memory, perhaps both in terms of the amount of information stored and the duration of memory, or the extent to which spatial cues, as opposed to other cues such as colour, are used.

Food storing species also have a specialized brain. In mammals, a special region of the cortex, called the hippocampus (so-called because in the eyes of some neuroanatomists it recalls the shape of a seahorse of the genus *Hippocampus*), is crucial in the formation of spatial and, perhaps, some other memories (Squire, 2004). Birds have an homologous structure that is essential for the recovery of stored food (Sherry & Vaccarino, 1989). Measurements of the brains of many different species of birds, including those that store food and those that do not, have shown that food storers have a larger hippocampus, relative to the rest of the brain, than do non-storers (Roth *et al.*, 2010) (Fig. 3.9a). Within one species, the black-capped chickadee (*Poecile atricapillus*), populations living in harsher winter conditions, and therefore more dependent on food storing for winter survival, have a larger relative hippocampus with more neurons than do birds living in less harsh conditions (Roth *et al.*, 2011; Fig. 3.9b).

Brain specializations in food hoarders and taxi drivers

The relative volume of the hippocampus, and/or the generation of new neurons, also varies with season in the black-capped chickadee, although the way in which this variation is linked to seasonal variations in food storing behaviour is not yet clear (Sherry & Hoshooley, 2010). One possibility is that new neurons are added in anticipation of the seasonal onset of storing, another is that the cause–effect arrow goes the other way round. Nicky Clayton and John Krebs (1994) found that in young marsh tits, experience of storing, or another spatial memory task, was necessary for growth of the hippocampus, a case of 'use it or lose it'! Either way, the reduction in the hippocampus during periods of little or no food hoarding suggests that brain tissue is costly to maintain.

Brain scans of humans also reveal plastic change in hippocampus volume in response to environmental demands. London taxi drivers face similar challenges of spatial memory to those of food hoarding birds. They have to undergo extensive training, learning to navigate between thousands of places in the city. This training, known colloquially as 'being on The Knowledge' takes about two years for a full licence. Magnetic resonance imaging showed that the taxi drivers had a larger posterior hippocampus (and a smaller anterior hippocampus) compared to control subjects, and the most experienced taxi drivers had the largest posterior hippocampus volume (Maguire *et al.*, 2000).

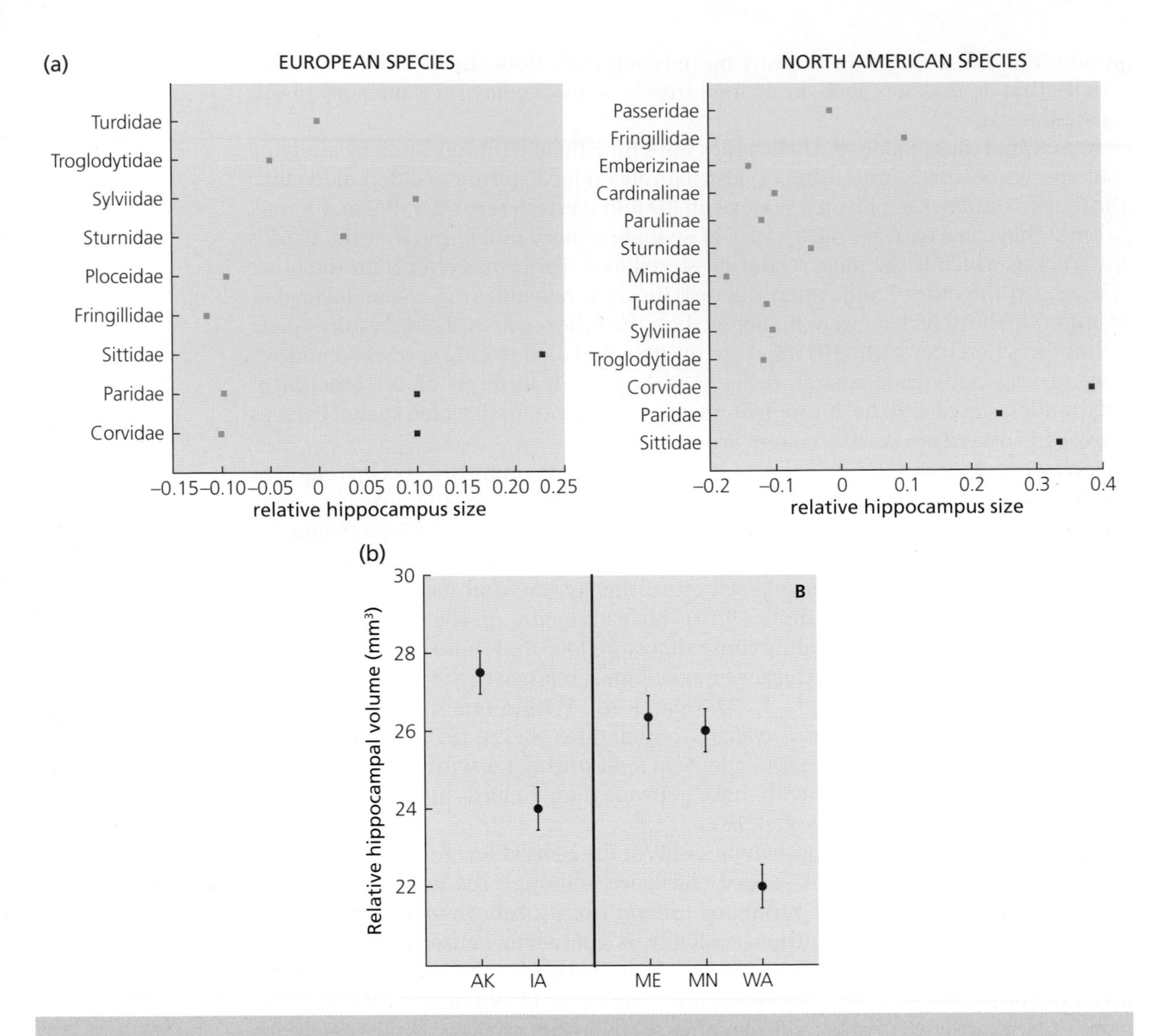

Fig. 3.9 (a) Two studies of interspecies comparisons of hippocampal size. Within families of birds, storing species have a larger relative hippocampus. The horizontal axis shows the relative size of the hippocampus for different families of birds, after correcting for the effects of body size and overall forebrain size (i.e. taking into account the fact that larger species will have larger brains). The pale blue points are averages for families that do not store food, whilst the dark blue points are families, or groups of species within families, that do store food. Storers have a larger relative hippocampus. After Krebs (1990). (b) Intraspecies comparisons: the volume of the hippocampus, relative to the rest of the forebrain, of black-capped chickadees (*Poecile atricapillus*) from Alaska (AK) and Iowa (IA) (left panel), and from Maine (ME), Minnesota (MN) and Washington (WA) (right panel). In the former comparison day length in the winter is shorter in Alaska than in Iowa, although both have very cold winters. In the latter, day length is similar in all three sites, but Washington has a milder winter climate than the other two places. The data show that in harsher winters, either as a result of shorter day length or colder temperatures, the birds have a larger relative hippocampus. After Roth *et al.* (2011).

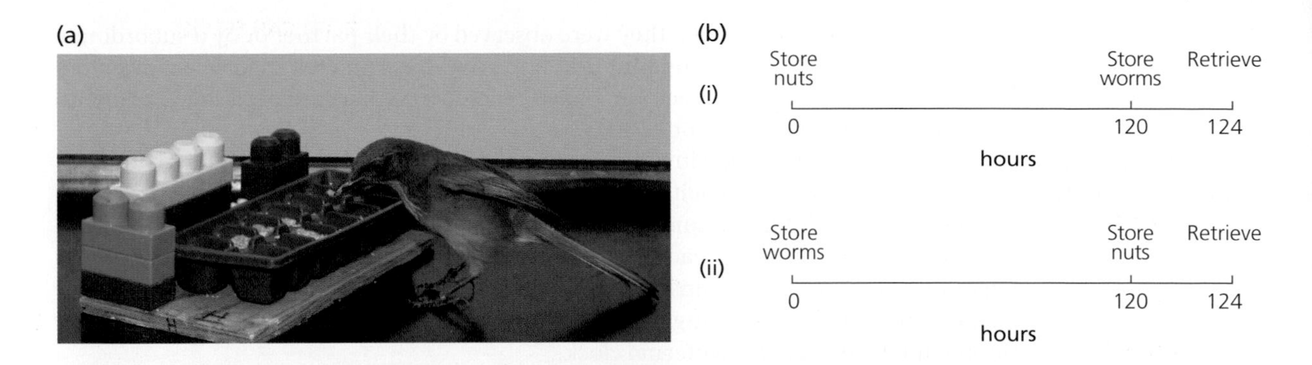

Fig. 3.10 (a) A scrub jay at a food storing tray in an experiment to investigate episodic memory. The coloured blocks provide spatial cues for the bird. Photo © Nicky Clayton. (b) Design of an experiment to test for episodic memory. In treatment (i) the birds are given nuts to store, then 120 h later they are given worms to store. Four hours after that, they are allowed to retrieve items. In treatment (ii), the order of storing is first worms, then nuts. From Clayton & Dickinson (1998). Reprinted with permission from the Nature Publishing Group. See text for explanation.

The evolution of cognition

Nicky Clayton and her colleagues have used the food storing behaviour of western scrub jays *Aphelocoma californica* in the laboratory to explore the mental capacities of these birds (Fig. 3.10a). She has investigated three aspects of behaviour that appear to demonstrate surprisingly complex mental representations.

Episodic memory: what, where and when?

One study showed that the scrub jays remember what they have stored, and when and where. This kind of memory for specific events in humans is called 'episodic memory' and is distinct from 'procedural memory' for learned skills such as riding a bike or playing the piano. The essence of the experiment was that the birds were allowed to store two kinds of food, nuts or worms, in a tray, with an interval of 120 hours between storing episodes. They were then allowed to retrieve both nuts and worms after another four hours (Fig. 3.10b). Normally scrub jays prefer worms, but worms decay after 124 hours whilst nuts do not. Thus, if the birds remember what they have stored, where it is and when it was stored, they should search in the nut part of the tray if they have stored worms 124 hours before (treatment (ii) in Fig. 3.10b) and in the worm part of the tray if they stored nuts 124 hours earlier (treatment (i)). The birds did this, even when both worms and nuts were removed during the retrieval period, to ensure that scent or another cue did not give the game away (Clayton & Dickinson, 1998). The birds appear to have an episodic-like memory.

Social cognition: recognizing potential thieves

Clayton's second study investigated 'social cognition': the ability of a bird to behave as though it could interpret the knowledge of another individual. Dally *et al.* (2006) found that when a scrub jay had been observed by another individual whilst caching food, it would, later on when in private, move its caches to new locations, as though it were aware of the fact that it had been observed and that the observer might pilfer the caches. Furthermore, the jays re-cached more of their food items when they had been observed by a dominant bird, than when they were observed by their partner or by a subordinate bird. There is also an intriguing hint that birds that have pilfered themselves are more likely to move their caches when they have been observed, a case of 'it takes a thief to know a thief' (Emery & Clayton, 2001).

Mental time travel: planning for the future

The third aspect of animal intelligence investigated in scrub jays is sometimes dubbed 'mental time travel', the ability to project into the future, independently of current physiological requirements, and plan appropriately, just as humans do when they make a weekly shopping list and head off to the supermarket even when they are not hungry. Experiments aimed at demonstrating this ability have to carefully control for other explanations, such as learning through repeated exposure to a sequence of events, or timing linked to the body's internal clock.

Raby *et al.* (2007) demonstrated behaviour in food storing scrub jays that could be interpreted as mental time travel. The birds lived in a set up with three compartments: a central area and a 'room' at either end. During a training phase, the birds were locked in one of the two end rooms overnight. In one room they were provided with food as soon as the lights came on in the morning, whilst in the other room, there was a delay of two hours before food was offered. After this training, the birds were allowed to store pine nuts placed in a bowl in the central compartment, and they preferred to hide the food in the room where they were not fed first thing in the morning. They did this on the first occasion they were allowed to store food, suggesting that they could project forward and anticipate in which room they would be hungry if locked in overnight. In a second experiment, birds were trained to experience breakfast in both rooms, but in one room the meal was always pine nuts and in the other it was always dog food. When given a choice of food to store, the birds hid pine nuts in the dog food room and dog food in the pine nuts room, as though anticipating the kind of food they would get for breakfast, and making their diet more varied by storing the opposite kind of food.

These studies show that food storing birds have surprising mental abilities that extend beyond spatial memory, but are they specialized adaptations that have arisen during evolution in association with food storing behaviour? At the moment we cannot say without a comparison of storing and non-storing species, such as those on spatial memory of corvids that were described earlier. The study of food storing birds does, however, provide a remarkable example of inter-relationship between ecology, behaviour and the brain.

The food storers also raise a more general question about 'animal intelligence'. Sarah Shettleworth (2010a, 2010b) asks whether the results from food storing birds, and other similar examples, could be the result of simple processes, such as associative learning, or whether they imply that non-human animals have more complex cognition involving a 'theory of mind', namely treating others as intentional beings, attributing to them knowledge, belief, desires and other intentional states.

Intelligence and simple rules

Shettleworth makes three important points. Firstly, apparently complex behaviours can be generated by very simple behaviour mechanisms: these are what Daniel Dennett (1983) has called 'killjoy' explanations. A famous example of this is the demonstration that pigeons, with appropriate pre-training, could solve a novel problem that involved moving a box to the correct place and then climbing onto it get to a reward that was otherwise out of reach (Epstein *et al.*, 1984). This experiment mimicked a classic report by Wolfgang Kohler (1929) who concluded that similar behaviour by chimpanzees showed that they had 'insight' (the 'aha' experience of suddenly figuring out the solution). Secondly, some kinds of 'intelligent' behaviour may be a specific adaptation to a particular ecological problem, such as the memory of food storing birds. Thirdly, it turns out that humans, more often than perhaps we appreciate, use subconscious rules of thumb rather than conscious calculations. The study by Melissa Bateson of paying for coffee, which is referred to in Chapter 12, is a good example. Experts in marketing capitalize on our subconscious biases in manipulating us into buying goods (Cialdini, 2001). Therefore, the Darwinian argument that there should be continuity between the intelligence of non-human animals and man can be turned on its head by saying that we are in some ways more like non-human animals than we sometimes recognize.

Feeding and danger: a trade-off

If you watch a squirrel eating chocolate chip cookies in the park, as Steve Lima and colleagues did (Lima *et al.*, 1985) you will notice that the squirrel generally comes to your picnic table, grabs a cookie and retreats to a tree to eat it. If you put out small fragments of cookie the squirrel will often make repeated sorties to the table and take each morsel back to the tree to eat it. This is obviously not a very efficient way to eat food: if maximizing net rate of energy intake or efficiency was the only important factor for a squirrel it would simply sit on the table and eat pieces of cookie until it was full. One interpretation of the squirrel's behaviour is that it is balancing the demands of feeding and safety from predators. It could feed at maximum rate and run a good chance of being killed by a cat by staying on the table, or it could be completely safe from cats but die of starvation in the trees. Neither of these is the best solution to maximize survival, so the squirrel does a mixture of the two. Lima *et al.* argued that the squirrel should be more prone to seek safety in the trees while feeding when this involves a smaller sacrifice in terms of feeding rate. Consistent with this they found that when the feeding table was close to the trees the squirrels were more likely to take each item to cover. Big pieces of cookie were more likely to be taken to cover than small ones; they take a long time to eat and are, therefore, more dangerous to handle out in the open and when handling time is long the relative cost of travelling back and forth is reduced.

Balancing foraging and safety

The balance between the benefits of feeding and of avoiding danger is also influenced by an animal's hunger. On a very cold day in winter, normally shy birds become quite tame at the garden bird table, presumably because their increased need for food overrides the danger of coming into the open. Manfred Milinski and Rolf Heller (Milinski & Heller, 1978; Heller & Milinski, 1979) studied a similar problem with sticklebacks (*Gasterosteus aculeatus*). They placed hungry fish in a small tank and offered them a simultaneous choice of different densities of water fleas, a favourite food. When the fish were very hungry they went for the highest density of prey where the potential feeding rate was high, but when they were less hungry the fish preferred lower densities of prey. Milinski and Heller hypothesized that when the fish feeds in a high density area it has to concentrate hard to pick out water fleas from the swarm darting around in its field of vision, so it is less able to keep watch for predators. This was later confirmed by Milinski (1984). A very hungry fish runs a relatively high chance of dying from starvation and so is willing to sacrifice vigilance in order to reduce its food deficit quickly. When the stickleback is not so hungry it places a higher premium on vigilance than on feeding

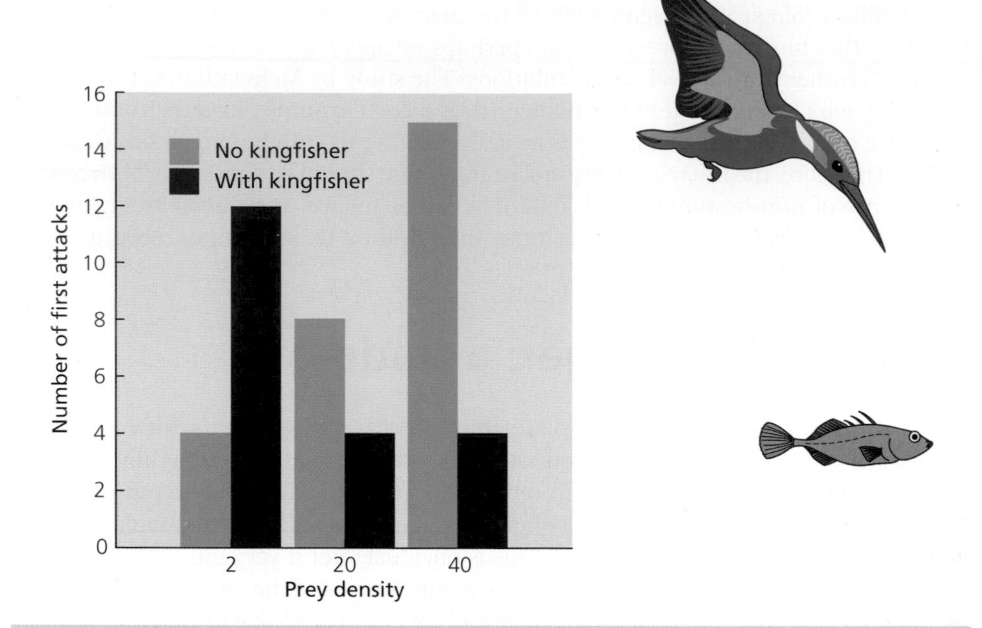

Fig. 3.11 Hungry sticklebacks normally prefer to attack high density areas of prey but after a model kingfisher was flown over the tank they preferred to attack low density areas. From Milinski and Heller (1978). Reprinted with permission from the Nature Publishing Group.

quickly, so it prefers the low density of prey. The balance of costs and benefits shifts from feeding to vigilance as the stickleback becomes less hungry.

Consistent with this hypothesis, Milinski and Heller found that predation risk influences choice of feeding rate. When they flew a model kingfisher (*Alecedo atthis*) (a predator on sticklebacks) over a tank containing hungry fish they found that the sticklebacks preferred to attack low rather than high prey densities (Fig. 3.11). This is to be expected if the hungry fish, in spite of its high chance of starvation, places a very high premium on vigilance when a predator is in the vicinity.

Hungry sticklebacks risk danger of predation to obtain high intake rates

An important difference between Milinski and Heller's analysis of foraging and those described earlier is that the cost–benefit calculations include the animal's hunger state. An optimization model in which the animal's state changes as a result of its behaviour (the fish becomes less hungry as a result of feeding) is referred to as a dynamic, as opposed to static, model. In fact, the traditional view that an animal's internal state controls its behaviour can be turned on its head and the animal can be seen as using its behavioural repertoire to control the internal state in an optimal way. The influence of the kingfisher on the stickleback is to alter the optimal allocation of time to feeding and vigilance, so that the fish decreases its hunger at a slower rate.

The idea of a dynamic feedback between foraging, body reserves and danger of predation has been used by Jim Gilliam (1982) to predict how an individual should shift between habitats as it grows up. His analysis applies to the bluegill sunfish (*Lepomis macrochirus*). In experimental ponds in Michigan, USA, Earl Werner found that these fish could obtain a higher rate of food intake by foraging on benthic invertebrates such as chironomid larvae than they could by foraging either on the plankton or near the emergent vegetation at the edge of the pond. As might be expected the fish spend most of their time (more than 75%) foraging on the benthos. However, when predators in the form of largemouth bass (*Micropterus salmoides*) were added to the pond, a significant change in habitat use by the sunfish was seen. The bass could eat only the smallest sunfish (the others were too big) and these fish now spent more than half their foraging time in the reeds feeding on plankton where they were relatively safe, even though as a result their food intake was reduced by about one-third and their seasonal growth rate by 27%. The bigger sunfish continued to forage with equanimity on the benthos (Werner *et al.*, 1983). The little fish thus face a trade-off: is it better to stay in the relative safety of the reeds and grow slowly, prolonging the period of vulnerability to predators, or is it better to gamble on rapid growth to a safe size in the benthos? Gilliam was able to show that the best thing (to maximize its total chance of survival) for the fish to do is to stay in the safety of the reeds until a certain size is reached and then to go for the benthic prey. This accords with observation: the young fish in the presence of predators tend to feed in safe places, and as they get bigger they shift to the better feeding areas.

Bluegill sunfish: age changes in habitat choice

Social learning

While individuals often sample alternatives themselves to determine which is the most profitable option (Krebs *et al.*, 1978; Lima, 1984), sometimes they use the behaviour of others as a source of information. Social learning is likely to be particularly beneficial when individual learning is costly, for example because it is time consuming or dangerous.

Social versus individual learning in sticklebacks

This is illustrated well by the contrast in sampling behaviour of two species of sticklebacks (Fig. 3.12). Three-spined sticklebacks *Gasterosteus aculeatus* have large spines and armoured body plates that protect them from predators and allow them to sample alternative food patches directly, in relative safety. By contrast, nine-spined sticklebacks *Pungitius pungitius* have weaker defences and spend much of their time hiding in weedy vegetation. Experiments show that nine-spined sticklebacks exploit 'public information' by watching conspecific or heterospecific demonstrator fish foraging and then choosing to approach the patch with better feeding rates. However, three-spined sticklebacks ignore public information and rely on their own experience to select rich over poor patches (Coolen *et al.*, 2003). These different strategies might reflect a trade-off between reliable but more costly self-acquired information and potentially less reliable but cheaper socially-transmitted information (Laland, 2008).

Social learning may involve several different learning mechanisms (Laland, 2008). For example, a naïve individual might simply have its attention drawn to a task by the action of others, and then learn by itself how to perform the task. Alternatively, naïve individuals might learn by copying the action of a demonstrator. Or naïve individuals might be taught by experienced individuals.

(a) (b)

Fig. 3.12 Two species of sticklebacks. (a) Nine-spined sticklebacks often rely on public information when choosing foraging sites, whereas (b) three-spined sticklebacks rely more on personal sampling of alternative options. Photos © Kevin Laland.

Copying and learning facilitation

Whiten *et al.* (2005) tested whether captive chimpanzees *Pan troglodytes* copied the actions of others. In one group, they trained one individual to obtain food by poking a stick into an apparatus to release a food item. In another group, one individual was trained to adopt a different method to obtain food from the same apparatus, by using the stick to lift up some hooks to release the food. When these two demonstrators were returned to their respective groups their companions observed them at work and then most of them adopted the particular technique seeded in their group (either poking or lifting). By contrast, no chimpanzees mastered either tool using technique in a third population that lacked a local expert. This group, with no demonstrator, severed as a control for possible effects of asocial learning.

Chimpanzees and meerkats learn foraging tricks from others

Thornton and Malapert (2009) tested whether a novel food acquisition technique could spread through a wild population in a study of meerkats *Suricata suricatta* in the Northern Cape, South Africa. Again, food could be obtained from an apparatus in one of two ways: either by climbing some steps and ripping the cover off a box, or by going through a cat-flap at the bottom of the box. In three groups, one individual was trained to perform the 'stairs' technique, in another three groups one individual was trained to perform the 'flap' technique, and in a further three control groups no individual was trained. The results showed that the two experimental groups were more likely to gain food from the box and that naïve individuals tended to adopt the technique of their demonstrator, either after observing it or scrounging food while the demonstrator fed from the box. In this case, naïve individuals might not necessarily have copied the actions of demonstrators but simply have had their attention drawn to either the stairs or the flap and then have learnt themselves how to access the food.

Local traditions

Social learning could lead to local traditions. For example, in some wild populations chimpanzees forage for ants by using a long wand. A chimp waits until many ants have gripped the wand, then it uses the other hand to wipe them off and transfer them to its mouth. In other populations, however, the chimps use a small stick to collect a few ants at a time and then they transfer them directly to the mouth with the stick (Whiten *et al.*, 1999). While these differences may well reflect cultural differences transmitted by social learning, it is difficult from observational studies alone to exclude the possibility that the differences might reflect genetic differences between populations, or differences in ecological conditions, such as resources available (Laland & Janik, 2006).

Local traditions in chimpanzees and fish

The best evidence for traditions in the wild maintained by social learning comes from translocation experiments with fish. Helfman and Schultz (1984) found that French grunts *Haemulon flavolineatum* had traditional daytime schooling sites on coral reefs and then also used traditional migration routes at dusk and dawn, to and from feeding grounds on nearby grass beds. These schooling sites and migration routes persisted for longer than the lives of individual fish. To test whether new recruits to the school learnt the local resting and migration routes from others, some fish were transplanted to other sites. These individuals quickly adopted the movement patterns of the residents at the new sites. By contrast, when residents were removed, newcomers used new sites or routes. Therefore, the local tradition broke down once the opportunity for social learning was removed.

In a similar experiment, Warner (1988) showed that mating sites of blueheaded wrasse *Thalassoma bifasciatum* on reefs could be maintained by socially transmitted traditions, rather than through individuals choosing the best sites based on resource quality. Mating sites remained in daily use for twelve years (four generations), yet when entire local populations were removed and replaced by new individuals, new sites were adopted and then maintained.

In future studies it would be interesting to examine the circumstances in which individuals benefit by following the social traditions of their group rather than investing in costly individual exploration, which might erode traditions or lead to new ones. In some cases, traditions persist even when they are maladaptive. In laboratory experiments with guppies *Poecilia reticulata*, Laland and Williams (1998) found that naïve fish would follow others that had been trained to adopt an energetically costly circuitous route to a feeder rather than select a less costly short route. Furthermore, once fish had been trained to follow the long route, they were slower to learn the quicker route than control fish. Therefore, socially-learned information can sometimes inhibit learning of the optimal behaviour pattern.

Teaching

In cases where individual learning or inadvertent social learning is very costly, or where opportunities for these are lacking, knowledgeable individuals may actively 'teach' others. Three criteria have been suggested for regarding an individual as a 'teacher': (a) it modifies its behaviour in presence of a naïve observer, (b) it incurs an initial cost from doing so and (c) the naïve observer acquires skills or knowledge more rapidly as a result (Caro & Hauser, 1992). Evidence for all three criteria in non-human animals is scarce.

Three criteria for teaching

Nigel Franks and Tom Richardson (Franks & Richardson, 2006; Richardson *et al.*, 2007) have shown that teaching, as defined above, does not require a large brain. When an ant, *Temnothorax albipennis*, finds food it leads another, naïve individual from the nest to the food source by 'tandem running' (Fig. 3.13). The leader runs ahead, with the follower maintaining frequent contact by tapping the leader's legs and abdomen with its antennae.

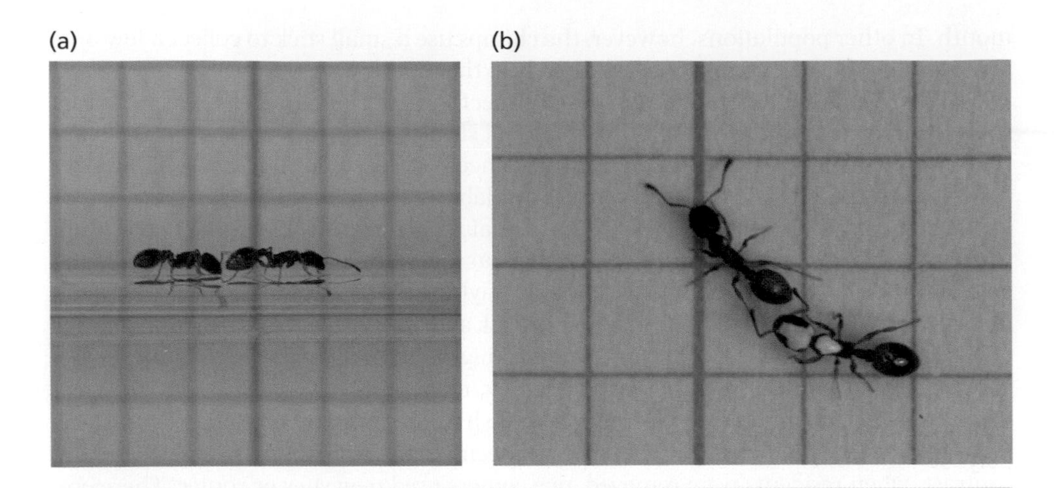

Fig. 3.13 (a) (b) Tandem running in the ant, *Temnothorax albipennis*. The leader teaches the follower where to go to find food. (Franks and Richardson, 2006). Photos © Tom Richardson.

Fig. 3.14 A meerkat pup with a scorpion. Adults teach the pups handling skills for these dangerous prey. (Thornton & McAuliffe, 2006). Photo © Sophie Lanfear.

The leader pays a cost when it is followed, travelling much more slowly because of regular stops to make sure the follower is still in tow, and to give the follower time to learn landmarks, which the follower itself will use to find its way back to the nest. When the gap between leader and follower grows too large, the leader slows down and the follower accelerates. Finally, followers find food more quickly when tandem running than when searching alone. These results show that the leader's behaviour fulfils all the criteria for teaching, with the follower acting as pupil. The lessons learned by followers are passed on when they become leaders, so although tandem runs are slow they propagate time-saving knowledge among the foragers in a colony.

Teaching in ants and meerkats

Alex Thornton and Katherine McAuliffe (2006) discovered teaching in meerkats, *Suricata suricatta*, which live in a harsh desert environment where they have to dig for prey (invertebrates, lizards). Among their favourite prey items are scorpions, which have deadly neurotoxins and powerful pincers (Fig. 3.14). Adult meerkats actively teach pups prey-handling skills by gradually introducing them to live prey and monitoring the pups' handling performance, retrieving prey that the pups have dropped or lost. Adults first provision pups with dead scorpions. Then, as the pups improve their skills, they bring live scorpions with the stings removed, before finally presenting intact scorpions. Hand feeding experiments confirmed that this teaching improved the pups' handling skills. Pups provided on three consecutive days with live, stingless scorpions were then more likely to handle a scorpion correctly and quickly than other pups who were provided with either dead scorpions or hard boiled egg, as a control.

Teaching using simple rules

As with the ants, the adult meerkat's teaching behaviour is likely to involve simple behavioural rules rather than complex cognitive mechanisms, such as recognizing the ignorance of pupils. For example, playback experiments showed that adults modified their presentation of prey in response to the maturation of the pups' begging calls. Playback of recorded calls of old pups caused adults to provision live scorpions, while playback of young pups' calls caused them to bring dead prey.

Optimality models and behaviour: an overview

In this chapter we have seen how optimality models can be used to analyse decisions about foraging and mating (Table 3.1). This approach is an extension of the idea of interpreting behaviour in terms of costs and benefits that we introduced in the last chapter. Let us now try to summarize some of the advantages and limitations of optimality modelling. Three main advantages illustrated by this chapter are:

(1) Optimality models often make testable, quantitative predictions so that it is relatively easy to tell whether the hypotheses that are represented in the model are right or wrong. For example, the honeybee workers were shown not to be maximizing net rate of energy delivery to the hive, but were maximizing efficiency in their foraging. The hypotheses that were tested in the bee study, and in all optimality studies, were hypotheses about the *currency* (what is being optimized) and about the *constraints* on the animal's performance (energy costs, handling times and so on). The currency is a hypothesis about the costs and benefits impinging on the animal; for example, for bees energetic costs and benefits seem to be much more important than, say, predation and other dangers. The constraints are hypotheses about the mechanisms of behaviour and the physiological limitations of the animal, whether it is able to recognize differences in nectar concentration, how fast it can fly and so on.

(2) A second advantage is that the assumptions underlying the currency and constraint hypotheses are made explicit. In the model used to analyse load size of starlings, for example, we had to make explicit assumptions about the loading curve, about the fact that the bird could encounter only one patch at a time, about the time taken to fly to the nest and so on. By making these things explicit in the model one is forced to think clearly about the problem.

Optimality models: testability, explicit assumptions, generality

(3) Finally, optimality models emphasize the generality of simple decisions facing animals. For example, foraging starlings and copulating dung flies face the same general problem of when to give up from a curve of diminishing returns.

Table 3.1 A summary of some of the decisions, currencies and constraints discussed

Animal	Decision	Currency	Some constraints	Test
Starling	Load size	Maximize net rate of gain	Travel time, loading curve, energetic costs	Load versus distance
Bee	Crop load	Maximize efficiency	Travel time, sucking time, energetic costs	Load versus flight time
Dung fly	Copulation time	Maximize fertilization rate	Travel time, guarding time, fertilization curve	Predict copula duration
Great tit	Size of worms	Maximize net rate of gain	Handling time, search time	Choice of large or small prey
Downy woodpecker	Patch time	Maximize net rate of gain	Travel time, recognition time	Number of holes inspected
Yellow-eyed junco	Where to feed	Minimize risk of starvation	Handling time, daily energy budget	Choice of variable or certain reward
Great tit/ marsh tit	Body reserves/ hoard size	Maximize survival	Energetic cost of carrying reserves	Body reserves/ hoard size in predictable and unpredictable environments
Squirrel	Where to eat	Maximize survival	Travel time, handling time	Vary size of food and distance
Stickleback	Where to feed	Minimize danger and starvation	Vigilance and foraging incompatible	Vary hunger and danger
Bluegill sunfish	Habitat choice	Maximize survival	Growth depends on food intake, danger related to size	Habitats used at different ages

What should we do when a model fails to predict observed behaviour?

Now for a difficulty with the optimality approach: deciding what to do when the model fails to predict what the animal does. Take the dung flies as an example: the model predicts reasonably well, but not exactly, the duration of copula. What should be done about the discrepancy? Should we ignore it, assuming it is within the acceptable range of error, or should we try to analyse if further? Assume for the moment that we wanted to take the latter course. One possibility is that the currency of the model was incorrect; dung flies may trade-off feeding and mating, or danger and mating, rather than simply maximizing rate of fertilization. A second possibility is that the currency is correct but that the constraints have not been identified correctly; perhaps males run out of energy reserves while in copula. Finally, the whole idea of dung flies or other animals maximizing a currency may be incorrect. Animals may simply not be that well tuned by the process of natural selection or they may be lagging behind when some aspect of the environment changes (as we saw in Chapter 1). The important point is that discrepancies between observed and predicted behaviour can be used to inspire further studies of currencies, constraints and the animal's environment, and so build up a better understanding of the animal's decision making. Another important step is to analyse more thoroughly the mechanisms underlying behavioural decisions.

Summary

Behaviour involves decision making (e.g. where to search, what to eat) which has costs as well as benefits. Individuals should be designed by natural selection to maximize their fitness. This idea can be used as a basis to formulate optimality models which specify hypotheses concerning: (i) the *currency* for maximum benefit (e.g. maximize rate of energy delivery to the nest, as in starlings, or rate of fertilizing eggs, as in dung flies) and (ii) the *constraints* on the animal's performance (e.g. search and handling time, energy costs, risks of predation). The emphasis of this approach is on quantitative, testable predictions about the choices that will maximize fitness. Often the observed behaviour deviates from the predictions of simple models: these discrepancies can then be used to refine the model to provide a better understanding of costs, benefits, currencies and constraints.

Foragers may minimize their risk of starvation by varying their risk-taking in response to their own hunger and the mean and variability of food rewards from different choices (juncos). Some animals store body fat as an insurance against hard times, while others hoard food. Food storing birds have remarkable spatial memory and a specialized brain (a relatively larger hippocampus than non-storers). Experiments with scrub jays reveal that they have episodic-like memory (for what they store, where and when), modify their storing in response to potential thieves and plan for the future.

Foragers often face a trade-off between feeding benefits and predation risk (fat stores in great tits, prey choice by sticklebacks, habitat choice by bluegill sunfish).

Individuals may sample alternatives themselves to determine the most profitable options, or they may use the behaviour of others as a cue (social learning), especially when individual learning is costly (nine-spined sticklebacks). Chimpanzees and meerkats learn foraging tricks from others in their group. Social learning may give rise to local traditions (fish foraging or mating sites). Sometimes knowledgeable individuals actively teach naïve individuals (ants, meerkats).

Both food storing and social learning (including teaching) may involve simple behavioural mechanisms rather than complex cognition based on the recognition of the intentions or knowledge of other individuals.

Further reading

Kacelnik and Bateson (1997) compare risk taking when foragers are faced with variable amounts of food versus variable delays in obtaining food. Shafir *et al.* (2008) compare risk-taking behaviour in honeybees and humans. Marsh *et al.* (2004) and Pompilio *et al.* (2006) show that an individual's energetic state during learning influences foraging choices in starlings and desert locusts respectively. These studies show how an understanding of behavioural mechanisms can illuminate a functional analysis in terms of costs and benefits.

Mangel and Clark (1988) and Houston *et al.* (1988) show how 'dynamic models' can be used to predict sequences of behaviour in response to changes in an individual's state.

Kendal *et al.* (2005) discuss the relative merits of social and asocial learning and Danchin *et al.* (2004) discuss how the use of public information can lead to cultural evolution. Taylor *et al.* (2009, 2010) and Kacelnik (2009) discusses cognition and tool use in animals while Tebbich and Bshary (2004) is an experimental study of tool use in the Galapagos woodpecker finch.

Sara Shettleworth's (2010a) book is a good reference for further studies of social learning and food storing.

TOPICS FOR DISCUSSION

1. Imagine that the observed load sizes in Fig. 3.2c did not give such a good fit to the model's predictions. What would you do next?
2. How are laboratory experiments on decision making in simple environments useful for understanding behaviour in the field?
3. Is average net rate of intake a sensible currency for foraging animals?
4. How might one investigate the mechanisms by which animals discriminate between fixed and variable amounts of food?
5. Why do some species store food while others do not? What might be the advantages of scatter hoarding? How would you test your hypotheses? Are there other situations in which you might predict differences in hippocampus size associated with ecological differences? (see Pravosudov *et al.*, 2006; Reboreda *et al.*, 1996).
6. Does tool using in animals show that they are intelligent? (see Taylor *et al.*, 2009, 2010; Kacelnik, 2009; Tebbich & Bshary, 2004).
7. Franks & Richardson (2006) suggest that a fourth criterion is needed to recognize teaching, namely bi-directional feedback between teacher and pupil. Do you agree?

CHAPTER 4

Predators versus Prey: Evolutionary Arms Races

Photo © Kyle Summers

In Chapter 3, we examined how predators search for and select their prey. We regarded various parameters, such as a predator's tactics for finding and handling prey, and a prey's tactics for avoiding predation, as more or less constant. Over evolutionary time, however, these may vary. Natural selection for efficient foraging by predators should select for improvements in prey defences. This, in turn, will select for improvements in predator tactics, leading to further changes in prey and predator lineages (Fig. 4.1). Antagonistic interaction leading to reciprocal evolutionary change is termed 'co-evolution', and any escalation of adaptations and counter-adaptations has been likened to an arms race (Dawkins & Krebs, 1979).

Red Queen evolution

Evolving to keep up with rivals

If both predators and prey improve over evolutionary time then, whereas their tactics may change, the relative success of each party may not do so. van Valen (1973) coined this kind of never-ending arms race 'Red Queen' evolution, after the Red Queen in Lewis Carroll's *Through the Looking Glass*. In this story, the Red Queen grabs Alice by the hand and they run together, faster and faster. To Alice's surprise, they never seem to move but remain in the same spot. 'In our country', she says, 'you'd generally get to somewhere else if you ran very fast for a long time'. The Queen replies, 'A slow sort of country! Here it takes all the all the running you can do to keep in the same place'.

Water fleas versus bacteria

At first, it seems unlikely that we would ever be able to test directly for Red Queen dynamics because this would involve comparing the success of past and future generations of predators against current prey populations (or, alternatively, pitting past and future prey populations against current predators). However, this has proved possible in a remarkable study of water fleas, *Daphnia magna*, and their bacterial parasite, *Pasteuria ramosa* (Decaestecker *et al.*, 2007).

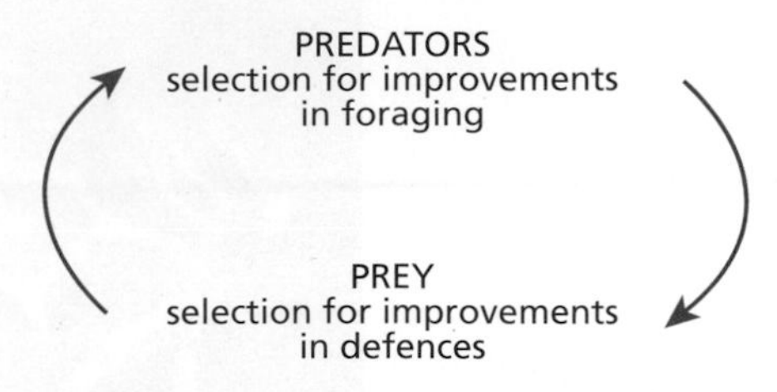

Fig. 4.1 Selection for efficient foraging by predators selects for better prey defences, which in turn selects for predator improvements, further prey improvements and so on.

Both the host and the parasite produce resting stages that accumulate in lake sediments, providing a living 'fossil record' of past generations. Host and parasite populations from different generations (over a 39-year period) were restored by reactivating dormant host eggs and parasite spores. *Daphnia* were then exposed to parasites from the same sediment layer (contemporary parasites) and from sediment layers with past and future parasite populations. Infectivity was higher with contemporary parasites than with parasites from previous growing seasons; therefore, the *Daphnia* evolved to beat past parasite genotypes while the parasites, in turn, rapidly evolved to adapt to the changing host genotypes. Infectivity was also lower with parasites from future growing seasons; therefore, parasite adaptations were specific to

Table 4.1 Examples of predator adaptations and counter-adaptations by prey.

Predator activity	Predator adaptations	Counter-adaptations by prey
Searching for prey	Improved visual acuity	Crypsis (background matching, disruptive patterns, countershading)
	Search image	Polymorphism
	Search limited area where prey abundant	Space out
Recognition of prey	Learning	Masquerade (resemble inedible object) Warning signals of toxicity (aposematism, Müllerian mimicry) Deceive predators by mimicking defended prey (Batesian mimicry)
Catching prey	Secretive approach, motor skills (speed, agility)	Signal to predator that it's been detected Escape flight Startle response: eyespots
	Weapons of offence	Deflect attack Weapons of defence
Handling prey	Subduing skills	Active defence, spines, tough integuments
	Detoxification ability	Toxins

(a) (i) (ii) (iii)

(b) (i) (ii)

(c) (d)

Fig. 4.2 Prey defences include: (a) Camouflage: examples from (i) a moth, (ii) a spider. Photos © Martin Stevens, and (iii) a leaf frog. Photo © Oliver Krüger. (b) Masquerade: (i) this moth caterpillar, *Biston betularia,* mimics a twig. Photo © Nicola Edmunds. (ii) This first instar alder moth, *Acronicta alni*, mimics a bird dropping. Photo © Eira Ihalainen. (c) Bright colouration signals toxicity, as in this poison frog, *Ranitomeya fantastica*. Photo © Kyle Summers. (d) Eyespots: as in this Peacock butterfly. Photo © iStockphoto.com/ Willem Dijkstra.

their current host populations. Finally, there was no overall change in parasite infectivity among contemporaneous populations through time, suggesting that (at least for this measure of fitness) the relative success of host and parasite remained the same over the generations. So the two parties were indeed 'running to stay in the same place'.

This kind of antagonistic evolution is likely to have played a key role in the evolution of biodiversity. Charles Darwin himself recognized this: 'Wonderful and admirable as most instincts are, yet they cannot be considered as absolutely perfect: there is a constant struggle going on throughout nature between the instinct of the one to escape its enemy and of the other to secure its prey' (quoted in Dawkins & Krebs, 1979). In this chapter, our aim is to explore first how predator–prey interactions lead to complex adaptations and counter-adaptations (Table 4.1; Fig. 4.2). Then we shall consider antagonistic interactions between cuckoos and their hosts. For both these systems we shall address two general questions:

Two questions about arms races

(1) *Adaptation or story telling?* Do the proposed adaptations of one party make functional sense given the adaptations by the other party in the arms race? This is not a trivial question. For example, Thayer (1909) suggested that flamingos are pink so as to blend in with the setting sun, so that lions find them difficult to see in the evening when out hunting. You may think this is an unlikely idea! But the proposal that moths blend in with tree trunks or mimic bird droppings may be equally unlikely. We need to perform experiments to test our hypotheses.

(2) *How can an arms race begin?* Prey cannot suddenly evolve perfect counter-adaptations any more than a vertebrate could instantly evolve a perfect complex structure, such as an eye. Our second question, therefore, is can even crude counter-adaptations decrease predation pressure and thus serve as a starting point for an evolutionary arms race?

Predators versus cryptic prey

Underwing moths ...

We begin by attempting to answer these two questions for one example of an arms race, namely birds as predators of cryptic prey. Alexandra Pietrewicz and Alan Kamil (1979, 1981) studied underwing moths (*Catocala* spp) in the deciduous woods of North America. There are up to 40 species living in a particular locality and they are hunted extensively by birds, including blue jays and flycatchers.

Testing hypotheses about adaptation

... 'cryptic' forewings and 'startling' hindwings?

The forewings of the moths appear cryptic, looking very much like the bark of the trees on which the moths rest (Fig. 4.3). The hindwings, on the other hand, are often strikingly coloured yellow, orange, red or pink (Fig. 4.3). The moths rest with their forewings covering the hindwings but when they are disturbed the hindwings are suddenly exposed. Our hypotheses, therefore, are that the forewings decrease detection and the hindwings may have a 'startle' effect on a predator that has found the moth, causing the bird to stop momentarily and thus giving the moth time to escape.

Fig. 4.3 Underwing moths, *Catocala* spp., have cryptic forewings and conspicuously coloured hind wings. This is *C. sponsa*. Photo © Martin Stevens.

Cryptic forewings

In support of the crypsis hypothesis for the forewings, different species of underwing moths select different backgrounds which match their own colour and thus maximize the cryptic effect. Furthermore, they orient themselves in particular ways, so that the wing patterns merge in with the fissure patterns on the bark. Pietrewicz and Kamil tested the importance of crypsis by giving a slideshow to blue jays in an aviary (Fig. 4.4a). Slides were projected on a screen; in some cases there was a moth present in the picture while in other cases there was no moth. If there was a moth, the jay was rewarded with a

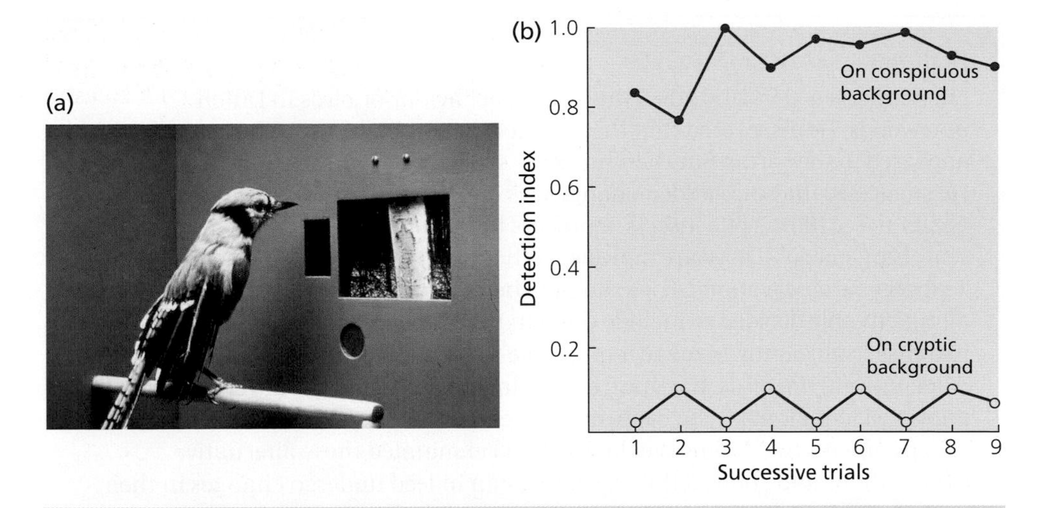

Fig. 4.4 (a) A blue jay in the testing apparatus. The slides are back projected on a screen in front of the bird. The advance key (see text) is to the left. A mealworm is delivered through the circular red hole if the jay makes a correct response. Photo by Alan Kamil. (b) The jays were more likely to detect *Catocala* moths on a conspicuous background. A jay pecking indiscriminately at all slides gets a low score on the detection index. From Pietrewicz and Kamil (1981).

mealworm for pecking at the slide, and the next slide was then shown after a short interval. If there was no moth in the picture then the jay could peck a second smaller 'advance' key to produce the next slide almost immediately. If the jay made either of two errors, namely pecking at the slide when no moth was present or pecking the advance key when a moth was present, it was 'punished' in the form of a delay to the next slide presentation.

Experimental test of crypsis: jays hunt for moth pictures on a screen

This experimental procedure is ingenious for two reasons. Firstly, the predator is faced with a perceptual problem only; there is no complication caused by other factors which might influence predation, such as prey taste, activity or escape efficiency. Secondly, because the predator is stationary and the prey are, in effect, moving past in front of it (in the form of a sequence of slides), it is easy to control the frequency and order in which the predator encounters prey. This would be more or less impossible if the predator was moving about the cage searching for real moths. It was found that the jay made many more mistakes if the moth was presented on a cryptic background than if presented on a conspicuous background (Fig. 4.4b). This provides direct support for the hypothesis about crypsis.

Polymorphic cryptic colouration

Polymorphic prey prevents search image use by predators

In many species of underwing moths the forewings are polymorphic, that is there are different colour forms coexisting within the same population. One hypothesis for this is that when a predator discovers a moth it may form a 'search image' for that particular colour pattern and concentrate on looking for another which looks the same (Box 4.1). If all the population were of exactly the same colour then all would be at risk, but if there was a polymorphism then a predator which had a search image for one morph may be more likely to miss the other morphs. Pietrewicz and Kamil were able to test this

BOX 4.1 SEARCH IMAGES

Luc Tinbergen (1960) studied the feeding behaviour of birds in Dutch pinewoods. He discovered that they did not eat certain insects when they first appeared in the spring but then suddenly started to include them in their diet. He suggested that the sudden change was due to an improvement in the birds' ability to see the cryptic insects, a process he called 'adopting a specific searching image'. There are, however, other hypotheses which could explain Tinbergen's observations. For example, the birds could have seen the insects all along but only decided to include them in their diet when their abundance increased sufficiently to make it profitable to search for them (Royama, 1970). Alternatively, the birds may have been reluctant at first to eat novel prey or have improved in their ability to capture the prey.

Experiments by Marian Dawkins (1971) eliminated these alternative explanations and showed that predators can indeed undergo changes in their ability to see cryptic prey. Her predator was the domestic chick and her prey were coloured rice grains. The clever design of the experiment was to keep the prey the same (therefore handling, acceptability constant) and to vary the background. Two examples are shown in Fig. B4.1.1. Chick (a) was presented with orange grains on a green background (green line) and on an orange background (yellow line). Chick (b) had green grains on an orange background (green line) and on a green background (yellow line). The two tests were run separately for each chick. In both cases the chicks found the prey quicker on a conspicuous background. On the cryptic background the chicks mainly pecked at background stones at first but after 3–4 min they eventually started to find the grain and by the end of the trial they had 'got their eye in' and were eating the cryptic prey at the same rate as when it was on the conspicuous background.

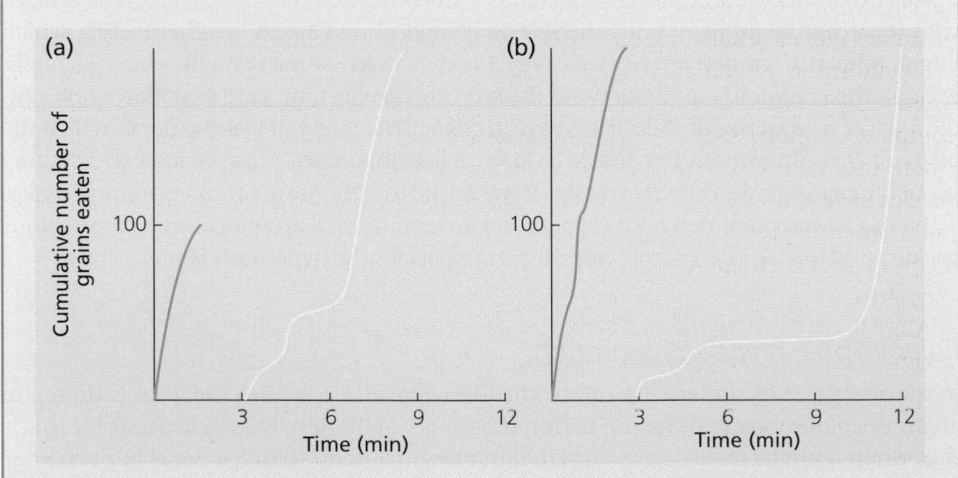

Fig. B4.1.1

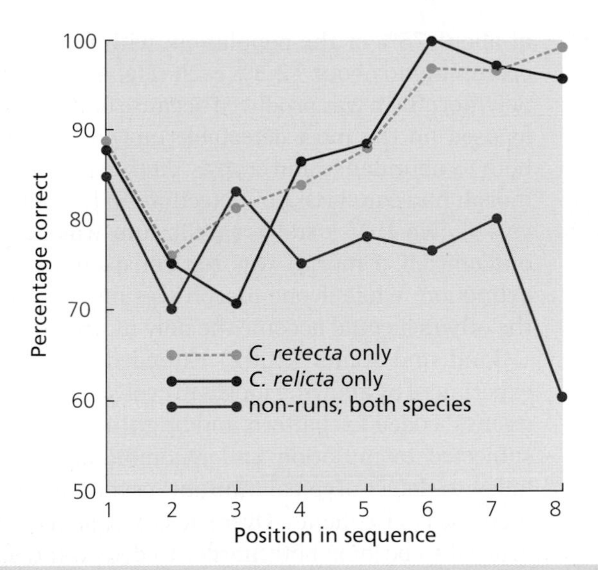

Fig. 4.5 The mean percentage of correct responses by jays when moths are presented in sequences of the same species (runs of either *Catocala retecta* or *Catocala relicta*) or in a sequence containing both species in random order. The jays improved their performance when runs of the same species were presented but not when the two species were presented in a mixed sequence. From Pietrewicz and Kamil (1979). Reprinted with permission from AAAS.

idea by presenting the jay with different sequences of slides. When, for example, runs were presented of 'all morph A' or 'all morph B', the jay quickly got its eye in and improved its success at pecking at the moth as the trial proceeded. When, however, morphs A and B were presented in random order, the jay did not improve its detection success with successive slides (Fig. 4.5). This shows that encountering polymorphic prey does indeed seem to prevent effective search image formation by the predator.

Apostatic selection: rarer prey types at an advantage

Polymorphic cryptic colouration occurs in many other prey species, including grasshoppers, homoptera, mantids and bivalve molluscs. However, it is particularly common in moth species that rest during the day on tree trunks and vegetation. About half of the North American *Catocala* moth species are polymorphic and some species have as many as nine different morphs. In theory, these polymorphisms could be maintained if predators focused on common prey types, so that individuals of rarer forms of prey were more likely to be overlooked. This is known as 'apostatic selection' (Fig. 4.6).

Evolution of prey polymorphisms

Alan Bond and Alan Kamil (1998) tested this idea using the same experimental design with the blue jays, but this time the jays hunted for digital images of *Catocala* moths projected onto a computer screen, with a patterned background (like a tree trunk) against which the moths were hard to detect. The experiments began with a founding population of equal numbers of three morphs of moth, one of which was more cryptic. At the end of each day, detected moths were considered as killed and were removed from the population. The population was then regenerated back to the initial size, maintaining the relative abundance of surviving morphs. Over thirty days ('generations') the abundance of the most cryptic morph increased but it then stabilized

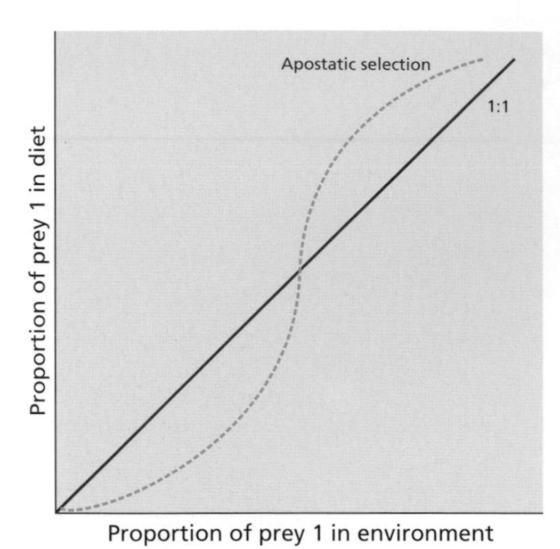

Fig. 4.6 Apostatic selection occurs when a predator eats more of the commoner prey types than expected from their relative frequency in the environment. This promotes polymorphism in prey, because rarer morphs are more likely to survive.

at about 75% of the population, with the other two morphs decreasing to about 12.5% each (Fig. 4.7). Therefore, a stable polymorphism was produced because of the way the predators focused on the most detectable morph, which depended on both its abundance and crypsis. At the stable equilibrium each morph had equal risks of detection. In further experiments, it was shown that a stable equilibrium was not the inevitable outcome; if a morph was too conspicuous it was eaten to extinction, while if one morph was much more cryptic than the others it could become the only morph in the population.

Bond and Kamil (2002) extended this 'virtual ecology' experiment by allowing moths to evolve via a genetic algorithm ('genes' coded for pattern and brightness, and offspring were subjected to mutation and recombination). The jays often failed to detect atypical 'mutant' cryptic moths and so these increased in frequency. Over successive generations the moths evolved to become both harder to detect and showed greater phenotypic variance. This 'virtual' experiment therefore mimics an arms race in nature, where prey evolve improved crypsis and polymorphism. It would be interesting to extend this by allowing the predator to evolve too. This would require both virtual predators and virtual prey.

Brightly coloured hindwings and eyespots

Experimental evidence for the startle effect

What about the brightly coloured hindwings? Debra Schlenoff (1985) tested the responses of blue jays to models which had variously patterned 'hindwings' concealed behind cardboard 'forewings'. The model moths were attached to a board and the jays were trained to remove them to get a food reward underneath. When the models were removed, the hindwings suddenly expanded from behind the forewings to mimic the reaction of the real moths. Jays which had been trained on models with grey hindwings showed a startle response when they were exposed to the brightly patterned hindwings typical of *Catocala*, whereas subjects trained on brightly patterned models did not startle to a novel grey hindwing. After repeated presentations the birds habituated to a particular *Catocala* pattern but a novel bright pattern elicited another startle response. These results provide good evidence for the startle hypothesis, and the habituation effect suggests an adaptive advantage for the great diversity in hindwing patterns of different sympatric species of *Catocala*.

Other cryptic moths and butterflies have eyespots on their wings (Fig. 4.2), which they expose when disturbed causing the predator to cease its approach. Experiments with peacock butterflies *Inachis io* have shown that birds are more likely to attack and eat butterflies whose eyespots have been painted over compared to controls (eyespots intact but painted elsewhere on the wing). Peacock butterflies are not distasteful, so the exposure of eyespots is a case of intimidating the predator by bluff (Vallin *et al.*, 2005).

Why are eyespots on prey an effective defence?

It has long been assumed that eyespots are effective deterrents because they mimic the eyes of the predator's own enemies, such as owls or birds of prey. However, experiments with artificial moths (mealworm bodies plus triangular paper wings) have shown that what makes a wing spot most effective in reducing predation is not eye mimicry itself, but rather high contrast and conspicuousness. Thus, stimuli with circles survived no better than those marked with other conspicuous shapes (such as bars) and circular spots more like real eyes (e.g. pale surrounds, dark centres) were no more effective than others (e.g. dark surrounds, pale centres). Therefore, eyespots seem to work simply by providing a conspicuous, novel stimulus that halts the predator's attack (Stevens *et al.*, 2007, 2008).

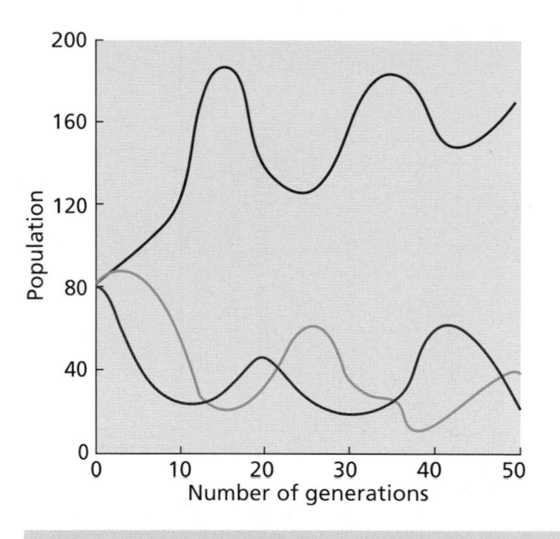

Fig. 4.7 In this experiment, blue jays hunted for digital images of moths on a computer screen (with rewards as in Fig. 4.4). The founding population had equal numbers (80 each) of three morphs, one (dark blue line) being more cryptic. The moth population 'evolved' over 50 generations (see text) to a stable distribution of the three morphs, with the more cryptic form becoming the most abundant. From Bond and Kamil (1998). Reprinted with permission from the Nature Publishing Group.

Does even slight concealment confer an advantage?

The experiments with the blue jays and moths show that crypsis and polymorphisms can indeed evolve as prey defences to thwart predators. But what was the starting point for the evolution of such marvellous camouflage? Could even a small increase in search time, caused by slight concealment, still bring a selective advantage?

Jon Erichsen *et al.* (1980) tested this by an experiment in which a great tit in a cage was presented with a 'cafeteria' of prey moving past on a conveyor belt, like the one in Fig. 3.5a. As with the jays, the clever design of this experiment is that it enables the observer to control precisely both the order in which the predator encounters prey and the rate of encounter. Three items came past the great tit on the belt.

(a) *Inedible twigs*. In fact, these were opaque pieces of drinking straw containing brown string.

(b) *Large cryptic prey*. These were also opaque pieces of straw but had a mealworm inside.

(c) *Small conspicuous prey*. These were clear pieces of drinking straw with half a mealworm inside, clearly visible.

The large prey were worth more energy per unit handling time than the small prey. However, the problem of selecting large prey is that of distinguishing them from the inedible twigs; the opaque straw which came past had to be picked up and examined to see whether it contained a mealworm or just inedible string. The experimental design therefore mimics the problem faced by a predator searching for a profitable, but concealed prey item.

Even slight concealment may bring an advantage

Now for the important point: the discrimination time for an inedible twig was only 3–4 seconds. Therefore, the tit, given time, could easily tell a twig from a large prey. Nevertheless, in theory the tit would maximize its rate of energy intake by ignoring the large prey altogether, provided conspicuous prey were encountered sufficiently frequently,

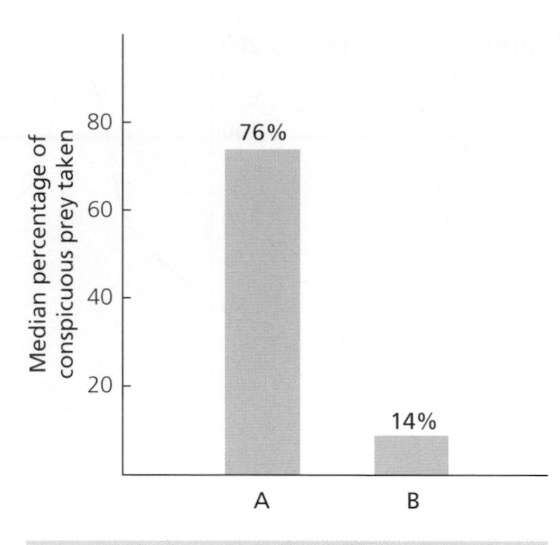

Fig. 4.8 Great tits foraging for artificial cryptic prey. The cost of distinguishing cryptic profitable prey may cause the predator to specialize on other more conspicuous prey. In treatment A, twigs were four times as common as the large prey that resembled them; in B the large prey were four times as common as the twigs. The abundance of conspicuous small prey was constant in A and B. According to an optimal foraging model (similar to those discussed in Chapter 3), it pays the predator to specialize on conspicuous prey in treatment A and on cryptic prey in B. From Erichsen *et al.* (1980).

or provided many of the large items were twigs rather than real prey. Experiments showed that the tits did indeed mainly ignore large prey under these conditions (Fig. 4.8). The conclusion is that even slight concealment, enough to impose only a second or two extra discrimination time, can still be of selective advantage to a prey. To gain an advantage the prey only has to be sufficiently better concealed to make another prey item more profitable for the predator. The results of this experiment support the idea that crude counter-adaptations can indeed provide a starting point for an evolutionary arms race.

Enhancing camouflage

Some animals can vary their crypsis to match the local background by changes in skin pigmentation (cephalopods: Hanlon, 2007; fish: Kelman *et al.*, 2006; chameleons: Stuart-Fox *et al.*, 2008; spiders: Théry & Casas, 2009). However, crypsis does not only involve background matching; other tricks also help to reduce detection.

Disruptive colouration

An experiment to test for disruptive colouration

Many cryptic moths, and other prey, have bold contrasting patterns on the periphery which seem to help break up the body outline (Fig. 4.9a). Innes Cuthill, Martin Stevens and colleagues tested the effectiveness of disruptive colouration by pinning artificial 'moths' to oak trees in a wood. The moths consisted of a dead mealworm for the body and triangular paper wings made from digital images of oak bark. The results showed that these cryptic moths survived attack from insectivorous birds much better than control moths which had uniform brown or black wings. However, disruptive patterns enhanced survival still further; when bold patterns were positioned on the wing edge (where they broke up the body outline) the moths survived significantly better than when the same bold patterns were placed inside the wings, with no overlap of the edges (Fig. 4.9b). Further experiments showed that disruptive patterns worked best if they matched bold colours on the background, but breaking up the body outline reduced predation even if the disruptive colours were conspicuous (Stevens *et al.*, 2006).

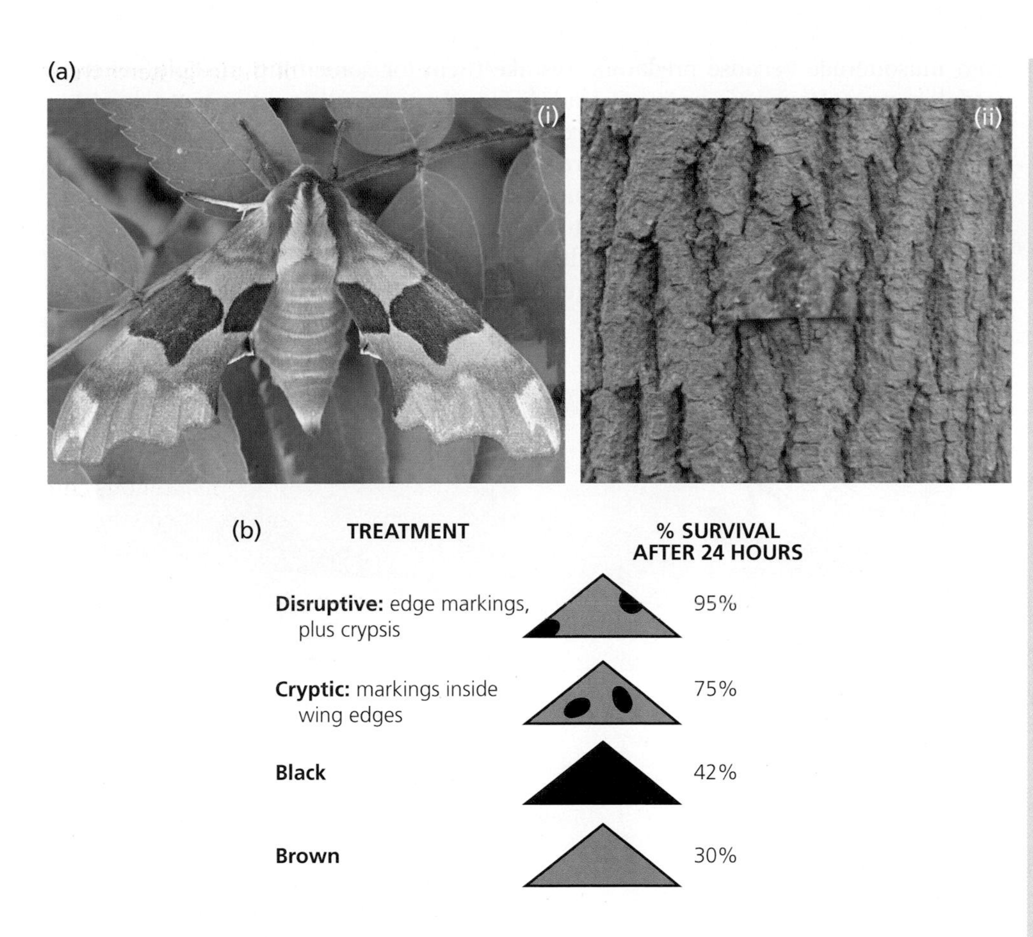

Fig. 4.9 (a) (i) Disruptive patterns on the wing margins of this lime hawk moth conceal the body outline. (ii) An artificial moth with cardboard wings and a mealworm body, pinned to a tree trunk to test the effects of disruptive patterns Photos © Martin Stevens. (b) Results of a field experiment with artificial moths (see text). Disruptive patterns increase survival beyond a cryptic pattern, which in turn survives better than plain black or brown wings. From Cuthill *et al.* (2005). Reprinted with permission from the Nature Publishing Group.

Countershading

Concealing shadows

When a uniformly coloured prey is illuminated by sunlight from above, its ventral shadow betrays its body shape, so even a cryptically coloured prey will be conspicuous to a visually-hunting predator. Many prey appear to solve this problem by countershading, namely darker colouring on the dorsal surface (or whichever surface is normally exposed to the light). The ventral shadow now combines with the darker dorsal surface to give the body a uniform brightness. For a caterpillar, countershading gives it a 'flat' appearance and enhances crypsis against the leaves on which it is foraging (Fig. 4.10a). Experiments with artificial caterpillars, made from pastry, show that countershaded prey do indeed survive better than uniformly coloured or reverse-shaded prey (Fig. 4.10b).

Masquerade

Resemblance of inedible objects

Masquerade involves resemblance of inedible objects such as twigs (stick insects, many caterpillars), leaves (many insects), bird droppings (some caterpillars), flowers (some spiders and mantids) or stones (plants of the genus *Lithops*). Prey presumably benefit from masquerade because predators mistake them for something inedible, whereas predators may also gain from masquerade because unsuspecting prey approach within striking range. Masquerade, therefore, involves camouflage without crypsis; the organism may be detected but it is not recognized as a prey or predator.

Masquerade versus crypsis

How could we distinguish whether an organism has been detected but misidentified (masquerade) rather than been undetected (crypsis)? John Skelhorn and colleagues (2010) did this by manipulating predators' experience of the inedible object while keeping their exposure to the masquerader constant. Their experiment involved

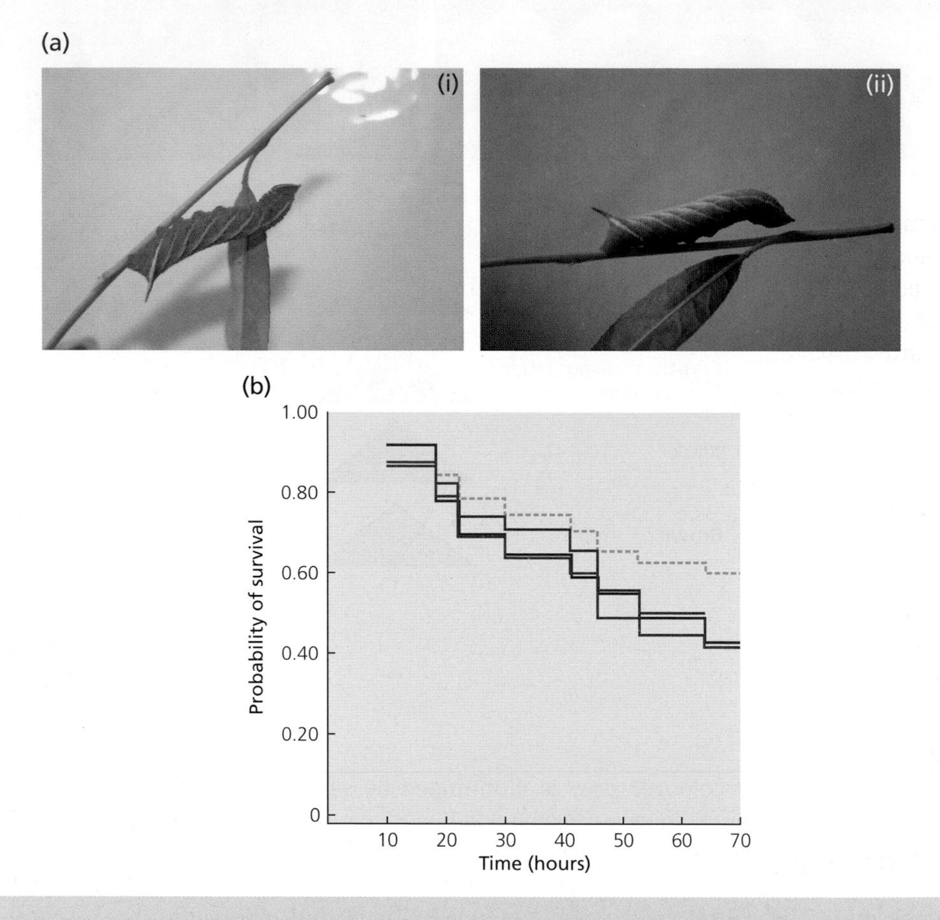

Fig. 4.10 Countershading. (a) (i) This eyed hawkmoth caterpillar, *Smerinthus ocellata*, feeds with its ventral surface uppermost. It has a darker ventral surface which, when lit from above combines with the shadow on the dorsal surface (below) to give it a uniform reflectance, which helps to conceal its body shape. (ii) When turned, so it is now illuminated dorsally, the lighter dorsal surface is highlighted and the darker ventral surface is now in shadow, creating a more pronounced gradient and rendering the caterpillar more conspicuous. Photos © Hannah Rowland. (b) An experiment with pastry 'caterpillars' pinned to the upper surface of branches in a wood. The countershaded caterpillars (dashed light blue line) survived better than plain dark (red line), plain light (purple line) or reverse-shaded prey (darker ventral surface; dark blue line). From Rowland *et al.* (2008).

domestic chicks as predators and twig-resembling caterpillars as prey, which were placed in clear view. The birds were divided into three groups: some encountered natural twigs; some encountered manipulated twigs, bound in purple cotton thread (to change their visual appearance without influencing their physical structure); and some experienced an empty arena. The birds were then presented with twig-resembling caterpillars. Birds with prior experience of natural twigs took longer to attack the caterpillars and handled them more cautiously compared to birds that had no experience of twigs, or that had experienced coloured twigs. Therefore, the caterpillars were detected but misidentified, a true case of masquerade.

Warning colouration: aposematism

Why bright colours?

Some prey are brightly coloured rather than cryptic. Fruit often becomes more brightly coloured when ripe, which increases the chance that it is eaten and so the seeds are dispersed. This is an example of a prey which is selected to be eaten by predators. On the other hand, many prey are also brightly coloured, yet they are presumably selected to *avoid* predation. Prey often have red, yellow or orange markings, often combined with black, which makes them especially conspicuous against the green vegetation (Fig. 4.11).

Brightly coloured prey are often toxic or have other defences

Bright colours in prey are often associated with repellent defences such as toxins, spines or stings. For example, the poison frogs (Dendrobatidae) of tropical central and south America are a monophyletic group with some 210 species. Some are cryptic and palatable and they tend to have a generalized diet. Others are brightly coloured (yellow, blue, red or lime green, often combined with black) and they are highly toxic, deriving at least some of their skin alkaloids from their specialized diet of ants, termites and mites (Daly *et al.* 2002). This association between diet specialization and bright colouration has evolved several times independently

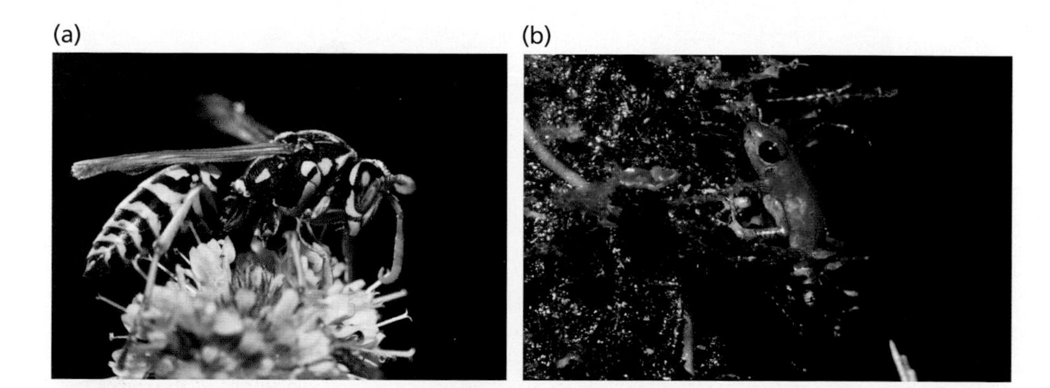

Fig. 4.11 Brightly coloured prey often have repellent defences. (a) A stinging wasp. Photo © iStockphoto.com/ElementalImaging (b) A red poison dart frog. Photo © Oliver Krüger.

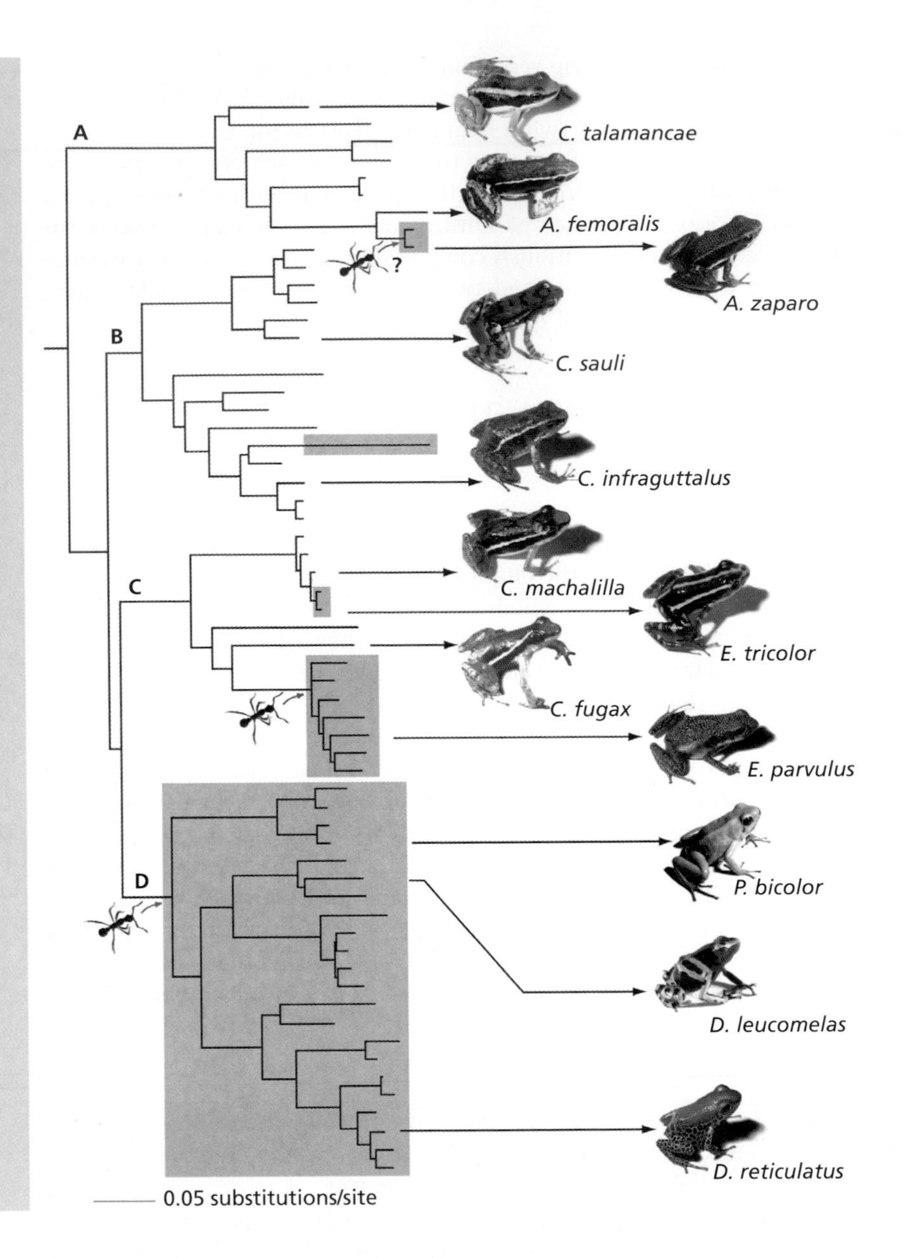

Fig. 4.12 Phylogeny of poison frogs (Dendrobatidae) based on molecular genetic analysis. The ant icons indicate two origins of a specialized diet, and a possible third origin is indicated by a question mark. The column of photos on the left shows representative cryptic and non-toxic species. The column on the right shows conspicuous and toxic species (the toxicity of *A. zaparo* is unknown). Figure from Santos *et al.* (2003); by courtesy of David Cannatella and Juan Carlos Santos.

from a generalist, cryptic ancestor within the Dendrobatidae family tree (Fig. 4.12; Santos *et al.*, 2003).

Wallace's hypothesis: warning signals to reduce predation

Why bright colours? In a letter to Charles Darwin, Alfred Russel Wallace suggested an answer: 'Some outward sign of distastefulness is necessary to indicate to its would-be destroyer that the prey is a disgusting morsel'. Wallace's hypothesis is that bright colours are best as warning signals. Such warning colouration is known as aposematism.

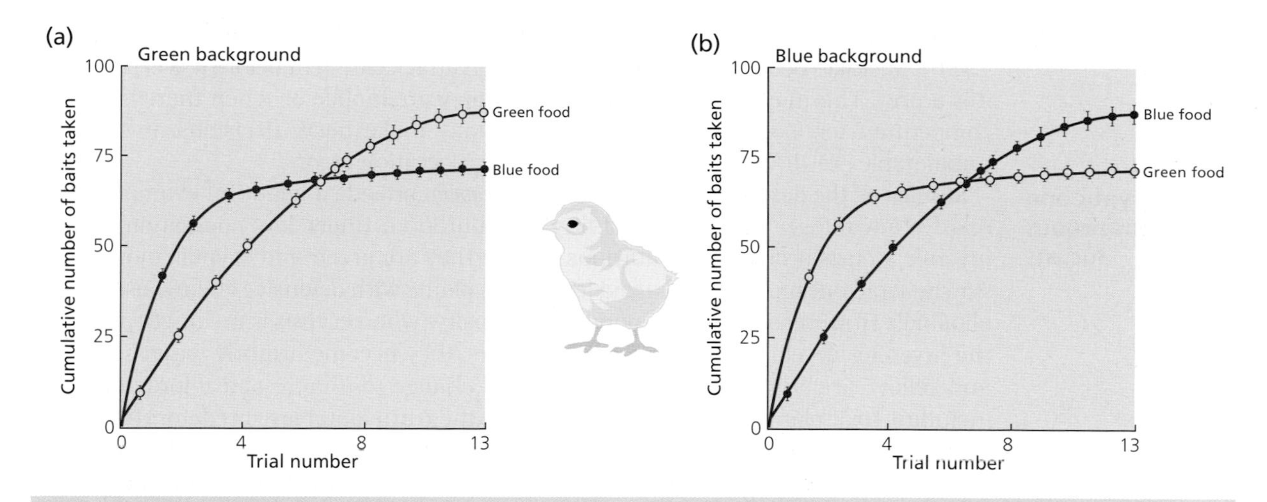

Fig. 4.13 Cumulative number of conspicuous and cryptic distasteful prey taken in successive trials by chicks. In (a) the green food is cryptic, in (b) the blue food is cryptic. In both experiments, by the end of the trial the distasteful prey has been eaten less when it is conspicuous. From Gittleman and Harvey (1980). Reprinted with permission from the Nature Publishing Group.

Conspicuous colours help predators to learn to avoid unpalatable prey

Do prey with repellent defences really benefit from being conspicuous? Gittleman and Harvey (1980) tested this idea by presenting domestic chicks with different coloured breadcrumbs. The chicks showed equal preference for blue or green crumbs. In the experiment, all the crumbs were made distasteful by dipping them in quinine sulfate and mustard powder. Four groups of chicks were used: (a) blue crumbs on a blue background or (b) green crumbs on a blue background, and (c) blue crumbs on a green background or (d) green crumbs on a green background. Whatever the background colour, the chicks took more of the conspicuous prey early on in the experiment. However, overall the cryptic prey suffered the greatest predation (Fig. 4.13). This suggests that it does indeed pay a distasteful prey to be conspicuous. Chicks may learn to avoid conspicuous prey more readily simply because they are more easily recognized on account of their bright colour (Roper & Redston, 1987), or the eating of a large number of prey in a short time may be a more powerful aversive experience than eating an even greater number of prey over a longer period.

Other experiments have shown that learning to avoid bright noxious prey can occur after just one trial and last for a long time. Miriam Rothschild had a pet starling which had just one experience of a brightly coloured distasteful caterpillar and it still refused to pick this species up a year later even though it had no further experience in the meantime. Sometimes predators will avoid brightly coloured and dangerous prey even though they have had no experience of them. For example, hand-reared 'naïve' flycatchers (the great Kiskadee *Pitangus sulphuratus*) will avoid coral snakes (*Micrurus* spp.), which are coloured with red and yellow bands (Smith, 1977).

... and may reduce recognition errors

There may be another advantage for conspicuous warning coloration. Not only can predators learn to avoid conspicuous colours more easily, but having learned to avoid a prey type they are probably less likely to attack by mistake if it is conspicuous (Guilford, 1986). Mistakes occur because a predator often attacks an item before it is certain that it is a prey. This may be advantageous when prey are mobile or when there is intense competition for food. Given that predators may make hasty decisions, it may pay unpalatable prey to be conspicuous to reduce recognition errors.

Cryptic and conspicuous locusts

Studies of the desert locust, *Schistocerca gregaria*, provide a beautiful example of the relative advantages of crypsis and bright colouration. Under low population density, juvenile locusts occur in a 'solitarious' phase; they are green and cryptic, move about slowly, avoid one another and avoid feeding on plants with defensive chemicals, such as alkaloids. However, under high population density (when crypsis is no longer possible), the juveniles develop into a 'gregarious phase'; they become brightly patterned black and yellow, are attracted to one another and change to a more active foraging style, including toxic plants in their diet which makes them unpalatable to predators (Despland and Simpson, 2005). Experiments show that predators learn faster to avoid unpalatable locusts when they are in this conspicuous black and yellow colouration rather than a cryptic green, so the change to the aposematic form at high density is adaptive in reducing predation (Sword *et al.*, 2000).

The evolution of warning colouration

Conspicuousness and distastefulness: which evolved first?

At the start of the chapter, two general questions for prey defences were posed. Clearly, bright colours are advantageous (question 1). But how did they evolve (question 2)? One possibility is that conspicuous colours evolved first, followed by distastefulness. For example, some brightly coloured birds like kingfishers are distasteful (Cott, 1940). Their colours may have been favoured for better mate attraction or territory defence and then, because they also increased conspicuousness to predators, this then favoured the evolution of distastefulness. The other possibility is that distastefulness came first. This may apply to those insects, such as caterpillars of the monarch butterfly, *Danaus plexippus*, which feed on plants containing toxins and incorporate the toxins in their bodies as a defence against predation. It is plausible that here unpalatability evolved first followed by conspicuousness. In this case, then, bright coloration evolves specifically as a warning device.

This last scenario poses an interesting problem. Imagine a population of unpalatable but cryptic larvae. A mutation arises in an adult which causes its larvae to be more conspicuous. These larvae would then surely be more obvious to predators and so more likely to perish. Although the predator may, as a result of its experience with the nasty taste, decide never to touch the brighter form again, because it is a rare mutant the predator is unlikely to encounter another one. Thus, the mutation goes extinct during the sampling and never has the chance to spread. How, then, can warning colours ever evolve?

Fisher's hypothesis: warning colours may have evolved through their effects on survival of relatives in the same group

R.A. Fisher (1930) was the first to propose a solution. He realized that distasteful, brightly coloured insects were often clumped in family groups (Table 4.2 gives an example). In this situation, because of the grouping, the predator does encounter others with the bright coloration, namely the siblings of the individual which perished in the sampling. Thus, their lives are saved and some copies of the gene for conspicuous

Dispersion	No. species of caterpillars	
	Aposematic	Cryptic
Family groups	9	0
Solitary	11	44

Table 4.2 Brightly coloured species of caterpillars of British butterflies are more likely to be aggregated in family groups than cryptic species (Harvey *et al.*, 1983).

coloration get through to the next generation. This process, where traits can be favoured because they benefit relatives, who share copies of the same genes, is termed kin selection; it will be explained in more detail in Chapter 11. Mathematical models have shown that with family grouping bright coloration can evolve in a distasteful species provided the brighter forms are not too conspicuous and provided the predator needs to sample fewer of them, compared to cryptic prey, in order to learn that they are distasteful (Harvey *et al.*, 1982).

But, individuals may benefit directly ... and grouping may have evolved after warning colours

Although Fisher's solution is ingenious, recent work has challenged two of its assumptions. Firstly, the assumption that the sampled individuals always perish may be wrong. Many brightly coloured insects have tough integuments which protect them against attacks by naive predators and they are released unharmed. Indeed, in some cases the defences themselves (e.g. hairs on a caterpillar) are brightly coloured. Thus, in some cases, there may be a direct advantage to the individual in being conspicuous; so long as the bright colouring is more easily remembered, a bright distasteful caterpillar is better protected than a cryptic one in subsequent encounters with the same predator (Sillén-Tullberg, 1985). Many predators are reluctant to attack a novel prey item ('neophobia') which would also promote the survival chances of a mutant, more conspicuous prey type (Marples & Kelly, 1999). Secondly, Fisher assumed that family grouping sets the stage for the evolution of warning coloration but a phylogenetic analysis (Chapter 2) of butterflies suggests that warning coloration evolved *before* gregariousness (Sillén-Tullberg, 1988). It seems likely, therefore, that warning coloration sometimes evolved because of the direct advantage it brought to individuals in decreasing their likelihood of attack, and grouping evolved afterwards through diluting predator attacks per individual (Chapter 6). Thus, grouping is not always critical for the evolution of bright colours. Note that many of the brightly coloured species in Table 4.2 are solitary.

This discussion highlights the central problem in aposematism, namely the trade-off between the costs of conspicuousness in increasing the probability of attack by naïve predators, and the benefits from increased protection against experienced predators through more memorable and detectable signalling. It seems likely that this balance of selection pressures will vary between different aposematic organisms, so one evolutionary pathway is unlikely to explain all cases.

Mimicry

The association between bright colours and repellent defences has led to the evolution of various forms of mimicry.

Müllerian mimicry: repellent species look alike

Fritz Müller (1878) was the first to notice the similarity in colour patterns between different repellent species (Fig. 4.14). For example, stinging wasps and distasteful beetles, bugs and moths living in one area may all share the same yellow and black colouration. Müller's idea was that if these repellent species look alike, then it will be easier for a predator to learn to avoid them all; it just has to learn the one pattern and then all the prey will benefit.

In theory, rarer morphs evolve to mimic commoner morphs

Müller suggested that colour patterns will evolve by a rarer morph converging on the colour of a commoner morph. Imagine two distasteful species with different colours; species A has a population of 10 000 individuals while species B is rarer with just 100 individuals. Assume a predator needs ten attacks to learn that a colour pattern is associated with distastefulness. A rare mutant in species A which resembles species B will suffer a disadvantage because it is more likely to be sampled in a population of 100. By contrast, a rare mutant is species B that resembles species A will be at an advantage because it is hidden in a larger population. Therefore, rarer morphs should evolve to match commoner morphs because they have a 'greater umbrella of protection'.

***Heliconius* butterflies**

A famous case of Müllerian mimicry involves the remarkable similarity between distasteful species of *Heliconius* butterflies in central and South America. Up to a dozen species living in the same area may share the same colour pattern, but colour patterns vary geographically; in some areas the butterflies are all blue, in other areas orange, red or of a tiger pattern (Fig. 4.14; Mallet & Gilbert, 1995). Two species, *H. melpomene* and *H. erato*, show parallel variation in pattern across south America and were the subject of one of the first experimental tests of Müllerian mimicry by Benson (1972). In Costa Rica, he painted some *erato* individuals so they were non-mimetic (in fact, they now resembled another race of *erato* in Colombia). These survived less well than controls, which were also painted but whose pattern remained mimetic. Reciprocal transfers of different colour morphs across a morph boundary produced the same result; non-mimetic forms survived less well and survivors had more beak marks on their wings, suggesting increased predation by birds such as jacamas (Mallet & Barton, 1989).

Convergent patterns may involve changes in the same genes

Crossing experiments within both *melpomene* and *erato* have identified the regions of the genome involved in controlling wing patterns in each species. Analysis of DNA sequences has revealed that producing convergent patterns in the two species involves changes in the same genes (Baxter *et al.*, 2010). It is not yet known whether the genes controlling colour patterns in mimicry rings involving different taxa (e.g. wasps and butterflies) are also the same, but they may well be.

Müllerian mimicry still needs more experimental work. How do predators learn from sampling? (Müller's idea of a fixed number of attacks to learn seems simplistic.) Why do different colour forms occur in different geographical regions? These may reflect differences in the signalling environment (different colours show best in different habitats) or perhaps simply chance differences in initial conditions (which prey species was the most abundant).

Fig. 4.14 Müllerian mimicry. In each case, populations of distantly related species converge on the same brightly coloured warning pattern within a single locality but the patterns vary across their range. (a) North American millipedes of the *Apheloria* clade (top row) and their mimics in the *Brachoria* clade (below) in three geographical regions. Photo © Paul Marek. (b) *Heliconius erato* (top row) and its mimic *Heliconius melpomene* (bottom row) in three geographical regions of the neotropics. Photo © Bernard D'Abrera and James Mallet. (c) Peruvian *Ranitomeya* (*Dendrobates*) frogs from two regions. *Ranitomeya imitator* (left in both panels) and its mimics *R. summersi* (left panel) and *R. ventrimaculata* (right panel). Photo © Jason Brown. From Merrill and Jiggins (2009).

Batesian mimicry: cheating by palatable species

Palatable species may mimic distasteful species

Henry Walter Bates (1862) spent eleven years exploring the Amazon rain forest. He noticed that palatable species of butterflies (mimics) sometimes closely resembled the appearance of distasteful species (models) and suggested that such mimicry evolved because of selective predation by birds. Darwin's *Origin of Species* had been published just three years before and Bates was thrilled by his discovery of mimicry: 'I believe this case offers a most beautiful proof of the theory of natural selection'.

There are some wonderful examples of Batesian mimicry (Fig. 4.15): for example, spiders mimic ants (Nelson & Jackson, 2006), hoverflies mimic wasps (Bain *et al.*, 2007) and harmless snakes mimic venomous snakes (Pfennig *et al.*, 2001). Many laboratory experiments have shown that predators learn to avoid noxious models and then subsequently avoid palatable mimics. Furthermore, mimics do better when they resemble the model more closely, when the model is more abundant, or more noxious, and when alternative palatable prey are available for the predator (Ruxton *et al.*, 2004).

Polymorphic Batesian mimics

Whereas Müllerian mimicry promotes uniformity of colour pattern, Batesian mimicry promotes polymorphism because mimetic patterns will be at an advantage when rare relative to the model (predators more likely to sample noxious models) and at a disadvantage when common (predators more likely to sample palatable mimics). Thus, as a mimic becomes relatively common, mutants resembling another noxious model may be favoured. There are several cases where one species has several morphs living in the same area, each resembling a different model species (e.g. African mocker swallowtails, *Papilio dardanus*; Vane-Wright *et al.*, 1999). In some species there is evidence that the frequency of each mimetic form is controlled by the frequency of the

Fig. 4.15 Batesian mimicry. (a) The highly venemous Sonoran coral snake, *Micruroides euryxanthus*, is the model for (b) its non-venemous Batesian mimic, the Sonoran mountain kingsnake, *Lampropeltis pyromelana*. These photographs were taken within 3 km of each other in Arizona. Photos © David W. Pfennig (c) An English pub sign fooled by a Batesian mimic; this is a hoverfly (*Syrphidae*)! Photo by Francis Gilbert.

models; in populations where a particular model is more common, a greater proportion of the palatable species mimics this model (Sheppard, 1959).

If predation on models increases as mimics become relatively more common, why don't models evolve to escape the mimics? A likely explanation is that whereas there is selection on mimics for improved mimicry, any mutant model which is different in colour will suffer increased predation because it will be rare and not recognized as noxious (Nur, 1970). Therefore, Batesian mimicry may be the outcome of an arms race that the mimic has won (Ruxton *et al.*, 2004).

Trade-offs in prey defences

Two costs

Prey defences have costs as well as benefits. Firstly, resources are limited, so there are *allocation costs*. Increased investment in defence means fewer resources for growth and reproduction. Secondly, alternative benefits may be forfeited, so there are *opportunity costs*. Improved crypsis on an oak tree may limit habitat choice because of increased conspicuousness on other backgrounds, and may restrict other activities, for example signalling to rivals or mates. An example of each is discussed, while recognizing that these two kinds of cost are often closely related.

Costs of aposematism

Theoretical studies suggest that investment in aposematic signals should vary depending on predation pressure (Endler & Mappes, 2004) and trade-offs between investing in repellent defences versus colourful signals (Speed & Ruxton, 2007; Blount *et al.*, 2009). This topic deserves more empirical studies and we focus on one example.

Wood tiger moths have warning colours ...

The wood tiger moth, *Parasemia plantaginis*, is aposematic at both the larval and adult stage. The caterpillars are hairy and black with an orange patch. The adult female has cryptic forewings and orange-red hindwings (Fig. 4.16). Both are moderately distasteful to insectivorous birds and the source of the toxins (iridoid glycosides) is the food plant of the caterpillars (ribwort plantain, *Plantago lanceolata*). The colouration of both

Fig. 4.16 Variation in aposematic colouration among individual wood tiger moths, *Parasemia plantaginis*. (a) The caterpillars vary in the size of the orange patch. (b) The hindwings of female moths vary from bright red to pale orange. Photos © Eira Ihalainen.

caterpillars and adult moths varies locally and geographically; the orange patch on the caterpillars may cover from 20 to 90% of the body, while the female moth's hindwing varies from pale orange to bright red. Studies in Finland by Carita Lindstedt, Johanna Mappes and colleagues have shown that more orange caterpillars and redder adult moths are more likely to be rejected by predatory birds, even though their levels of toxins are no greater than those of less bright individuals (Lindstedt *et al.*, 2008). Bright colours are, therefore, more effective as warning signals. Why, then, aren't all the caterpillars and moths brightly coloured? Experiments reveal that the aposematic colours are both costly to wear and costly to make.

... both costly to wear, and costly to make

Costly to wear: The size of the caterpillar's orange patch is heritable, so laboratory selection lines were used to produce caterpillars with large and small orange patches. At low temperatures, caterpillars with smaller orange patches (hence more black) grew faster because they could absorb heat more effectively and, hence, forage for longer each day. Caterpillars reared at low temperatures developed smaller and darker orange patches, suggesting that the costs of thermoregulation have selected for phenotypic plasticity in aposematic colouration (Lindstedt *et al.*, 2009).

Costly to make: In another experiment, caterpillars were reared on plants with either a low or a high concentration of glycosides. Both groups of caterpillars and subsequent adult moths had equal levels of toxins in their bodies, so excess toxins were disposed of effectively. However, detoxification was costly because moths reared on the high concentrations produced fewer offspring. Furthermore, although the diet did not affect caterpillar colouration, female moths reared on high toxin concentrations developed less bright hindwings. This suggests that resources devoted to getting rid of toxins left fewer resources for pigment production (Lindstedt *et al.*, 2010).

Conspicuousness versus crypsis

As a defence against predators it may pay to be cryptic, but this may conflict with the advantage of being conspicuous for other activities such as territory defence or mate attraction. As an example of this trade-off, in many species of birds the males are brightly coloured in the breeding season but moult into duller female-like plumage after breeding.

In guppies, brighter males have a mating advantage ...

John Endler's (1980, 1983) work on the coloration of guppies (*Poecilia reticulata*) provides an illuminating experimental study of this trade-off. Endler studied several isolated populations of these little fish in the streams of Trinidad and Venezuela. Males are more colourful than females. Three types of colour can be distinguished. (a) *Pigment colours* (carotenoids — red, orange and yellow), which are obtained from the diet. If fish are fed carotenoid-free food then these colours fade within a few weeks (Kodric-Brown, 1989). (b) *Structural colours* (iridescent blue and bronze) produced by reflection of light from scales. (c) *Black spots* (melanin), which are partly under nervous control and can increase or decrease in size. Laboratory experiments showed that brighter colours brought a mating advantage. Females were particularly attracted to the orange spots (Houde, 1988).

... but suffer increased predation

To test whether there was counter-selection against bright colours due to predation, Endler sampled streams with different predator communities. He found that males living in streams with greater predation pressure were duller in colour, having both fewer colour spots per fish and also smaller spots (Fig. 4.17a). Not only did predation intensity influence guppy coloration but the type of predator was also important. In

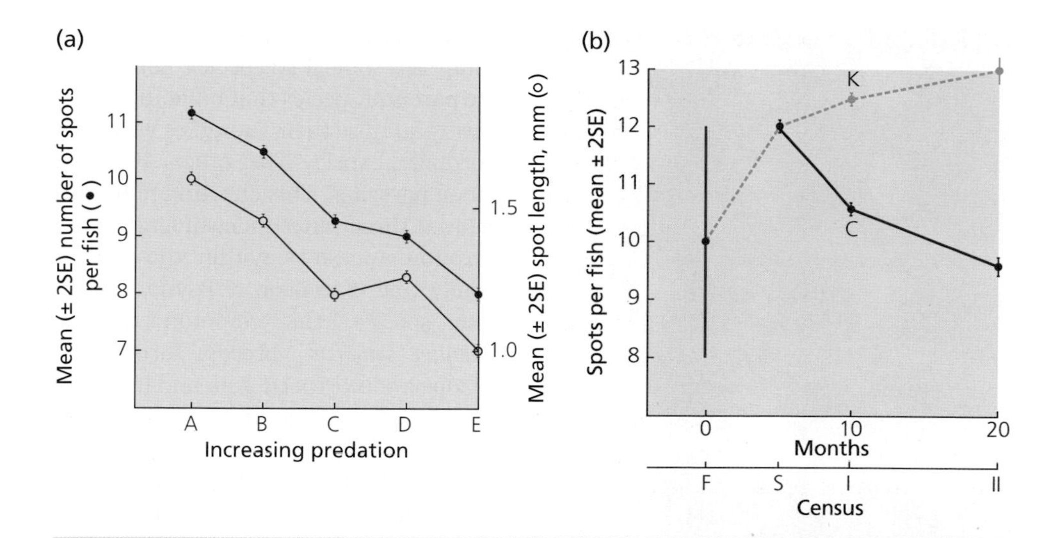

Fig. 4.17 Influence of predation on colour pattern of male guppies. (a) Both number of colour spots per fish and spot size are smaller in streams with greater predation. The main predators are other fish and prawns. Data from five streams in Venezuela with increasing levels of predation from A to E. From Endler (1983). (b) A selection experiment in the laboratory. F, foundation population of guppies kept with no predators. S, start of the experiment; predators added to population C but not to population K. Note the rapid change in population C after predation began. I and II are the dates of two censuses. From Endler (1980).

some streams the main predator was a prawn, which was red blind, and here the male guppies were significantly redder.

An evolutionary experiment in the field

Finally, Endler showed how colours could change in response to changed selection pressure. In controlled laboratory populations studied over many generations, males kept with predators evolved to be duller, while those kept isolated from predators evolved to be brighter, with both larger and more spots (Fig. 4.17b). Similar effects could be shown under field conditions: when 200 dull males from a high predation stream were introduced into a new isolated stream in Trinidad, which had no guppies and no predators, the population evolved during a two-year period into one with much more colourful males. Endler's experiments provide a convincing example of how natural selection can change colour patterns in relation to a shifting balance between different selection pressures.

Cuckoos versus hosts

Brood parasites trick hosts into raising the parasite's offspring

Some species of birds, fish and insects are brood parasites; they lay their eggs in the nests of other species (hosts) and so get their young raised for free. As with predator and prey, one party (the brood parasite) gains a benefit while the other (the host) suffers a cost. Clearly, then, we would expect selection to favour host defences. This, in turn, should select for counter-adaptations by the parasite. We shall focus on one example to show how field experiments can help to uncover the various stages of this arms race.

Fig. 4.18 The common cuckoo, *Cuculus canorus*, has several genetically different host-races each of which lays a distinctive egg (central column) which matches, to varying extents, the eggs of its particular host species (left-hand column). In the examples here, the host species and their corresponding cuckoo host races are, from top to bottom: robin, pied wagtail, dunnock, reed warbler, meadow pipit and great reed warbler. The right-hand column is of variously coloured model eggs used to test host discrimination. From Brooke and Davies (1988).

The cuckoo family (Cuculidae) comprises some 140 species. Some 60% are parental species that build their own nests and raise their young, as we might normally expect. The other 40% are brood parasites. This cheating habit has evolved three times independently from parental ancestors within the cuckoo family tree (Sorenson & Payne, 2005). One species, the common cuckoo, *Cuculus canorus*, breeds throughout Europe and northern Asia and has been particularly well studied. This species has several genetically distinct host races, each of which specializes on one host species and lays a distinctive egg type that tends to match the egg of its particular host (Fig. 4.18).

How common cuckoos lay their eggs

The female cuckoo adopts a particular procedure when parasitizing a host nest (Fig. 4.19a). She usually finds nests by watching the hosts build. Then she waits until the hosts have begun their clutch and parasitizes the nest one afternoon during the hosts' laying period, laying just one egg per host nest. Before laying, she remains quietly on a perch nearby, sometimes for an hour or more. Then she suddenly glides down to the nest, removes one host egg and, holding it in her bill, she then lays her own egg and flies off, later swallowing the host egg whole. She is incredibly quick; the time spent at the host nest is often less than ten seconds! Having laid her egg, the cuckoo then abandons it, leaving all subsequent care to the hosts.

Hosts sometimes reject the cuckoo egg, but often they accept it (Fig. 4.19b). The cuckoo chick usually hatches first, needing an unusually short period of incubation, whereupon, just a few hours old and still naked and blind, it balances the host eggs on its back, one by one, and ejects them from the nest (Fig. 4.19c). Any newly hatched host young suffer the same gruesome fate and so the cuckoo chick becomes the sole occupant of the nest and the hosts slave away, raising it as if it was their own (Fig. 4.19d).

To what extent has each party evolved in response to selection pressures from the other? Experiments provide evidence for coevolution.

Fig. 4.19 (a) A female common cuckoo parasitizing a reed warbler nest. Firstly she removes a host egg. Then, holding it in her bill, she sits briefly in the nest and lays her own egg in its place. Photo by Ian Wyllie. (b) A cuckoo egg (right) in a reed warbler nest. (c) A newly-hatched cuckoo chick ejecting host eggs one by one. Photo © Paul Van Gaalen / ardea.com (d) The reed warbler hosts continue to feed the young cuckoo even as it grows to seven times their own weight. Photo © osf.co.uk. All rights reserved.

Cuckoos have evolved in response to hosts

Laying tactics to beat host defences

There are two sources of evidence for this. Firstly, the female cuckoo's egg-laying tactics are specifically designed to circumvent host defences. Experimental 'parasitism' of reed warbler nests with model cuckoo eggs reveals that the hosts are more likely to reject the model egg if it is a poorer match of the hosts' own eggs, if it is laid too early (before the hosts themselves begin to lay), if it is laid at dawn (when the hosts themselves lay) or if the hosts are alerted by the sight of a cuckoo on their nest. Therefore, the cuckoo's mimetic egg, timing of laying and unusually fast laying are all adapted to increase the success of parasitism (Davies & Brooke, 1988).

Cuckoos have evolved host-egg mimicry

Secondly, egg mimicry by cuckoos evolves in response to host discrimination. To quantify egg mimicry it is important to do so as through a bird's eye. Birds have better colour vision than humans, with four cone types sensitive, respectively, to very short (ultraviolet), short (blue), medium (green) and long (red) wavelengths. Mary Caswell Stoddard and Martin Stevens (2011) measured the reflectance spectra of cuckoo and host eggs from the various cuckoo host races (Fig. 4.20a), calculated how these wavelengths would be captured by the four cone types, and then plotted the colours in tetrahedral colour space (Fig. 4.20b). They found that the match between cuckoo and host eggs varied across the different host races of cuckoo (Fig. 4.20c), with a better match when the hosts were stronger discriminators (Fig. 4.20d).

A question that is still unresolved is whether these differences in host discrimination represent: (i) different equilibria set by differences in the costs and benefits of egg rejection or (ii) evolution in progress, with older hosts having stronger rejection. For example, dunnock hosts may reject foreign eggs less than great reed warbler hosts either because: (i) rejection is more costly or of less benefit (lower parasitism risk) or (ii) because dunnocks are more recent hosts and have not had sufficient time to evolve defences.

Hosts have evolved in response to cuckoos

Defences evolve in response to cuckoo parasitism

If host discrimination evolves specifically in response to brood parasitism, then we would predict that species untainted by cuckoos would show no rejection of odd eggs. This is indeed what is found; species that are unsuitable as hosts, either because of a seed diet (young cuckoos need to be raised on invertebrates) or because they nest in holes (inaccessible to a female cuckoo), show little, if any, rejection of eggs unlike their own and, in contrast to hosts, they also show little aggression to adult cuckoos near their nests (Davies & Brooke, 1989a, 1989b; Moksnes *et al.*, 1991).

Host egg signatures and cuckoo forgeries

Hosts not only evolve egg rejection as a defence, their egg patterns evolve, too, providing more distinctive 'signatures' to signify 'this is my egg'. Compared to species with no history of brood parasitism, species exploited by cuckoos have less variation in the appearance of eggs within a clutch and more variation between clutches of different females (Stokke *et al.*, 2002). This makes life harder for the cuckoo, since it is easier for a host to spot a foreign egg if all its own eggs look exactly the same, and distinctive markings for individual host females makes it harder for the cuckoo to evolve a convincing forgery of that species' eggs.

The diversity of egg markings within one host species is especially remarkable in an African warbler, the tawny-flanked prinia, *Prinia subflava*, which is parasitized by the cuckoo finch, *Anomalospiza imberbis* (Fig. 4.21). The host eggs vary in background colour, marking size, variation in markings and in marking dispersion. These four parameters vary independently, which is exactly what would be predicted if hosts had been selected to maximize the individual distinctiveness of their egg signatures.

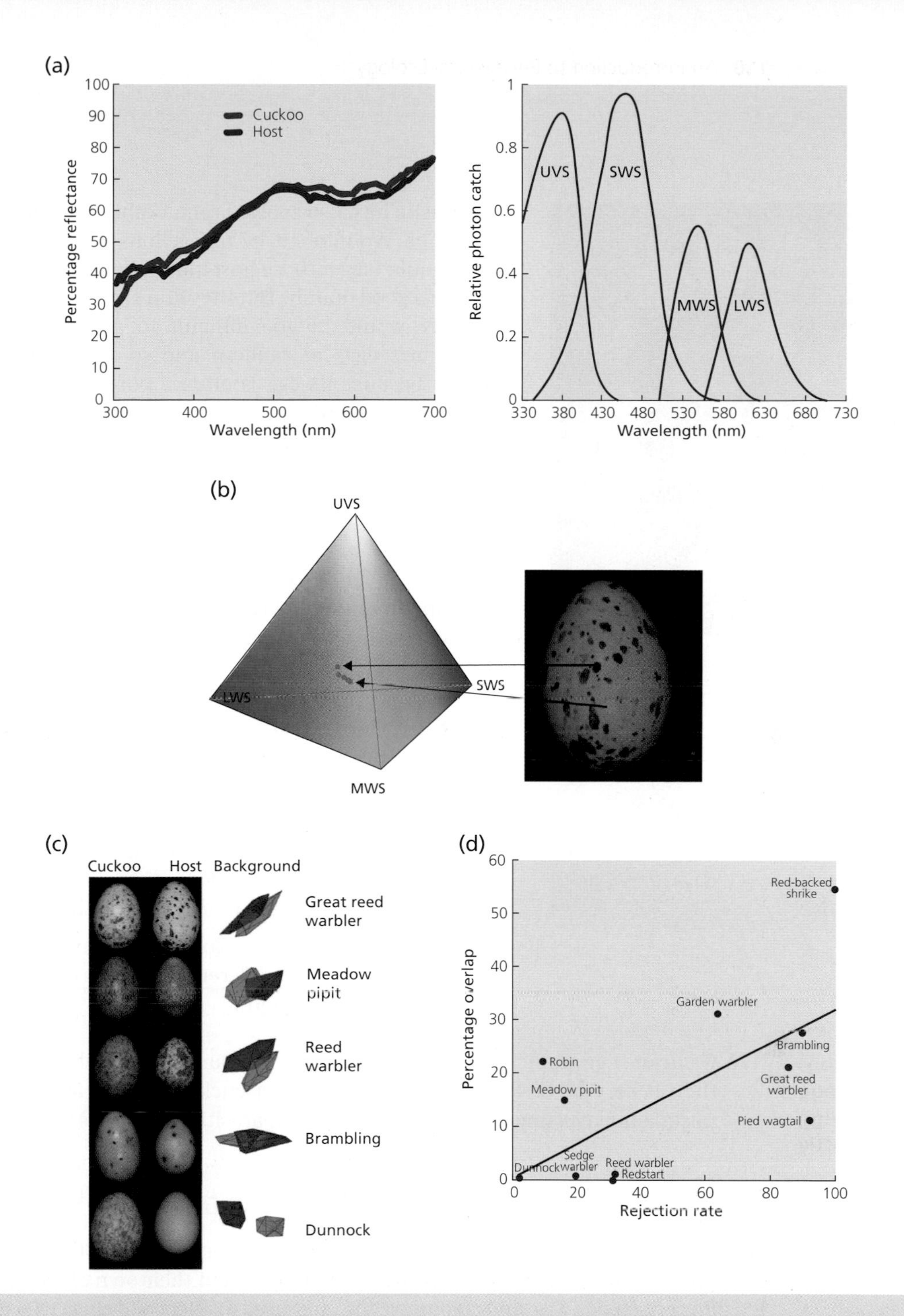

Fig. 4.20 Seeing eggs through a bird's eyes (Stoddard & Stevens, 2011). (a) The relative stimulation of the four avian cone types (ultraviolet sensitive, short-, medium- and long-wave sensitive) is determined from reflectance spectra and the spectral sensitivity functions for each cone type. From Hart (2001). With permission from Elsevier (b) Egg colours are mapped in avian tetrahedral colour space; the position of a colour is determined by the relative stimulation of the four retinal cones. (c) Background colours and spot colours are then compared for cuckoo and host eggs. Only background colours are shown here. The overlap between common cuckoo (red) and host (blue) distributions in tetrahedral colour space is shown for various host races of cuckoo. (d) The relationship between host rejection rate of non-mimetic eggs and background colour overlap for eleven cuckoo host races. Colour overlap is expressed as the percentage of the host volume overlapped by the cuckoo volume. Egg photographs by Mary Caswell Stoddard © Natural History Museum, London.

Fig. 4.21 Tawny-flanked prinia eggs (outer circle) and cuckoo finch eggs (inner circle). The diversity of egg 'signatures' in the host leads to a signature-forgery arms race between host and cuckoo and remarkable diversity in egg colours and markings within a species. From Spottiswoode and Stevens (2012). Photo by Claire Spottiswoode.

The parasite has also evolved remarkable variation in its eggs (Fig. 4.21). Wouldn't it be marvellous if individual cuckoo finches could target those host individuals for whom their egg would be a good match? But they don't, perhaps because such selectivity would be too difficult to achieve. Instead, the parasite lays eggs at random and so suffers high rates of rejection because its egg is often a poor match for the host eggs (Spottiswoode & Stevens, 2010). Thus, host egg signature variation can be an effective defence.

The egg arms race: a coevolutionary sequence

These experiments enable us to reconstruct the likely sequence of the cuckoo–host arms race at the egg stage (Davies & Brooke, 1989a, 1989b).

(i) At the start, before small birds are parasitized, they show little, if any, rejection of foreign eggs (small birds with no history of cuckoo parasitism, because they are unsuitable as hosts, do not reject).

(ii) In response to parasitism, hosts evolve egg rejection (hosts do reject) and more distinctive individual egg signatures (hosts have more egg variation).

(iii) In response to host rejection, cuckoos evolve egg mimicry (egg mimicry in the different host races reflects the degree of host discrimination).

Once cuckoos evolve egg mimicry, host egg rejection is costly

(iv) If cuckoo egg mimicry is sufficiently good, and if parasitism levels are not too high, then it may be best for hosts to accept most cuckoo eggs to avoid the costs of mistakenly rejecting their own eggs from unparasitized clutches (Box 4.2).

Egg rejection versus chick rejection as a host defence

Hosts occasionally abandon common cuckoo chicks (Grim *et al.*, 2003) but why don't they always do so? The cuckoo chick is much larger than their own chicks and its gape is a different colour. Size and colour are the cues used to reject odd eggs; why not pick up on these same cues at the chick stage too?

Why accept the cuckoo chick? A cognitive challenge?

Arnon Lotem (1993) suggested an ingenious solution to this puzzle. At the egg stage, experiments have shown that hosts imprint on their eggs the first time they breed and then they reject eggs that differ from this learnt set (Rothstein, 1982; Lotem *et* al., 1995). Such imprinting is not a foolproof defence. Firstly, if hosts are unlucky and are parasitized in their first clutch then they may misimprint on the foreign egg too and so regard it as one of their own. Secondly, once the parasite has evolved egg mimicry then hosts may make recognition errors and reject one of their own eggs rather than the parasite's egg (Box 4.2). Therefore, as for prey, host defences are costly. However, the costs of misimprinting would become prohibitively large at the chick stage if hosts are faced with parasitic chicks, like the common cuckoo, which hatch early and eject the host eggs. Then, if hosts are parasitized in their first breeding attempt, all they see in the nest at the chick stage is a cuckoo chick. If they imprint on this, then they'd reject their own chicks in future, unparasitized clutches. Faced with

BOX 4.2 *SIGNAL DETECTION* (REEVE 1989)

Imagine a world with both 'desirable' and 'undesirable' signallers. For a cuckoo host, desirable signallers are their own eggs, undesirable signallers are cuckoo eggs. For a predator, desirable signallers are tasty prey, undesirable signallers are inedible objects or noxious prey. Now imagine there is some overlap in the signal from desirable and undesirable signallers (host eggs are mimicked by cuckoo eggs; tasty prey resemble inedible objects or noxious prey), as shown in Fig. B4.2.1.

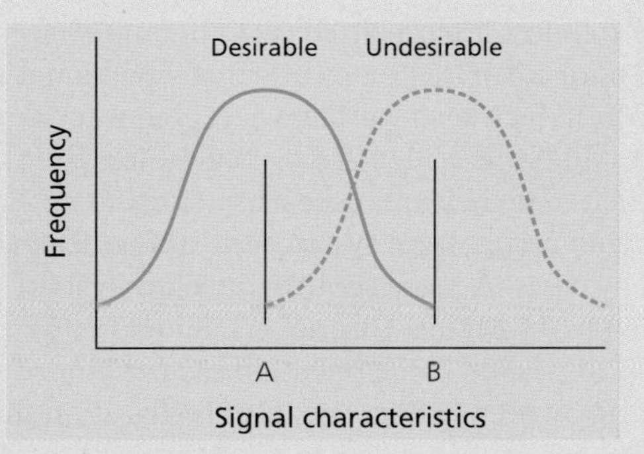

Fig. B4.2.1

Where should the host or predator set its rejection threshold? If it rejected signals to the right of A (a strict threshold), it would reject all undesirable signallers but at the cost of rejecting many desirable signallers too. At the other extreme, if it rejected signals to the right of B (a lax threshold), all desirable signallers would now be accepted but at the cost of accepting many undesirable signallers too. The optimal threshold depends on: (i) the relative frequency of desirable and undesirable signallers, and (ii) the economics of the four outcomes: accepting versus rejecting desirable signallers and accepting versus rejecting undesirable signallers.

For reed warbler hosts whose clutch may be parasitized by a common cuckoo, the following pay-offs (explained in the footnotes) apply (modified from Davies *et al.*, 1996a):

	Reproductive success of host	
Host response	**Not parasitized**	**Parasitized**
	(4 host eggs)	**(3 host + 1 cuckoo egg)** [a]
Accept	Correct acceptance [b]	Missed rejection [c]
	(4)	(0)
Reject	False alarm [d]	True alarm [e]
	(3)	$(0.7 \times 3) + (0.3 \times 0) = 2.1$

[a] Average clutch is four eggs. The cuckoo removes one host egg and replaces it with her own egg. [b] Hosts raise all four eggs. [c] The cuckoo egg remains and the cuckoo chick ejects all host eggs. [d] Assume hosts reject one egg. [e] Hosts again reject one egg but field observations show that they make recognition errors: in 70% cases they reject the cuckoo egg and so save their three remaining eggs but in 30% cases they reject one of their own eggs, so the cuckoo egg remains and the cuckoo chick ejects their eggs.

It's clearly better to reject if parasitized (2.1>0) but to accept if not parasitized (4>3). The critical parasitism frequency at which rejection does better than acceptance is 32.2%. In fact, reed warblers accept most cuckoo eggs (parasitism is usually much less than this critical frequency) unless they have seen a cuckoo at their nest, in which case they are more likely to reject an egg.

The key point is that recognition systems are not perfect, so the best response will be influenced by costs. A more sophisticated analysis of rejection of cuckoo eggs is given in Rodriguez-Girones & Lotem (1999). For a signal-detection approach to cuckoo chick rejection see Langmore *et al.* (2009), and for alarm calling by prey (where predator recognition may not be perfect) see Getty (2002).

this conundrum, the rule 'accept everything' is better at the chick stage than 'imprint and reject' (Lotem, 1993).

but, some Australian hosts do reject foreign chicks ... and their cuckoos have evolved mimetic chicks

This clever argument may help to explain why rejection of common cuckoo chicks is rare. However, misimprinting cannot be a general constraint because in Australia, hosts of bronze-cuckoos *Chalcites* spp. (where the cuckoo chick also ejects host eggs) do regularly reject a cuckoo chick, either by abandoning it (superb fairy wrens abandon 38% of Horsfield bronze cuckoos; Langmore *et al.*, 2003) or by picking it up and tossing it out of the nest (Sato *et al.*, 2010). Bronze-cuckoo chicks have evolved visual mimicry of their host's chicks, presumably in response to such host rejection (Fig. 4.22).

Experiments show that superb fairy-wrens do not reject their own young after raising a cuckoo, and with increased experience of raising their own young they are more likely

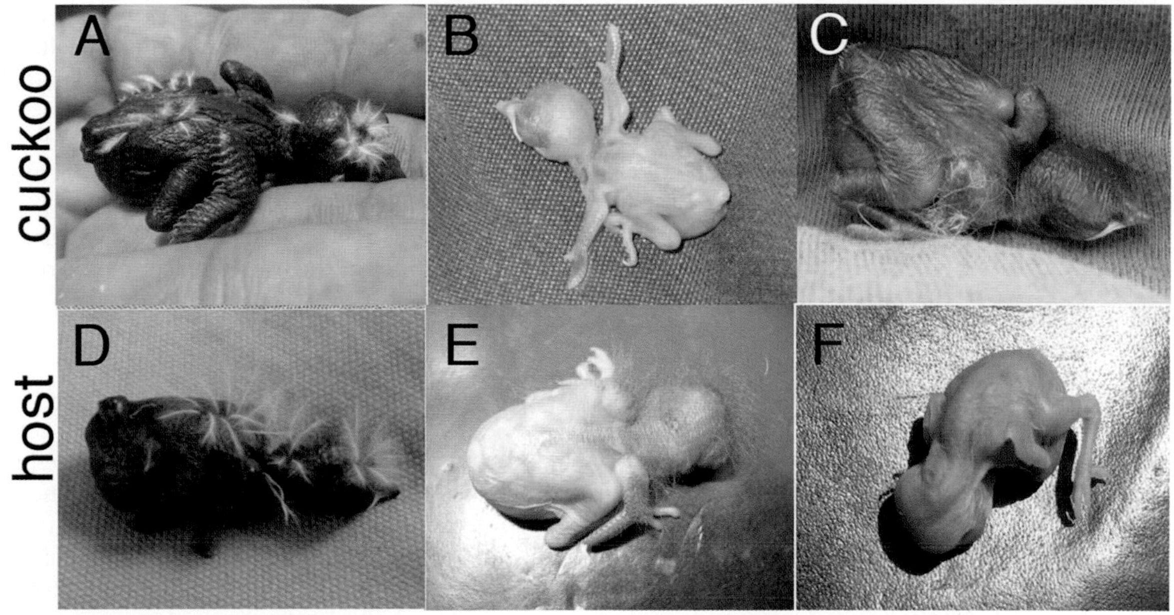

Fig. 4.22 Visual mimicry of host young by Australian bronze-cuckoo (*Chalcites*) chicks. From left to right : A. Little bronze-cuckoo chick (above) and D. the host's chick (below), the large-billed gerygone. B. Shining bronze-cuckoo (above) and E. the host's chick (below), the yellow-rumped thornbill. C. Horsfield's bronze-cuckoo (above) and F. the host's chick (below), the superb fairy-wren. From Langmore *et al.* (2011).

to reject a cuckoo (Langmore *et al.*, 2009). Therefore, these hosts are likely to avoid misimprinting by means of an innate template to guide selective learning, akin to that involved in selective learning of own-species songs in songbirds.

Summary

Antagonistic interactions between predators and prey lead to the evolution of adaptations and counter-adaptations. The result may be 'Red Queen' coevolution, where tactics change over time but the relative success of each party remains the same. Experiments involving birds searching for cryptic prey show that background matching does indeed reduce predation, and that it can be enhanced by disruptive colouration and counter-shading. Predation may also promote the evolution of polymorphism in prey, which reduces the efficiency of search image formation by the predator. Even slight crypsis, sufficient to impose a few seconds extra discrimination time, can bring an advantage to prey and so serve as the starting point for an evolutionary arms race.

Some prey are brightly coloured. Bright colours and eyespots may startle predators. Bright colours are also effective warning signals for noxious prey (aposematism). The evolution of aposematism involves a trade-off between increased conspicuousness to naïve predators and better protection against experienced predators through more detectable and memorable warning signals. Aposematism has promoted the evolution of both Müllerian mimicry (repellent species look alike) and Batesian mimicry (palatable species mimic noxious species).

Prey defences are costly and involve trade-offs in both their production and use. Experiments with guppies reveal that male colouration reflects a balance between the advantages of bright colours for mating and dull colours for avoiding predation.

Cuckoo–host interactions involve coevolution of cuckoo tricks to get their eggs and chicks accepted by hosts and of host defences to reject them. Field experiments reveal the successive stages of coevolution.

Further reading

We have not attempted to give a comprehensive review, rather to take a few well-defined problems and to show how these can be tackled by field and laboratory experiments. A wonderful review of predator–prey interactions is given in the book by Ruxton, Sherratt and Speed (2004). Animal camouflage is reviewed in the issue of the Philosophical Transactions of the Royal Society, edited by Stevens & Merilaita (2009). Mappes *et al.* (2005) review aposematism. Marshall (2000) shows that conspicuous colours in reef fishes can sometimes look cryptic from a distance. Mallet and Joron (1999) review warning colours and mimicry. Rowland *et al.* (2007) test by experiment whether there is a mutualistic or a parasitic relationship between unequally defended co-mimic species. Darst *et al.* (2006) compare investment in warning signals and toxicity in poison frogs.

Krüger (2011) shows how acceptance of a non-mimetic cuckoo egg can sometimes be adaptive for hosts. Rothstein (2001) and Lahti (2005, 2006) consider what happens to host defences after brood parasitism ceases. Davies (2011) compares cuckoo adaptations involving 'trickery' (to beat host defences) and 'tuning' (to improve host care through tuning into host life histories). Hauber *et al.* (2006) show, by experiment, how hosts vary their acceptance of foreign eggs in response to parasitism threat.

The papers by Susanne Foitzik and colleagues explore the antagonistic interactions between slavemaker ants and their hosts, and provide many interesting parallels with cuckoo–host interactions (Foitzik & Herbers, 2001; Foitzik *et al.*, 2001, 2003). Kilner and Langmore (2011) compare cuckoo–host interactions in birds and insects.

Buckling and Rainey (2002) is a classic demonstration of a coevolutionary arms race, using laboratory cultures of a bacterium and a bacteriophage.

TOPICS FOR DISCUSSION

1. Does grouping in prey promote the evolution of warning signals? (Read these three papers, which provide contrasting views from a clever experimental design with a 'novel world' of prey: Alatalo & Mappes, 1996; Tullberg *et al.*, 2000; Riipi *et al.*, 2001).
2. In butterflies, eyespots may be signals for mate choice as well as for deterring predators (read Robertson & Monteiro 2005). How would you investigate the influence of these two selection pressures?

3. Why be a Batesian mimic rather than cryptic (no warning colours) or aposematic (with repellent defences too)?
4. Why are prey sometimes (but not always) polymorphic?
5. Some predators mimic non-threatening, or even inviting species, or objects, to gain access to prey (Heiling *et al.*, 2003). Compare this 'aggressive mimicry' with masquerade, camouflage and Batesian mimicry in prey.
6. Some Australian hosts of bronze-cuckoos often reject cuckoo chicks whereas European hosts of common cuckoos rarely do so. In both cases, the cuckoo chicks eject the host eggs or host young from the nest, so hosts lose all their reproductive success from a successfully parasitized nest. What could explain this difference in host rejection? (Read Kilner & Langmore, 2011).

Photo © Douglas Emlen

CHAPTER 5 Competing for Resources

Our discussion in Chapter 3 of how individuals exploit resources omitted a crucial factor: competition. When many individuals exploit the same limited resources, they become competitors and the best way for one individual to behave often depends on what its competitors are doing. In other words, the pay-offs for various strategies are frequency dependent. We need, therefore, to consider what might be the stable outcome of competition. John Maynard Smith and George Price (1973) introduced the concept of the *Evolutionarily Stable Strategy* or *ESS*, namely a strategy that, if all members of a population adopt it, cannot be bettered by an alternative strategy. They originally introduced the idea to model the evolution of fighting strategies, but it is widely applicable to all cases where individuals interact. The key question to ask is: could a mutant strategy do better?

Evolutionarily stable strategies

A human example helps to explain this idea. Imagine a crowd of people sitting on the floor to watch a concert. Someone stands to get a better view. Those behind now have to stand in order to see and so a wave, from sitting to standing, begins to spread through the crowd. Eventually everyone is standing, with the end result that no-one can see any better than before! In this example, sitting is not an ESS (in a crowd of sitters, someone standing does better) whereas standing is an ESS (once everyone is standing, it then doesn't pay anyone to sit). This example also makes the point that the ESS is often not what would be best for everyone; if only everyone had agreed to sit, they would all have been more comfortable.

The Hawk–Dove game

To illustrate 'ESS thinking' more formally, we shall consider the Hawk–Dove game (Maynard Smith, 1982). Assume contestants meet randomly to compete for a resource. Imagine a simple world in which there are just two possible strategies. *Hawks* always

An Introduction to Behavioural Ecology, Fourth Edition. Nicholas B. Davies, John R. Krebs and Stuart A. West.

(a) *Pay-offs: change in fitness from a contest*		
Winner gains, V = 50. Loser gains 0. Injury cost, C = loses 100.		
Assume that: (i) When a Hawk meets a Hawk, on half the occasions it wins and on half the occasions it suffers injury. (ii) Hawks always beat Doves. (iii) Doves immediately retreat when they meet a Hawk. (iv) When a Dove meets a Dove, they share the resource.		
(b) *Pay-off matrix: pay-offs to attacker*		
	Opponent	
Attacker	***Hawk***	***Dove***
Hawk	½V – ½C = –25	V = +50
Dove	0	½V = +25

Table 5.1 The game between Hawk and Dove (Maynard Smith, 1982).

fight and may injure their opponents, though in the process they risk injury themselves. *Doves* never engage in fights. These two strategies are chosen to represent the two extremes we may see in nature. The pay-offs are explained in Table 5.1a. (The exact values do not matter for the moment, as long as $V<C$, and are chosen simply because this game is easier to explain with numbers rather than algebra.) It is important to note that these are simply the *changes* in fitness resulting from a contest. The individual that does not obtain the resource need not have zero overall fitness. For example, if the resource is a territory in a good habitat, then the loser may still get to breed in a poor habitat. So the value of winning the contest is the difference between the reproductive success in the good and poor habitats.

We now draw up a two by two matrix, with the average pay-offs for the four possible encounters, as explained in Table 5.1b. How would evolution proceed in this game? Consider what would happen if all individuals in the population are Doves. Every contest is between a Dove and another Dove and the pay-off is +25. In this population, any mutant Hawk would soon spread because when a Hawk meets a Dove it gets +50. Therefore, Dove is not an ESS.

However, Hawks would not spread to take over the entire population. In a population of all Hawks the pay-offs per contest are –25, and any mutant Dove would do better because it retreats immediately with a payoff of 0. (Remember that this doesn't mean that Doves have zero fitness in a population of Hawks. It means that the fitness of a Dove does not alter as a result of a contest with a Hawk.) Therefore, Hawk is not an ESS either.

The Hawk–Dove game helps us to think about the evolutionary stability of contest behaviour

The key point in this game is that each strategy does best when it is relatively rare: in a population of Doves, Hawks prosper, while in a population of Hawks, Doves prosper. This leads to frequency dependent selection; the outcome will be a stable equilibrium where the frequencies of Hawks and Doves are such that their average pay-offs are equal. If the population moves away from this equilibrium, then one of the strategies will be doing better, so it will increase in frequency, suffer reduced success as a consequence and drive the population back to the equilibrium once more. For the values in Table 5.1, the stable mixture can be calculated as follows: Let h be the proportion of

Hawks in the population. Therefore, the proportion of Doves must be $(1 - h)$. The average pay-off for a Hawk is the pay-off for each type of contest multiplied by the probability of meeting each type of contestant. Therefore:

$$H\ average = -25h + 50\,(1 - h)$$

Similarly, for Dove the average pay-off will be:

$$D\ average = 0h + 25\,(1 - h)$$

At the stable equilibrium, *H average* must equal *D average*. Solving the equation above by setting *H average* = *D average* gives h = ½; therefore, the proportion of Doves $(1 - h)$ must also be ½. In general, if V < C, the stable proportion of Hawks in this game is given by V/C.

A mixture of Hawk and Dove is evolutionarily stable

The ESS in Table 5.1 could come about in two ways.

(1) There is an evolutionarily stable polymorphic state, with individuals all playing pure strategies, half of them Hawk and half of them Dove.

(2) Individuals all adopt a mixed strategy, playing Hawk randomly with probability ½ and Dove with probability ½.

It is instructive to note that at the ESS, the average pay-off per contest is +12.5. If only everyone had agreed to be Doves, the pay-off would be +25! As with our human crowd example, the optimal strategy to maximize everyone's fitness is often higher than the pay-offs at the ESS. Nevertheless, we expect evolution to lead to stable strategies because, in the words of Richard Dawkins, 'they are immune to treachery from within'. The Hawk˜Dove game makes another general point. At the stable equilibrium there is often variation in the population; either between or within individuals. Variation is, therefore, not always noise about a population norm. Rather, it is often the expected stable outcome when individuals compete.

The ESS solution does not maximize every individual's fitness

The assumption that V < C will apply to many contests in nature. For example, it will rarely be worth the risks of serious injury from a fight simply to win an item of food or access to a shelter. However, if V > C then Hawk is an ESS. Intuitively, it is easy to see why this is so. We need to consider the consequences of the current contest for lifetime reproductive success. This will involve a balance between the value of the contested resource and the expected value from future contests. When the value of the resource is similar to, or greater than the value of the future, we would expect individuals to risk more in contests, even at the cost of serious injury or death. Indeed, if the value of the future is close to zero then contestants should, in theory, never give up after starting a fight, so the contest should be fatal for at least one of the opponents (Enquist & Leimar, 1990).

If V > C, Hawk is an ESS

As predicted, fatal fighting has been reported in cases where individuals have a short lifespan and few reproductive opportunities. For example, male bowl and doily spiders, *Frontinella pyramitela*, (Austad, 1983) and male fig wasps (Hamilton, 1979; Murray, 1987) often fight to the death for their once-in-a-lifetime chance to win a receptive female.

The Hawk–Dove game is clearly too simple to apply to any real cases in nature; there are likely to be more than just two strategies, strategies will vary with individual strength, encounters will not be random and so on. In Chapter 14, we will show how displays often involve assessment of an opponent's fighting potential, so individuals vary their fighting tactics from contest to contest. The fixed strategies in the hawk–dove game ('sealed bids') are, therefore, also rather simplistic. Nevertheless, the game illustrates a valuable way of thinking about how evolution will proceed whenever there is competition. The key question is: what will be the evolutionarily *stable* outcome? In this chapter, we will now use 'ESS thinking' to examine how individuals compete for scarce resources, focusing on the two problems: where to search and how to behave. We will start by discussing the simplest form of competition, *exploitation*, which simply means 'using up resources', and then go on to describe another form of competition, *resource defence*, in which individuals keep others away from resources through dominance or territoriality. Next, we will show how competition for either food or mates often leads to individuals in a population showing variability in competitive behaviour, with mixtures of producers and scroungers or fighters and sneaks. Finally, we will link this variation to the concept of animal personalities.

Competition by exploitation: the ideal free distribution

The ideal free model

Let us start with a simple model. Imagine there are two habitats, a rich one containing a lot of resources and a poor one containing few, and that there is no territoriality or fighting, so each individual is free to exploit the habitat in which it can achieve the higher pay-off, measured as rate of consumption of resource. With no competitors, an individual would simply go to the better of the two habitats and this is what we assume the first arrivals will do. But what about the later arrivals? As more competitors occupy the rich habitat, the resource will be depleted, and so less profitable for further newcomers. Eventually a point will be reached where the next arrivals will do better by occupying the poorer quality habitat where, although the resource is in shorter supply, there will be less competition (Fig. 5.1). Thereafter, the two habitats should be filled so that the profitability for an individual is the same in each one. In other words, competitors should adjust their distribution in relation to habitat quality so that each individual enjoys the same rate of acquisition of resources. This theoretical pattern of distribution of competitors between resources was termed the 'ideal free' distribution by Stephen Fretwell (1972) because it assumes that animals are free to go where they will do best (there is no exclusion of weaker competitors by stronger ones) and that the animals are ideal in having complete information about the availability of resources.

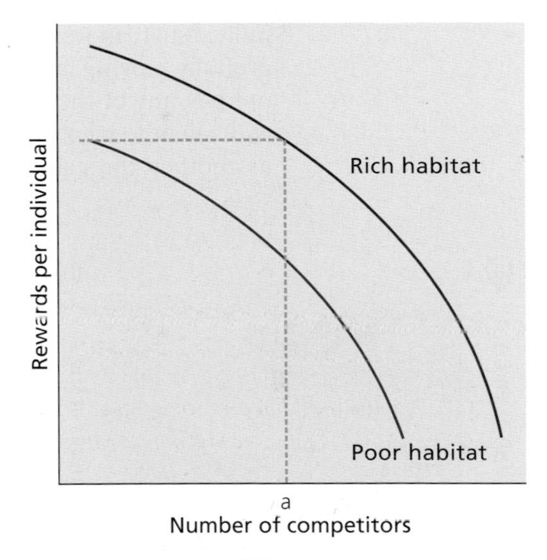

Fig. 5.1 The ideal free distribution. There is no limit to the number of competitors that can exploit the resource. Every individual is free to choose where to go. The first arrivals will go to the rich habitat. Because of resource depletion, the more competitors the lower the rewards per individual so at point *a* the poor habitat will be equally attractive. Thereafter, the two habitats should both be filled so that the rewards per individual are the same in each. After Fretwell (1972).

A simple model of how competitors should distribute between habitats

We can see an example of this in action when people stand in line at the counters of a supermarket. If all the serving clerks are equally efficient and all the customers are equal in the time they require for service, then the lengths of all the lines should end up being equal. If one line gets shorter, then customers would profit by joining it until its length becomes the same as the others. Because everyone is free to join whichever line they like, each person goes to the best place at the time, and the lines fill up in an ideal free way with the result that every customer should have the same waiting time for service.

Competing for food: sticklebacks and ducks

An equivalent in animals is Manfred Milinski's (1979) experiment with sticklebacks. Six fish were put in a tank, and prey (*Daphnia*) were dropped into the water from a pipette at either end. At one end prey were dropped into the tank at twice the rate of the other end. The best place for one fish to go clearly depends on where all the others go. There was no resource defence and Milinski found that the fish distributed themselves in the ratio of the patch profitabilities, with four fish at the fast-rate end and two at the slow-rate end. When the feeding regimes were reversed, the fish quickly redistributed themselves so that four were again at the fast end (Fig. 5.2a). This is the only stable distribution under ideal free conditions. With any other distribution it would pay an individual to move. For example, if there were three fish at each end then one fish would profit by moving from the slow to the fast-rate end. Once it had done so, it would then not pay any of the other fish to move. In our supermarket analogy, this experiment is equivalent to what should happen if one clerk is twice as efficient at serving customers as another; the stable distribution would be for this line to be twice as long.

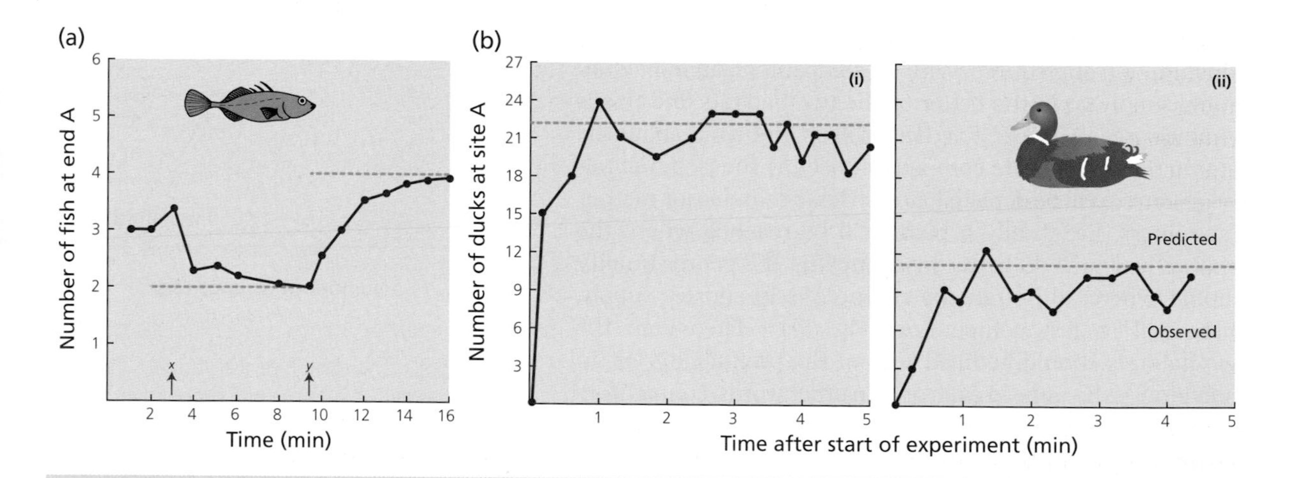

Fig. 5.2 (a) Milinski's (1979) feeding experiment with six sticklebacks. At time *x*, end B of the tank had twice the amount of food as end A. At time *y* the profitabilities were reversed. The pale blue lines indicate the number of fish predicted at end A according to ideal free theory, and the points and dark blue line are the observed numbers (mean of several experiments). (b) Harper's (1982) feeding experiment with a flock of 33 mallard ducks. In (i) food was thrown into the pond at twice the rate at site A compared to site B. In (ii) the profitabilities were reversed. Pale blue lines indicate the predicted numbers at site A according to ideal free theory. The points and dark blue line are the observed numbers (means of many experiments).

David Harper (1982) performed a similar experiment to test how a flock of mallard ducks *Anas platyrhynchos* divided up between two foraging sites. Two people stood on either side of a pond. Each person threw pieces of bread into the water, with one throwing in bread at twice the rate of the other. Just as with the sticklebacks, the relative numbers at each site matched the ratio of rates at which bread was thrown in, with twice as many ducks at the site where bread was being thrown in twice as fast (Fig. 5.2b).

Fish and ducks settle in a stable distribution between feeding patches

The stable distribution of sticklebacks and ducks could be achieved in two ways. For example, if one habitat was twice as profitable as another then stability could come about by: (i) competitor numbers adjusting so that twice as many individuals go to the good habitat as to the poor one; (ii) all individuals visiting both habitats, but each spending twice as much time in the good habitat as the poor one. These two experiments tested the ideal free distribution by examining the *numerical prediction* (number of predators at each site); as predicted the ratio of competitors numbers match the ratio of food input rates. Two other predictions could also have been tested: the *equal intake prediction*, that is the intake rate should be the same at both sites; and the *prey risk prediction*, that is prey mortality should have been the same at both sites (Kacelnik *et al.*, 1992).

These examples are of a 'continuous input' system, in which prey density does not change with time because prey arrive at a constant rate and are eaten as soon as they arrive. Whilst this may be a realistic representation of some natural foraging environments, such as streams with insects drifting past waiting predators, more often prey (or other resources) are likely to be gradually depleted. The predictions of the ideal free model are then more complicated (Kacelnik *et al.*, 1992).

Competing for mates: dung flies

Female dung flies, *Scatophaga stercoraria*, come to fresh cowpats in order to lay their eggs. Swarms of males are waiting for them on and around the dung (Fig. 5.3a), and whenever a female arrives the first male to encounter her copulates with her and then guards her while she lays her eggs (Chapter 3). Females prefer to lay in fresh dung and as the pat gets older, and a crust forms over it (thus making it less suitable for egg laying), fewer females arrive. The male's problem is: what is the optimum time to spend waiting for females at each cowpat?

A competitive game for male dung flies: how long to wait for a female?

Just like the sticklebacks and ducks, the best decision for one individual depends on what other competitors are doing. For example, if most males wait for short times then a male who stayed a little longer would have high mating success because he could claim all the late arriving females. If, on the other hand, most males were staying a long time then it would pay our male to move quickly to a new pat to claim the early arriving females there. This competitive situation is analogous to the one faced by the sticklebacks and ducks, except that now we have frequency dependent pay-offs for different times rather than at different places. In theory, we would again expect the outcome of competition to be an ideal free, or stable, distribution. This is where the relative numbers of males at a pat matches the expected relative numbers of arriving females, so no waiting times are either over-or under-exploited compared to the rest.

Males adopt evolutionarily stable waiting times

What do male dung flies do? Geoff Parker (1970b) counted males on cowpats and found that numbers declined exponentially with time (Fig. 5.3b). He then calculated the expected male mating success at different waiting times, given this observed temporal distribution of males. He found that expected mating success was indeed equal across the different times (Fig. 5.3c). Therefore, male dung flies achieve the predicted stable

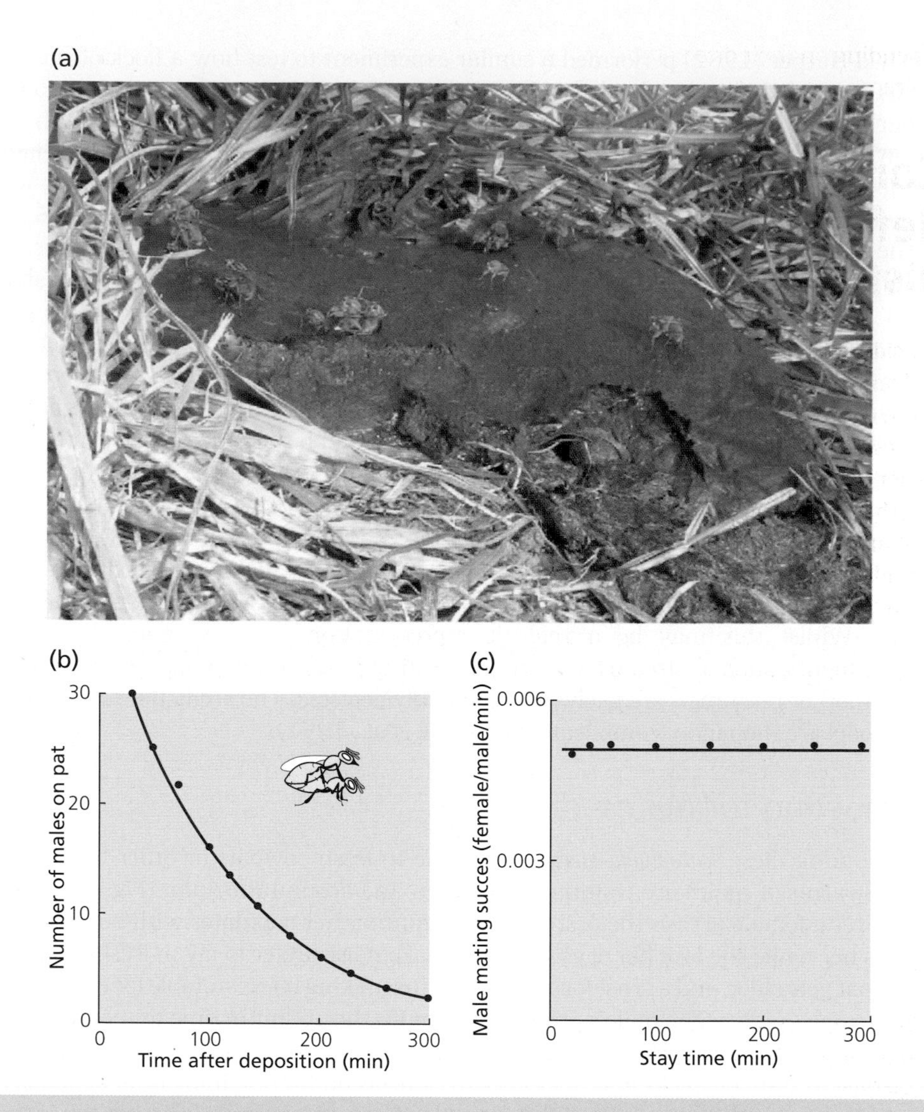

Fig. 5.3 (a) Male dung flies on a cowpat, waiting to mate with females that come to lay their eggs in the dung. In this photo, there are six searching males. Two pairs are being attacked by another male while the male is guarding his egg-laying female (centre and left), and there is a struggle for possession of a single female (top margin of the pat in the centre). Photo © G. A. Parker. (b) The number of males declines exponentially with time after pat deposition. (c) Given this distribution of stay times, the result is that mating success of males adopting different stay times is about equal, as predicted by the ideal free model. From Parker (1970).

distribution in time just as the sticklebacks and ducks did so in space. It is not yet known how the stable distribution is achieved. In theory, it could come about because different individual dung flies have different stay times (some males are short stayers, others are long stayers), or because individuals are variable (sometimes staying a short time, sometimes a long time). The latter seems more likely; perhaps males vary their stay time depending on their direct assessment of female arrival rate, or indirect assessment based on pat age and competitor numbers.

Competition by resource defence: the despotic distribution

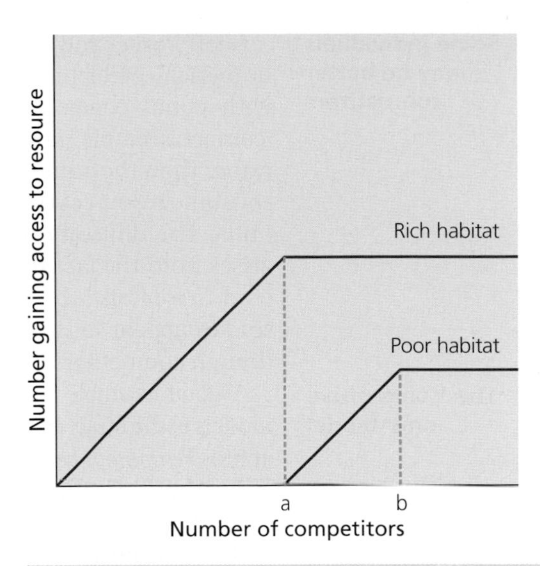

Fig. 5.4 Resource defence. Competitors occupy the rich habitat first of all. At point a this becomes full and newcomers are now forced to occupy the poor habitat. When this is also full (point b), further competitors are excluded from the resource altogether and become 'floaters'. After Brown (1969).

Consider the same situation as before: two habitats, one rich and one poor. This time, though, the first competitors to settle in the rich habitat defend resources by establishing territories (pieces of ground containing the resource), so later arrivals are forced to occupy the poor habitat even though they do less well there than the individuals in the rich area. When the poor habitat fills up with territory-defending individuals the latest arrivals of all may end up being excluded from the resource altogether (Fig. 5.4). This kind of situation is very common in nature. In Wytham Woods, near Oxford, UK, the best breeding habitat for great tits is in oak woodland. This is quickly occupied in the spring and becomes completely filled with territories. Some individuals are excluded from the oak wood and have to occupy the hedgerows nearby where there is less food and, consequently, lower breeding success. If great tits are removed from the best habitat then birds rapidly move in from the hedgerows to fill the vacancies (Krebs, 1971). Similarly, in red grouse (*Lagopus lagopus scoticus*) territorial birds defend the richest areas of the heather moors as breeding and feeding territories. Excluded birds have to go about in flocks and exploit poor habitats where their chances of survival are low. Once again, if a territory owner is removed its place is quickly taken by a bird from the flock (Watson, 1967).

Removal experiments show that territorial behaviour may exclude some competitors from good habitats

In these examples the strongest individuals are despots, grabbing the best quality resources and forcing others into low quality areas or excluding them from the resource altogether.

The ideal free distribution with unequal competitors

Most examples in nature will have features of both the simple models we have discussed above. Perhaps the commonest situation will be where the best place to search depends on where all the other competitors are, but within a habitat some individuals get more of the resource than others. In the duck experiment, for example, population counts showed a stable distribution of individuals but some ducks were better competitors than others and grabbed most of the food (Harper, 1982). The stable distribution could come about because of the way the subordinates distribute themselves in relation to the despots. In effect, the despots are part of the habitat to which the subordinates respond when deciding where to search.

Some individuals may be better competitors

Geoff Parker and Bill Sutherland (1986) pointed out that it might be difficult from the numerical prediction alone to distinguish between the simple ideal free distribution with equal competitors and one with unequal competitors, which they call the 'competitive unit' model, because it hypothesizes that the number of 'competitive units', rather than the number of individuals, is equalized across patches. If one individual can consume resources twice as rapidly as another it scores twice the number of competitive units. The difficulty in distinguishing the two versions of the ideal free distribution arises from the fact that, by chance alone, the competitive unit distribution will often tend to look like the simple ideal free distribution (Fig. 5.5). This may be why many studies appear to support the numerical prediction of the ideal free distribution, even though competitors in these studies were unequal.

The 'competitive unit' model

A good example which shows features of both the resource defence and ideal free models is the study by Thomas Whitham (1978, 1979, 1980) of habitat selection in the aphid (*Pemphigus betae*). In the spring, females known as 'stem mothers', settle on leaves of narrowleaf cottonwood (*Populus angustifolia*) to feed and they become entombed by expanding leaf tissue, so forming a gall. A stem mother reproduces parthenogenetically and the number of progeny she produces depends on the quantity and quality of the juices she can tap from the leaf. The largest leaves provide the richest supplies of vascular sap and result in the greatest reproductive success, with up to seven times the number of progeny that are produced by settling on a small leaf. As we would expect, all the large leaves are quickly occupied, so additional settlers have the problem of whether to settle on large leaves and share the resources or occupy smaller leaves alone.

Gall aphids: a test of the competitive unit model

Whitham made measurements of reproductive success which enabled him to plot a family of fitness curves for habitats of varying quality (leaves of different sizes) and with different densities of competitors (number of females (= galls) per leaf). Figure 5.6a shows the results, which enable us to draw three conclusions. Firstly, for any competitor density, the average reproductive success increases with habitat quality. Secondly, within a habitat of a certain quality, success decreases as the number of competitors increases. This shows that stem mothers settling on the same leaf must compete with each other for resources. Thirdly, if the *average* reproductive success is calculated for aphids which are alone on a leaf, those who share a leaf with one other and those who share with two others, no significant differences are found. There was also no significant difference in average success on leaves with different numbers of competitors when other fitness measures were used, such as body weight of the stem mother, abortion rate, development rate or predation. The results support the predictions of the ideal free model. The conclusion, therefore, is that the stem mothers settle on leaves of different sizes such that the average success in good habitats with a high density of competitors is the same as in poor habitats with fewer competitors.

Average success is equal on leaves of different quality, but individuals near the leaf base do better

However, although the results of average success on different sized leaves are in accord with ideal free predictions, within a habitat not all individuals get equal rewards. This is because a leaf is not a homogeneous habitat. The best place to be is on the mid-rib at the base of the leaf blade because everything translocated into and out of the leaf must flow past this point. Basal galls on a leaf give rise to more young than distal galls and the stem mothers spar with each other, like boxers in a ring, for occupancy of these prime positions (Fig. 5.6b). As we would predict from the defence model, if a basal individual is removed her place is quickly occupied by another aphid from a distal site.

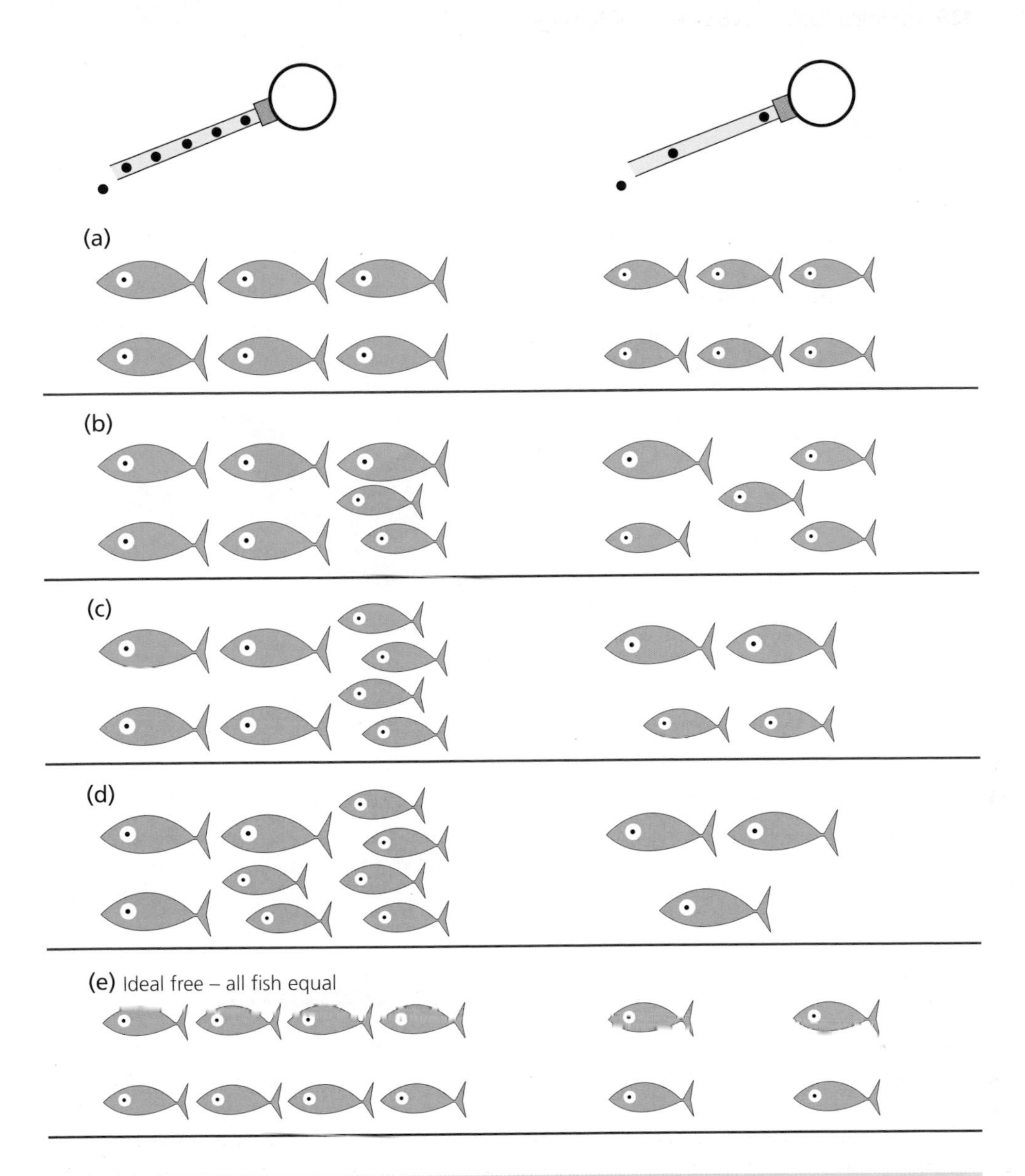

Fig. 5.5 An illustration of how in theory it is difficult to distinguish between a numerical distribution based on the simple ideal free distribution with equal competitors (e) and a distribution with unequal competitors (a–d). The left-hand patch has twice the input rate of that on the right, so the ideal free distribution (e) of 12 equal competitors is 8:4. To illustrate unequal competitors, we imagine that six of the fish (drawn as twice the size) are capable of eating twice as many prey per unit time than the other six. There are now four possible ways of distributing the 12 fish so that the average intake at the two ends is equal (a–d). However, the number of different possible combinations of individuals which achieves each of these distributions varies. Imagine each fish has a name. The 12 fish can be arranged in only one pattern to achieve distribution (a). However, for (b), (c) and (d), there are many ways of arranging the individual fish to achieve the distribution; the numbers of ways are 90, 225 and 20. In short, by chance alone, (c) is the most likely to be observed. Note that it has the same numerical pattern as (e). After Milinski and Parker (1991).

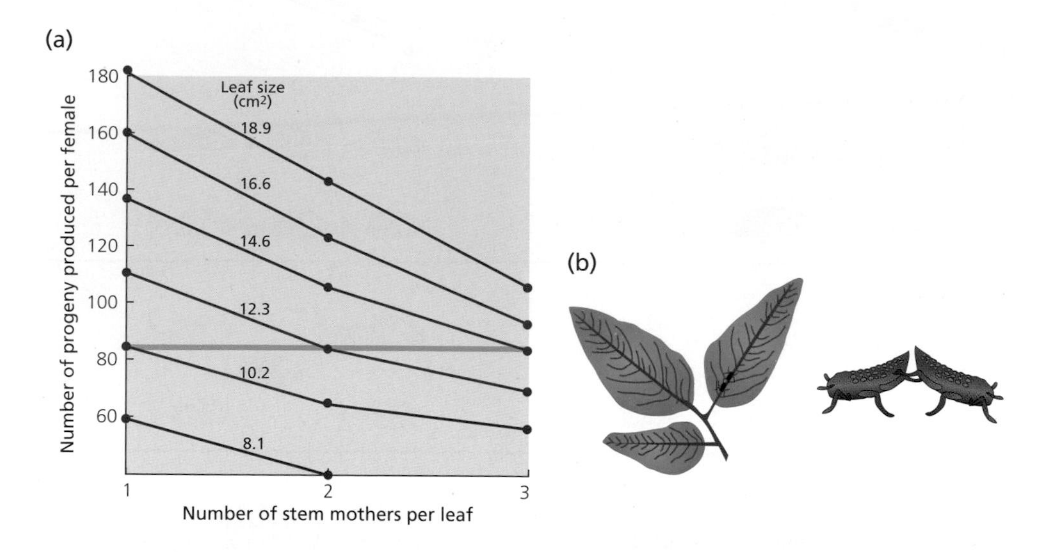

Fig. 5.6 (a) The thin lines are a family of fitness curves for habitats of varying quality (leaf size) and competitor density (no. stem mothers per leaf) in the aphid *Pemphigus betae*. The pale blue horizontal line is the average success for one, two and three stem mothers per leaf. See text for explanation. From Whitham (1980). (b) Stem mother aphids fight for prime positions on a leaf by kicking and pushing. The winner will settle at the base of the mid-rib where food is richest. From Whitham (1979). Reprinted with permission from the Nature Publishing Group.

The economics of resource defence

Some animals, as we have seen, compete for resources by exploitation, others by defence. Is it possible to predict when the latter form of competition should be adopted instead of the former?

Economic defendability

Jerram Brown (1964) first introduced the idea of economic defendability. He pointed out that defence of a resource has costs (energy expenditure, risk of injury and so on) as well as the benefits of priority of access to the resource. Territorial behaviour should be favoured by selection whenever the benefits are greater than the costs (Fig. 5.7). This idea led field workers to look in more detail at the time budgets of territorial animals.

Costs and benefits of territory defence

Frank Gill and Larry Wolf (1975), measured the nectar content of territories of the golden-winged sunbird (*Nectarinia reichenowi*) in East Africa, where it defends patches of *Leonotis* flowers outside the breeding season. They also calculated from time budget studies and laboratory measurements of the energetic costs of different activities, such as flight, sitting and fighting, how much energy a sunbird expends in a day. When the daily costs were compared with the extra nectar gained by defending a territory and excluding competitors, it turned out that the territorial birds were making a net energetic profit. Therefore, the resource was economically defendable (Box 5.1).

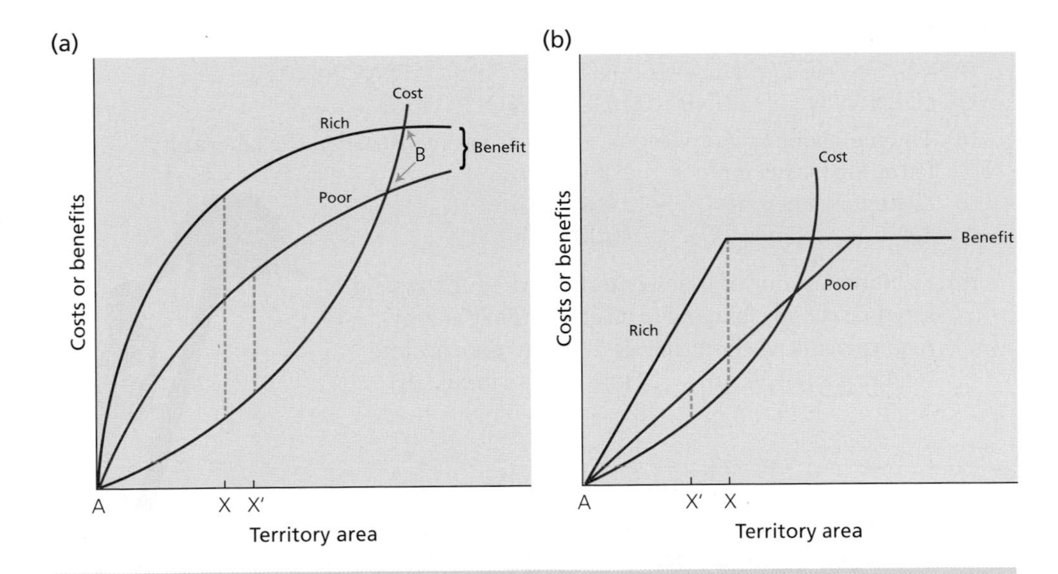

Fig. 5.7 (a) The idea of economic defendability. As the amount of resource defended (or territory size) increases, so do the costs of defence. The benefits (e.g. amount of food available) are assumed to increase at first but level off as the resource becomes superabundant in relation to the animal's capacity to process the resource. Two benefit curves are shown, one for a rich environment and one for a poor environment: the benefit curve rises more steeply in the former because the density of resources is higher. The resource is economically defendable between A and B. Within this range the optimal territory size depends on the currency: for maximizing net gain (B–C) the optimal size is smaller for the rich environment (X) than for the poor one (X′) (note that this is where the slopes of the cost and benefit curves are equal). (b) The same model, but with slightly different shaped curves. In this case the optimal territory size to maximize net gain is predicted to *increase* in a rich environment (X now greater than X′). Therefore, the shapes of the cost and benefit curves are crucial for the predictions that are made! After Schoener (1983).

Time and energy budgets of sunbirds predict when resources should be defended

The idea of economic defendability has also been used to predict the levels of resource availability which could lead to territorial defence (Fig. 5.7). If resources are at low density, the gains from excluding others may not be sufficient to pay for the cost of territorial defence. Instead, the animal might abandon its territory and move elsewhere. There may also be an upper threshold of resource availability beyond which defence is not economical. This upper boundary could arise because there may be so many intruders trying to invade rich areas that defence costs would be prohibitively high. Alternatively, there may be no advantage of territoriality at high resource levels if the owner cannot make use of the additional resources made available by defence.

Resource defence does not pay when nectar is too scarce or too abundant

In Gill and Wolf's sunbirds, one advantage of territorial defence was that it raised the amount of nectar per flower (by exclusion of nectar thieves) and, hence, saved foraging time (Box 5.1). But if nectar levels are already high, the extra increment resulting from territorial defence saves hardly any foraging time. This is because the bird's rate of food intake at high nectar levels is limited by the time it takes to probe its beak into the flower (the handling time). For example, Gill and Wolf calculate that an increase from 4 to 6 μl

BOX 5.1 THE ECONOMICS OF TERRITORY DEFENCE IN THE GOLDEN-WINGED SUNBIRD (GILL & WOLF, 1975)

(a) The metabolic cost of various activities was measured in the laboratory:

Foraging for nectar	1000 cal/h
Sitting on a perch	400 cal/h
Territory defence	3000 cal/h

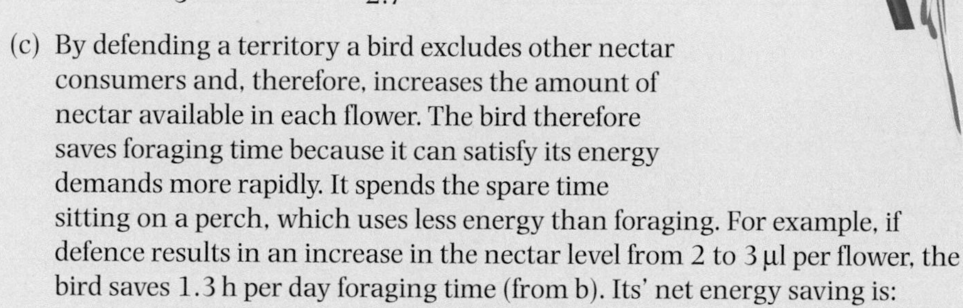

(b) Field studies showed that territorial birds need to spend less time per day foraging to meet their daily energy requirements when the flowers contain more nectar:

Nectar per flower (μl)	Time need to forage (h)
1	8
2	4
3	2.7

(c) By defending a territory a bird excludes other nectar consumers and, therefore, increases the amount of nectar available in each flower. The bird therefore saves foraging time because it can satisfy its energy demands more rapidly. It spends the spare time sitting on a perch, which uses less energy than foraging. For example, if defence results in an increase in the nectar level from 2 to 3 μl per flower, the bird saves 1.3 h per day foraging time (from b). Its' net energy saving is:

$$(1000 \times 1.3) - (400 \times 1.3) = 780\,\text{cal}$$

foraging resting

(d) But this saving has to be weighed against the cost of defence. Measurements in the field show that the birds spend about 0.28 h per day on defence. This time could otherwise be spent sitting, so the extra cost of defence is:

$$(3000 \times 0.28) - (400 \times 0.28) = 728\,\text{cal}$$

In other words, the sunbirds make an energetic profit when the nectar levels are raised from 2 to 3 μl as a result of defence. Gill and Wolf found that most of their sunbirds were territorial when the flowers were economically defendable.

per flower would save the birds less than 0.5 h of foraging time while, as shown in Box 5.1, an increase from 1 to 2 μl saves four hours. Thus, when nectar levels are already high, territorial exclusion of nectar thieves does not pay for itself in savings of foraging time.

Shared resource defence

Sometimes the economics of resource competition may favour shared defence. An example is the winter feeding territories of the pied wagtail (*Motacilla alba*) along stretches of the River Thames near Oxford, UK (Davies & Houston, 1981). Here, pied wagtails feed on insects washed up by the river onto the bank. After a bird has foraged in a particular stretch and depleted the insects, the numbers gradually build up as new insects are washed ashore.

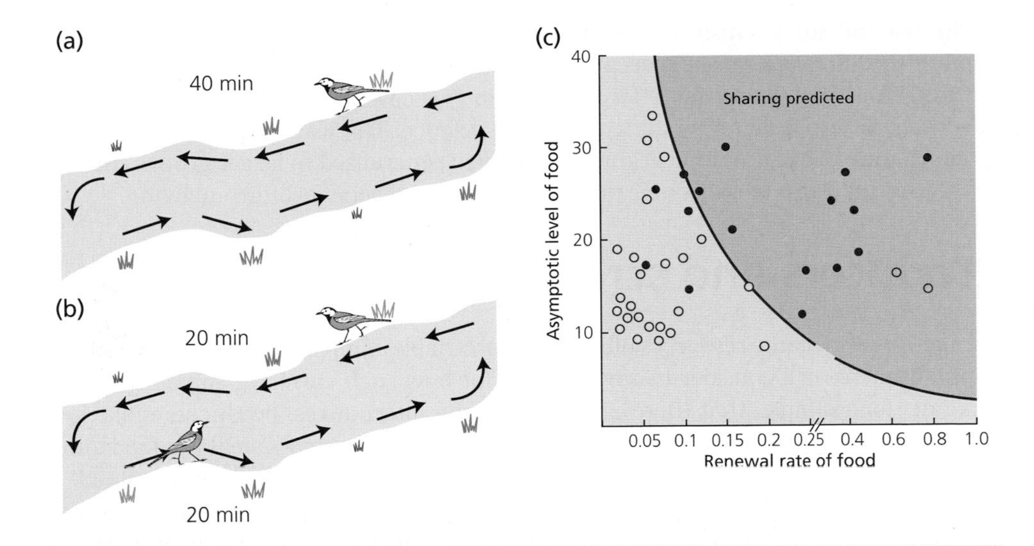

Fig. 5.8 (a) A pied wagtail territory owner exploits its riverside territory systematically, feeding along one bank and then back down the other bank (a circuit of about 40 min) to allow time for prey renewal between successive visits to the same stretch. (b) Sharing the territory with a satellite brings benefits in terms of help with defence, but costs in reducing prey renewal times by half when each bird walks half a circuit behind the other. (c) An owner was predicted to share its territory when the rate of renewal of food and the asymptotic abundance of food were above the curve. These combinations represent instances where the owner gains a net benefit in feeding rate because costs of sharing are outweighed by the benefits. The observed outcomes are shown as dots, each one representing a single day: solid dots – satellite accepted, open dots – satellite chased off. From Davies and Houston (1981).

The territory owner works systematically around its territory and revisits each stretch on average about once every 40 minutes, which gives time for prey renewal (Fig. 5.8a).

It is not hard to see why territorial defence pays: without exclusive use of the river bank, a wagtail's strategy for harvesting the renewing food could easily collapse because of prey depletion by intruders. One might expect, therefore, that wagtails would always defend exclusive territories, but this is not so. Sometimes the territory owner tolerates a second bird, a so-called 'satellite'. The two sharers tend to move around the territory out of phase with one another, so that the average return time to a site is halved to 20 minutes, resulting in a lower feeding rate for the owner (Fig. 5.8b). To counterbalance this cost there is a benefit of sharing because the satellite chases away intruders. The owner therefore saves defence time, leaving more time to feed. The net effect of these costs and benefits on feeding rate depends on the food supply. On days when the rate of renewal of food is high the cost of sharing is relatively low, and the benefit of sharing is relatively large because intrusion rates increase with increased food on the territory. By calculating how these costs and benefits influence an owner's feeding rate it was possible to predict on which days it would pay a territory owner to share and on which days it should evict a satellite. The predictions were right for 34 out of 40 days (Fig. 5.8c).

Pied wagtails gain from territory sharing when food renewal is rapid and intrusion rate is high

The wagtail study illustrates two general points. Firstly, it is an example of how apparently different kinds of costs and benefits (defence and feeding) can sometimes be reduced to a single currency – feeding rate in this case. Secondly, it shows that one advantage of group living is shared resource defence. The wagtail groups were never larger than two, but the same argument could be generalized to larger groups (Brown, 1982). In Chapter 6 we will come back to the costs and benefits of group living.

Producers and scroungers

Competition for scarce resources often leads to variable competitive behaviour within a population. We will consider two hypotheses for how such variation can come about. Imagine, for example, that there are two foraging alternatives: producers make food available by digging or otherwise exposing prey, while scroungers steal the food found by the producers. How could a mixture of producers and scroungers be maintained?

One possibility is that producers are the better quality competitors and scroungers have to 'make the best of a bad job', settling for a technique with a poorer pay-off because of their inferior competitive ability. As the proportion of scroungers in the population increases, the fitness of a producer will decline (more of its food is stolen) and the fitness of a scrounger will also decline (more competition for stealing). However, producers always do better than scroungers (Fig. 5.9a).

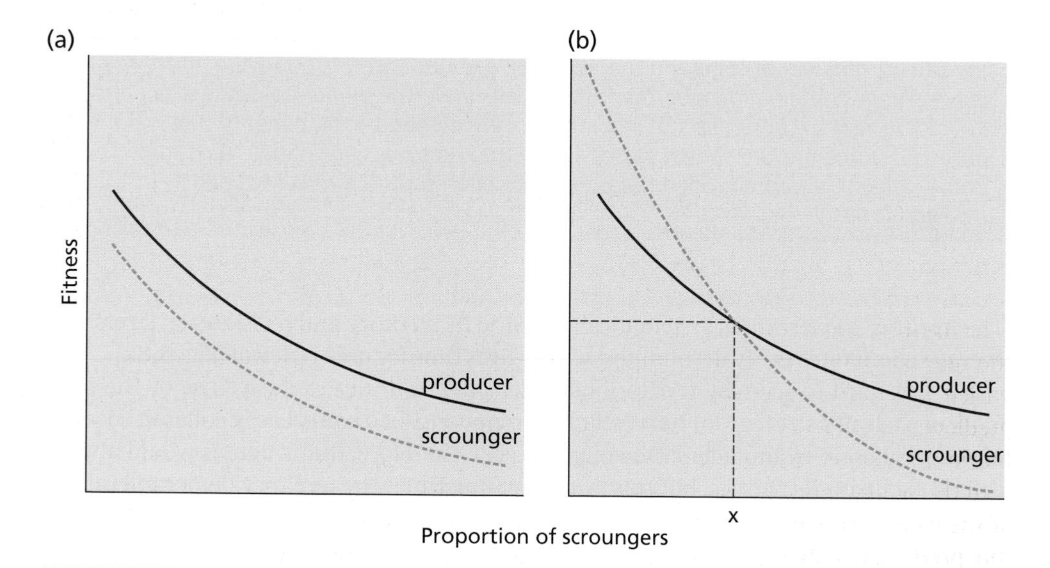

Fig. 5.9 Two models for how a mixture of producers and scroungers may be maintained in a population. (a) As the proportion of scroungers increases, both producers and scroungers have declining fitness, but producers always do best. Scroungers are poorer quality competitors. (b) There may be no difference in competitive ability. Each behaviour does best when rare. The stable equilibrium frequency of the two is at x, where producers and scroungers have equal fitness. Model (b) after Barnard and Sibly (1981). With permission from Elsevier.

The ruddy turnstone *Arenaria interpres* provides an example of variable foraging techniques which fits this hypothesis, though there are several techniques involved rather than just two. This shorebird feeds in small wintering flocks on rocky shores where it searches for small invertebrates, especially crustaceans and molluscs. Philip Whitfield (1990) studied colour-banded birds in south-east Scotland and discovered that they used six distinct techniques: routing (flicking or bulldozing seaweed to expose prey), stone turning, digging, probing, surface pecking and hammering barnacles. Individuals showed varying degrees of specialization and their predilections were maintained over two successive winters. Dominant individuals tended to rout and often displaced subordinates from good seaweed patches. Subordinates sometimes stole food exposed by the dominants. When dominants were removed temporarily (and kept in aviaries), some subordinates increased their use of routing. However, others persisted with their own favourite techniques.

Producers could do best, with scroungers 'making the best of a bad job'

Variation could also be maintained in a population even if there was no difference in competitive ability between individuals. Imagine individuals were free to choose between producer and scrounger. If most of the population were producers, then scroungers might do best (plenty of food to steal). On the other hand, if most were scroungers then there would be intense competition for stolen food and a producer might do best. This would lead to frequency dependent selection and, in theory, the result would be a stable mixture of producers and scroungers, where each behaviour enjoyed the same pay-off (Fig. 5.9b).

Or, there could be a stable mixture with producers and scroungers doing equally well

Kieron Mottley and Luc-Alain Giraldeau (2000) performed aviary experiments with captive flocks of spice finches, *Lonchura punctulata*, to test whether producers and scroungers would reach this predicted stable equilibrium frequency when individuals were free to choose which behaviour to adopt. They studied flocks of six birds. Each aviary was divided into two sections (Fig. 5.10a). On the 'producer' side individuals had access to a string next to each perch. By pulling on the string, a producer released seeds into a dish on the scrounger side opposite. The producer could feed on the seeds by stretching its neck through a small hole in the division between the compartments. Individuals on the scrounger side had no string, so they searched for patches made available by the producers. Two treatments were tested: scroungers could gain easy access to the seeds (dish uncovered) and only partial access (dish covered).

In the first part of the experiment, birds were unable to move between the two sides of the aviary and the numbers of the flock on the producer and scrounger sides were varied. As predicted, scroungers did better when there were more producers and the predicted equilibrium frequency of scroungers was lower when scroungers found it harder to access the food (Fig. 5.10b).

An experimental test with spice finches

In the second part of the experiment, all six birds were given free access to both sides of the aviary. The numbers choosing the producer versus scrounger sides converged on the predicted stable frequencies after a few days of testing with each treatment (Fig. 5.10c). Therefore, variability in a population can come about because of frequency dependent pay-offs from different choices.

Alternative mating strategies and tactics

The two hypotheses in Fig. 5.9 can also help us to explain why individuals within a population often vary in the way they compete for mates. It is useful to distinguish two terms:

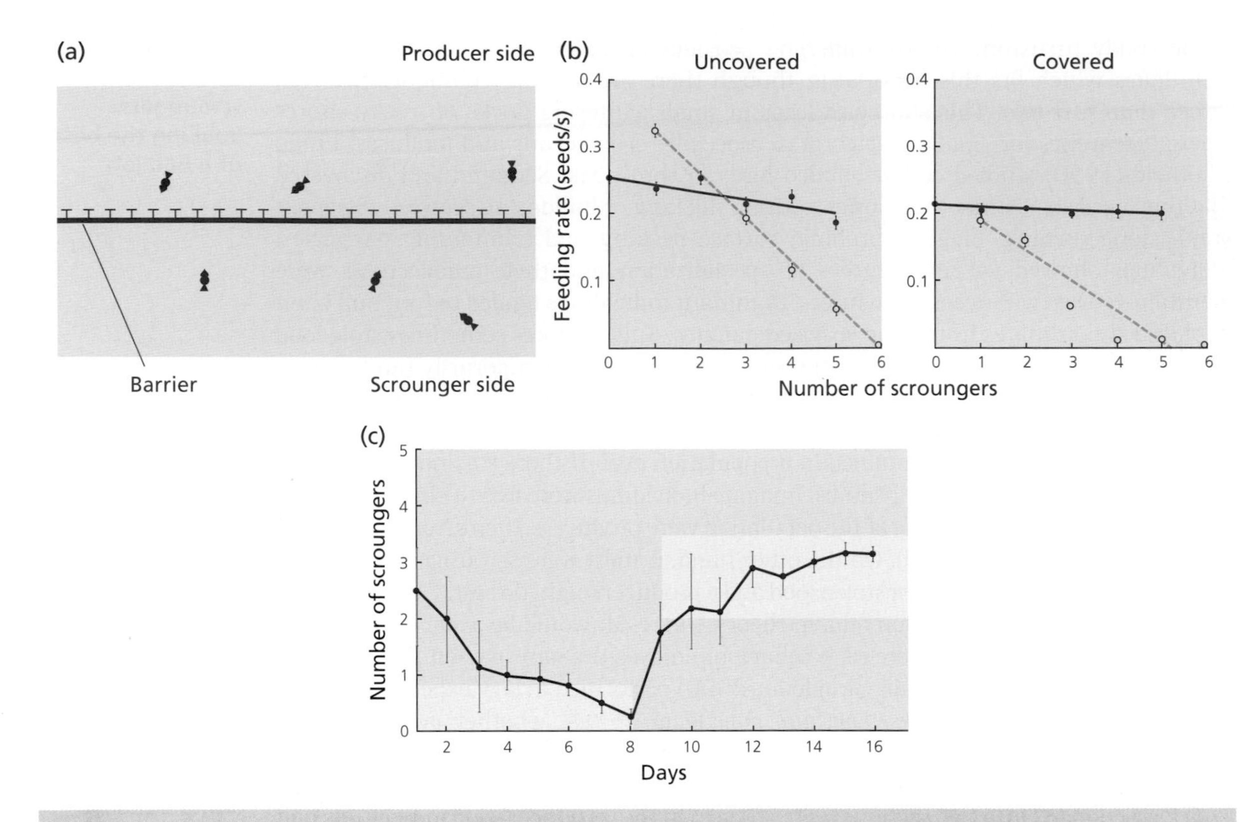

Fig. 5.10 Mottley and Giraldeau's (2000) experiment with spice finches to test the model in Fig. 5.9b. (a) Experimental aviary. On the producer side, birds can sit on little perches (the T shapes) and pull a string to release food into a dish. On the scrounger side, birds have to wait for producers to make food available (see text for details). (b) Results from one flock, showing how foraging rate of producers (p – solid dots) and scroungers (s – open circles) varies with the number of birds in the scrounger compartment. There are six birds in total in the flock. In the graph on the left, the feeding dishes were uncovered and scrounger = producer feeding rate when 2–3 birds of the flock are scroungers. In the right-hand graph, the dishes were covered, which reduced scrounger (but not producer) food access; here equal success occurs when 0–1 birds are scroungers. (c) When all six birds have free access to either side, the numbers converge on the predicted stable equilibrium (shaded areas) over successive days of the experiment. On days 1–8, dishes were covered and on days 9–16 they were uncovered.

Strategy. This is a genetically based decision rule, so differences between strategies are due to differences in genes. For example one strategy might be 'always fight' and another strategy might be 'always sneak'. Strategies might also be *conditional strategies*, where individuals vary their competitive behaviour depending on their body size ('fight if larger than size x, sneak if smaller than x'), or depending on the environment ('fight above threshold cue y, sneak below threshold cue y'). Here, the genetic variation between strategies would involve differences in response thresholds.

Strategy differences reflect genetic differences

Tactic. This is a behaviour pattern played as part of a strategy. For example, if the strategy is 'fight above threshold x, sneak below x', then 'fight' and 'sneak' are now alternative tactics within the same overall genetic strategy.

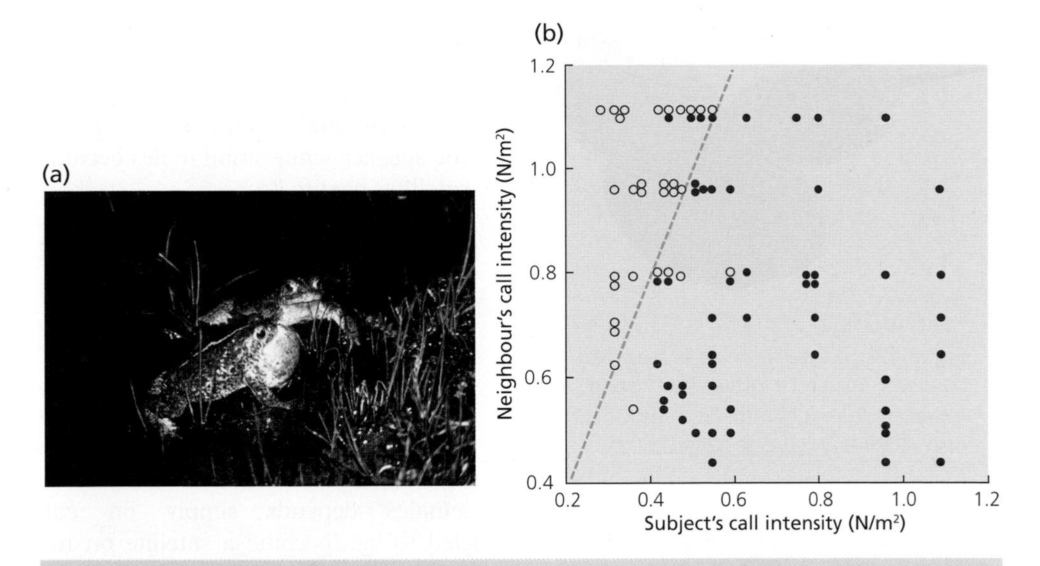

Fig. 5.11 (a) A calling male natterjack toad with a silent satellite male next to him, waiting to intercept any females that are attracted. Photo © Nick Davies. (b) How a male natterjack toad decides whether to be a caller or a satellite. The subject's call intensity is plotted against the call intensity of his nearest neighbour. Males were predicted to become satellites when their neighbours produced calls twice as loud as their own calls (the area to the left of the dashed line). The open circles refer to males who were satellites and the closed circles males who called. From Arak (1988). With permission from Elsevier.

Tactics within a strategy

Don't spend time agonizing over these differences now; they will become clearer once we discuss some examples which will show why the strategy–tactic distinction is useful.

Conditional strategies with alternative tactics

Natterjack toads: callers and satellites

Large male natterjack toads are callers, small males are satellites

In spring, male natterjack toads, *Bufo calamita*, migrate to ponds and call to attract females for mating. The calls are very loud; on still evenings they can be heard from over a mile away and from one metre the call is louder than that legally permitted from a car engine as heard from the sidewalk! Anthony Arak (1983) broadcast male calls from loudspeakers and showed that females moved passively down sound gradients towards the loudest call. Therefore, the largest males, who had the loudest calls, attracted the most females. Small males, unlikely to make themselves heard in the chorus, adopted satellite behaviour in an attempt to intercept females on their way to the callers (Fig. 5.11a). Callers clearly did better than satellites; on average 60% of the males were callers yet they gained 80% of the matings. Therefore, small males made the best of a bad job until they grew larger and had louder calls. Nevertheless, they varied their behaviour depending on the degree of competition from the larger males. If large males

Fig. 5.12 A satellite male horseshoe crab (left) next to a female (front) and guarding male (behind). From Brockmann *et al.* (1994) and Brockmann (2002).

were removed from the chorus, then the small males began to call. If calls were broadcast from a loudspeaker, then large males came over to attack the speaker while small males became satellites next to it.

How do satellites decide which callers to parasitize? Observations showed that when two males were together, one calling and the other a satellite, then they had about an equal chance of capturing the female. (Overall, callers did better because not all callers had satellites.) If a satellite gains 50% of a caller's females, and attraction of females depends simply on call intensity, then a male's decision rule is predicted to be: 'become a satellite on my neighbour if his call is at least twice as loud as my own'. Fig. 5.11b shows that 89% of males adopted the behaviour predicted by this simple rule.

Satellite males make adaptive choices concerning which callers to parasitize

Therefore caller and satellite are two tactics within the conditional strategy 'call above threshold x, sneak below x', where the threshold depends on the male's body size (and hence call loudness) relative to that of his competitors. This is an example of case (a) in Fig. 5.9; satellites (scroungers) are poorer competitors, nevertheless, they maximize their chance of gaining a female by their choice of which callers (producers) to parasitize.

There are many examples in nature like this, where poorer competitors have to choose alternative tactics to make the best of a bad job. For example, small male elephant seals attempt to sneak matings while a large male is busy defending other females in his harem. Male horseshoe crabs, *Limulus polyphemus*, in poor body condition attempt to sneak fertilizations while a female is spawning with a guarding male (Fig. 5.12).

Tactics may vary with age, condition or body size

When small body size is correlated with age, individuals may change to more profitable tactics when they become older and stronger. A male toad or seal, for example, may change tactics depending on its assessment of competitor intensity. In some cases, however, body size is fixed throughout an individual's lifetime and reflects feeding success in the immature stages. Here, a poor quality competitor may have to make the best of a bad job throughout its lifetime. An example is the bee, *Centris pallida*, where the largest males, which had good food as a larva, are three times the mass of the smallest males, who had poor larval food. Large males fight to gain access to virgin females just as they emerge from the ground. Small males, who are poor fighters, have to settle for the less profitable tactic of searching for the few airborne females who have emerged unmated (Alcock *et al.*, 1977).

Morphological switches with body size: dung beetles and earwigs

In some cases the alternative tactics of a conditional strategy involve morphological specializations. Male dung beetles in the genus *Onthophagus* come in two morphs: large males ('majors') have long horns on their heads while small males ('minors') are hornless (Fig. 5.13a). The development of horns is facultative and depends on the

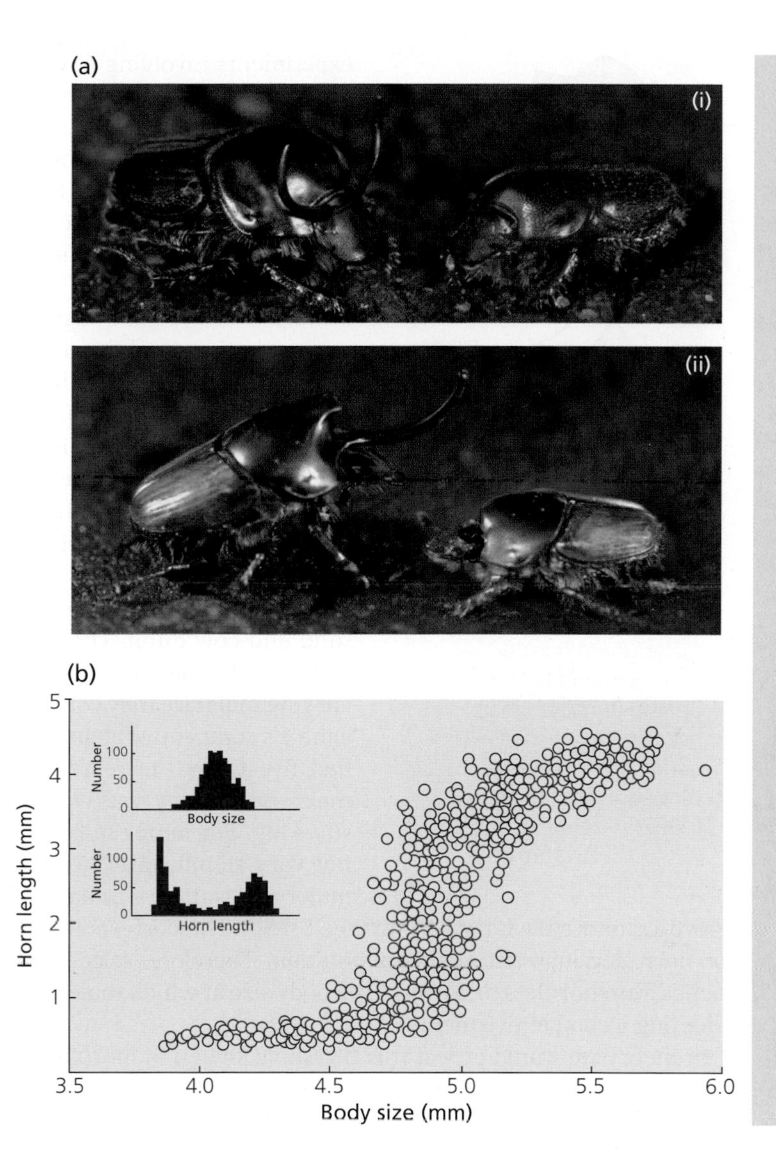

Fig. 5.13 (a) Horned (left) and hornless (right) males of the dung beetles: (i) *Onthophagus taurus* and (ii) *O. nigriventris*. Photos © Douglas Emlen. (b) Scaling relationship between horn length and body size (thorax width) for 810 *taurus* males collected from pastures in Durham County, North Carolina. Inserts illustrate the frequency distribution of body sizes and horn length. From Moczek and Emlen (2000). With permission from Elsevier.

amount of dung available to a growing larva. A hormonal switch during larval development (Emlen & Nijhout, 1999) leads to horn growth only above a critical threshold body size (Fig. 5.13b).

Large male dung beetles have horns and fight, small males are hornless and sneak

Males of the two morphs compete for females in different ways. Female *Onthophagus* are attracted to vertebrate dung. In the neotropics *O. acuminatus* exploits the dung of howler monkeys, while *O. taurus* exploits cattle and horse dung in many temperate regions of the world. Females dig tunnels beneath the dung piles and lay their eggs in brood chambers, which they provision with pieces of dung to feed their larvae (Fig. 5.14). Male competition for females is intense. Major males fight to defend a female's tunnel and then guard the entrance. Minor males attempt to sneak matings through side tunnels and they scuttle off to safety if attacked by a major male (Fig. 5.14). In laboratory

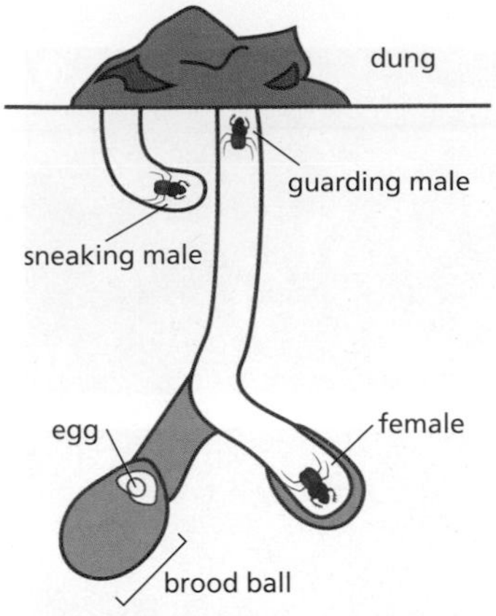

Fig. 5.14 Alternative mating tactics in *Onthophagus* dung beetles. Large, horned males guard burrow entrances and fight to defend females. Small, hornless males sneak matings through side tunnels. From Emlen (1997).

experiments involving fights between two males of the same size, those with larger horns were more likely to win. On the other hand, horns reduced a male's running speed and agility in the tunnels. Therefore, horns are good for fighting while hornless is good for sneaking (Moczek & Emlen, 2000). But what would we expect the threshold size to be for horn development?

Mart Gross (1996) suggested that a conditional strategy's threshold for a switch in tactics should occur at the point where high status individuals (in this case major males) would benefit by investing in a competitive phenotype (in this case horns). Fig. 5.15a explains his model. John Hunt and Leigh Simmons (2001) tested this model for a population of *Onthophagus taurus* in Western Australia. In laboratory experiments they placed groups of ten males and ten females together in buckets with moist sand and cow dung. They then collected broods of larvae and measured male fertilization success. By varying male size they could test how size influenced a male's competitive ability. On average, major males had five times the reproductive success of minor males, so fighting was clearly more profitable than sneaking. For minor males, fertilization success did not vary significantly with body size. However, for major males there was a marked increase in success above a pronotum width of 5 mm (Fig. 5.15b). This corresponded well with the threshold for horn development in this population. Therefore, selection has led to a switch in tactics from hornless to horns at the body size at which males gains increased fitness by adopting a competitive tactic.

Predicting the size threshold for horn development

In theory we would expect this threshold switch in morphology to occur at different body sizes in different populations, depending on how local conditions influence competition for mates. Douglas Emlen (1996) showed by selection experiments that the threshold has a genetic basis. He selected males that produced unusually long horns and males that produced unusually short horns for their respective body sizes. After seven generations of selection he had shifted the position of the sigmoid allometry switch along the body size axis (Fig. 5.16). These results show that populations have the potential to evolve in response to selection on horn length, through modifying the threshold for horn development.

Size thresholds for morphological switches are subject to natural selection

A study of the European earwig *Forficula auricularia* provides an excellent example of how selection has shifted a threshold switch in morphology (Tomkins & Brown, 2004). Just like the dung beetles, males come in two morphs. 'Macrolabic' males have long abdominal forceps which they use in fights for females under stones and logs. 'Brachylabic' males have short forceps and they attempt to steal copulations. Again, just like the dung beetles, there is a threshold body size for development of long forceps

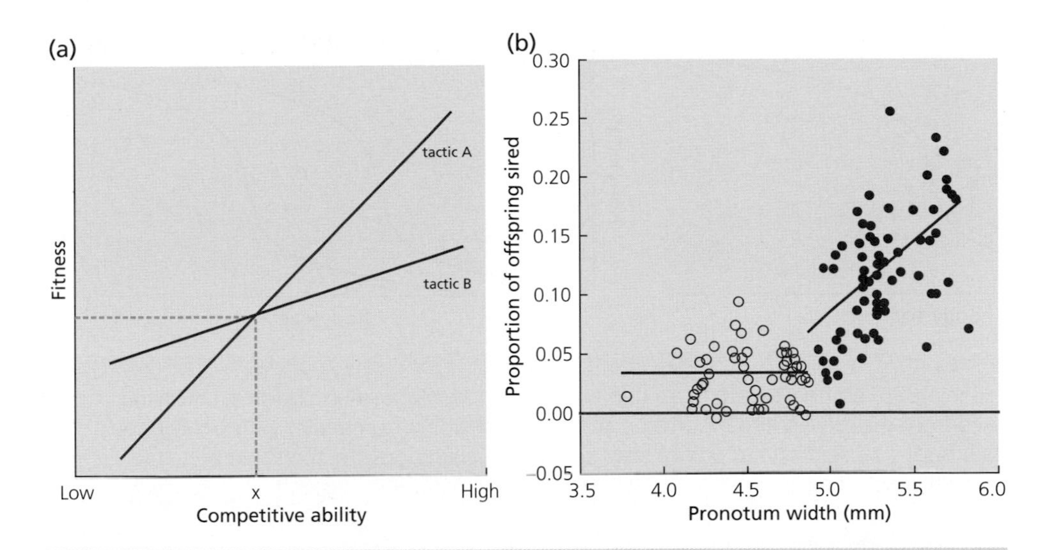

Fig. 5.15 (a) Gross's (1996) model for the threshold morphological switch between alternative tactics within a conditional strategy. The fitness of each alternative tactic A and B varies with an individual's competitive ability. On average tactic B has the lower pay-off. However, below threshold x, B does best while above the threshold A does best. With permission from Elsevier. (b) The reproductive success of horned (solid symbols) and hornless (open symbols) male dung beetles *Onthophagus taurus*. Horned males begin to do better at a body size of about 5 mm pronotum width, which corresponds to the threshold for horn development in the population under study. Hunt and Simmons (2001).

(a sigmoidal relationship; Fig. 5.17a). However, this threshold varies between isolated island populations just 40 km apart. On islands where earwig population density is higher, and where there are higher pay-offs for fighter morphs (more females can be defended), the switch to long forceps occurs at smaller body sizes, so a greater proportion of males in the population has long forceps (Fig. 5.17b).

Alternative strategies: equilibria and cycles

The examples we have discussed so far all involve one conditional strategy with alternative tactics. Individuals employing tactics with poor pay-offs are doing the best they can given their poor competitive ability. Tactic choice is a case of bad luck, not bad genes. However, in theory, different competitive behaviours could result from alternative genetic strategies, so there would be a genetic polymorphism in the population. In this case, if one strategy always had poorer pay-offs then this *would* be a case of bad genes, and we would expect natural selection to eliminate it from the population. Therefore, if alternative genetic strategies persist in population we would expect them to have, on average, equal success.

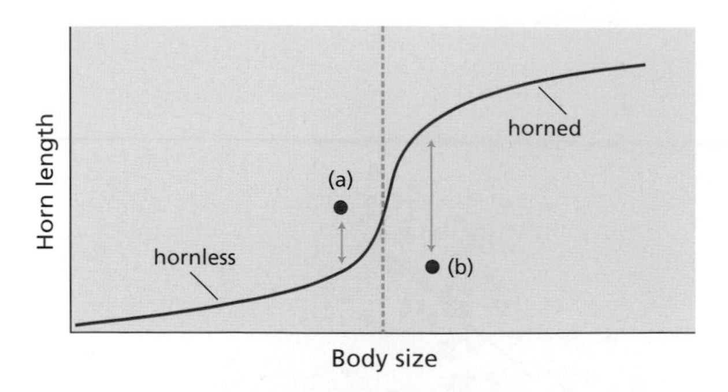

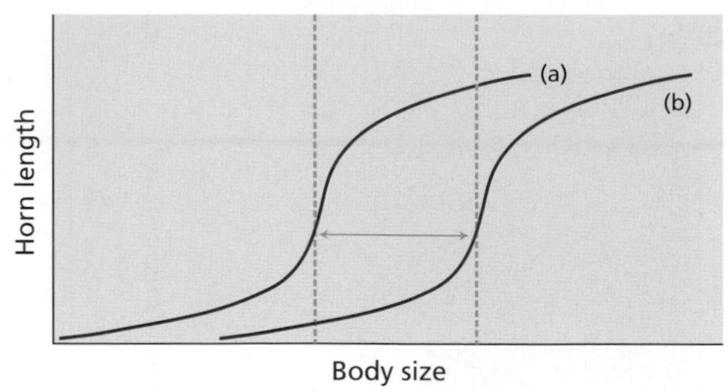

Fig. 5.16 Douglas Emlen's (1996) artificial selection experiment with *Onthophagus* dung beetles. Left-hand graph: in one line, he bred from males with larger horns (a) and in the other line from males with smaller horns (b) than expected for their body size. Right-hand graph: after just seven generations the threshold switch to horns differed between the two selection lines.

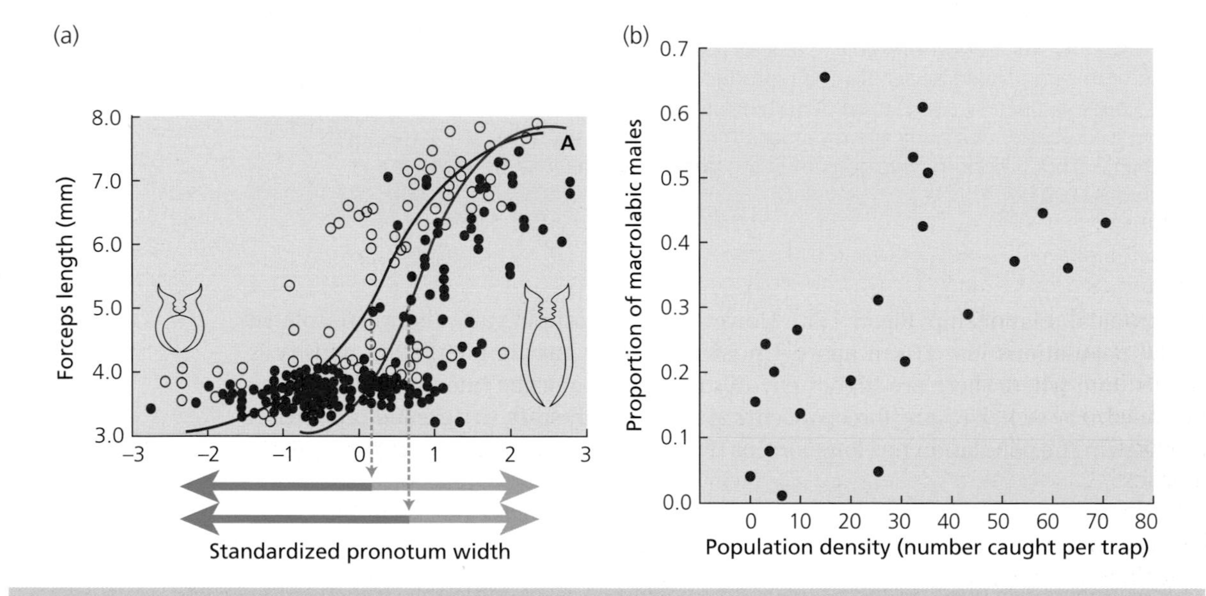

Fig. 5.17 A threshold morphological switch for male forceps length in European earwigs. From Tomkins and Brown (2004). Reprinted with permission from the Nature Publishing Group. (a) The sigmoidal relationship between male forceps length and body size (measured as pronotum width) for two island populations in the North Sea, UK (Bass Rock, open circles; Knoxes Reef, Farne Islands, solid circles). The x axis is the standardized pronotum width for each population (mean zero ± 1, 2, 3 standard deviations). The switch to long forceps occurs at a relatively (and absolutely) smaller body size on the Bass Rock, so a greater proportion of the males here has large forceps (proportion with large forceps indicated by length of the yellow bars in the lines below the x axis). (b) Relationship between population density of earwigs and proportion of macrolabic (long forceps) males in the population for 22 islands in the North Sea.

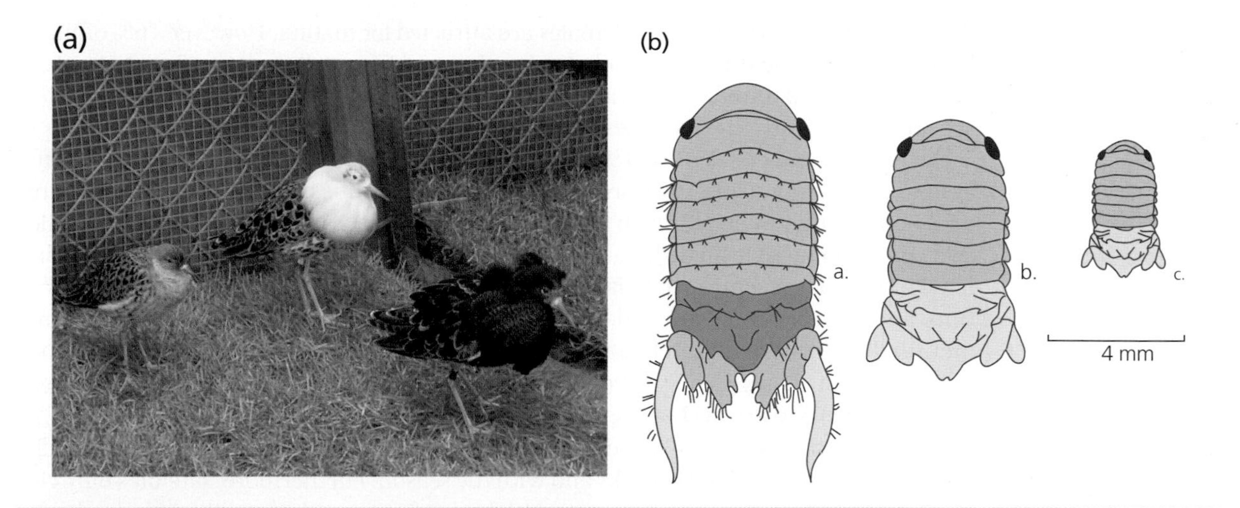

Fig. 5.18 Alternative genetic mating strategies. (a) In the ruff there are three male strategies: territorial males (right) have dark ruffs, satellite males (centre) have white ruffs and female mimics (left) have no ruffs. Photo © Susan McRae. (b) In the marine isopod *Paracerceis sculpta* there are three male morphs which differ in size and behaviour; from left to right: alpha, beta and gamma males. From Shuster (1989).

Frequency dependent selection can lead to equal success for alternative strategies

The most likely way for this to come about is by frequency dependent selection. Let us return to Fig. 5.9b, where producers and scroungers could coexist at a stable frequency. In the spice finches the equilibrium was achieved by behavioural decision making. In an evolutionary game involving genetic alternatives the equilibrium would come about by natural selection. Imagine, for example, that the proportion of the genetic alternative 'scrounger' was below x in Fig. 5.9b. The scrounger morph has higher fitness than the producer morph, so it will be favoured by selection and will increase in frequency across the generations. As it does so, the pay-off for scrounging will decrease (more and more competition between scroungers for resources produced by producers). On the other hand, beyond x producers do better so they now increase in frequency (and so the proportion of scroungers decreases). At x, there is a stable mixture of the two genetic morphs because here each has the same reproductive success.

In theory, we might expect this kind of genetic polymorphism to be rare in nature because conditional strategies with alternative tactics would enable competitors to fine tune their behaviour to fit local environmental conditions and so would be favoured by selection. Nevertheless, there are some marvellous cases of alternative genetic strategies; some examples are now discussed.

Ruffs: fighters, satellites and female mimics

The ruff is a shorebird with a remarkable difference between males and females (Fig. 5.18a). Males are much larger and in the spring they develop ornamental neck ruffs and head tufts and aggregate on display grounds (leks – see Chapter 10) to compete for females. The scientific name *Philomachus pugnax* signifies 'love of fighting' and it describes well the behaviour of the majority of the males; they have dark ruffs and tufts and they fight to defend small territories to which females are attracted for mating. However 16% of males have white ruffs and tufts and their behaviour is very different; they do not fight but instead act as satellites on the edge of territories and attempt to steal copulations while the fighters are busy defending their territories (Hogan-Warburg, 1966; van Rhijn, 1973).

David Lank and colleagues (1995) reared chicks from eggs collected in the wild and assigned paternity using DNA profiles. They also bred ruffs in captivity. These pedigree data showed that male plumage and behaviour differences were genetically controlled and consistent with a single locus, two allele autosomal genetic polymorphism. A third male morph has recently been discovered (Jukema & Piersma, 2006); this has no ruff or tufts and looks just like a female (though it is a little larger). Indeed, it is sometimes mounted by fighter and satellite males. These female mimics occur at a low frequency (1% or less of males). They have enormous testes (2.5 times the volume of those of fighter males) and are likely to behave as sneakers.

Three male morphs in a shorebird

Assessing the success of these three male strategies is a challenging task; satellite mating success varies between leks and with the season. Furthermore, satellites may visit several leks and a female may copulate with both a fighter and a satellite male, leading to multiple paternity in half the clutches (Lank *et al.*, 2002). So it is not yet known whether the three morphs are maintained by frequency dependent advantage, or whether the strategies persist because each does best in different places or at different times.

A marine isopod with three male morphs

Stephen Shuster and colleagues studied a marine isopod crustacean, *Paracerceis sculpta*, in the northern part of the Gulf of California. This breeds in the body cavities of an intertidal sponge. Males come in three discrete morphs (Fig. 5.18b). These are determined by a single autosomal locus with three alleles which influence male growth and maturation rate (Shuster & Sassaman, 1997).

(i) *Large 'alpha' males* position themselves at the entrance of the sponge, with their head pointing down into the cavity (the spongocoel) and their large abdominal spines (uropods) sticking out as a defence. Females are attracted to spongocoels with an alpha male and the male grabs them and feels them carefully before allowing them to enter. An alpha male may accumulate harems of up to 19 females in a single spongocoel and defends them against other alpha males who come and try to evict him. The male mates with a female when she moults. The female then broods her eggs and the young leave the spongocoel to feed on intertidal algae.

(ii) *Small 'beta' males* lack spines and resemble sexually mature females. They mimic female behaviour and so trick the alpha male into allowing them to enter the spongocoel. Once inside, they have a chance to mate with the females.

(iii) *Tiny 'gamma' males* invade spongocoels by stealth, using their small size and rapid movements to rush through the entrance. Alpha males try to block their entrance or to grab them and fling them away. However, gamma males sometimes sneak through and once inside the sponge they, too, have a chance to gain matings with the females.

Each male morph of a marine isopod gains equal success

In laboratory experiments with artificial spongocoels made from a synthetic polymer, Shuster (1989) varied the number of females and males per spongocoel and then

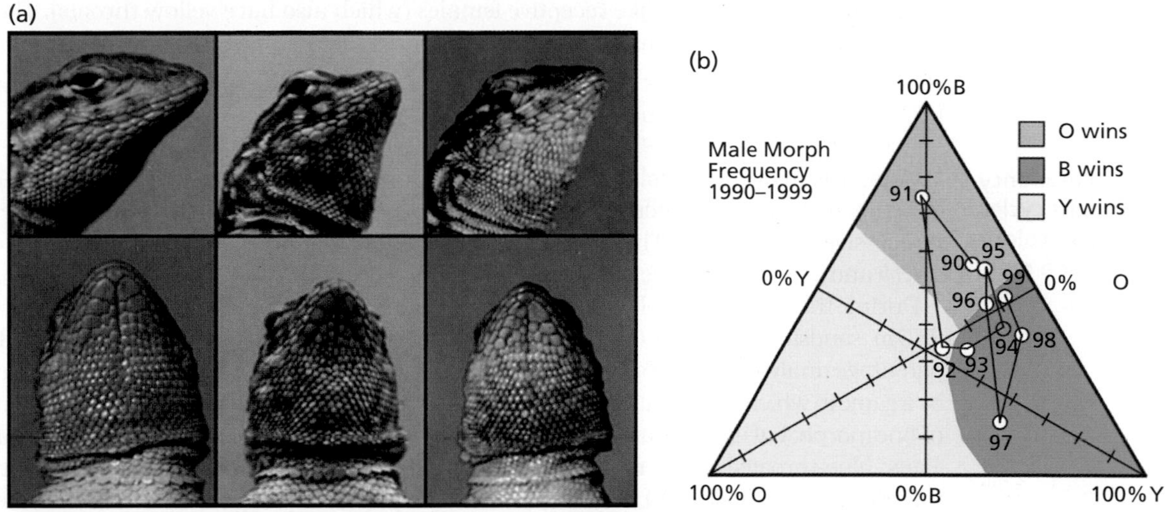

(a)

(b)

Fig. 5.19 (a) The three male colour morphs of the side-blotched lizard: orange, blue and yellow. Each has a different mating strategy (see text). Photo © Barry Sinervo. (b) Observed frequencies of each male strategy (O, Y, B) in a Californian population from 1990 to 1999. The triangle plots the frequencies as follows: 0–100% blue from base to apex, 0–100% orange from right side to left vertex, and 0–100% yellow from left side to right vertex. Shaded areas indicate the zones where each morph has highest fitness. From Alonzo and Sinervo (2001).

measured the fertilization success of the three male morphs using genetic markers. He found that alpha males guarded effectively and sired all the young when there was just one female. However, beta and gamma males did increasingly well when there were two or three females per spongocoel, perhaps because the alpha male could not guard them all at once, especially when they moulted (and so were receptive for mating) at the same time. These laboratory results were then used to calculate the average reproductive success of the three male morphs in nature, given the frequencies of females and males per spongocoel in field samples. The calculations showed that each male morph gains equal success (Shuster & Wade, 1991). It is not yet known how success is equalized; there may be frequency dependent pay-offs or different morphs may do better at different female densities and fluctuations in female abundance may help to maintain the three morphs.

Side-blotched lizards: cycles of orange, blue and yellow

Our final example is a clear case where strategies have frequency dependent pay-offs but here there is no stable equilibrium. Instead, the frequencies of the strategies cycle over time. Barry Sinervo and colleagues have studied side-blotched lizards, *Uta stansburiana*, in California (Sinervo & Lively, 1996; Zamudio & Sinervo, 2000; Alonzo & Sinervo, 2001). Males have one of three throat colours and each type differs in the way it competes for females (Fig. 5.19a):

Orange-throated males are aggressive and defend large territories within which live several females.

Yellow-throated males look like receptive females (which also have yellow throats). They do not defend territories. Instead, they attempt to sneak matings.

Blue-throated males are less aggressive than orange-throated males. They defend small territories in which they guard a single female.

Frequency dependent cycles in three colour morphs of a male lizard

Comparison of throat colours of fathers and sons indicates that these differences are genetic and involve one locus with three alleles: the *o* allele is dominant and the *b* allele is recessive to the *y* allele. Therefore, orange males have genotype *oo*, *ob* or *oy*; blue males are *bb* and yellow males are *yy* or *by*.

During the years 1990–1999, the frequencies of the three morphs changed along a 250 m sandstone outcrop (Fig. 5.19b). Detailed studies of behaviour, combined with parentage analysis using DNA profiles, revealed that these frequency changes were driven by a game in which each strategy had a strength, which enabled it to outcompete neighbours of one morph, but also a weakness which left it vulnerable to neighbours of another morph. Thus, blue-throated males mate-guard their females and so avoid cuckoldry by yellow-throated sneaker males, but they can become overpowered by aggressive orange-throated males. Orange-throated males, on the other hand, cannot guard all their females at once and so they are vulnerable to cuckoldry by yellow-throated males. Finally, yellow-throated males are outcompeted by mate-guarding blue males. The game is, therefore:

rare orange beats common blue

rare yellow beats common orange

rare blue beats common yellow

and so on, as we repeat the cycle again.

This is just like the rock–paper–scissors game that children play: paper covers rock, scissors cuts paper, rock crushes scissors, and so on.

We hope the reader is now convinced that the strategy–tactic distinction is crucial. Where alternative tactics are involved, as part of one conditional strategy, it is usual for one tactic to have a higher pay-off (callers in natterjack toads, horned male dung beetles). However, where the alternatives are genetic strategies then for them to coexist we would expect them to have equal average pay-offs at a stable equilibrium frequency, or to exhibit frequency dependent cycles.

ESS thinking

In this chapter, we have used 'ESS thinking' to examine how competitors search for resources and have seen that individuals tend to adopt stable distributions in response to resources varying in space and time. We have also shown how the stable outcome of competition is often for there to be variability in behaviour. This can occur within individuals in the form of conditional strategies such as: 'play tactic A below threshold x, and tactic B above threshold x'. In this case, we need to analyse which threshold forms the ESS. Or, the variability could be between individuals, with some playing A and others playing B as alternative strategies. In this case there may be an evolutionarily stable polymorphic state or cycles involving frequency dependent replacement of strategies over time.

We recommend 'ESS thinking' throughout the book, because evolution is the outcome of competitive games. In Chapter 1, for example, we argued that restricting the birth rate for the good of the group was not stable because selfish mutants, which optimized their production of young, would invade and spread. Later in the book we will use ESS thinking to consider whether optimal group sizes are stable, the honesty of signalling systems, the evolution of sex ratios and many other problems.

ESS thinking: Could a mutant strategy do better? What is the stable outcome of competition?

Animal personalities

We are all familiar with differences in human personalities; for example, some individuals are bolder, more sociable or more aggressive than others. These differences have a genetic basis and they have consequences for our health, social relationships and sexual behaviour (Carere & Eens, 2005). In this chapter, we have seen that animals, too, often differ in their behaviour, with individual differences in tactics or strategies arising as an evolutionary outcome of competition for resources. Sometimes these differences are discrete, for example caller versus satellite behaviour in toads, horned versus hornless beetles or orange, blue and yellow-throated lizards. Sometimes the differences are continuous, for example waiting times in dung flies. Recent studies have begun to investigate continuous differences in more detail and have shown that they often involve suites of correlated traits. For example, in birds, rodents and fish, individuals that are relatively aggressive to conspecifics are also often bolder in their approach to predators and quicker to explore novel environments. Such consistent differences in behaviour, both over time and across different situations, are comparable to those in humans and are now being referred to with the same terminology, namely animal personalities, temperaments, coping styles or behavioural syndromes. Personalities are being recognized not only in vertebrates but even in insects and spiders. (Sih *et al.*, 2004; Dingemanse & Réale, 2005; Réale *et al.*, 2007).

Studies in The Netherlands have shown that great tit personalities can be assessed by a remarkably simple technique. In one four-year study (Dingemanse *et al.*, 2002), 1342 wild great tits from a nest-box population were caught and kept overnight in aviaries, where they were housed individually. The following morning, before they were released back into the wild, their exploratory behaviour was measured by placing each bird individually in a small room with five artificial trees. The total number of flights and hops within the first two minutes was used as an index of their exploratory behaviour. This measure was repeatable (individuals were consistent when re-tested) and heritable (offspring scores correlated with those of their parents). In another study, great tits were bred in aviaries and two selection lines were created: one breeding from juveniles who had the highest exploration scores and one from those with the lowest scores. Over four generations, there were strong responses to selection in both lines showing that there is a genetic basis to exploratory behaviour (Drent *et al.*, 2003).

Suites of correlated behavioural traits and consistent individual differences

The index of exploratory behaviour was correlated with various behavioural traits. More exploratory individuals were more aggressive towards conspecifics, bolder in their approach of novel objects, more likely to scrounge food from others and showed lower physiological signs of stress when handled (Dingemanse & Réale, 2005).

Personalities in great tits are heritable

Two interesting questions arise from these and other studies of animal personalities. Firstly, what maintains the genetic variation in exploration, or other personality traits? One possibility is that there is frequency dependent selection, and the outcome is a stable mixture of traits, for example various degrees of exploratory behaviour. For example, fast explorers do best in a population of slow explorers and vice versa, leading to a stable mixture of the two (see the example of 'rover' and 'sitter' *Drosophila* larvae in Chapter 1). This has been a familiar theme throughout this chapter. Another possibility is that different personalities do better under different circumstances (e.g. different ecological or social environments) and that a varying environment in space and time leads to persistence of the different traits because of fluctuating selection pressures.

In great tits, different personality types do better under different ecological conditions

Studies of great tits support this last possibility but the link between personality and fitness is complex (Dingemanse *et al.*, 2004). In years with high winter food supplies (beech nuts), relaxed competition for food led to better survival for slow exploring females (no benefit to fast explorers from seeking out novel feeding sites). However, the resulting high survival of the tits led to more intense male competition for breeding sites, in which fast exploring males did best. By contrast, in years with low winter food supplies, fast exploring females survived better (they found novel food sources more quickly) but, curiously, among the males it was the slow explorers who did best. The causes of these differences are not yet fully understood, but the message is that long-term studies are needed to elucidate their effects.

Why are different behavioural traits correlated? ... A constraint?

The second question is: why are different behavioural traits often correlated to form personality types? One possibility is that they are controlled by the same hormones or genes and so the traits are constrained to occur together, boldness towards predators with risk taking in foraging and aggressiveness towards conspecifics, for example. If traits were not independent then there might be trade-offs, where one behaviour can be optimized only at the expense of another. Therefore, to understand the evolution of correlated behaviours we could not study their costs and benefits in isolation (Sih *et al.*, 2004).

... or adaptation?

While different behaviours may sometimes be constrained to occur together because of common proximate control, in three-spined sticklebacks (*Gasterosteus aculeatus*) the correlation between intraspecific aggressiveness and exploratory behaviour or boldness towards predators differs between populations, suggesting that these can evolve as separate traits (Bell, 2005; Dingemanse *et al.*, 2007). The more likely possibility, therefore, is that personality traits form suites of adaptive traits which are selected to go together. For example, a slow life history, which favours investment in adult survival and breeding over many seasons, may select for less risk taking across several contexts (foraging, approaching predators) than a fast life history, which favours increased investment in the current brood at the expense of adult survival (Wolf *et al.*, 2007; Biro & Stamps, 2008). Consistent with this view is the finding that in a number of species bolder individuals tend to have higher reproductive success but lower survival than shyer conspecifics (Smith & Blumstein, 2008).

Summary

When individuals compete for scarce resources, such as food or mates, their best options will be influenced by what their competitors are doing. In these cases, we need to consider what is the stable outcome of competition, namely the evolutionarily stable strategy (ESS). Animals may compete by exploitation or by resource defence, or by a mixture of both. A simple model of exploitation is the 'ideal free' distribution, which predicts the stable distribution of individuals across good and poor habitats. Experiments reveal that sticklebacks and ducks distribute in an ideal free manner between foraging sites of different profitability. Observations show that male dung flies, waiting for females at cowpats, also adopt a stable distribution across different waiting times.

Resource defence is influenced by both benefits and costs. The idea of economic defendability can predict territory defence in sunbirds and territory sharing in pied wagtails.

Competition can lead to variable foraging behaviour within a population, for example a mixture of producers and scroungers. Producers may be better competitors with scroungers 'making the best of a bad job', or the two may coexist in a stable mixture with equal pay-offs. These same two alternatives apply to variable mating behaviour within a population. This may result from alternative tactics within one conditional strategy; for example, poorer competitors sneak matings while better competitors display or fight for females (natterjack toads). Conditional strategies may involve morphological switches at threshold body sizes (horns in dung beetles, long forceps in earwigs). Variable mating behaviour can also reflect alternative genetic strategies within a population (ruffs, isopods). At equilibrium, we would expect equal average success of the alternative strategies. However, in side-blotched lizards there is no stable equilibrium; three colour morphs of males show cycles in frequency because each morph has an advantage over one morph but is vulnerable to exploitation by another morph.

Many of the examples discussed in this chapter involve discrete traits (sneaker versus fighter, horned versus hornless). Recent studies have recognized individual differences in continuous traits, such as exploratory behaviour or aggression. Consistent differences in individual behaviour, both over time and across different contexts (and often involving suites of correlated traits) are referred to as personalities. These may involve differences in risk taking, aggression or sociality. In great tits, for example, there are genetic differences in individual exploratory behaviour which are correlated with many other behavioural traits. In theory, personality variation may be maintained by frequency dependent selection as an ESS. Or, different personalities may do best under different social and ecological conditions.

Further reading

Tom Tregenza (1995) discusses the ideal free distribution. Giraldeau and Dubois (2008) discuss producer–scrounger games in foraging. Jane Brockmann (2001) reviews alternative strategies and tactics. Gross (1996) and Tomkins and Hazel (2007) consider models and evidence for condition dependent switches in tactics. Simmons and Emlen (2006) show how the size threshold for horns in dung beetles also influences investment in testis mass (sneaks have large testes). Müller *et* al. (2006) describe alternative tactics in both male (satellite behaviour) and female (brood parasitism) burying beetles, where there is competition for carcasses, which provide food for their larvae. Hori (1993) shows how frequency dependent selection leads to a stable frequency of 50% left 'handed' and 50% right 'handed' scale-eating cichlid fish (where 'handedness' refers to which side the mouth twists to, facilitating scale-eating from one side of a host fish). Raymond *et al.* (1996) suggest left-handedness in humans has been maintained by frequency dependent advantage in fights; in interactive sports, which depend on fighting ability either directly (boxing, fencing) or indirectly (tennis), there are more left-handers in the sporting elite than expected from the population at large. However, in non-interactive sports (gymnastics, dart, snooker, javelin) left-handers do not have an unusual advantage.

TOPICS FOR DISCUSSION

1. In David Harper's experiment with ducks (Fig. 5.2b), the equilibrium flock sizes occur within a minute, before each duck has had time to visit both patches. How could ducks reach the stable distribution so quickly? What rules could individuals use?
2. How can the effects of predation differences between patches be incorporated into the ideal free distribution? (see Abrahams & Dill, 1989).
3. How would you apply the idea of economic defendability to resources other than food (e.g. nest sites, mates)?
4. Condition dependent tactics are more common in nature than alternative genetic strategies. Why?
5. Would you expect all populations of dung beetles to switch to horned males at the same body size threshold? If not, why?
6. How would you measure the reproductive success of the three male morphs of the ruff (territorial, satellite, female mimic)?
7. Another example of a 'rock–paper–scissors' game is provided by strains of the bacterium *Escherichia coli*. These strains can produce a toxin or not, and they can be resistant to a toxin or not. Experiments by Kerr *et al.* (2002) reveal which conditions lead to the coexistence of several strains (see also the commentary by Nowak & Sigmund, 2002). Compare these bacterial games with the lizards we discussed in this chapter.
8. Should we expect all animals to have personalities?

CHAPTER 6
Living in Groups

Photo © iStockphoto.com/stevedeneef

Animal groups provide some of the most remarkable spectacles of the natural world. In winter, as dusk falls, flocks of starlings, *Sturnus vulgaris*, fly from the fields where they have been foraging all day towards their night-time roosts, such as buildings in city centres or woods in the countryside. From all directions they come, congregating into a vast flock, often with tens or hundreds of thousands of individuals wheeling in ever-changing smoke-like formations before they go to roost. In the seas, vast shoals of fish perform similar, spectacular coordinated movements, circling in a tight sphere, forming a torus (a doughnut-like shape with a hollow centre), or suddenly scattering in all directions with flashes of silver (Fig. 6.1).

Questions about grouping behaviour

In this chapter we shall first consider why individuals often form groups like these, despite the potential costs of increased competition for resources and infection by pathogens. There are many potential benefits of group living, including protection from climatic extremes (e.g. huddling for warmth) and aerodynamic or hydrodynamic advantages in locomotion, but we shall focus on two main benefits, namely reducing predation and improving foraging success (Table 6.1). We shall also analyse the costs and benefits of different group sizes. Is there an optimal group size to maximize individual fitness? How can grouping be stable despite individual conflicts of interest? Finally, we shall consider the mechanisms of group decision making. How do groups make such wonderful, coordinated movements? How do groups decide when and where to go next? Are there leaders and followers? Or do individuals reach a consensus by voting?

An Introduction to Behavioural Ecology, Fourth Edition. Nicholas B. Davies, John R. Krebs and Stuart A. West.

(a)

(b)

Fig. 6.1 Living in groups: (a) In winter, tens of thousands of starlings gather in spectacular amoeba-like flocks at dusk, prior to their night-time roost. Photo © osf.co.uk. All rights reserved (b) When a predator approaches, many fish form tight shoals. Why do individuals form groups? How are group movements coordinated? Photo © iStockphoto.com/stevedeneef.

Table 6.1 Summary of some anti-predator and foraging benefits of grouping

Benefits	How grouping may benefit individuals
Anti-predator	Dilute risk of attack (swamping predators; selfish herding).
	Predator confusion.
	Communal defence.
	Improved vigilance for predators.
Foraging	Better food finding (information centres).
	Better food capture (group hunting).

Predation and food influence the costs and benefits of group living

How grouping can reduce predation

Diluting the risk of attack

Dilution: theory

G.C. Williams (1966) and W.D. Hamilton (1971) were pioneers in rejecting the idea that gregarious behaviour evolves through benefits to the population or species. Instead, they sought individual benefits from grouping. They proposed what is perhaps the simplest idea of all, namely individuals associate with others as a form of cover-seeking, to reduce their personal risk of attack. If an individual is alone, clearly it is at risk of attack if a predator encounters it. On the other hand, if it is in a group of N individuals, it now has a 1/N chance of being the victim. So individuals may join others in the hope that someone else gets attacked rather than them! This dilution advantage will favour grouping provided attack rate does not increase proportionately with group size; if a group of N individuals suffer N times as many attacks as a singleton, then clearly individuals will be no safer in a group than on their own. However, if there are less than N times as many attacks at a group of N, then individuals will still be safer in a group. The key point of this idea is that even when there are no other advantages from grouping, for example better detection of predators or communal defence, individuals will often still be safer in a group through dilution of the risk of attack.

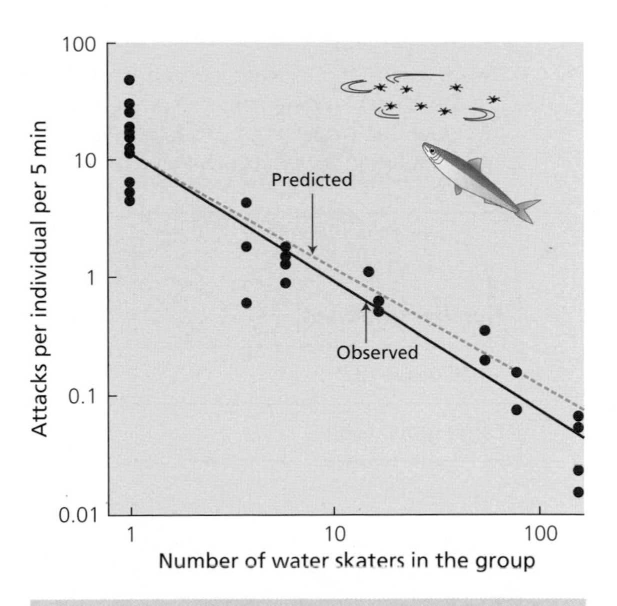

Fig. 6.2 An example of the dilution effect. The prey are insects called water skaters (*Halobates robustus*) that sit on the water surface; their predators are small fish (*Sardinops sagax*). The fish snap the insects from below, so there is little possibility that vigilance increases with group size. The attack rate by the fish was similar for groups of different sizes, so the attack rate per individual varies only because of dilution. The 'predicted' line is what would be expected if the decline in attack rate with group size is entirely caused by dilution; this line is very close to the observed. From Foster and Treherne (1981). Reprinted with permission from the Nature Publishing Group.

Dilution: evidence

Grouping can dilute an individual prey's risk of being attacked

So much for the theory – what about the evidence for dilution? Figure 6.2 provides an example where predator attack rate does not vary with prey group size, so here grouping by the prey results in perfect dilution. In a group of 100, an individual suffers 1/100th the attack rate compared to being alone. More often, however, predator attack rate will increase with group size because larger groups are more conspicuous. Nevertheless, grouping usually still brings a net dilution advantage. For example, in the Camargue marshes of the South of France, wild horses are attacked by blood sucking flies (Tabanidae), which not only remove blood but also transmit bacterial and viral diseases. During the weeks when these flies are most active, the horses aggregate into larger groups. Duncan and Vigne (1979) varied group size experimentally and found that more flies were attracted to larger groups of horses; nevertheless, attack rate per horse was still lower in a larger group (Table 6.2).

... even when attack rate per group increases with group size

A spectacular example of dilution is the winter aggregations of monarch butterflies in Mexico, where thousands or millions of individuals assemble in enormous communal roosts, clothing the trees over an area of up to 3 ha. The monarch is not a very palatable butterfly but some birds attack them in these roosts. Counts of the remains of depredated butterflies showed that although larger colonies attracted more predators, predation rate per individual was lower in a larger colony, so the advantage of dilution outweighed the disadvantage of greater conspicuousness in a larger roost (Calvert *et al.*, 1979).

Table 6.2 Dilution advantage from grouping. Wild horses in the Camargue, southern France, were kept in a group of three or in a group of 36. Although the larger group attracted more biting tabanid flies, individual horses suffered fewer attacks in the larger group (Duncan & Vigne, 1979).

	Mean number of biting flies	
Number of horses	**per group**	**per horse**
Small group (3)	30	10
Large group (36)	108	3

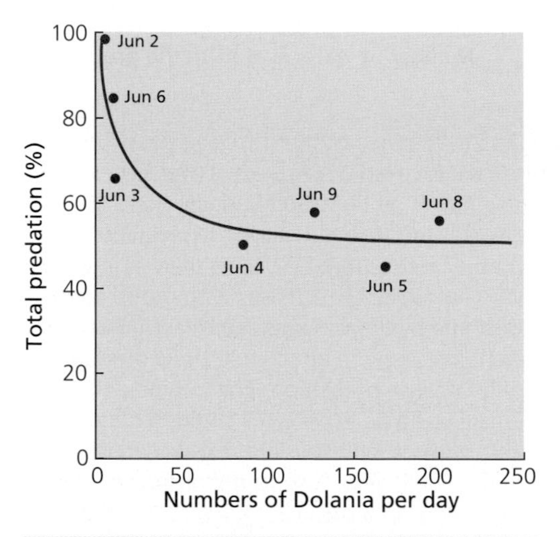

Fig. 6.3 Synchronous emergence swamps predators. The percentage of adult female mayflies *Dolania americana* preyed upon by aquatic and aerial predators combined, during seven days in June. Individual mayfly are safest on days where more females emerge. From Sweeney and Vannote (1982).

Synchrony in time can also dilute predator attacks per individual prey

Synchrony in time: predator swamping

Dilution may also be achieved by synchrony in time, which swamps the capacity of predators to capture prey. The synchronous emergence of mayflies provides a dramatic example of this effect. In North America, the emergence of *Dolania americana* occurs over a two week period in late May to early June, when the larvae transform into winged adults at the water surface, just before sunrise each day. The adults immediately mate, lay eggs and then die all within an hour or so. They are preyed upon at the water surface by beetles, and after emergence by dragonflies, bats and birds. Individual mayfly are safest from predation during days when more adults emerge (Fig. 6.3). It could be argued that this synchronous emergence has primarily evolved to enhance mating success. However, parthenogenetic mayflies have similar synchrony in emergence, which suggests that reducing predation through predator swamping is a major selective pressure favouring synchrony (Sweeney & Vannote, 1982).

Selfish herds

Just as a mayfly in the middle of the emergence period is safer than one at either end, so individuals in the middle of a group may enjoy greater security than those at the edge. W.D. Hamilton (1971) suggested that individuals should approach others to reduce their domain of danger, and this might explain the continuous movements in swarms, flocks and shoals as individuals try to gain the safest positions in the group. He called this the 'selfish herd' effect (Fig. 6.4a).

Reducing the domain of danger

Alta De Vos and Justin O'Riain (2010) tested Hamilton's assumption that an individual's risk of attack was related to its domain of danger. They studied Cape fur seals, *Arctocephalus pusillus*, in False Bay, by the Cape Peninsula of South Africa. The seals suffered heavy predation from great white sharks, *Carcharodon carcharias*, which detected seals as silhouettes on the surface and then attacked at tremendous speed from below, resulting in the whole body of the shark breaching the water as it grabbed the prey. De Vos and O'Riain made groups of decoy seals from styrofoam boards, and fixed these to a raft using reed poles attached to the dorsal surface of each decoy. By varying the distance between decoys on the raft, they could present prey with different domains of danger. The rafts were towed behind a boat near the seal colony and attracted the sharks. As assumed in Hamilton's model, an individual decoy's risk of shark attack increased with an increase in its domain of danger (Figs 6.4b, 6.4c). This is a particularly neat test of Hamilton's model because it distinguishes the selfish herd effect from other factors that may reduce attacks on groups of prey, such as improved prey vigilance or predator confusion (see next section).

Individuals in the middle of a group may be safer than those at the edge ...

Jens Krause (1993a) tested Hamilton's prediction, that alarmed individuals would seek safety amongst companions, in laboratory experiments with dace *Leuciscus leuciscus* and minnows *Phoxinus phoxinus*. These cyprinid fish live in shoals and if they detect chemicals from the damaged skin of a companion wounded by a predator, they form tighter shoals. A shoal of fourteen dace was habituated to this chemical stimulus by repeated exposure and then single naïve minnows were added to the shoal. Before the chemical was added to the water there were no differences in the positioning of the minnows and dace in the shoal. However, after the introduction of the chemical, the naïve minnows moved closer to the other fish and positioned themselves so that they were surrounded by near neighbours on all sides, just as predicted by the selfish herd model.

... but position choice may reflect a trade-off between predation and foraging

An individual's position in a group is likely to reflect a trade-off between foraging benefits and predation risk. Experiments with fish, for example, show that hungry individuals tend to occupy positions at the front of the shoal, where they will be the first to encounter food (Krause, 1993b). However, front positions are also more vulnerable to predation, so satiated fish tend to seek more central positions (Bumann *et al.*, 1997).

Predator confusion

Individuals in groups may also be safer from attack because the predator has difficulty in focusing on one target as different individuals in the group continually move across its line of sight. We can experience this confusion effect ourselves when we are thrown a group of balls; it is much more difficult to catch one of the group than when the balls are thrown one at a time.

Grouping may confuse predator attacks

Neill and Cullen (1974) tested the hunting success of four aquatic predators in laboratory experiments in aquaria, where the prey were small fish. Squid, cuttlefish and pike are ambush predators; they have a slow, stalking approach and then make a rapid strike after a short chase. The perch is a chasing predator; it dashes after prey, often with a long pursuit. For all four predators the success of an attack declined with increasing prey shoal size (Fig. 6.5) and Neill and Cullen suggest that this is mainly because of increasing predator confusion. This was most evident with the perch, where shoals of prey disrupted attacks

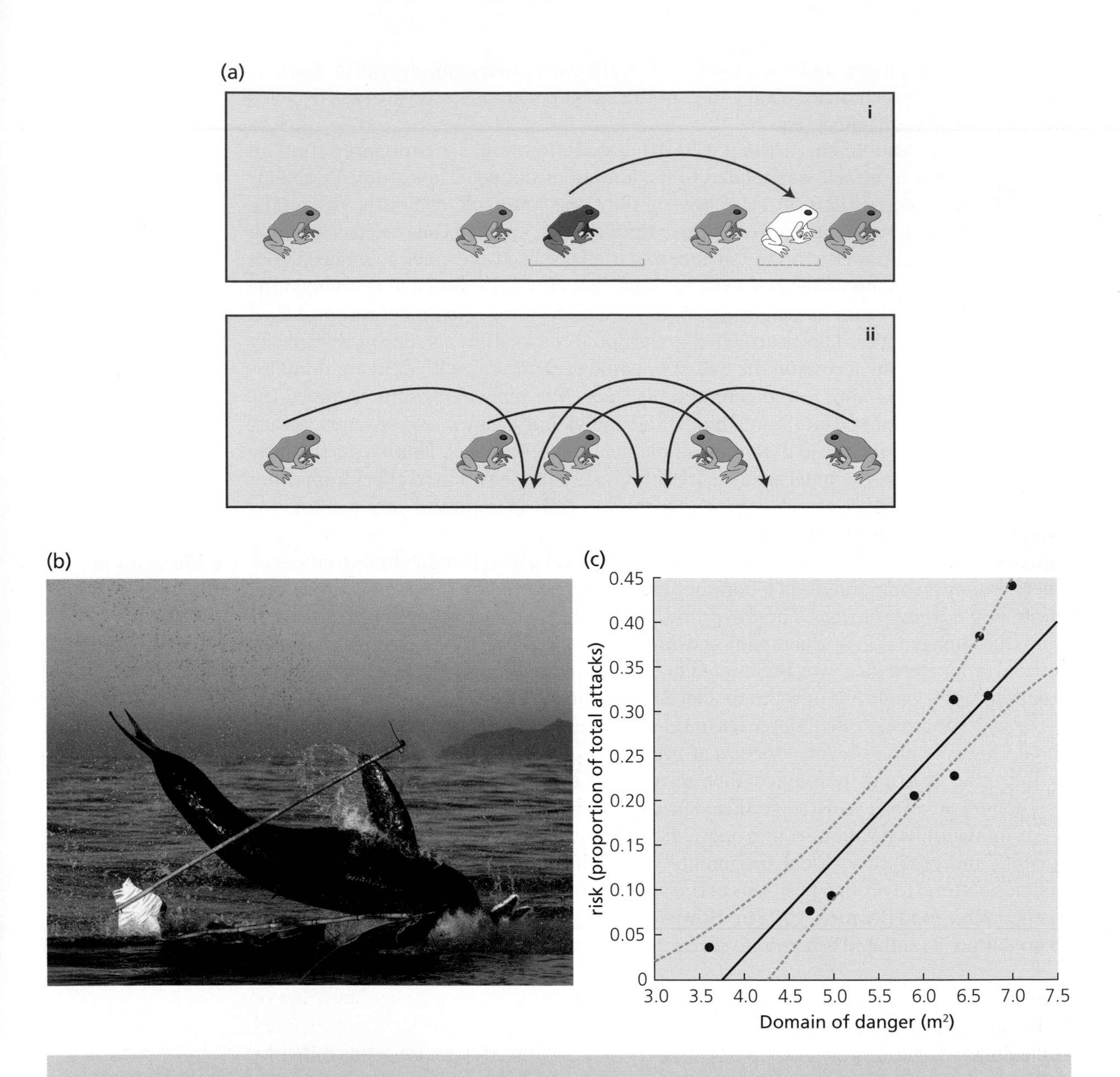

Fig. 6.4 (a) W.D. Hamilton's (1971) model of the selfish herd. Each frog has a 'domain of danger' in which it is likely to be selected for attack if a predator appears: this is shown for one individual in (i) as the solid bar, and is the zone stretching half way to the neighbour on either side. Any predator approaching this zone will select this frog as the nearest potential victim. The frog can reduce its zone of danger by jumping to settle in between two closer neighbours (arrowed movement, new domain of danger shown by dashed line). If all frogs follow this principle, the result will be increased aggregation (ii). (b) A test of Hamilton's model. An experiment with groups of decoy styrofoam seals, attached to a raft using reed poles, and then presented to great white sharks (see text). A shark attacking the decoys. Photo © Claudio Velasquez Rojas/Homebrew Films. (c) An individual seal decoy's risk of shark attack increased with its domain of danger, as assumed in Hamilton's model. De Vos and O'Riain (2010).

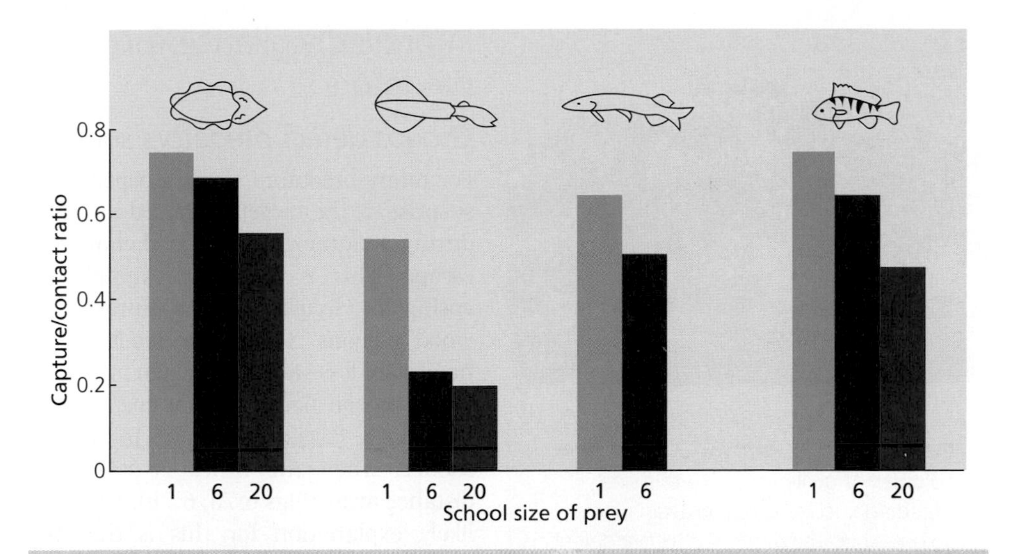

Fig. 6.5 The confusion effect of grouping by prey. The capture success per attack of, from left to right: squid, cuttlefish, pike and perch, when attacking small prey fish in singles, groups of six, or groups of twenty. In all cases, capture success declines with increasing prey group size. From Neill and Cullen (1974).

by forcing the predator to continually switch targets during the pursuit. Milinski (1984) also found that sticklebacks preferred to attack single water fleas *Daphnia* on the edge of a swarm, perhaps because these were easier to target.

Indirect evidence for the confusion effect is provided by experiments showing that predators often target odd-coloured individuals in a group (Ohguchi, 1978; Landeau & Terborgh, 1986). By concentrating on an individual that looks different from the rest, a predator may be able to counteract confusion.

Communal defence

Prey are often not just passive victims but may actively defend themselves by attacking or mobbing a predator and grouping may enhance prey defence. Nesting black-headed gulls, *Larus ridibundus*, will mob crows who come to their colony in search of eggs and chicks. In the centre of a dense colony many gulls mob a crow at the same time because it is close to many nests, and group defence reduces crow success (Kruuk, 1964). Similarly, breeding success is greater in denser parts of the colonies of guillemots (= common murres) *Uria aalgae* because the defence of many, closely-packed incubating birds is more effective in deterring nest predators such as gulls (Fig. 6.6; Birkhead, 1977).

Communal mobbing of predators

Andersson and Wicklund (1978) demonstrated by experiment the increased effectiveness of group defence by fieldfares *Turdus pilaris*, a colonial-nesting thrush which breeds in Scandinavian boreal forests. Fieldfares vigorously mob and defecate on crows and other predators, and artificial nests placed near fieldfare colonies survived better than those placed near solitary fieldfare nests.

Fig. 6.6 Group defence. In dense colonies of guillemots, like this one, breeding success is higher than in sparse colonies because of more effective defence against nest predators such as gulls. From Birkhead (1977). Photo © T. R. Birkhead.

Improved vigilance for predators

Groups detect predators sooner

For many predators success depends on surprise; if the target is alerted too soon during an attack it has a good chance of escape. This is true, for example, of goshawks, *Accipiter gentilis*, hunting for wood pigeons, *Columba palumbus*. The hawks are less successful when attacking larger pigeon flocks mainly because the birds in a larger flock take to the air sooner, when the hawk is still some distance away (Figs 6.7a, 6.7b). The most likely explanation for this is that the bigger the flock, the more likely it is that one bird will be alert when the hawk looms over the horizon (Pulliam, 1973). Once one pigeon takes off in alarm, the others follow. Water skaters in larger groups also respond sooner to an approaching model predator and begin their escape movements when the predator is further away (Fig. 6.7c).

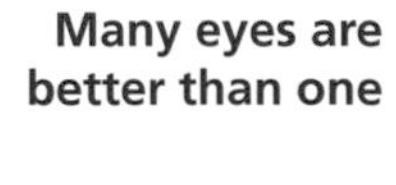

Many eyes are better than one

A problem in interpreting these results is that there may be a delay between prey detecting the predator and their escape response. Therefore, escape reactions are not necessarily a good measure of detection. However, in general individuals should be safer in larger groups so if anything there should be a longer delay between detection and response in larger groups. This would lead to response times increasing with group size,

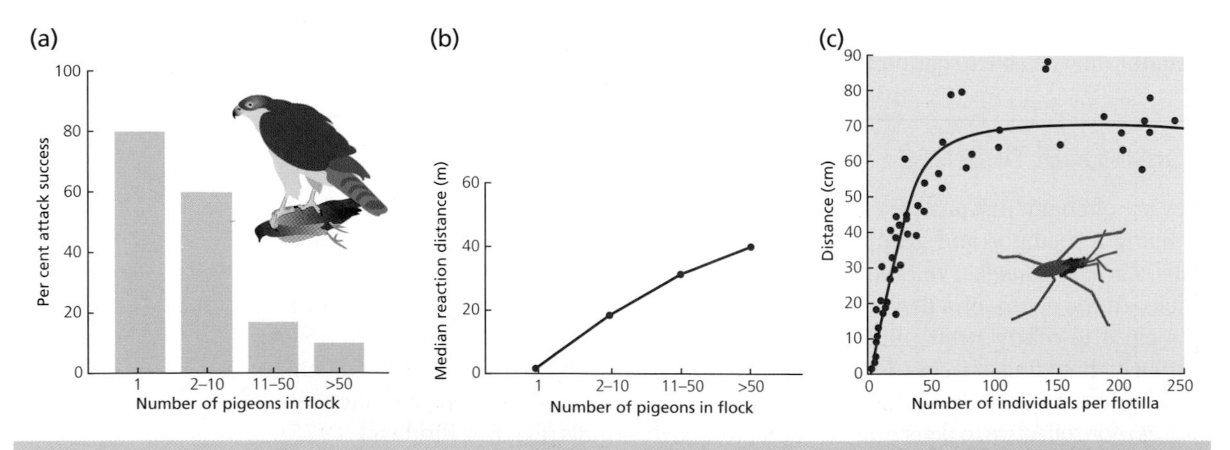

Fig. 6.7 (a) Goshawks are less successful when they attack larger flocks of wood pigeons. (b) This is largely because bigger flocks take flight at greater distances from the hawk. The experiments involved releasing a trained hawk from a standard distance. From Kenward (1978). (c) Water skaters, *Halobates robustus*, in larger groups also respond sooner to an approaching model predator, by agitated movements on the water surface when the predator is further away. From Treherne and Foster (1980).

the opposite of the results in Figs 6.7b and 6.7c. This suggests that individuals in larger groups do indeed detect predators sooner (Krause & Ruxton, 2002).

Are larger groups more vigilant? Brian Bertram (1980) studied small groups of ostriches *Struthio camelus* in Tsavo National Park, Kenya, where they feed out on the open plains and are vulnerable to attack by lions. He found that as group size increased, individual ostriches reduced the proportion of time they had their head up, scanning the environment (Fig. 6.8a). Nevertheless, the overall vigilance of the group (at least one individual scanning) increased with group size. This increase was as predicted if each bird raised its head independently of the others (Fig. 6.8b). The ostriches also raised their heads at random time intervals, which makes it impossible for a stalking lion to predict how much time it has to creep forward undetected in between look-ups by its victim. Any predictable pattern of looking could be exploited by the lion in its tactics of approach.

Ostriches scan at random

These data suggest that prey could gain a double advantage in vigilance by grouping; firstly, individuals in groups could scan less and so devote more time to feeding and, secondly, many eyes improve overall vigilance in the group. However, the relationship between individual vigilance and group size can be complex and involve many effects (Roberts, 1996). For example, as group size increases an individual's risk of predation declines due to dilution as well as through many eyes and both these effects may lead to reduced individual vigilance. Furthermore, increased group size may increase feeding competition, so individuals may have to devote more time to feeding, in which case reduced scanning could sometimes be a cost of larger groups rather than a benefit.

Response to others' alarms

Do individuals benefit by responding to the alarms of others in the group or do they have to wait to respond to the predator itself? Magurran and Higham (1988) used a

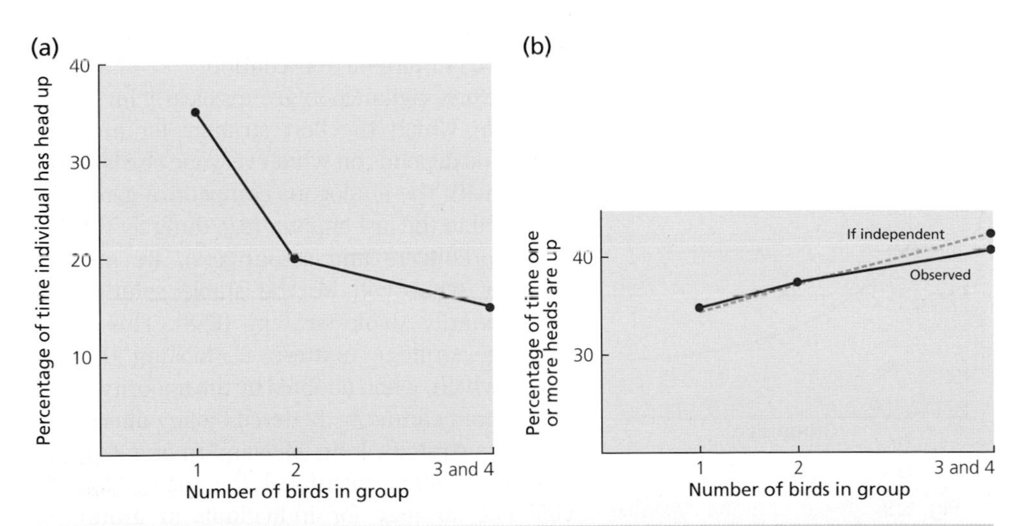

Fig. 6.8 Vigilance in groups. (a) An ostrich spends a smaller proportion of its time scanning (head up) when it is in a larger group. (b) The overall vigilance of the group (at least one bird scanning) increases with group size (solid line) and follows the relationship expected if each individual looks up independently of others in the group (broken line). From Bertram (1980). With permission from Elsevier.

one-way mirror in an experiment with minnows, so that some of the shoal could not observe the predator themselves but they could observe the reaction of other minnows who were threatened by the stalk of a model pike *Esox lucius*. Although the naïve minnows couldn't see the pike, they nevertheless hid in response to the alarm behaviour of the other fish. Similarly, Treherne and Foster (1981) showed that the approach of a model predator caused anti-predator movements to spread across flotillas of water skaters, *Halobates robustus*, as individuals bumped into their neighbours. They coined this wave of alarm the 'Trafalgar Effect', after the Battle of Trafalgar when signals were transmitted across chains of ships, allowing Admiral Nelson to know of the approaching enemy even when they were over the horizon and he could not yet detect them himself. So there's good evidence that individuals can benefit from the alarm of others in the group.

Evidence that individuals respond to the alarm of others

Cheating

Imagine an ostrich on its own. If it spent all its time with its head down, foraging, then it would never starve to death, but it would eventually be killed by a lion. If it spent all its time with its head up, scanning, it would always detect the predator but it would eventually starve to death. So scanning involves a trade-off and, in theory, there will be some optimum pattern of foraging and scanning that maximizes the individual's overall chance of survival through avoiding both starvation and predation.

In theory, in scanning groups it may pay individuals to cheat

Now imagine a large group in which everyone occasionally scans and occasionally feeds. Any individual that cheated, by reducing its scanning, would gain an advantage; it could now devote more time to feeding and its selfish behaviour would have little, if any adverse effect on overall group vigilance, so it could still benefit from any alarm raised by its vigilant companions. The problem, of course, is that if everyone cheated there would be little vigilance and everyone would be vulnerable to predation.

An ESS for vigilance

Therefore, vigilance in groups clearly involves a game, in which the best strategy for any one individual depends on what everyone else is doing. Just as with the analogous competitive games we discussed in the last chapter (e.g. dung fly waiting times, producers and scroungers), we need to consider what will be the stable solution, or evolutionarily stable strategy (ESS). This is the vigilance strategy (pattern of looking up and down) which, when adopted by the majority of the population, cannot be bettered by any alternative vigilance strategy. John McNamara and Alasdair Houston (1992) calculated the theoretical ESS vigilance strategy for individuals in groups of different sizes. The precise pattern varied depending on various parameters, such as relative risks of starvation and predation, but the general result was for vigilance to decline with increasing group size, just as we observed in the ostriches (Fig. 6.9).

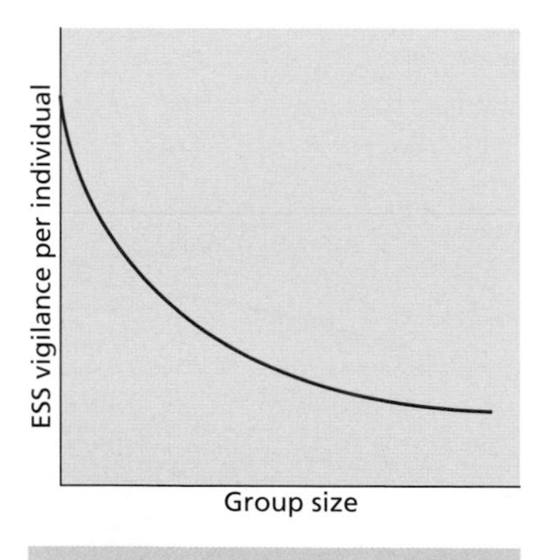

Fig. 6.9 The evolutionarily stable vigilance (ESS) for individuals in groups of different sizes (see text for explanation). From McNamara and Houston (1992). With permission from Elsevier.

(a)

(b)

Fig. 6.10 A cheetah (a) is less likely to attack the more vigilant Thomson's gazelles (b) in a group. Photos © Oliver Krüger.

Cheating may be reduced if more vigilant individuals are safer from attack

In theory, the temptation to cheat will be reduced if individuals who spot the predator gain an extra advantage over their non-vigilant companions. Observations suggest that vigilant individuals do indeed gain a personal advantage, and that this accrues through two effects. Firstly, predators may be less likely to target vigilant individuals. Claire FitzGibbon (1989) observed cheetahs, *Acinonyx jubatus*, hunting pairs of Thomson's gazelles, *Gazella thomsoni*, on the Seregeti plains, Tanzania (Fig. 6.10). She measured the vigilance of each gazelle during the predator's stalk, and found that the cheetah targeted the least vigilant gazelle in 14 out of 16 cases. Furthermore, gazelles that were successfully caught by the cheetah had been less vigilant during the stalk than those that escaped, because the latter detected the predator sooner.

Secondly, studies of small birds at feeders have shown that individuals who are more vigilant at the initiation of an alarm flight depart more quickly to safety than non-vigilant individuals (Elgar *et al.*, 1986). Lima (1994) demonstrated this experimentally by rolling a ball silently down a ramp towards a single target bird in a flock of sparrows and juncos. The sides of the ramp were high enough so only the target bird could see the ball. Lima found that the target individual fled to cover first, followed by individuals who had been vigilant during the target's departure, and finally by individuals who had not been vigilant. Therefore, vigilant individuals gain a double advantage; they detect the predator sooner and detect alarmed companions sooner, too.

Sentinels

Sentinels warn others ...

In some animal groups, for example meerkats, *Suricata suricatta*, and various species of babblers, individuals often act as sentinels, watching for predators from prominent look-out perches while the rest of the group forages on the ground below. When the sentinel spots an approaching predator (a raptor, snake or carnivorous mammals) it gives an alarm and everyone rushes to safety (Fig. 6.11).

... but sentinel behaviour could be best for selfish individuals

At first sight sentinels seem to be behaving altruistically, risking their own welfare for the good of the group (Chapter 11). However Peter Bednekoff (1997) showed, in a theoretical model, that sentinel behaviour could arise through selfish individual actions. His model's key assumptions are that individuals are only likely to go on sentinel duty when they are satiated, and that acting as a sentinel may be beneficial to the sentinel itself because it can detect a predator sooner. The outcome can then be a series of change-overs in the group, as sentinels become hungry and are replaced by satiated

Fig. 6.11 Sentinels in: (a) meerkats and (b) pied babblers. These individuals watch for predators from look-out perches while the rest of the group forages. Is the sentinel altruistic or selfish? Photos © Tom Flower.

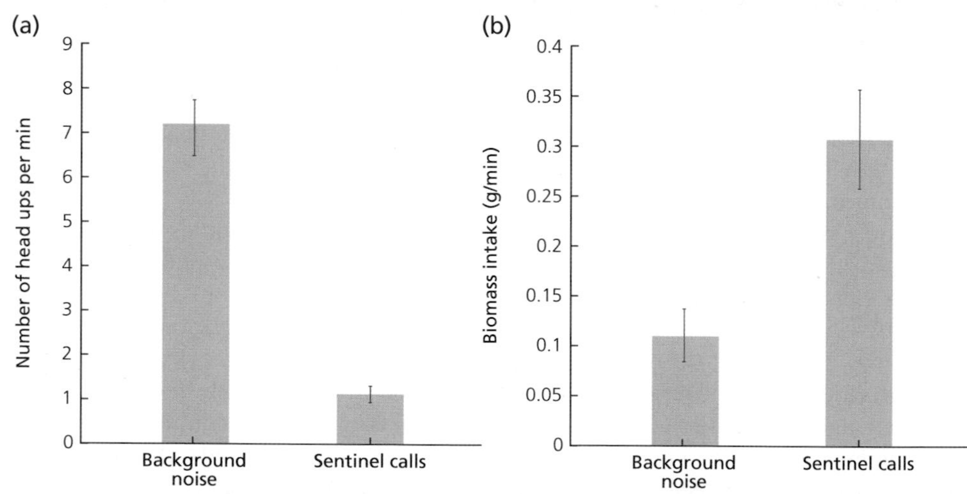

Fig. 6.12 The watchman's song. Sentinels give quiet vocalizations while on guard. Playback of sentinel calls to foraging pied babblers leads to lower vigilance (a), and improved foraging success (b) by the foragers in the group, compared to control playback of background noise. From Hollen, Bell and Radford (2008). With permission from Elsevier.

foragers. Each individual is simply choosing the best option for itself, depending on its own state and the action of others in the group.

Selfish sentinels in meerkats and babblers

Observations of both meerkats and Arabian babblers, *Turdoides squamiceps*, support this selfish view of sentinel duties (Clutton-Brock *et al.*, 1999a; Wright *et al.*, 2001). Both species feed in the open desert, digging for invertebrates in the sand. While foraging, it is difficult to spot approaching predators. Individuals are more likely to become sentinels when they are satiated, either after a natural feeding bout or after experimental provisioning with extra food. Furthermore, sentinels are not at increased risk of predation; in fact, they usually detect predators first and are often the first to flee to safety (a burrow for a meerkat, a bush for a babbler). Finally, even solitary individuals become sentinels when they are satiated, supporting the idea that sentinel behaviour benefits individuals. Alarm calling by the sentinel is a case of mutualism (Chapter 12); both caller and responders gain because individuals benefit from group living.

The 'watchman's song'

One aspect of sentinel behaviour is particularly fascinating; sentinels give quiet vocalizations while on guard. Wolfgang Wickler (1985) suggested this might function as a 'watchman's song', announcing to the rest of the group that they were safe because someone was on guard duty. Playback experiments to both meerkats (Manser, 1999) and pied babblers (*Turdoides bicolor*) (Hollen *et al.*, 2008) revealed that foragers decreased their own vigilance when they heard sentinel calls. In the babblers, this also led to a marked increase in foraging success (Fig. 6.12). Sentinels may experience little cost from these quiet vocalizations and may gain a benefit if improved foraging by others leads to better coordination of sentinel activities in the group.

How grouping can improve foraging

Better food finding

The comparative studies described in Chapter 2 revealed that species which feed on large ephemeral clumps of food, such as seeds or fruits, often live in groups. For these animals, the limiting stage in feeding is the problem of finding a good site; once the patch has been found there is usually plenty of food, at least for a short while. Peter Ward and Amotz Zahavi (1973) developed the idea that communal roosts and nesting colonies of birds may act as 'information centres', in which individuals find out about the location of good feeding sites by following others. The idea is that unsuccessful birds return to the colony or roost and wait for the chance to follow others who have had more success on their last feeding trip. Unsuccessful birds might recognize successful ones by, for example, the speed with which they fly out from the colony on their next trip.

Colonies and roosts may act as information centres

The phrase 'information centre' has the connotation of mutual cooperation in the transfer of information, as for example in a honeybee or ant colony. As we shall see in Chapter 13, there are special reasons to expect cooperation in social hymenopteran colonies because individuals interact with close relatives. However, bird roosts or nesting colonies often involve large anonymous, assemblages of unrelated individuals. So the key question to ask is: what would informers gain from telling others where the food is? Richner and Heeb (1996) suggested that the informers may simply gain the advantages of group foraging (e.g. better protection from predators), and if this is the main benefit, rather than information sharing, then a better term would be 'the recruitment centre' hypothesis. Indeed, if predation pressure favours grouping, then a consequence may be that successful individuals could be followed to food sources without any active signalling on their part. In this case the group would simply be an 'eavesdropping centre'.

Nevertheless, game theory models show that the opportunity to share foraging information can, in theory, also drive the evolution of communal roosts. If pooling the independent search effort of many individuals is the most effective way of locating rare food bonanzas (e.g. animal carcasses for carrion eaters such as crows and ravens), and if there is little cost to foraging in a group (because the food patch is large or only temporarily available), then the strategy 'search independently and recruit others from the colony or roost' can be an ESS (Mesterton-Gibbons & Dugatkin,

1999; Dall, 2002). Any additional advantage of group foraging would, of course, further enhance communal roosting or coloniality.

Information transfer in raven roosts

Two studies of ravens, *Corvus corax*, provide strong evidence that communal roosts act as information centres in the manner envisaged by Ward and Zahavi (1973); namely, successful foragers actively share information about the location of good feeding sites. In Maine, USA, Bernd Heinrich and colleagues found that a territorial pair of ravens could defend a mammal carcass against intrusions from one to two juveniles, but gave way if there was a gang of six or more juveniles. Upon discovering a large, defended carcass juveniles called loudly to recruit others (Heinrich *et al.*, 1993). However, the most effective recruitment to carcasses occurred at the juvenile communal roost sites. This was demonstrated experimentally by keeping some birds in captivity for a short time so they were unaware of the location of food in the wild. When these naïve birds were released at roosts, they followed knowledgeable birds to carcasses. By contrast, when naïve birds were released away from roosts they were rarely sighted at the carcasses (Marzluff *et al.*, 1996). Intriguingly, knowledgeable juveniles initiated the pre-roost soaring displays at dusk, and also initiated the departure of feeding groups from the roost the following dawn, suggesting that they actively advertised their discovery of the food sources.

Jonathan Wright and colleagues (2003) studied a large winter roost of up to 1500 ravens (mainly unpaired juveniles) on the isle of Anglesey, North Wales, UK. At distances of 2–30 km from the roost, they put out sheep and hare carcasses embedded with small, colour-coded plastic beads. The ravens ingested these beads at the carcass and then regurgitated them in pellets back at the roost. Beads from each carcass tended to appear at specific sites within the roost, showing that birds which fed together also slept together. Aggregations of beads at the roost grew daily, with an increasing radius centred on the first beads from a carcass, showing that there was an increasing number of naïve birds joining the feeding and roosting group. Furthermore, groups of birds would leave the roost together led by the bird that roosted centrally in the aggregation, and they flew directly towards a specific carcass.

In rats, information is transferred by smell

Laboratory studies have also shown that naïve individuals can learn from experienced individuals about the location of food. Geoff Galef and Stephen Wigmore (1983) trained rats (*Rattus norvegicus*) to search for food in a three-arm maze. Each arm had food with a different flavour, cocoa in one, cinnamon in another and cheese in the third. In the first part of the experiment the rats learned that on any particular day only one of the three sites contained food, but the site was unpredictable. Then on the days of the actual experiment each of the seven test rats was allowed to sniff a 'demonstrator' rat in a neighbouring cage. The demonstrator had been allowed to feed on whatever randomly chosen food was available for that day, and four of the seven test rats, having sniffed the demonstrator, went to the correct site on their first choice of the day. 'Sniff' is the operative word, because other experiments showed that the cue the test rat picks up from the demonstrator is the smell of the food it has eaten.

Information transfer may be tempered by the costs of food sharing

Learning about potential food sources in a more direct way, by seeing others exploiting them, is important in both flocks of birds and shoals of fish (Krebs *et al.*, 1972; Laland & Williams, 1997). In some cases, the decision to recruit others to a feeding site depends on the costs of food sharing. In experiments with house sparrows, *Passer domesticus*, Mark Elgar (1986a) found that birds who discovered a food source would give recruitment 'chirrup' calls to attract others if the food was divisible (bread crumbs), but not if the food was indivisible (same amount of bread but in one piece). Therefore, there is a trade-off between flock benefits (safety from predators) and costs (food sharing).

(a)

(b)

Fig. 6.13 Group hunting. When they hunt in a group, (a) spotted hyenas and (b) lions can successfully attack prey which are larger than themselves. Photo (a) © Hans Kruuk and (b) © Craig Packer.

Better prey capture

Predators may sometimes improve their ability to capture prey by hunting in a group. Hans Kruuk (1972) studied spotted hyenas, *Crocuta crocuta*, in the Serengeti National Park and Ngorongoro Crater area of Tanzania. He found that solitary hyenas could easily kill a Thomson's gazelle fawn, which was usually simply picked up from its hiding place in the grass. However, zebras, *Equus burchelli*, could defend themselves by biting and kicking and they showed no fear when a single hyena wandered past. A pack of hyenas, however, was an entirely different proposition. When hunting zebra, the hyenas assembled in packs of 10–25 individuals before the hunt (Fig. 6.13a). The zebra group

Hyenas and lions may benefit from group hunting

Fig. 6.14 A pack of fourteen spotted hyenas hunting a group of zebras. The zebra mares and foals run in a tight formation, followed by the stallion, who repeatedly charges at the hyenas. From Kruuk (1972).

walked or ran in a tight bunch and the hyenas usually followed in a crescent formation behind, attempting to bite any zebra on the edge of the group, while dodging its defensive kicks (Fig. 6.14). The chase lasted for up to 3 km, by which time the hyenas often had singled out one zebra that had fallen behind, either because it was in poorer condition or because one hyena had managed to get a grip on it. All the other hyenas would then immediately concentrate on this one zebra and help to pull it down.

Individual lionesses may favour particular positions in a group hunt

Lions, *Panthera leo*, also benefit from hunting in groups when tackling large prey, such as buffalo, which are difficult for one lion to kill alone, or when making surprise attacks on prey which can outrun them (Fig. 6.13b). Philip Stander (1992) studied lions in Etosha National Park, Namibia, where the main prey were zebra, springbok (*Antidorcas marsupialis*) and wildebeest (*Connochaetes taurinus*). He could recognize individual lionesses by brand marks and some were radio-collared, so he could track the group during a hunt. He found remarkable individual specializations, rather like those of a football team. The pride usually adopted a formation where some lionesses ('wings') circled the prey while others ('centres') waited in hiding. Individuals tended to specialize as either left or right wings or centres (Fig. 6.15). The wings tended to stalk the prey and initiate the attack, while the centres, which were usually the largest and heaviest lionesses in the group, most often made the kill as the prey was driven towards them. Not only did individual lionesses have regular positions in the hunting formation, but hunts were also more successful when most lionesses were playing in their preferred positions.

Not all hunting groups have this level of sophistication. Often a group is more successful simply because the simultaneous attack by many predators is more effective in causing panic in the prey, and as they flee in all directions the predators can more easily pick them off, one by one. This applies for example, to black-headed gulls (*Larus ridibundus*) and jack (a predatory fish, *Caranx ignobilis*), where an individual's hunting success against shoals of small fish prey increases with the number of predators in the hunting group (Götmark *et al.*, 1986; Major, 1978).

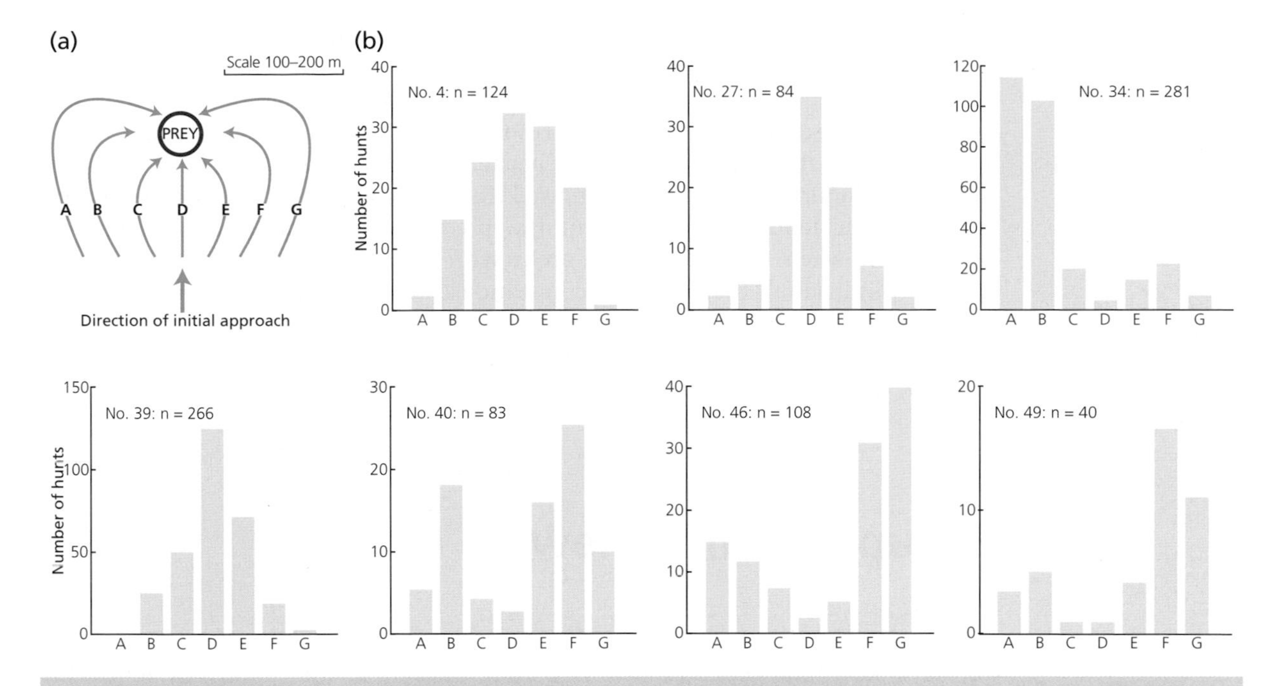

Fig. 6.15 Cooperative hunting in lions may involve individual specializations. (a) Seven stalking roles taken by lionesses towards a prey: A–B positions are 'left wings', C–E are 'centres' and F–G are 'right wings'. (b) Stalking roles of seven lionesses of the Okondeka pride in Etosha National Park, Namibia. The number of hunts is shown above each set of histograms. Individuals clearly differed in their positions: for example, number 34 preferred the 'left wing', numbers 4, 27 and 39 were 'centres', while numbers 46 and 49 preferred the 'right wing'. From Stander (1992).

Evolution of group living: shoaling in guppies

The studies described so far in this chapter are based on measurement of short-term costs and benefits of group living, such as food intake and predation risk. In one study, it has been possible to show how these costs and benefits translate into evolutionary change. In 1957, the American ichthtyologist C.P. Haskins moved 200 guppies (*Poecilia reticulata*) from a predator-rich river system in Trinidad (the Caroni system) to the almost predator-free headwaters of another river system, the Oropuche. The transferred fish subsequently colonized the downstream parts of their new river system and again encountered predators.

Rapid evolution of shoaling in response to changing predation

In 1989–1991, more than 30 years after the original transfer, Anne Magurran (Magurran *et al.*, 1992) collected guppies from several streams, including the original predator-rich source from which Haskins took the guppies, the predator-free site whether they were first introduced and the predator-rich streams they later colonized from the introduction site. In guppies, shoaling reduces predation (Magurran,

1990). However, shoaling also has a cost: Magurran and Seghers (1991) showed that guppies with a high propensity to shoal are less aggressive in competition over food. Therefore, there is a behavioural trade-off; selection for increased shoaling decreases competitiveness for scarce resources. By placing single guppies in an aquarium with the choice of an empty beaker at one end and a shoal in a beaker at the other, Magurran *et al.* (1992) showed that the fish transferred by Haskins first reduced their tendency to shoal in the predator-free streams but secondarily increased it when they re-invaded areas with predators. These differences were shown by individuals bred and raised in standard conditions in the laboratory. Thus, in guppies, the short-term costs and benefits of shoaling translated into an evolutionary response to decrease and later increase shoaling behaviour. These changes took place within 100 generations (guppies probably have about three generations per year).

Group size and skew

Optimal versus stable group sizes

We have seen that larger groups may bring benefits to individuals, by improving protection from predators or foraging success. However, larger groups may also bring increasing costs, from resource competition or disease (Fig. 6.16). In theory then, there might be an optimal group size to maximize an individual's fitness (Fig. 6.17).

In theory, there may be an optimal group size ...

Using George Schaller's (1972) data on the hunting behaviour of lions in the Serengeti, Caraco and Wolf (1975) predicted the optimum hunting group size to maximize an individual's food intake. Capture success increased with lion group size (a benefit) but if a prey was caught then food per lion per kill decreased with increasing group size (a cost). Caraco and Wolf calculated that for wildebeest hunts, the optimal lion group size to maximize food per lion per hunt was two lions. However, observed group sizes were, on average, larger with three to four lions per hunt (Fig. 6.18).

Packer *et al.* (1990) suggested that lion group sizes were unlikely to be predicted by hunting needs alone because larger groups were beneficial in other ways, for example for territory defence against neighbouring prides or defence of cubs against infanticidal males. This is an argument concerning the appropriate currency for maximizing individual fitness. But even if we measured all the relevant factors influencing the costs and benefits, would we expect individuals to live in an optimal group size?

... but optimal group sizes may not be stable

Richard Sibly (1983) pointed out that optimal group sizes may not be stable. His idea is illustrated in Fig. 6.19. In this example the optimal group size is seven, but newcomers will benefit from continuing to join the group until the group size is 14. At this point, the next individual would do better alone (note that this settlement pattern is like that in the ideal free model discussed in Chapter 5; Fig. 5.1). The same principle would apply if one imagined groups splitting and reforming as smaller units, but the argument is more complicated (Kramer, 1985). A group of 12, for example would split into two units of six, but then one individual would migrate to form a seven- and a five-membered group, the seven would be joined by another because eight is better than five and so on.

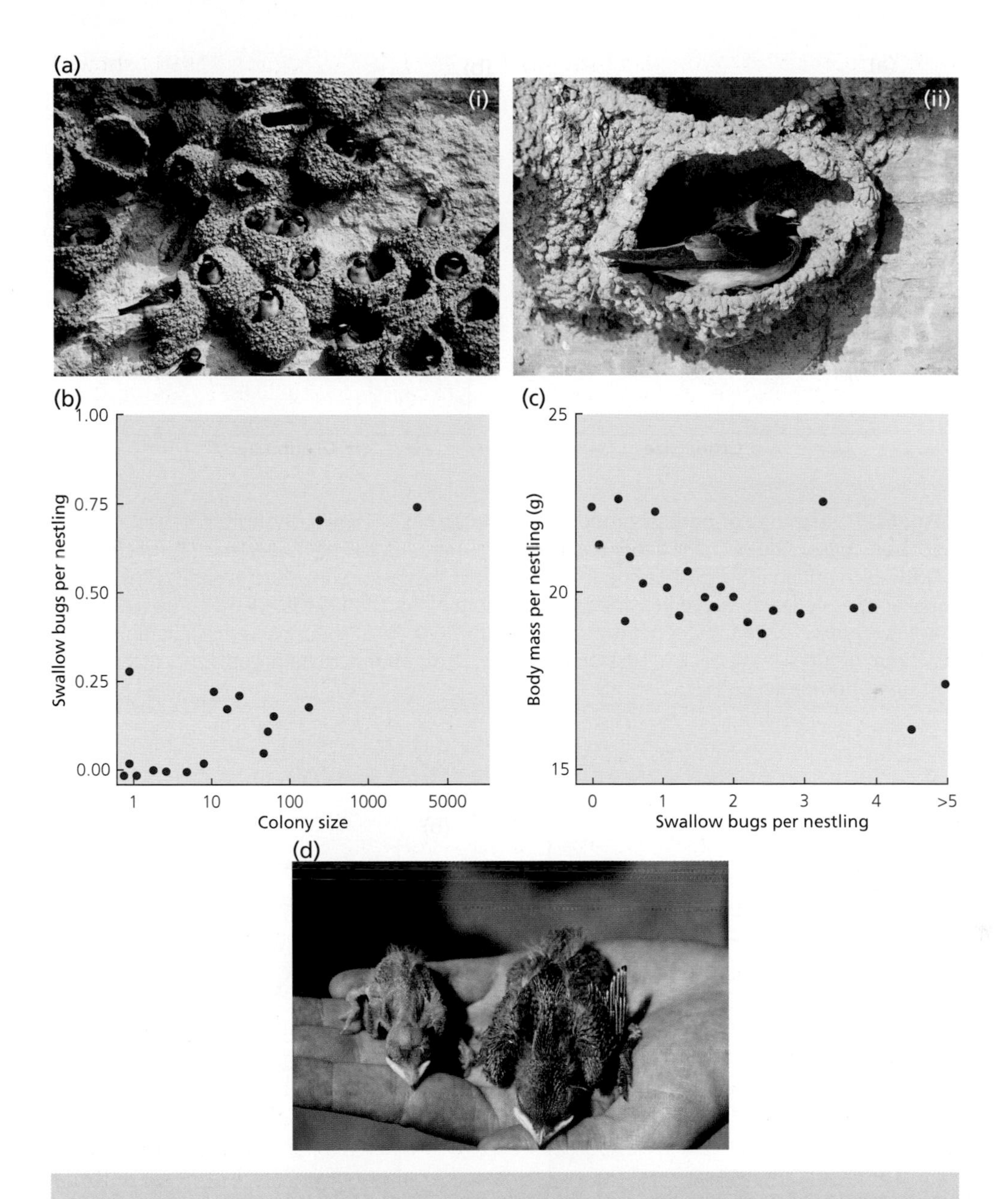

Fig. 6.16 An example of the benefits and costs of group living. (a) (i) A cliff swallow breeding colony. (ii) A bird at its nest entrance. Individuals gain benefit from the colony, which is an information centre enhancing food finding (ephemeral insect swarms; Brown 1988). (b) However, there are also costs; the number of blood-sucking hemipterans (swallow bugs, *Oeciacus vicarius*) on nestlings increases with colony size and (c) nestling body mass declines with increasing numbers of these ectoparasites. (d) Two nestlings, both 10 days old. The one on the right is from a fumigated nest, where the bugs were removed. The one on the left is from a naturally infested nest in the same colony. Photos (a) and (d) © Charles R. Brown. Figures (b) and (c) from Brown and Brown (1986). With permission of the Ecological Society of America.

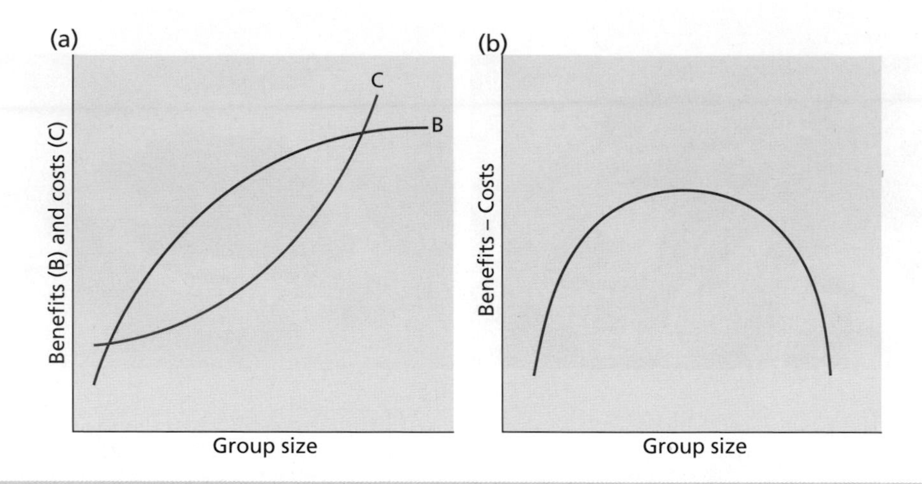

Fig. 6.17 The idea of optimal group size. (a) As group size increases both benefits and costs increase. However, in theory the increase in benefits will be a decelerating function (with each added individual having less effect than the last), while the increase in costs will accelerate (each added individual having more effect than the last). Therefore, costs will eventually exceed the benefits at large group sizes. (b) In theory, there will be an optimal group size (benefits – costs a maximum) at an intermediate group size. After Krause and Ruxton (2002).

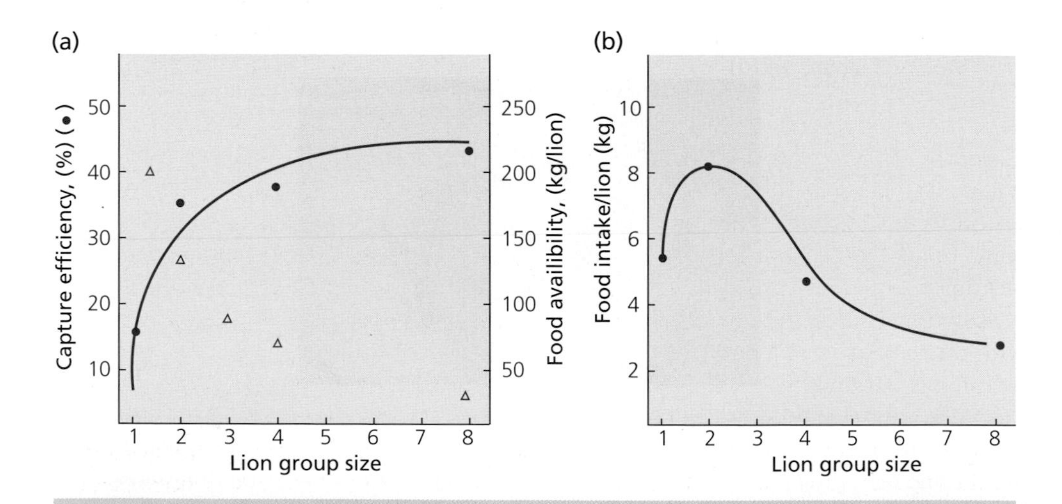

Fig. 6.18 Caraco and Wolf's calculation of optimal hunting group size for lions when hunting wildebeest in the Serengeti. (a) With increasing lion group size, capture success increases (solid circles) but if a kill is made, then food per lion decreases (open triangles). (b) This results in an optimal group size of two lions, to maximize food per lion per chase. Observed group sizes, however, are larger, on average three to four lions per hunt. From Caraco and Wolf (1975).

The exact outcome depends on the shape of the curve in Fig. 6.19, and it is possible to draw fitness curves which will result in a stable optimal group size (Giraldeau & Gillis, 1985).

The general point from Sibly's model is that unless groups of optimal size can prevent further newcomers from settling, groups in nature will often be larger than the optimum because it will pay individuals to join groups until a stable distribution is achieved.

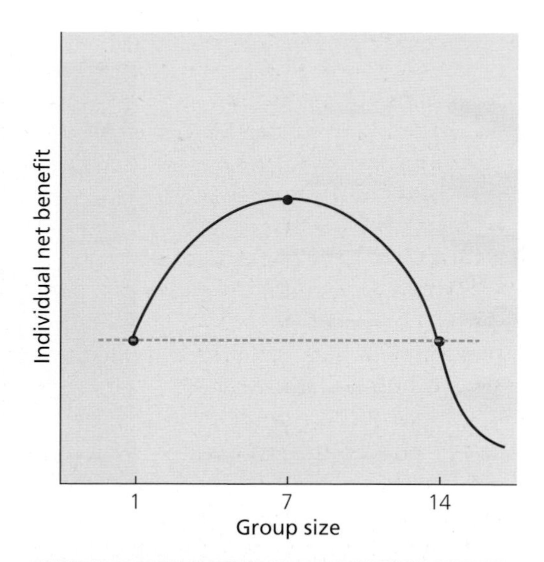

Fig. 6.19 Optimal group sizes may not be stable. In this example individual fitness is maximal at a group size of seven, but a new arrival would do better to join this group than to be solitary because individual fitness in a group of eight is higher than in a group of one. Further individuals should continue to join until the group size is 14. Only after this would the next newcomer do better alone. After Sibly (1983). With permission from Elsevier.

Individual differences in a group

Skew Theory

Fig. 6.19 considers average individual fitness in groups of different sizes. However, in most groups there will be individual differences in the net benefit from group living. In foraging flocks, for example, individuals at the front of the group may do better than those at the back (Major, 1978), and when a predator arrives individuals who can get to the centre of the group will be safer than those at the edge. Often older, larger or more experienced individuals will commandeer the best positions and force others to settle for poorer places in the group. In theory, subordinate individuals will put up with lower pay-offs so long as they could not do better by moving elsewhere. This idea forms the basis of what has come to be known as 'skew theory' (Vehrencamp, 1983).

Skew models consider the effects of group size on individual reproductive success, which will be the outcome of all the potential costs and benefits of group living, including foraging and protection from predators. One possibility is that dominant individuals have complete control of reproduction in the group. In this case, if a subordinate tries to take too great a share of the reproduction, to the detriment of the dominant, then the dominant can do something about it, such as evict the subordinate from the group. As a consequence, subordinates may restrain their reproduction to avoid eviction by the dominant. Alternatively, if dominants do not have complete control, there may be 'tug of war' conflicts over reproductive shares.

Individual differences in benefits: who is in control of skew?

This is a topic where empirical studies have lagged behind a proliferation of theoretical models (Keller & Reeve, 1994; Reeve, 2000; Johnstone, 2000; Kokko, 2003). We shall consider two studies of how grouping is maintained despite a skew in benefits within the group.

(a)

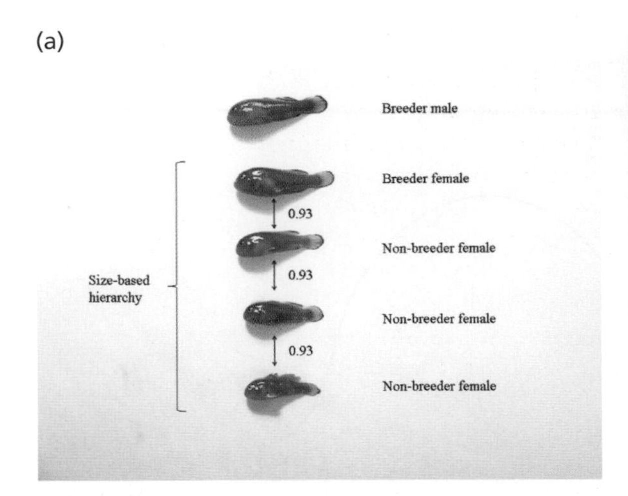

(b)

Fig. 6.20 Reproductive skew. (a) Size hierarchy in a group of the coral-dwelling goby, *Paragobiodon xanthosomus*. These individuals were anaesthetised to enable this photo to show the size ratios within the group. Subordinates restrain their growth to avoid being evicted by the dominants. Photo © Marian Wong. (b) A banded mongoose group. Dominant females evict pregnant subordinates when the group size exceeds the optimum from the dominant's point of view. Photo © Hazel Nichols.

Queuing by size in groups of coral reef fish

Coral reef fish often live in groups, with larger individuals dominant over smaller individuals. In many species, this dominance hierarchy forms a social queue; only the largest individuals in the group breed and when they die, or are removed experimentally, the next largest individuals take over as breeders. It might be thought that subordinates grow more slowly simply because they are out-competed by the dominants for food. However, in both clownfish, *Amphiprion percula* (Buston, 2003), and the coral-dwelling goby, *Paragobiodon xanthosomus* (Wong *et al.*, 2007), the size difference between individuals adjacent in rank is more regularly stepped than would be expected by chance. In the gobies, subordinates tend to be 90–95% of the length of the next-sized fish in the hierarchy, and as they reach this size threshold their growth rate usually declines (Fig. 6.20a). In laboratory experiments with artificially constructed groups kept in aquaria, it was found that subordinates were never evicted from the group by their immediate dominant when their size ratio was less than 0.95, but above this ratio the dominant was increasingly likely to evict them.

Subordinates may restrain their growth to avoid eviction by dominants

These observations suggest that subordinates often restrict their growth so they can continue to remain in the group and await a breeding vacancy without risk of eviction. Wong *et al.* (2008) tested this by offering subordinate gobies supplemental food. Some refused the food, despite no feeding interference from dominants. They remained small and were tolerated in the group. Others continued to eat, grew beyond the 95% threshold ratio and were evicted by the dominant. Therefore, subordinates have to choose between restraint, with the benefits of group living while they wait to become a breeder, or rapid growth, which may hasten their breeding chances but at the risk they will be evicted from the group.

Banded mongooses

Banded mongooses, *Mungos mungo*, live in mixed-sex groups of 8–40 adults. Each group contains one to five older, dominant females and varying numbers of young, subordinate females (Fig. 6.20b). In a study in Uganda, Michael Cant and colleagues (2010) found that a dominant female had greatest individual reproductive success in a group of four to five breeding females. Beyond this, pup survival declined, most likely because of competition for food. There was no evidence for reproductive restraint by subordinates; all females in the group entered oestrus around the same time and mated. However, dominant females were increasingly likely to evict pregnant subordinates when there were more than five breeding females in the group. Some of these evicted subordinates left for good, while others aborted their litters and were then re-accepted in the group. This option of abortion and re-acceptance may explain why subordinates showed no initial pre-emptive restraint.

Dominants may not be able to control subordinate reproduction

The main message from the fish hierarchies and mongooses is that we need to assess individual pay-offs from group living and to consider how grouping can be stable despite a skew in benefits within the group. In some cases, dominant individuals can control subordinates by threat (of attack or eviction) or by action (egg eating, infanticide, interference with mating, aggression). In other cases, the conflict is not resolved and subordinates fight back, so groups may not be stable. For example, in banded mongooses there are sometimes mass evictions or departures of subordinates and, hence, large fluctuations in group size (Cant *et al.*, 2010).

Group decision making

Local rules and self-organized groups

We now turn from the costs and benefits of grouping to a discussion of the mechanisms involved in group movements. The complex, coordinated movements of bird flocks and fish shoals have been a source of wonder to human observers ever since ancient times. Early philosophers imagined some unseen leader must be controlling the group, or perhaps coordination was achieved by 'thought transference' or other mystical powers. Recent work has shown how complex group movements and leadership can emerge from individuals adopting simple local movement rules (Couzin & Krause, 2003).

Traffic lanes in humans and ants

Local rules may lead to complex group movements

Consider a crowd of humans moving in opposite directions along a street or corridor. When no-one else is nearby, an individual will accelerate to a desired speed and orientate towards its destination. If it meets another individual head on, it slows down and moves aside to avoid collision. At first, this occurs frequently but then individuals who begin to follow others find that they are less likely to have to perform time-wasting manoeuvres and they, in turn, protect others who follow them. These local rules produce a crowd which soon self-organizes into lanes; when new individuals arrive, they tend to fall behind others who move in 'their' direction.

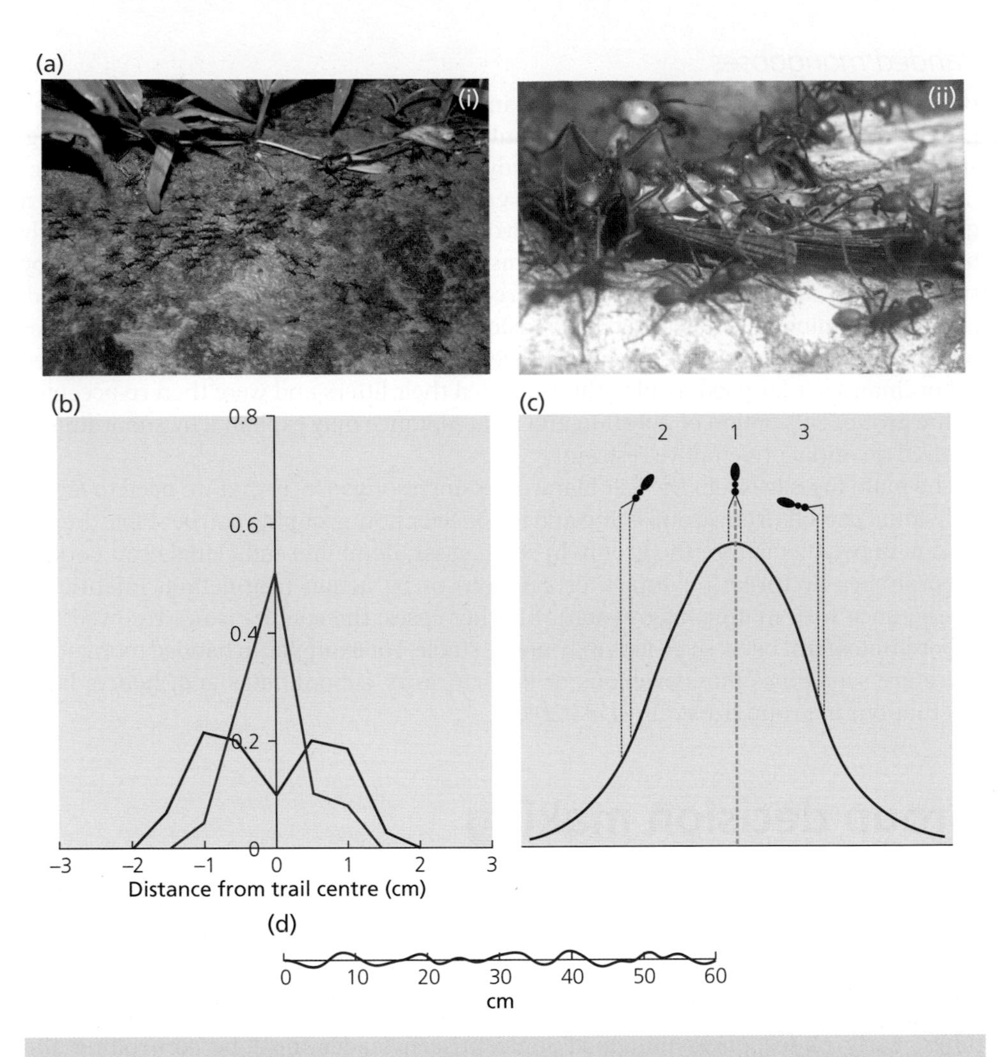

Fig. 6.21 (a) (i) A colony of army ants *Eciton burchelli* on the trail. Photo © Stefanie Berghoff. (ii) Some are taking a large insect prey back to the nest. Photo © Nigel Franks. (b) Ants returning to the nest with food (red curve) tend to occupy the centre of the trail with outbound foragers (blue curve) on the periphery, either side. (c) Individual ants detect the pheromone gradient (shown as a normal distribution about the centre of the trail) with the tips of their two antennae, moving towards the side with the strongest concentration. (d) The result is a straight line movement along the trail or, with some error in detection of pheromone concentration, a sinusoidal trail around the centre. From Couzin and Franks (2003).

Ian Couzin and Nigel Franks (2003) have studied similar traffic lanes in New World army ants *Eciton burchelli*. Colonies of this species have up to 200 000 foragers and form spectacular trails up to 20 m wide and 100 m long, with ants leaving the nest predominantly using both margins of the trail, and ants returning to the nest with arthropod prey using the centre (Fig. 6.21a,b). Couzin and Franks showed that this remarkable group organization arises from three simple movement rules followed by individuals:

(i) Ants follow pheromone trails, which they detect with the tips of their two antennae, moving towards the side with the strongest pheromone concentration (Fig. 6.21c). When the ant is near the middle of the trail, there will be only a slight difference in concentration between the antennae. As it leaves the middle, the concentration gradient increases and the ant makes a turn back to the centre again. The result is that it follows the trail centre in a straight line if there is no error in pheromone detection, or with a sinusoidal trajectory if there is greater error when the difference in concentration between the antennae is small (Fig. 6.21d).

(ii) Individuals have a preferred direction: away from the colony when searching for food, back home when they have found a prey item.

(iii) Individuals turn to avoid collisions with other ants and outbound ants have a higher avoidance turning rate than inbound ants. This may arise simply because inbound ants are less manoeuvrable because they are burdened with prey.

Computer simulations of ants adopting these three rules, with parameters estimated from individual army ant behaviour, resulted in the emergence of distinct spatial structuring on the trail, with returning ants in the centre and outbound ants on the periphery either side, just as observed in nature. Therefore, local rules can influence traffic organization on a large spatial scale, reducing the number of collisions between ants moving in different directions and increasing the efficient flow of ants and food.

Army ants: traffic lanes as an outcome of three simple rules for individuals

Local rules in fish shoals

Complex three-dimensional movements in fish shoals can also emerge from local decision making by individuals (Couzin *et al.*, 2002). Imagine an individual fish has three behavioural zones: a zone of repulsion, a zone of orientation and a zone of attraction, as explained in Fig. 6.22a. Computer simulations show that groups will form three distinctive collective organizations simply as a result of changes in these zones, and that the transitions between the three organizations is sharp. If individuals exhibit attraction to others but little or no orientation, the group forms into a 'swarm', with individuals moving in many different directions (Fig. 6.22b). As the size of the zone of orientation increases, the group forms a 'torus', moving around an empty core (Fig. 6.22c). As the zone of orientation increases further still, the group begins to move in a single direction (Fig. 6.22d). All these patterns are seen in real fish shoals (Fig. 6.22e-g). The key point is that in theory large changes in group behaviour can emerge simply from minor changes in an individual's local response to neighbours.

Large scale changes in shoaling as an outcome of changes in individual local responses ...

In further simulations Couzin *et al.* (2002) added a model predator which moved towards the highest perceived density of prey. The prey now had a further rule: 'if detect a predator, move away'. This led to the kind of collective behaviour often seen in real groups, such as waves of escape spreading through the shoal (the Trafalgar effect), fragmentation of the shoal or sudden 'flash expansions' in all directions.

... and changes in group size

Finally, in theory simple changes in an individual's local responses could also result in changes in group size. Hoare *et al.* (2004) found that group size in the banded killifish *Fundulus diaphanus* varied as might be predicted from our previous discussion of costs and benefits. When broken skin extract from an injured fish was added to the aquarium, the fish formed larger shoals; when food odour was added, the fish tended to swim

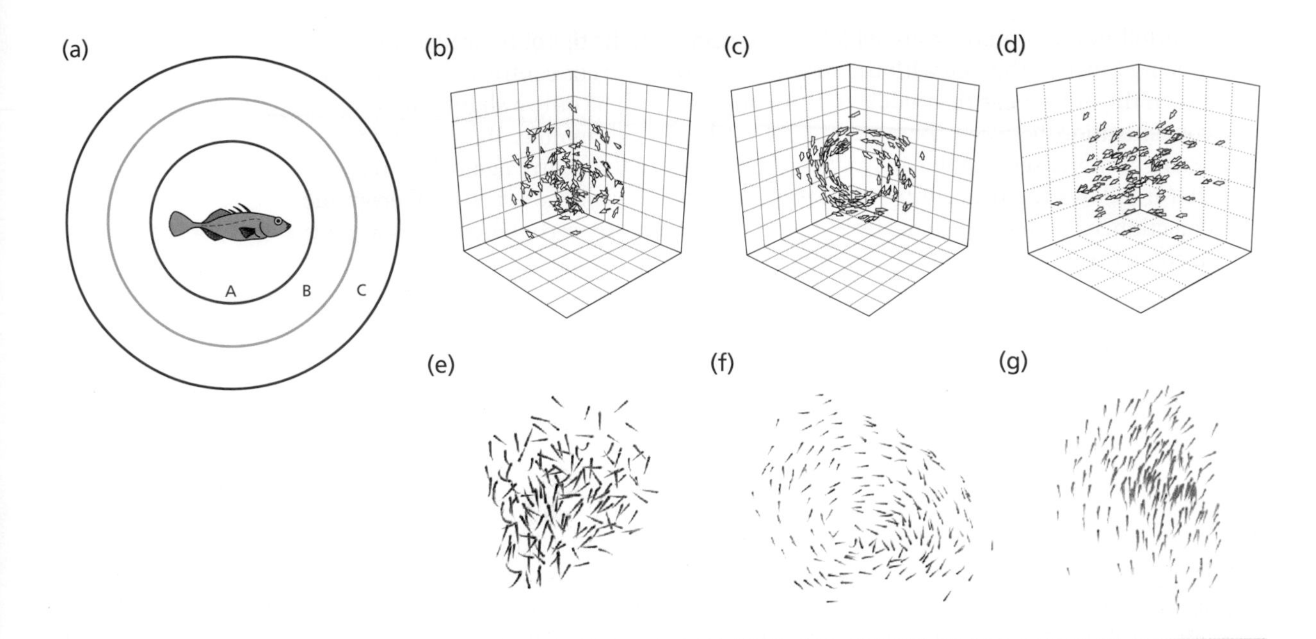

Fig. 6.22 (a) Three behavioural zones of an individual fish: A is a zone of repulsion (here individuals move away from others to avoid collision). B is a zone of orientation (here individuals align with the movement of others). C is a zone of attraction (here individuals move towards others). Computer simulations show that different group structures can emerge simply as a result of changes in these zones. (b) With little or no orientation zone, a swarm forms. (c) As the size of the orientation zone increases, the group forms a torus. (d) As the zone of orientation increases still further, the group begins to move in one direction. From Couzin *et al.* 2002. (e) (f) (g): The three states in golden shinerfish *Notemigonus crysoleucas*. These are the only states observed in many hours of tracking these fish. Photos courtesy of Iain D. Couzin.

around as individuals; and when both food and skin extract were introduced together, shoals were intermediate in size. Fig. 6.23a summarizes these results in terms of median group size. Do individuals vary group size by counting their companions? Not necessarily. Different grouping may emerge as the outcomes of variation in local interactions. Hoare *et al.* simulated the consequences of varying the radius of the zones of alignment and attraction in Fig. 6.22a. They hypothesized that food odours would decrease these zones (individuals less attracted to neighbours) while signals of a predator would increase them (more attraction to neighbours). Simulations in which the zones were varied in this way gave an excellent fit not only to the median observed group sizes in the different treatments, but also to the variation in group sizes (Fig. 6.22b).

Future work needs to investigate the local rules actually used by individual fish to test whether these predicted group consequences follow from real behaviour.

Leadership and voting

Leaders and followers

Consider a group of individuals, each with the three behavioural zones depicted in Fig. 6.22a. Now imagine that a small proportion has information about a preferred direction, for example to a food source. All the others are naïve, with no preferred direction, and

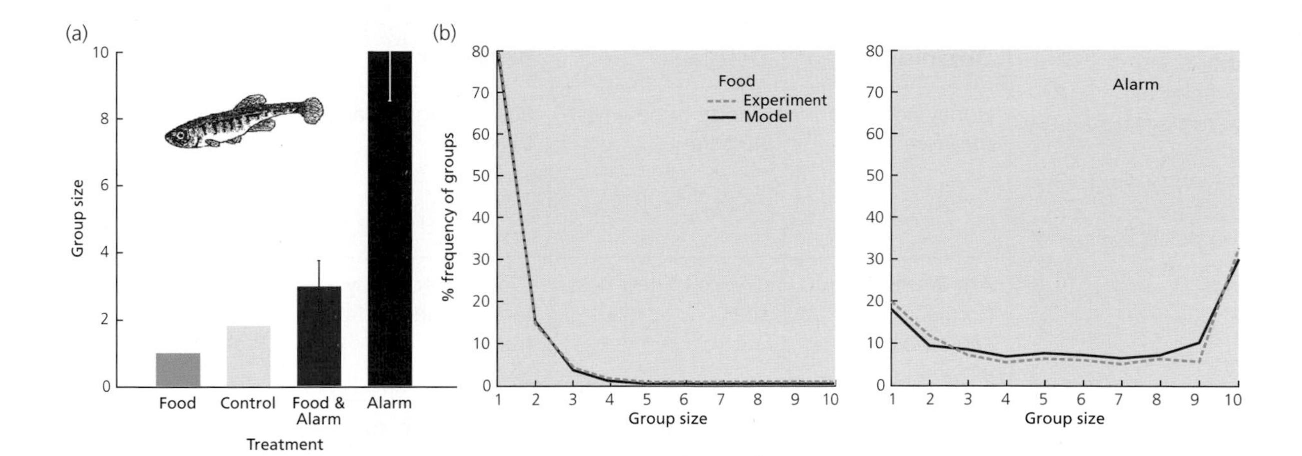

Fig. 6.23 Shoal size in banded killifish. (a) Median group size in experiments where the treatments were: food odour added to aquarium, control, food odour and alarm odour (extract of broken skin of killifish), and alarm odour only. (b) The variation in observed group sizes across two of the treatments, compared with outcomes of a simulation which varied an individual's response to neighbours (see text). From Hoare *et al.* (2004). With permission from Elsevier.

they also do not know which are the informed individuals in the group. Computer simulations reveal that even a small proportion of knowledgeable individuals (less than 5% in a group of 200) can influence the direction of movement of the whole group and guide everyone to a target location. Therefore, group decision making can occur without explicit signals or complex communication (Couzin *et al.*, 2005).

Experiments with groups of humans provide evidence for this effect (Dyer *et al.*, 2008). Some individuals (naïve) were given a piece of paper with the instruction 'stay with the group'; others were given the instruction 'move to a specific target'. The results showed that a small informed minority could guide a group of naïve individuals to the target without verbal communication or obvious signalling. When informed individuals had conflicting information about the target, then the majority dictated the group direction.

A small informed minority can lead the group

In animal groups, too, individuals who move in specific directions to resources, for example because they are hungrier (Rands *et al.*, 2003) or bolder than the rest (Harcourt *et al.*, 2009), become leaders of the group's movements.

Voting

In some cases groups reach a consensus about when and where to go by opinion polling (voting), without any control by leaders (Conradt & Roper, 2005). The pooling of information is common in human groups. Francis Galton (1822–1911), mathematician and cousin of Charles Darwin, attended a cattle fair in which people placed wagers on the weight of an ox. Galton noted down all the wagers (nearly 800). Individual wagers varied widely but the overall average was remarkably close to the true weight of the ox. Therefore, pooling information can reduce extreme errors and result in the best overall solution.

Table 6.3 Examples of opinion polling and consensus decision making in groups (Conradt & Roper, 2005).

Organism	Decision	Behaviour	References
Honey bee *Apis mellifera*	Choice of new nest nest site	Scouts waggle dance to recruit	Visscher & Camazine (1999) Seeley & Buhrman (2001) Seeley (2003)
Ant *Temnothorax* (formerly *Leptothorax*) *albipennis*	Choice of new nest site	Workers recruit by tandem runs and then transport	Franks *et al.* (2002) Pratt *et al.* (2002) Pratt (2005)
Whooper swan *Cygnus cygnus*	When to move after a rest	Head movements	Black (1988)
Mountain gorilla *Gorilla gorilla beringei*	When to move after a rest	Calls ('grunts') signalling impending departure	Stewart & Harcourt (1994)
African buffalo *Syncercus caffer*	Which direction to move after a rest	Stand up and gaze in a particular direction.	Prins (1996)

Buffalo herds: voting when and where to go

Animal groups also sometimes make decisions based on how individuals vote (Table 6.3). Herbert Prins (1996) studied African buffalo in Manyara, Tanzania. Females and their offspring lived in large, stable herds of several hundred individuals. During the heat of the day the herd usually lay down to rest, and then towards dusk they all went off as a group to feed in one place. How does the herd decide where to go? Before the herd sets off, individuals occasionally stand up and then lie down again. At first sight, this seems to be individuals simply stretching their legs. However, Prins noticed that individuals adopted a special posture when they stood up; they gazed steadily in one direction and held their head up high. He took note of the directions of gaze and discovered that the herd's eventual direction of movement could be predicted by the vector of the individuals' gaze directions. He suggested that individuals were voting and that the group moved off once there was a consensus about where to go.

Ants: voting for a new nest site

The best evidence for voting comes from studies of how social insect colonies choose a new nest site (Franks *et al.*, 2002; Table 6.3). The ant *Temnothorax* (formerly *Leptothorax*) *albipennis* lives in small colonies (less than 500 workers) that nest in rock crevices. When a crevice crumbles, the colony has to find a new nest site. These ants are tiny (workers are 2.5 mm long) and the area they search is just a few square metres. When a scout finds a potential new site, it returns to the colony and begins to lead other ants to the site, one by one, by tandem running. This involves walking ahead of the follower, who signals its continued presence by tapping the leader's body with its antennae from time to time (Fig. 6.24a). If the recruit approves of the new site it, in turn, will lead further recruits to the site.

During this search phase, several sites may be inspected and recruits may build up and then decline at various sites depending on how individuals assess their suitability.

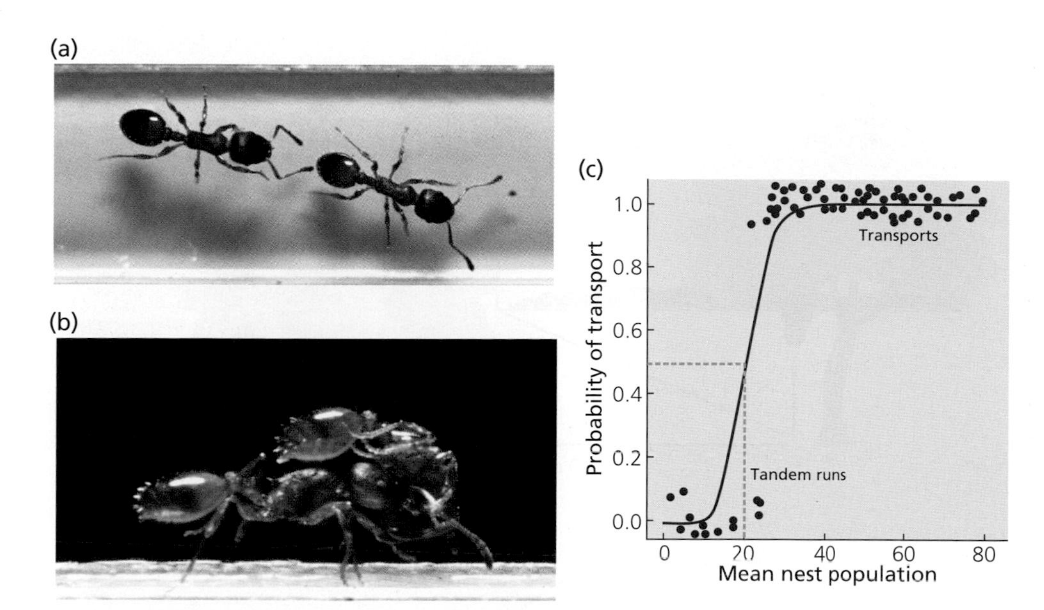

Fig. 6.24 Opinion polling in the ant *Temnothorax albipennis*. (a) Recruitment to potential new nest sites occurs first by tandem running, with the follower signalling its presence by tapping the body of the leader with its antennae. (b) After a sufficient number of workers has voted for a new site, there is more rapid recruitment involving transport; one worker simply carries another one to the site. Photos from Franks *et al.* (2002). (c) In one study, the switch from tandem runs to transport occurred when there were 20 recruits at a new site. From Pratt (2005).

Eventually a critical quorum of individuals builds up at the new site, at which there is a switch in recruitment method from slow tandem running ('follow me') to quicker transport, where others are simply picked up and carried to the new site (Fig. 6.24b). In one laboratory study, where colonies nested between glass plates, the switch to rapid transport occurred when about 20 workers had assembled at a new site (Fig. 6.24c).

An optimum quorum size

This method of slow recruitment at first is a case of 'adaptive procrastination'; it allows the colony to inspect several potential sites before sufficient votes at one site leads to a final decision. In theory, the optimal quorum size will reflect a trade-off between speed and accuracy in decision making. Experiments show that the quorum size for the switch to rapid transport to a new site is lower when conditions have deteriorated rapidly at the old nest, so that a quicker decision to move is needed (Franks *et al.*, 2003).

Voting for a new nest site in honeybees: scouts lead the way

Honeybees, *Apis mellifera*, also set off in search for new nest sites, but their search is on a grander scale (references in Table 6.3). A queen and some 15 000 loyal workers wait in a swarm on the branch of a tree while scouts search up to 10 km away for a new home. The scouts assess potential cavities and then return to the swarm and waggle dance on the surface to advertise the new site's direction and distance (Fig. 6.25). The vigour of the dance signals their enthusiasm for the new site. After an extended period of signalling, in which many scouts may dance for many potential sites, a consensus is

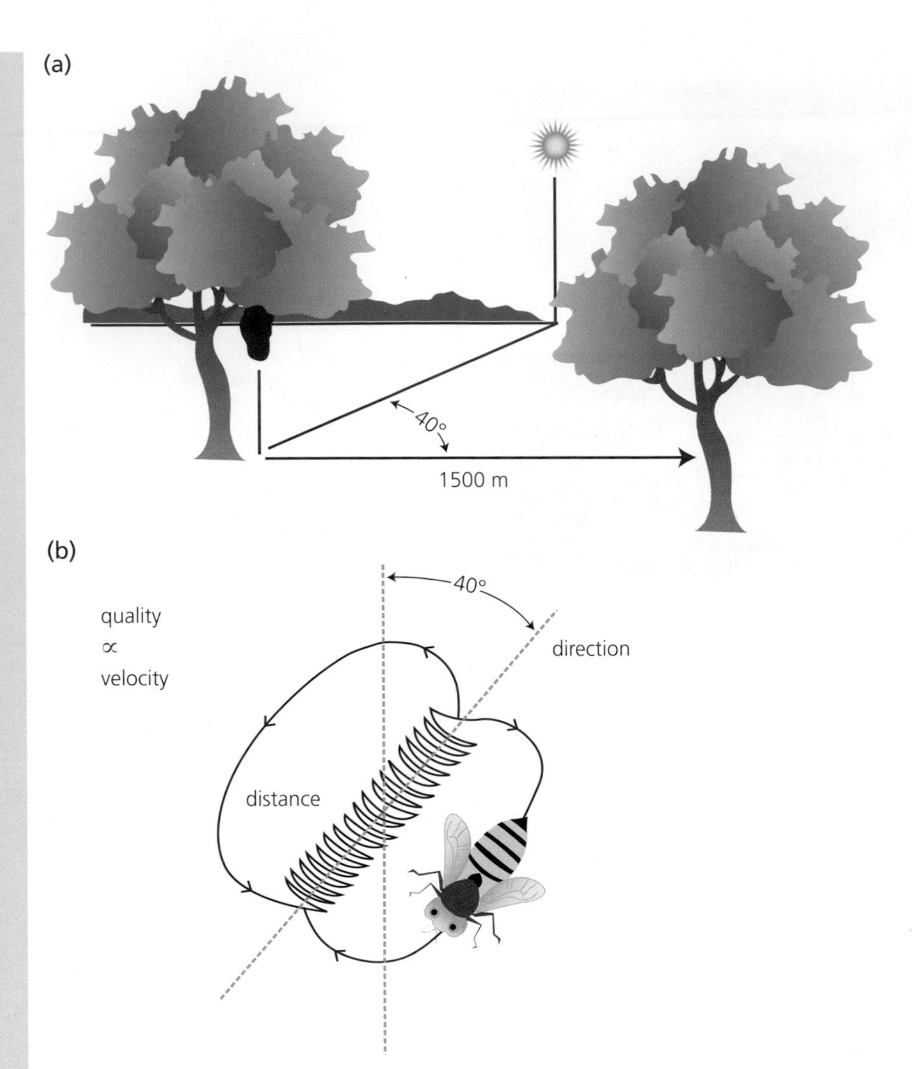

Fig. 6.25 Recruitment to a new nest site by the waggle dance of the honeybee. (a) A scout from a swarm waiting in the left-hand tree has found a potential new nest site in the tree on the right, 1500 m away and 40° clockwise from the sun. (b) The scout performs a waggle dance on the surface of the swarm. She signals the direction of the new nest site by the angle of the waggle run from the vertical, the distance by the duration of the straight waggle run and the quality of the site by the speed with which she repeats the waggle run. From Franks *et al.* (2002), after Seeley (1995).

reached and just one site is danced for vigorously. Colony consensus occurs through scouts examining the dances of other bees and switches in allegiance from one site to another. It is not yet known what determines when voting ceases and a final decision is made; this may be when there is a critical threshold of scouts signalling one direction back at the swarm, or of scouts assembling at a new site. However, once the decision is made the swarm speeds off in tight formation like a missile, with the scouts (about 5% of the swarm) guiding the rest. Experiments reveal that scouts whose Nasanov glands are sealed (and which, therefore, cannot release pheromones) can still guide the swarm just as effectively to a new site, so the cue is a visual one (Beekman *et al.*, 2006). The scouts fly fast through the swarm and the rest follow, presumably using local rules, like those discussed for fish shoaling.

Summary

Grouping can reduce individual predation in several ways: by diluting the risk of attack (water skaters, mayflies), by reducing an individual's domain of danger (selfish herd: seals attacked by sharks), by confusing predator attacks (fish shoals), by communal defence (gull and guillemot colonies) and by improved vigilance for predators (bird flocks). Within groups, individuals behave in their own selfish interests. When hungry, they may go to the front of the group where they will encounter food first, while when satiated they may seek safety in the centre (selfish herd effect).

Individual vigilance levels are maintained in larger groups because more vigilant individuals are less likely to be attacked (gazelles) and they can flee to cover faster, because they are sooner to spot the predator or their alarmed companions. In some groups, satiated individuals may benefit from acting as sentinels (meerkats, babblers).

Grouping can also improve foraging through better food finding (information centres; e.g. raven roosts) or better prey capture, either through coordinated group hunting (lions) or through scattering prey groups (gulls, some predatory fish). A study of guppies in Trinidad has revealed the evolution of reduced and then increased shoaling tendency in response to changing predation pressure over a 30-year period.

In theory, there may be an optimal group size but this may not be stable. There is often 'skew' in the net benefits to individuals from grouping. Subordinates may restrain their resource intake to avoid being evicted by dominants (fish size hierarchies). Dominants may evict subordinates who try to increase their share of resources (banded mongooses).

Complex group movements emerge from individuals adopting simple local movement rules in response to neighbours (e.g. traffic lanes in army ants, fish shoaling). Leadership may emerge if a small proportion of the group has a preferred direction of movement. A group may move after individuals have voted for their preferred choice (e.g. house hunting in ants and bees).

Further reading

Krause and Ruxton (2002) is an excellent review of the benefits and costs of group living. Caro (2005) reviews anti predator defences in birds and mammals, including grouping. Cresswell (1994) is a fine field study of raptors hunting flocks of redshank. Giraldeau and Caraco (2000) review models and empirical studies of social foraging. Lima (1998) considers individual decision making under the risk of predation.

Packer *et al.* (1990) and Mosser and Packer (2009) show that a major benefit of lion sociality is through joint defence of a high quality territory versus other lion prides.

Cant (2011) reviews how reproductive conflict and skew in social groups might be resolved by threats from dominants (attack, eviction), action by dominants (infanticide, interference with mating, aggression) or by threats or action by subordinates in the group (departure, interference). Cant and Johnstone (2009) model how these threats to exercising alternative options (leaving the group or evicting others from the group) might influence conflict resolution. Hamilton (2000) uses skew theory to predict how recruiters and joiners should behave in groups. Radford and Ridley (2008) show how calling regulates spacing and food competition in babbler flocks.

Couzin and Krause (2003) review how group movements arise from local individual decision making. Conradt and Roper (2005, 2007) discuss how groups reach a consensus and King *et al.* (2009) review how leadership emerges in groups. King *et al.* (2008) is a field study of baboons, revealing how subordinates follow dominants even when they suffer short-term foraging costs. Franks *et al.* (2002) review opinion polling and house hunting in social insects.

TOPICS FOR DISCUSSION

1. Devise an experiment which would provide a convincing test of the 'confusion effect'.
2. How would you test the hypothesis that groups are of stable rather than optimal size?
3. In fish size hierarchies, when would you expect subordinates to grow to challenge dominants rather than wait for a breeding vacancy?
4. How would you test whether individuals really have the three behavioural zones in Fig. 6.22a?
5. Discuss how individual specializations in hunting roles, as in lions, might develop in a group.

Chapter 7

Sexual Selection, Sperm Competition and Sexual Conflict

Photo © Joah Madden

In the second part of *'The Descent of Man and Selection in Relation to Sex'*, Charles Darwin (1871) proposed his theory of 'Sexual Selection'. He had been puzzling over the extravagant traits which often occurred in one sex only, usually the males. Why, for example, is it only male kudu (an African antelope) that have enormous horns, and only male birds of paradise that have such remarkable, ornamented plumage (Fig. 7.1)? Darwin argued that these structures could not be essential for survival, otherwise surely the females would have them too. Instead, he proposed that these traits had evolved simply because they were of advantage in competition for mates, a process he called 'Sexual Selection'.

Darwin's theory of sexual selection: selection for traits that increase mating success

Darwin suggested that: 'the sexual struggle is of two kinds; in the one it is between the individuals of the same sex, generally the males, in order to drive away or kill their rivals, the females remaining passive; whilst in the other, the struggle is likewise between the individuals of the same sex, in order to excite or charm those of the opposite sex, generally the females, which no longer remain passive, but select the more agreeable partners'. The first process, where victory goes to the more powerful competitor, might explain the evolution of male weapons (such as kudu horns) and other male attributes (e.g. large size and strength) which give a male an advantage in direct combat with rival males. The second process, where victory goes to the most charming competitor, might explain the evolution of male ornaments (such as those of a bird of paradise). Somewhat confusingly, the first process is now often referred to as 'intrasexual selection' or 'male–male competition', and the second as 'intersexual selection' or 'female choice'. This is confusing because, as Darwin recognized, both processes involve intrasexual competition, in the first case to win mates by force and in the second to win them by charm.

Competing for mates by force or by charm

Before examining the evidence for sexual selection, and its consequences, we need to consider why it is usually the males who compete for females, rather than vice versa. This takes us right back to the beginning, to the fundamental differences between male and female.

An Introduction to Behavioural Ecology, Fourth Edition. Nicholas B. Davies, John R. Krebs and Stuart A. West.

Fig. 7.1 Darwin's theory of sexual selection was proposed to explain the evolution of traits, usually found in males, concerned with competition for mates, either by force or by charm. (a) Horns of the male kudu *Tragelaphus strepsiceros*. Photo © Oliver Krüger. (b) Ornaments of the male Raggiana bird of paradise *Paradisaea raggiana*, left, on display to a female, right. Photo © Tim Laman/naturepl.com

Males and females

Differences in gamete size

Females: the sex that produces large gametes

Sexual reproduction entails gamete formation by meiosis and the fusion of genetic material from two individuals. It almost always, but not invariably, involves two sexes called male and female. In higher animals the sexes are often most readily distinguished by external features such as genitalia, plumage, size or colour, but these are not the fundamental differences. In all plants and animals the fundamental difference between the sexes is the size of their gametes: females produce large, immobile, food-rich gametes called eggs, while male gametes or sperm are tiny, mobile and consist of little more than a piece of self-propelled DNA. Sexual reproduction without males and females occurs in many simple, unicellular organisms, such as *Paramecium*, where the 'gametes' which fuse during sex are of the same size. This is referred to as *isogamous* sexual reproduction. The fusion of two gametes of unequal size, one large and one small is, however, much commoner and occurs in virtually all sexually reproducing multicellular plants and animals. It is called *anisogamous* sex.

Evolution of anisogamy

Parker *et al.* (1972) proposed an elegant model to explain how anisogamy might have evolved from isogamy. Assume that the survival of a zygote depends on its size; the larger the zygote the more food to sustain its development and so the better its chance of surviving. Larger gametes would be favoured by selection if this increase in survival more than compensated for the fact that larger gametes could be produced in smaller numbers. For example, if an individual's resources could be divided into two large or four small gametes, then the zygote produced by large gametes would have to be more than twice as good at surviving for large gametes to be favoured by selection.

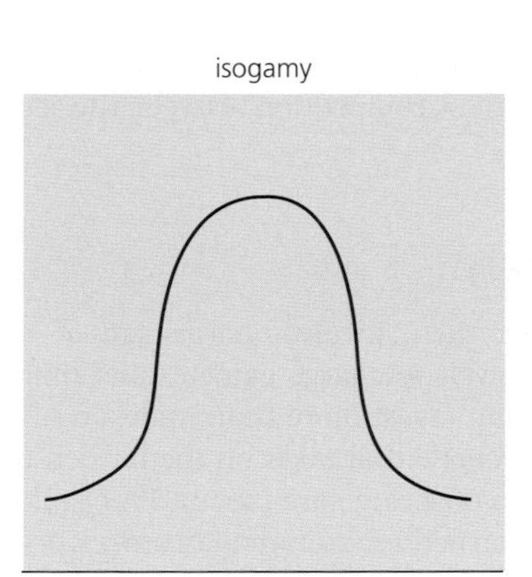

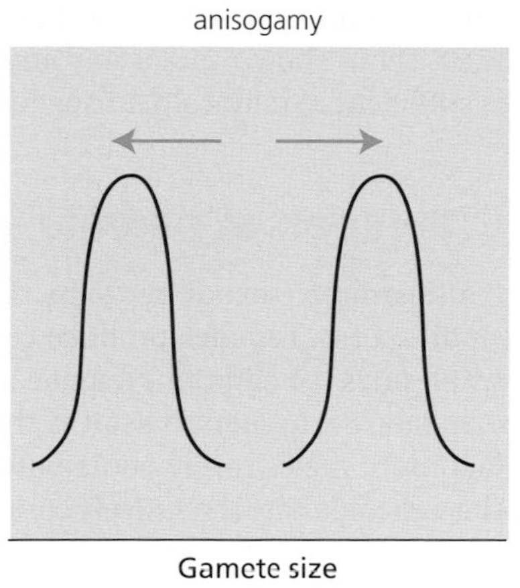

Fig. 7.2 Parker *et al.* (1972) proposed that anisogamy evolved from isogamy by disruptive selection for two gamete sizes: large gametes with food reserves (eggs) and small gametes (sperm) which parasitize the investment of the large gametes.

Now, once larger gametes evolve there is immediately selection on smaller gametes to seek out larger gametes to fuse with, in order to parasitize their food reserves. At the same time, larger gametes will be selected to resist the small ones since the most viable zygote of all will be one resulting from the fusion of two large gametes. However, fusion between small and large gametes will predominate, simply because small gametes are produced in larger numbers. Furthermore, the penalty for a large gamete failing to resist a small gamete will not be so severe as that paid by a small gamete if it fails to find a large partner; a medium sized zygote has a chance of survival, a small one has none. This means that selection acts more strongly on the small gametes. Also, because smaller gametes are made in larger numbers, they are likely to occur in a greater variety of genotypes, and suffer a higher rate of mortality (both a consequence of numerical superiority). Therefore, they will evolve faster and during evolutionary time will 'outwit' the defences of the larger gametes.

Small gametes parasitize the investment of large gametes

Parker *et al.* show that, in theory, the end result of this ancient evolutionary arms race will be individuals specializing on either the production of many, small gametes (males) or the production of fewer, large gametes (females). Producers of intermediate sized gametes lost out because they enjoyed neither the advantage of great numbers nor the benefits of large food reserves for the zygote (Fig. 7.2).

One of the basic assumptions of Parker *et al.*'s model is supported by a comparative survey of the family of algae called Volvocales. Unicellular genera, in which food reserves for zygote growth are relatively unimportant, tend to be isogamous, while multicellular genera, where food reserves play a role in zygote survival, are mostly anisogamous (Knowlton, 1974); other families of algae show similar, but less clear trends (Bell, 1978).

Sperm competition maintains anisogamy

A primordial form of gamete competition may, therefore, account for why there are two gamete-producing morphs, namely males and females. Gamete competition can also explain why the two sexes are maintained (Parker, 1982). Consider a cow's ovum, which is roughly 20 000 times the size of a bull's sperm. Now imagine a mutant bull which doubles its sperm size but with the consequence that its sperm numbers are halved (because resources are limited). The larger sperm would increase the cytoplasmic reserves of the zygote by just one unit in 20 000, which is a trivial benefit. There would, however, be a huge cost whenever two bulls mated with the same cow because the mutant bull's relative fertilization success drops from one half to one third. Parker (1982) has shown that a tiny amount of competition between the sperm of rival males is sufficient to maintain anisogamy.

Differences in parental care

Anisogamous sexual reproduction, then, involves parasitism of a large egg by a small sperm. Females produce relatively few large gametes and males produce many small ones. In addition, females often invest more than males in other forms of care (Chapter 8). In mammals, it is the female that takes on the burden of pregnancy and lactation; males rarely contribute to offspring care (about 5% of all mammal species, though male care is relatively common in three mammalian orders: primates, carnivores and perissodactyls = the odd-toed ungulates). In birds, bi-parental care is the norm but females often invest more in care. In other taxa, when parental care occurs female care is commoner in both reptiles and invertebrates, male and female care are about equally common in amphibians, while only in fish is male care more common than female care. In fish, male care is particularly common in territorial species where males can continue to attract new mates while guarding eggs, so male care involves fewer opportunity costs (lost matings) than in other taxa (Chapter 8).

Parental investment and sexual competition

The theory of Robert Trivers

Robert Trivers (1972) was the first to recognize the link between sex differences in investment in resources for gametes and other forms of care (parental investment), and sexual competition. He wrote: 'Where one sex invests considerably more than the other, members of the latter will compete among themselves to mate with members of the former'. The key point is that the sex with the least parental investment has a greater potential rate of reproduction (Clutton-Brock & Parker, 1992). Thus, in general, a male can potentially fertilize eggs at a much faster rate than a female can produce them. Even in species where males temporarily deplete their sperm supply when offered a surfeit of females, their potential for producing offspring is greater than that of females (Nakatsuru & Kramer, 1982). This means that while a female can usually best increase her reproductive success by increasing the rate of converting resources into eggs and offspring, a male can best increase his success by finding and fertilizing many different females.

Male reproductive success is often limited by access to females ...

The different effects of mating rate on reproductive success of the two sexes were neatly demonstrated by A.J. Bateman (1948) with experiments on *Drosophila* and it was his results in particular which helped to inspire Trivers's theory (Fig. 7.3). The greater reproductive potential of males is graphically demonstrated by mammals, such as man, in which a female spends many months producing a single child, during which time a male could potentially fertilize the eggs of hundreds of other mates (Table 7.1).

... whilst females are often limited by resources

As we shall show in the rest of this chapter, the asymmetry in parental investment, and consequent difference in reproductive potential of the two sexes, has far-reaching

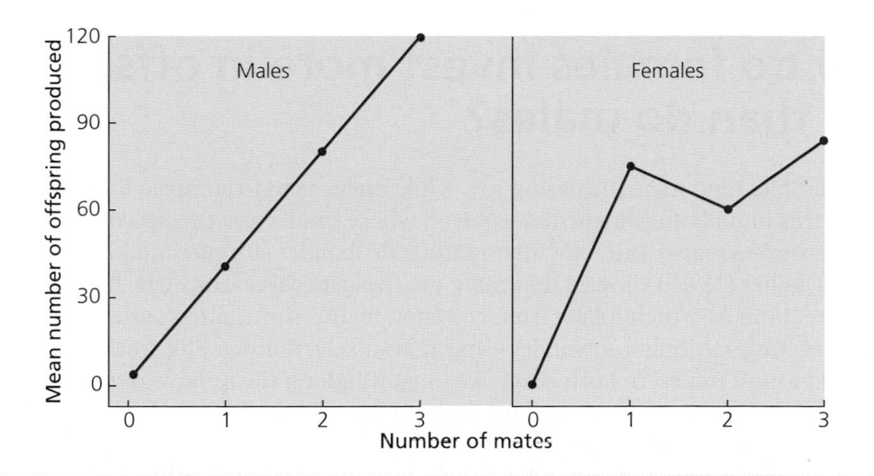

Fig 7.3 A.J. Bateman (1948) put equal numbers of male and female fruit flies (*Drosophila melanogaster*) in bottles and scored the number of matings and offspring produced by each individual, using genetic markers to assign parentage. For males reproductive success goes up with number of matings, for females it does not, beyond the first mating.

Table 7.1 In polygamous or promiscuous species some males have a much higher potential reproductive rate than females

	Maximum number of offspring produced during lifetime	
Species	**Male**	**Female**
Elephant seal	100	8
Red deer	24	14
Man	888	69
Kittiwake gull	26	28

The data for man came from the Guinness Book of Records: the male was Moulay Ismail the Bloodthirsty, Emperor of Morocco, the woman had her children in 27 pregnancies. The data for elephant seals are from Le Boeuf and Reiter (1988), for red deer from Clutton-Brock *et al.* (1982). In the monogamous kittiwake, where male and female invest similarly in each offspring, the difference in maximum reproductive output is negligible (Clutton-Brock, 1983).

consequences for sexual behaviour. Where females invest more in each offspring than do males, male courtship and mating behaviour is to a large extent directed towards competing for and exploiting female investment, while females are expected to choose those males who offer the best resources or genes.

Why do females invest more in offspring care than do males?

Clearly females tend to be the caring sex, while males tend to compete for matings. We can link this initially to gamete dimorphism, where small gametes (sperm) compete for female resources (eggs). But why, after mating, do females still care more?

David Queller (1997) showed that there are two general reasons why females should care more, both of which follow from common features of mating systems. These are females mating multiply (polyandry) and sexual selection leading to skewed mating success amongst males. In both cases, we must weigh up the relative returns to a male from putting its resources (time and energy) into either helping raise a brood, or attempting to obtain other mates.

Firstly, consider a mated pair, where there is a chance that another male (or males) has fathered some of the offspring in the brood. When the paired male has to share parentage of the brood, this reduces their genetic value to him, and so reduces the relative benefit of staying to help. Put simply, there is little benefit from staying to help raise the offspring of other males! Consequently, we would expect males to be less likely to care in species where paternity is shared between multiple males.

For males, parental care is often of less benefit because of paternity sharing ...

Secondly, consider a species in which sexual selection leads to a skewed mating success amongst males, with some males obtaining an above average share of the matings. Queller (1997) realized that no matter how much greater their *potential* reproductive rate, the *average actual* reproductive rate of males must be exactly the same as that of females, if the sex ratio is 1:1. This must be true, because each offspring has exactly one mother and one father. However, consider the subset of males and females who have succeeded in mating and so who have offspring available for care. If most or all of the females breed, while only a fraction of the males are successful (namely the strongest competitors) then the expected future reproductive success of these few successful males will indeed be greater than that of the females. So these males have more to gain by not helping and instead putting more resources into obtaining other mates. Consider, for example, a group of 20 displaying male grouse in which the largest and most vigorous male does all the mating (which will be with 20 females, if the sex ratio is 1:1). The expected reproductive success of this male is, therefore, 20 times that of a female. This male should be less inclined to care than a female because care would detract more from his potential future reproductive success than it would from her success.

... and the opportunity costs of care are often also greater for males

Therefore, the logic of Queller's (1997) second point is as follows: if females invest more prior to mating, there will be competition among males for female investment; this will lead to greater variance in male success (some males, the best competitors, will have great success while others will fail to mate); these successful males will then be less inclined to care after mating. As females provide more care, sexual selection on males intensifies, so there is positive feedback, making it even less likely that males will care (Kokko & Jennions, 2009). Box 7.1 summarizes some of the problems in measuring the strength of sexual selection.

BOX 7.1 MEASURING THE STRENGTH OF SEXUAL SELECTION (KLUG *ET AL.*, 2010)

A common measure of the maximum potential strength of directional sexual selection (the 'opportunity for sexual selection') is *Is*, which is a standardized measure of intrasexual variation in mating success, measured as the square of the coefficient of variation in mating success for a given sex (Wade, 1979; Shuster & Wade, 2003). It has often been assumed that *Is* will be determined by the sex ratio of individuals capable of mating at any one time (the 'Operational Sex Ratio', or OSR; Emlen & Oring, 1977). Males usually invest less in gametes and parental care, so they are more often available in the mating pool, competing for mates. Thus, the OSR will most often be male-biased.

The simple example in Fig. B7.1.1 shows that it is not possible to make general predictions about the relationship between the OSR and sexual selection without also knowing how the OSR influences the ability of males to monopolize mates. In A, the addition of a new male to the mating pool may lead to an increase in mate monopolization by the topmost male (B) or it may make mate monopolization more difficult (C). In D, the OSR changes without changing absolute density of competitors (a male arrives, a female leaves), again leading either to increased (E) or decreased mate monopolization (F). In the first diagram (below), *Is* behaves as predicted: it increases as mate monopolization increases (A to B) and decreases as mate monopolization decreases (A to C). In the second diagram (next page), *Is* also increases as more males remain unmated (D to E), but it increases too when mating success is as egalitarian as possible (D to F).

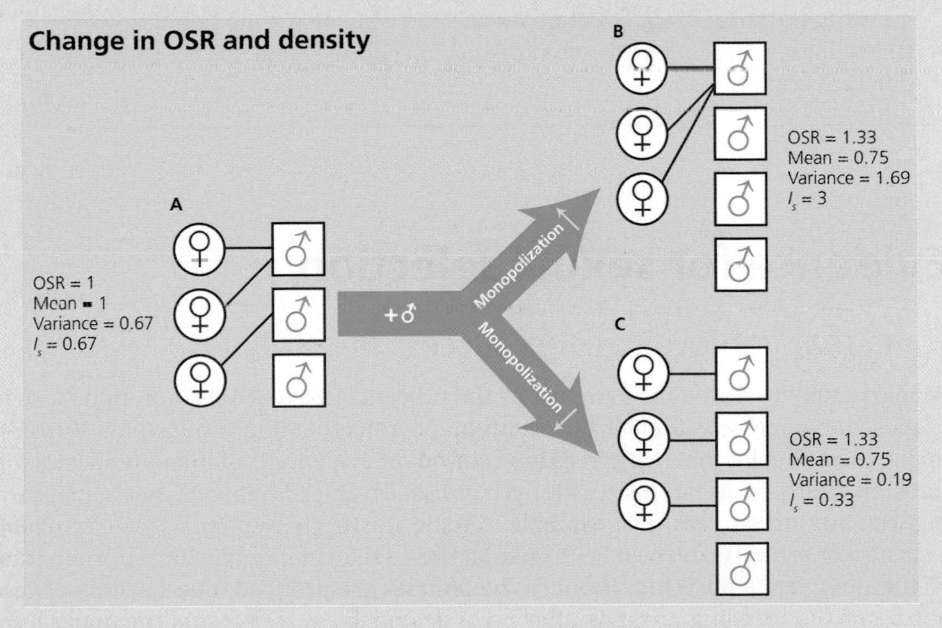

Fig B7.1.1

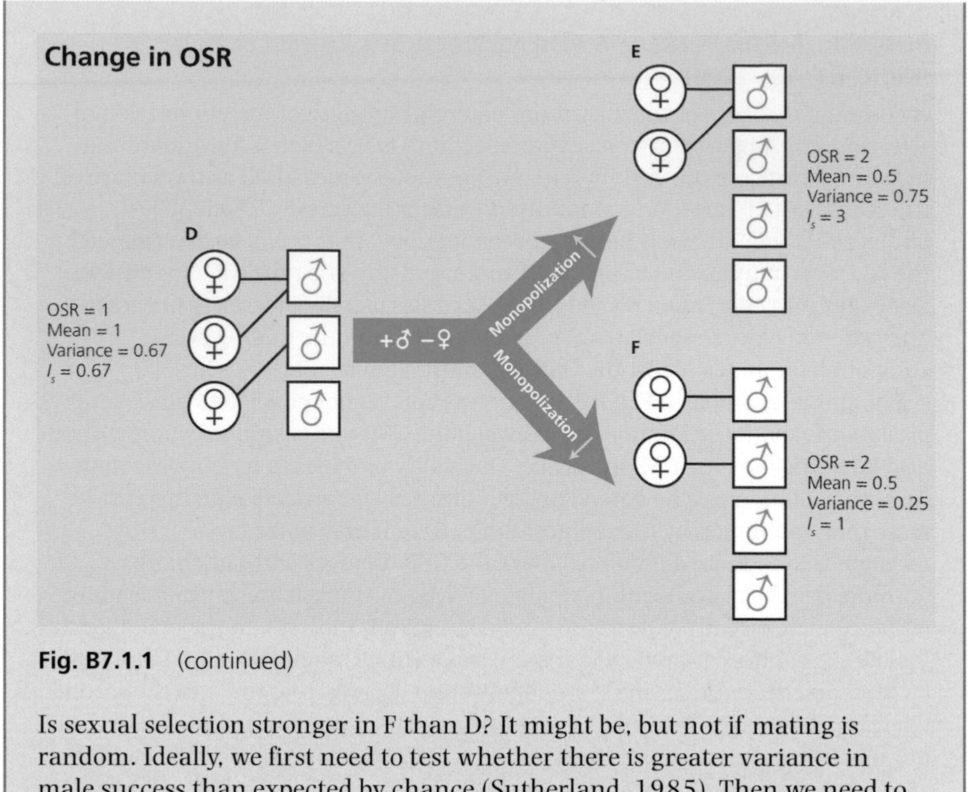

Fig. B7.1.1 (continued)

Is sexual selection stronger in F than D? It might be, but not if mating is random. Ideally, we first need to test whether there is greater variance in male success than expected by chance (Sutherland, 1985). Then we need to quantify the strength of selection on specific traits (e.g. long tails). However, this will make cross-species comparisons harder (different species may have different traits).

Evidence for sexual selection

Traits that improve a male's success in combat

People readily accepted the first part of the process of Sexual Selection that Darwin proposed, namely selection for the evolution of traits that improved a male's success in direct combat for mates. There is now abundant evidence that in many species it is the males with the greatest size, strength or best developed weapons that achieve the highest mating success. For example, female northern elephant seals, *Mirounga angustirostris*, haul up on beaches to give birth and then mate again for the production of the next year's offspring. Because the females are grouped, due to the localized nature of the breeding grounds, they are a defendable resource and the males fight with each other to monopolize harems (Fig. 7.4). The largest and strongest males win

Fig. 7.4 (a) Male southern elephant seal. (b) Two males fighting for harems of females. (c) The male is huge compared to the female. These are two subadults. Photos © Oliver Krüger.

the largest harems; in one long-term study in California by Burney Le Boeuf and Joanne Reiter (1988), each year as few as five out of 180 competing males were responsible for 48–92% of the matings with 470 females. Adult males were from three to seven and a half times as heavy as adult females. While most females first bred at four years of age, males delayed attempting to breed until 6–9 years old, building up their size and strength so they had a chance in the intense fights for females. Whereas a female's reproductive success depended on her ability to nurture offspring, a male's depended on his mating success resulting from fights and defence of harems.

Larger and stronger males often gain the most females

The evolution of sexual dimorphism leading to larger male size, strength or weaponry is, therefore, easy to explain because these traits increase a male's success in male–male combat for mates.

Female choice

The second process suggested by Darwin, namely female choice for the most 'charming or agreeable' males took longer to gain acceptance simply because Darwin had little, if any, firm evidence that females did choose. In fact, we had to wait for 90 years since Darwin first proposed his theory before anyone thought to test it experimentally. Malte Andersson's (1982) classic experiments on the long-tailed widowbird (*Euplectes progne*) in Kenya, took advantage of the invention of Superglue. The males of this sparrow-sized bird have a remarkable tail which is often more than half a metre long (females have normal, short tails). Males defend territories in grassland to which they may attract several females for nesting. Their tails are not displayed in contests with other males. However, whenever a female flies past the male performs a slow, cruising flight in which his tail is displayed and expanded into a deep keel. Could this extraordinary tail have evolved by female choice?

Malte Andersson's classic experiment with long-tailed widowbirds

Early in the breeding season, Andersson mapped the territories of 36 males and scored the number of females each had attracted. He then divided them at random into four groups. In one group, he cut the tails to about 14 cm; he then used these severed pieces to increase the tails of another group, which became elongated by an average of 25 cm. The remaining two groups were controls: one was left untouched while the other had their tails cut and then re-glued without any change in length. By counting the number of nests in each territory (which measured the number of females attracted), Andersson showed that before the experimental manipulations there was no difference in mating success of the different groups, while afterwards the long-tailed males did significantly better than the controls or the shortened-tail group (Fig. 7.5). Shortened-tail males did not become less active in courtship displays, nor were they more likely to give up their territories. Therefore, the increased success of the elongated-tail group reflected female choice. Andersson's key finding was that females preferred tails that were even longer than normal-sized tails. This suggests there must be a balance between the forces of sexual selection (favouring longer tails) and natural selection (limiting tail length, perhaps because longer tails hinder survival).

Females prefer elaborate ornaments ...

The costs of sexual ornaments have been demonstrated in similar tail manipulation experiments with barn swallows, *Hirundo rustica*. Males with experimentally elongated tails paired up more quickly and were also preferred by females seeking extra-pair matings (Møller, 1988). However, these males were handicapped in their foraging; they caught smaller prey and grew poorer quality feathers and shorter tails at the next moult. As a result, they were slower to attract a mate the following year and suffered reduced reproductive success (Møller, 1989).

... but elaborate ornaments may be costly to male survival

Another early elegant experimental study demonstrating female choice is that of Clive Catchpole (1980; Catchpole *et al*., 1984) on the song of the European sedge warbler, *Acrocephalus schoenobaenus*. The song consists of a long stream of trills, whistles and buzzes and is sung by the male after arriving back on its breeding territory from the winter quarters in Africa. As soon as the male has paired, it stops singing (this species is monogamous). Catchpole's measurements showed that males with the most elaborate songs were the first to acquire mates (Fig. 7.6a). Furthermore, when females were brought into the laboratory and treated with

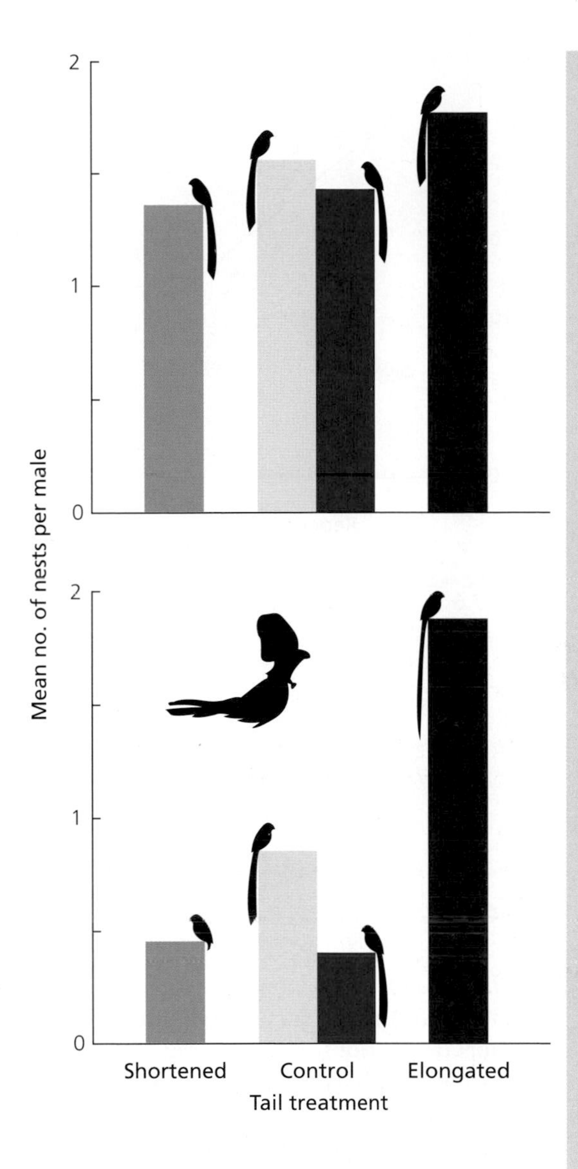

Fig. 7.5 Malte Andersson's classic experiment showing how male tail length in long-tailed widowbirds influences female choice. Top: Before experimental manipulations, there was no significant difference across the four groups in the mean number of nests per male territory (a measure of the number of females attracted). Bottom: After experimental manipulation, males with elongated tails attracted more females compared to shortened-tail males and controls. The drawing shows the male display flight. After Andersson (1982). Reprinted with permission from the Nature Publishing Group.

oestradiol to make them sexually active, they were more responsive to playbacks of larger repertoires than to small repertoires (Fig. 7.6b).

Female choice for more complex songs

In some cases, males have multiple ornaments – some for attracting females and some for deterring rival males (Box 7.2).

Why are females choosy?

Many other studies have now demonstrated female choice (Andersson, 1994). But why do females choose? Once again, Darwin did not provide an answer other than suggesting that animals, like humans, had an aesthetic sense. During the last four decades many

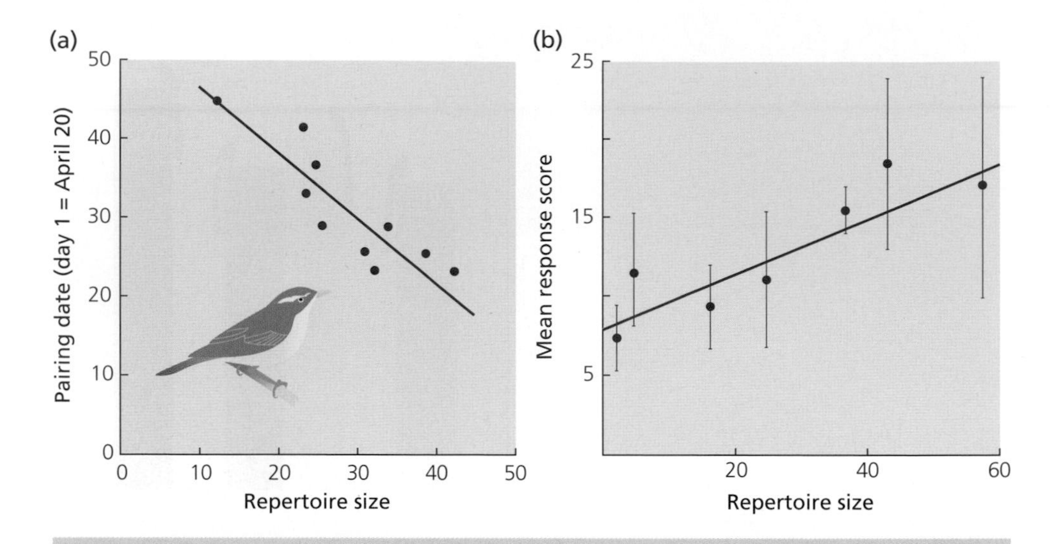

Fig. 7.6 (a) Male sedge warblers with the largest song repertoires are the first to acquire females in the spring. The size of song repertoire is estimated from sample tape recordings of each male. The results were collected in such a way as to control for the possibilities that older males, or males in better territories, both mate first and have larger repertoires. From Catchpole (1980). (b) The mean ± s.e. response score of five females to repertoires of different sizes. The response score measures sexual behaviour. From Catchpole *et al.* (1984). Reprinted with permission from the Nature Publishing Group.

BOX 7.2 MULTIPLE ORNAMENTS IN WIDOWBIRDS AND BISHOPS (*EUPLECTES* SPP.)

The red-collared widowbird, *Euplectes ardens*, lives in grasslands of eastern and southern Africa. Females and non-breeding males are drab brown and have short tails. Prior to the breeding season, males moult into a striking black plumage, with a crescent-shaped red collar on the chest and they grow a long tail, too (Fig. B7.2.1). Why have multiple ornaments?

During courtship, males display their tails in a slow flight over their territories. Males with longer tails pair up sooner and attract most females (up to nine). The colour and size of the collar does not correlate with mating success, but is correlated with a male's success in territory defence against rival males (Pryke *et al.*, 2001, 2002). Therefore, the red collar functions in male–male competition while the long tail functions in female choice. Furthermore, there is a trade-off between investment in the two ornaments; males with larger or redder collars have shorter tails. This may reflect a physiological trade-off between resources allocated to tail growth and to the red patch, which is produced by carotenoid pigments acquired entirely from the diet.

Across the *Eupletes* species, too, there is an inverse correlation between tail length and carotenoid colouration. Widowbirds tend to have elongated tails and little carotenoid pigmentation while the bishops have short tails but extensive yellow or red plumage (Fig. B7.2.2). An energetic trade-off between the two signals may, therefore, have influenced evolution of plumage diversity in the genus, perhaps reflecting the balance between the forces of male–male competition and female choice. Bishops tend to breed in colonies, where males defend small, densely-packed territories, leading to intense male–male competition (hence emphasis on carotenoid colour displays). By contrast, widowbirds have larger, dispersed territories, where female choice of males may be more important (hence emphasis on long tails) (Andersson *et al.*, 2002).

Fig. B7.2.1 Red-collared Widowbird. Photo © Warwick Tarboton.

Fig. B7.2.2 Red bishop. Photo © Oliver Krüger.

Finally, a phylogeny of 33 subspecies (all 17 species) of *Euplectes* shows that long tails and brighter red colours have evolved at least twice from ancestors with short tails and yellow colour signals. This suggests that during evolution there has been sexual selection for increased advertisements. In addition, long-tailed widowbirds (the species with the longest tail) have shown recent losses of carotenoid pigments, further supporting the idea of a trade-off between colour and tails (Prager & Andersson, 2009).

Fig. 7.7 Male bullfrogs compete for good egg-laying territories by wrestling (left). The winners then advertise for females by calling (middle) and females lay their eggs in the male's territory. From Howard (1978a,b).

theoretical and empirical studies have tackled the problem of what benefits females may gain from choice. It is useful to distinguish two broad types of benefit.

Good resources

In both the long-tailed widowbird and the sedge warbler, females nest in the territories of their chosen male, so choice may enhance their access to good resources, such as nest sites or food. Male long-tailed widowbirds do not help to feed their young, but male sedge warblers do help and males with larger song repertoires are better parents (Buchanan & Catchpole, 2000). Therefore, in both these cases choice of a male trait (tail or song) could be a cue to resources that improve a female's reproductive success.

In many animals, males compete to control resources which females need for breeding. For example, male North American bullfrogs (*Rana catesbeiana*) defend territories in ponds and small lakes where females come to lay their eggs (Fig. 7.7). Some territories are much better for survival of eggs than others (warmer, hence faster development, and less predation by leeches *Macrobdella decora*). Females prefer these good laying sites and the preferred territories are hotly contested by males, with the largest, strongest males gaining the best sites (Howard 1978a, 1978b). Thus, female choice and male–male competition often go hand in hand.

Sometimes female choice of males increases a female's access to resources

In other cases, females may choose males on the basis of their ability to provide food. Female hanging flies (*Hylobittacus apicalis*) will mate with a male only if he provides a large insect for her to eat during copulation. The larger the insect, the longer the male is allowed to copulate and the more eggs he fertilizes (Fig. 7.8). The female gains from a large insect by having more food to put into her eggs. Gifts provided by insects during courtship may help to protect, rather than nourish, the eggs. In the moth *Utethesia ornatrix*, the male transfers alkaloids to the female during mating, which the female uses in anti-predator defence. Further, the same alkaloids are used by the male as a pheromonal attractant. The female is able to assess the quantity of poison she will receive by the concentration in the pheromone (Dussourd *et al.*, 1991).

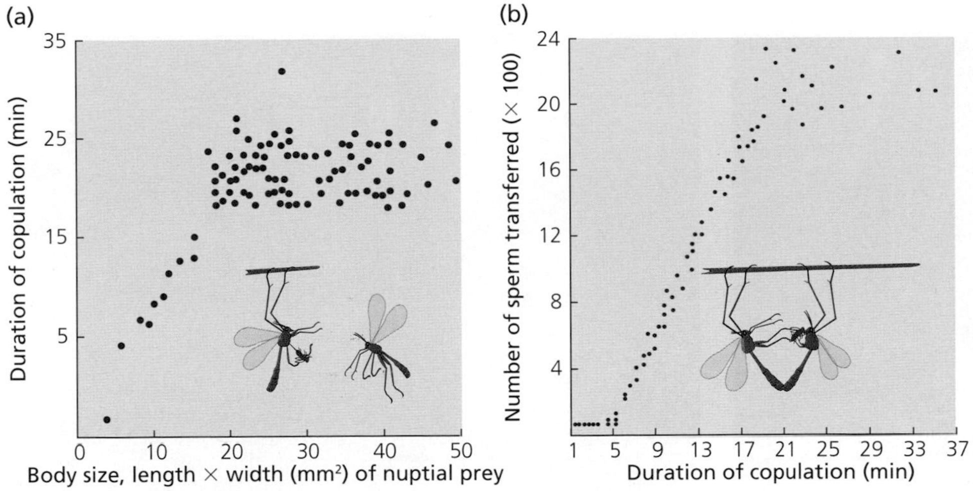

Fig. 7.8 Female choice for good resources. (a) Female hanging flies (*Hylobittacus apicalis*) mate with males for longer if the male brings a larger prey item for her to eat during copulation. (b) The male benefits from long copulation because he fertilizes more eggs. From Thornhill (1976).

Good genes

But sometimes all the female gains from a male is sperm

In some cases, all the female gets from a male is sperm to fertilize her eggs. Nevertheless, even here females are often very choosy. Bowerbirds provide a wonderful example. There are twenty species, all inhabiting New Guinea and Australia. In most, the males play no part in parental care; all their reproductive effort is put into display. They use sticks, grasses and stems to construct bowers, of various structures depending on the species; some are in the form of little avenues, others are towers, others are little roofed huts. The males then decorate their bowers with colourful fruits, flowers, feathers, bones, stones, shells and insect skeletons. Some will collect man-made objects too, such as pen tops, bottle tops, clothes pegs, car keys, jewellery and even, in one case, an old man's glass eye! The colour varies with the species; male satin bowerbirds, *Ptilonorhynchus violaceus*, prefer blue objects, while male spotted bowerbirds, *Chlamydera maculata*, prefer white and green objects (Fig. 7.9). So elaborate is the hut-like bower of the Vogelkop bowerbird *Amblyornis inornatus*, with its carefully arranged separate piles of pink flowers, green moss and shiny beetle elytra, that early explorers in New Guinea assumed these must be religious shrines built by local tribes people (Fig. 14.8c; Frith & Frith, 2004).

Female choice for more ornamental bowers in bowerbirds

Females visit bowers solely for mating, and in both satin bowerbirds (Borgia, 1985) and spotted bowerbirds (Madden, 2003a) it has been shown that males with the best decorated bowers gain the most matings. In satin bowerbirds, experimental removal of the decorations reduced the male's mating success (Borgia, 1985). In spotted bowerbirds, choice experiments revealed that males preferred to add to their bowers those objects which were best predictors of mate attraction, especially green *Solanum* berries (Madden, 2003b). Males often steal decorations from neighbouring bowers and they also sometimes destroy other males' bowers, so a male's decorations may signal his ability to defend his treasures as well as his ability to collect them. When Joah Madden (2002) added *Solanum* berries to spotted bowerbird bowers, this attracted increased bower destruction from rival males. When he removed berries and offered males excess berries to re-decorate their

Fig. 7.9 The bowers of: (a) Satin bowerbird. Photo by Michael & Patricia Fogden/Minden Pictures/FLPA. (b) Spotted bowerbird. Satin bowerbirds prefer blue objects. This bower has feathers and human debris including pens, pieces of plastic and toothbrushes. Spotted bowerbirds prefer green objects, especially *Solanum* berries. Photo © Joah Madden.

bowers, males tended to add sufficient berries to restore their previous number of decorations, and no more. This suggests that males modulate their decorations in relation to their own social status, so females may be able to assess a male's competitive ability from his bower.

When females gain only sperm from matings, as in the bowerbirds, could they be gaining genetic benefits from mate choice? Two hypotheses have been proposed. We shall discuss each in turn before we consider how we might test them.

Genetic benefits from female choice: two hypotheses

Fisher's hypothesis: females gain attractive sons

R.A. Fisher (1930) was the first to clearly formulate the idea that elaborate male displays may be sexually selected simply because it makes males attractive to females. This may sound circular, and indeed it is, but that is the elegance of Fisher's argument. At the beginning, he supposed, females preferred a particular male trait (let us take long tails as an example) because it indicated something about male quality. Perhaps males with longer tails were better at flying and, therefore, at collecting food or avoiding predators. An alternative starting point is to suppose that longer tails were simply easier to detect or that females had a pre-existing sensory bias to respond to certain stimuli (Ryan *et al.*, 1990; Chapter 14). If there is some genetic basis for differences between males in tail length the advantage will be passed on to the female's sons. At the same time, a gene which causes females to prefer longer than average tails will also be favoured, since these females will have sons better able to fly or more readily detected by potential mates.

Selection for attractiveness alone

Now, once the female preference for longer tails starts to spread, longer-tailed males will gain a double advantage: they will be better at flying and be more likely to get a mate. The female similarly gets a double advantage from choosing: she will have sons that are both good fliers and attractive to females. As the positive feedback between female preference and longer tails develops, gradually the benefit of attractive sons will become the more important reason for female choice, and the favoured trait might eventually decrease the survival ability of males. When the decrease in survival counterbalances sexual attractiveness, selection for increasing tail length will grind to a halt (Lande, 1981; Kirkpatrick, 1982). Box 7.3 describes some aspects of Fisher's hypothesis in more detail.

Good genes for sons and daughters

Amotz Zahavi's handicap hypothesis: only good quality males can afford elaborate ornaments and display

Amotz Zahavi (1975, 1977) suggested an alternative view of elaborate male sexual displays. He pointed out that the peacock's long tail is a handicap in day-to-day survival. He then went on to suggest that females prefer long tails (or other equivalent traits) precisely *because* they are handicaps and, therefore, act as a reliable signal of a male's genetic quality. The tail demonstrates a male's ability to survive in spite of the handicap, which means that he must be extra good in other respects. If any of this ability is heritable, then the tendency to be 'good' at surviving will be passed on to offspring. Thus, females select for good genes by selecting to mate only with males whose displays honestly indicate their genetic quality. Note that in this hypothesis the 'good genes' are genes for the utilitarian aspects of survival and reproduction, rather than genes purely for attracting females, as assumed in Fisher's hypothesis.

Selection for increased genetic quality of offspring

When it was first published, Zahavi's idea was not accepted, but subsequent theoretical papers have shown that the handicap hypothesis is a plausible explanation for the evolution of elaborate sexual displays, and perhaps of animal signals in general (Chapter 14). The most important feature of theoretical models of the handicap principle that 'work' (i.e. show that females could benefit from choosing males because of their handicaps) is that males only express the handicap, in other words develop the full sexual display, when they are in good condition (Grafen, 1990a, 1990b). This gets around the difficulty some critics saw in Zahavi's original idea, that males were forced to carry the handicap whether or not they could afford it, because it was viewed as a fixed trait. There are different variants of the flexible handicap idea (some authors refer to 'revealing handicaps' that reveal a male's current vigour, others to 'condition dependent handicaps' expressed in proportion to the male's condition), but the essential feature of all these models is that the degree of expression of the male sexual display tells the female about his genetic quality.

Another problem that initially bothered theoreticians is this; if there is strong selection by females for males with the 'best genes' then genetic variation might rapidly decline, as only a few male genotypes would be successful in breeding. In the same way, artificial selection for traits tends to have strong effects at first but then becomes less effective as genetic variation for the trait gets 'used up'. If female choice is costly, for example in time searching for a mate, then, in theory at least, females should stop choosing as genetic variation declines (Andersson, 1994).

How is genetic variation for quality maintained over the generations?

Current thinking suggests that this is not so much of a problem in practice, for four reasons. Firstly, populations suffer from a continuous input of deleterious mutations and so it may pay females to always be choosy to avoid these in potential mates (Kondrashov, 1988; Agrawal, 2001). Secondly, if only males in the best condition can afford to grow elaborate ornaments, or have vigorous displays, then female choice is unlikely to deplete the genetic variation available simply because so many genes are involved in influencing male condition. In fact, just about every physiological process going on in a male's body will have some influence on condition, so the genetic variation will be enormous (Rowe & Houle, 1996). Thirdly, females may choose different male traits in different years, which could maintain genetic variation underlying multiple sexual ornaments. In lark buntings, *Calamospiza melanocorys*, for example, in some years females prefer the blackest males, in others those with the largest wing patches and in others those with the largest beaks, and so on (Chaine & Lyon, 2008). These rapidly changing patterns of sexual selection parallel the oscillating patterns of natural selection described for Darwin's finches, where a rapidly changing food supply selects for small beaks in some years and large beaks in others (Grant & Grant, 2002).

Host–parasite arms races: female choice for disease resistance

The fourth factor that might maintain genetic variation in fitness is host–parasite arms races. Bill Hamilton, who lived from 1936 to 2000, was one of the most influential evolutionary biologists since Darwin. He once commented that two of the most dramatic changes in the English countryside during his lifetime were caused by disease: the myxoma virus that periodically killed the rabbit population and led to vegetation change, and Dutch elm disease, a fungus which killed most of the elm trees that once graced lowland England. This convinced him that disease must be a powerful selection pressure on organisms in nature. Together with Marlene Zuk (Hamilton & Zuk, 1982), he suggested that sexual displays were reliable indicators of genetic resistance to disease. According to this view, by choosing for elaborate displays females are acting as diagnostic veterinarians, selecting males which are genetically equipped to resist current infection. Because the parasites and hosts are engaged in a never-ending arms race of adaptation and counter-adaptation, involving genetic changes in both sides, the 'good genes' are changing all the time and so it always pays females to be choosy.

Testing the hypotheses for genetic benefits

To demonstrate that a trait could have evolved by the Fisher process, it would be necessary to show that there is genetic variation in both female preference and the male trait, and that preference and trait genes covary (Box 7.3). Several studies have now shown this predicted genetic correlation between trait and preference.

Covariance of male trait and female preference in stalk-eyed flies ...

Wilkinson and Reillo (1994) studied stalk-eyed flies, *Cyrtodiopsis dalmanni*. These small flies have their eyes held out on stalks which are particularly long in the males,

BOX 7.3 SEXUAL SELECTION FOR NOSE LENGTH: THE IMPORTANCE OF GENETIC COVARIANCE FOR FISHER'S HYPOTHESIS (LANDE, 1981)

(1) Imagine that at the start there was a range of nose lengths and of female preferences in the population. Females with a preference for slightly longer than average noses would be mated to males with longer noses and vice versa. The crucial fact to note is that offspring of these matings would have *both* the nose and preference genes: either genes for long nose plus long preference or short nose and short preference. The preference is expressed only in females and the nose in males, but everyone carries both kinds of gene. In short, there will arise an association or *covariance* between nose and preference genes. You could look at a female's preference and predict what kind of nose genes she carries to give to her sons (Fig. B7.3.1).

(2) How will evolution proceed, given this covariance? If equal numbers of females have preference above and below the mean nose length (x), there will be no change. But if by chance there was a slight predominance of females on one side of the mean (it could be long or short but let us take long), then positive feedback will start. This is shown by the arrows in the figure. Females select for long noses (long-nosed males have a higher chance of mating) and, thereby, *because of the covariance*, select for long preference. This in turn produces a further push to long noses, and hence an increase in preference.

(3) The final outcome of sexual selection in quantitative models of this hypothesis depends on the exact assumptions made in the model, for example whether or not there is a cost of female choice (Pomiankowski *et al.*, 1991). However, the important general point is that covariance between the male trait and female preference underlies Fisher's hypothesis.

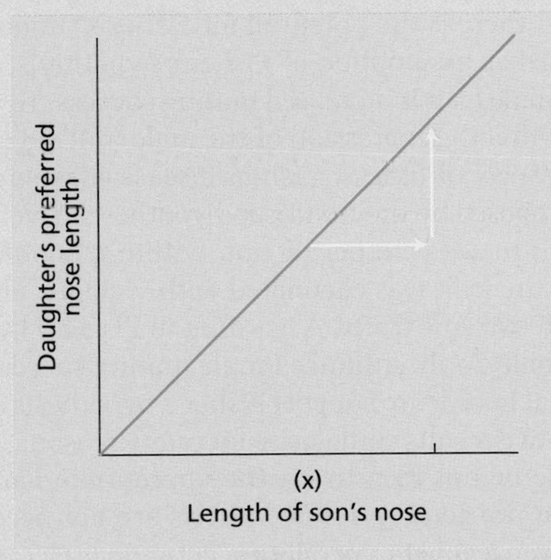

Fig. B7.3.1 Genes for long nose and long preference go together in the offspring. The slope of the line represents the degree of association or covariance.

Fig. 7.10 A male (top) and a group of three female stalk-eyed flies roosting on a root hair. Note the much greater eye span of the male. From Wilkinson and Reillo (1994).

where the eye span can exceed their body length (Fig. 7.10). Along stream banks in peninsular Malaysia, they roosted on plant roots and males with the largest eye spans were accompanied by more females. In laboratory choices it was clear that females preferred males with the largest eye spans. Wilkinson and Reillo then conducted artificial selection experiments: in one line they selected for males with the longest eye spans and in the other for males with the shortest. After 13 generations they found that female choice had changed too, as a correlated response: in the long eye span male line females preferred long eye span males, while in the short eye span line they preferred males with short eye spans.

... in guppies

In the streams of Trinidad and north eastern Venezuela, there is considerable variation in the colouration of male guppies. In streams with high predation pressure, the males tend to be dull whereas in streams with few predators they are brightly coloured, with large orange spots (Chapter 4). Houde and Endler (1990) found that there was a correlation between the amount of orange in males in a population and the strength of female preference for orange. Sticklebacks, too, vary between populations in the brightness of the male's red nuptial colouration. Females prefer redder males (Milinski & Bakker, 1990) and breeding experiments revealed that a son's intensity of red was correlated with that of his father. Furthermore, daughters of red males preferred red males while daughters of dull males showed no preference for red (Bakker, 1993). Once again, therefore, there was a positive genetic correlation between trait and preference.

... and in sticklebacks

These results suggest there is the potential for Fisher's runaway process (Box 7.3) but they do not test the key assumption of Fisher's hypothesis, namely that the only benefit of the selected male trait is increased mating success. To show this, it would be necessary to show also that the expression of the male trait did not correlate with any inherited 'utilitarian' aspect of fitness, such as disease resistance or ability to gather scarce resources, as proposed by the handicap hypothesis. Two ways to examine this prediction would be: (i) to see whether or not, within a population of males, more extreme expression of the trait was correlated with viability, and (ii) to examine the offspring of males with extreme traits. According to Fisher's hypothesis they should have no enhanced viability, only enhanced male mating success. The difficulty with both of these tests is that in order to support Fisher's hypothesis one would need to see a negative result. Negative results could arise for many reasons, including not having a large enough sample or not measuring the appropriate variables. Furthermore, Fisher's hypothesis and the good genes hypothesis are not necessarily incompatible (Iwasa *et al.*, 1991). Given a genetic correlation between preference and trait, Fisher's runaway process has the potential to operate even if the trait is an honest signal of male genetic quality. We now turn to two case studies that have tested the 'good genes' hypothesis.

(a)

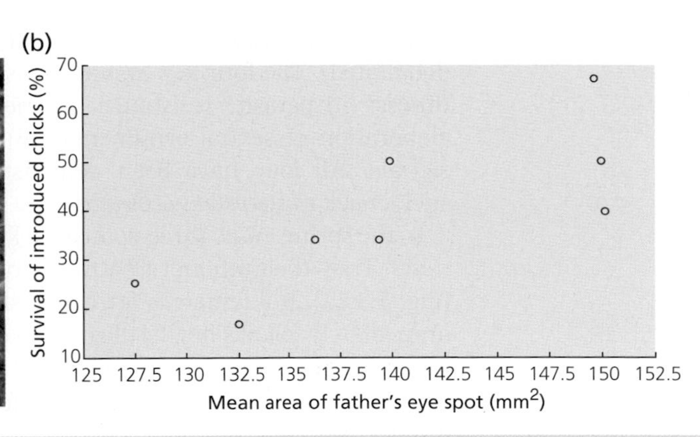

Fig. 7.11 (a) Male peacock displaying. Photo © Marion Petrie. (b) Males with larger eyespots on their tails sired offspring that survived better. From Petrie (1994).

Peacocks

By choosing more ornamented males, females gain more viable offspring

The extraordinary tail of the male Indian peafowl *Pavo cristatus* has often been regarded as the pinnacle of sexual selection. The train comprises numerous elongated upper tail coverts, with blue-centred green and copper ocelli, and males erect these to form a shimmering fan in their display to females (Fig. 7.11a). Marion Petrie tested whether this spectacular display signalled a male's genetic quality. She studied a feral population of peafowl in Whipsnade wildlife park, southern England. Firstly, she showed that the number of eyespots in the train predicted a male's mating success and that experimental reduction of eyespot numbers led to a reduction in male mating success between years (Petrie *et al.*, 1991). Then she performed an experiment in which females were paired at random with males of different natural tail ornamentation. The eggs were all collected and raised by chickens under standard conditions, and the peafowl chicks were then given food ad libitum in aviaries. Petrie (1994) found that both the sons and the daughters of males with the more ornamented tails grew better. Furthermore, when they were released into the park they survived better to two years of age (Fig. 7.11b). This suggests that in this population females obtained more viable young through choosing the most ornamented males. In another study in Japan, however, there was no evidence that females preferred males with more elaborate trains (Takahashi *et al.*, 2008). Perhaps, like the guppies, female choice varies in different ecological conditions.

Sticklebacks

Difficulties of testing female choice for disease resistance

Testing the Hamilton–Zuk hypothesis is by no means straightforward. For example, it is not sufficient to show merely that females prefer males with lower parasite burdens. They may do this not because they are shopping for good genes for their offspring but simply because they want to avoid infection during the act of mating, or because they want a partner able to provide efficient parental care (heavily parasitized males may be debilitated). The four key assumptions that need testing are: (i) parasites reduce host fitness; (ii) parasite resistance is genetic; (iii) parasite resistance is signalled by the elaboration of sexual ornaments; (iv) females prefer males with the most elaborate signals. All four have been demonstrated in detailed studies of the three-spined stickleback *Gasterosteus aculeatus*.

In the spring, male three-spined sticklebacks develop bright red colouration and build nests. They then attempt to attract gravid females by displaying with a zig-zag dance (Fig. 7.12a). If a female is attracted, she enters the nest to lay her eggs and the male immediately follows her, fertilizing the eggs externally. The male then cares for the eggs and young fry for a period of about 10 days. Manfred Milinski and Theo Bakker (1990) showed, by a clever experiment, that females preferred to mate with males with more intense red colouration. When given a choice between two males with nests in a laboratory tank, the females preferred the redder male under normal white light condition, but under green light (where the red colour was not visible) the females showed no strong preference (Fig. 7.12b). The males courtship display was unaffected by the lighting conditions, so this experiment showed neatly that it was the male red colouration that influenced female choice.

Female sticklebacks prefer redder males, who are in better condition ...

Why do females prefer redder males? Milinski and Bakker showed that a male's red colouration signalled his physical condition (redder males had greater mass per unit length). When the preferred bright males were infected with a parasite, both their condition and their red intensity declined. Under white light females now no longer preferred them, while under green light the males' parasitization had no effect on female choice.

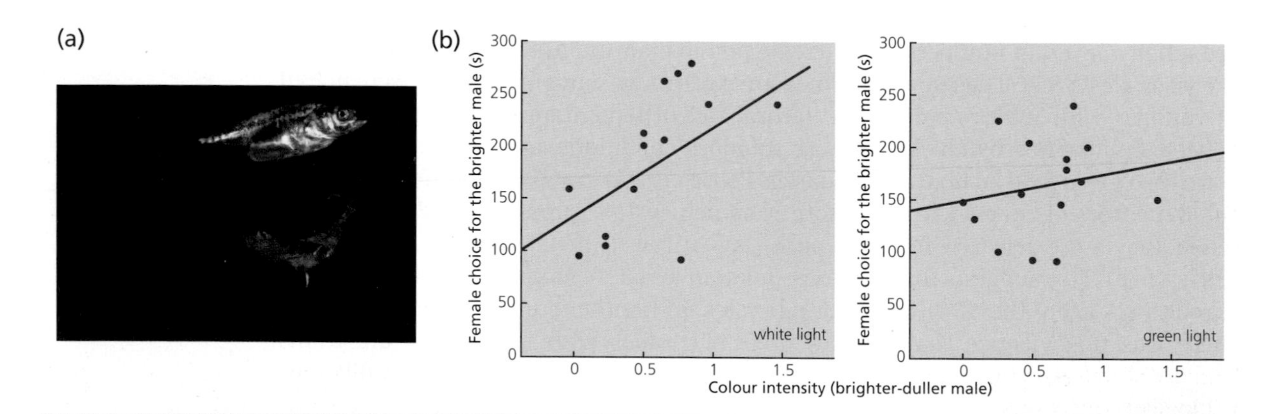

Fig. 7.12 (a) Male three-spined stickleback in breeding colouration zig-zagging around a gravid female. Photo © Manfred Milinski. (b) Under white light (left graph) females prefer the brighter of two males. The experiment was done with 15 pairs of males and plots their difference in red intensity against the strength of the female preference for the brighter male. The greater the difference in red intensity, the stronger the preference for the brighter male. Under green light (where females are unable to distinguish differences in red) there is no significant preference (right graph). From Milinski and Bakker (1990). Reprinted with permission from the Nature Publishing Group.

Therefore, one advantage of choosing redder males is that females get a healthy male, better able to fan and protect her eggs and young. However, further experiments showed that healthy males were also likely to provide 'good genes' for the female's offspring, namely alleles which confer resistance to current infections, just as hypothesized by Hamilton and Zuk (1982). The Major Histocompatibility Complex (MHC for short) is a cluster of genes which codes for MHC molecules (glycoproteins) which help fight parasite infections. Different MHC molecules recognize and bind different foreign peptides, which are then destroyed (the details are complex and Milinski (2006a) provides a good review). The MHC contains the most polymorphic gene loci known in vertebrates, so individuals vary enormously in their MHC profile. Females are able to assess a male's MHC profile by odour, because MHC gene products are fragrant. In laboratory experiments where females were given choices between the odours of different males, they preferred the odours of males whose MHC alleles provided the optimal complement to their own alleles, including alleles that provided resistance against the current prevailing infections (Reusch *et al.*, 2001; Eizaguirre *et al.*, 2009). Therefore, by choosing healthy males, females gain genetic benefits for their offspring as well as better parental care.

... and males whose MHC alleles provide a good compliment to their own. So female choice leads to genetic benefits as well as better parental care

Sexual selection in females and male choice

Female ornaments

When males make a large contribution to parental investment, males may be choosy about whom they mate with; this can lead to sexual selection in females, who evolve traits to increase their access to males (Fig. 7.13a). For example, in monogamous birds

(a)

(b)

Fig. 7.13 Sexual ornaments in females. (a) In great crested grebes, both sexes have head-feather ornaments, likely to have evolved through mutual mate choice. Photo © osf.co.uk. All rights reserved. (b) Sexual swellings in a female Chacma baboon, *Papio ursinus*, from the Cape Peninsula, South Africa. Photo © Esme Beamish.

both sexes often invest heavily in parental care and it pays both to choose high quality partners. In some cases, both sexes have similar ornaments (e.g. the head plumes of some grebes). Darwin (1871) suggested these evolved through 'mutual mate choice'. An alternative hypothesis is that female ornamentation arises simply as a non-adaptive correlated response when males evolve ornaments. However, selection experiments show that such correlated responses are weak and ornament expression is often limited to one sex (Wiens, 2001). Furthermore, comparative studies reveal frequent changes in female ornamentation during evolution (Kraaijeveld *et al.*, 2007). There is now good experimental evidence for mutual mate choice. For example, in the breeding season both male and female crested auklets, *Aethia cristatella*, develop spectacular forehead crests and experiments reveal that both sexes display most vigorously to members of the opposite sex with longer crests (Jones & Hunter, 1993).

Males may also be choosy: mutual mate choice

Female advertisement takes on an extreme form in some primates which live in multimale groups, where a female has access to several males. Here, females develop large sexual swellings, the size and colour varying during the menstrual cycle and signalling changes in fecundity (Fig. 7.13b). In some species these swellings become so large that the female finds it difficult to sit down comfortably, so the swellings must be as much a handicap as a male peacock's tail. In baboons, *Papio cynocephalus anubis*, individual differences in the size of the swellings are correlated with female quality (ability to rear offspring; Domb & Pagel, 2001). Females may compete for matings with the dominant male (who is best able to protect them or their offspring), or they may compete to mate with several males in the group to give each a sufficient paternity chance that they will desist from infanticide (Chapter 2). Similar sexual swellings occur in alpine accentors, *Prunella collaris*, a montane songbird which also lives in multimale groups. A male will help to feed a female's brood only if he has mated with her and so has a chance of paternity. By sharing matings (and paternity) among several males a female increases the chance that one or more males will help (Davies *et al.*, 1996b). Females compete for male attention by presenting bright red cloacas (Nakamura, 1990) and by singing to attract males during their fertile period (Langmore *et al.*, 1996).

Sex role reversal

Females may compete for males

In some cases, female competition for males becomes so strong that there is reversal of the usual sex roles. In the pipefish, *Syngnathus typhle*, it is the male who becomes pregnant; he has a brood pouch in which a clutch of fertilized eggs are kept safe and provided with nutrients and oxygen (Fig. 7.14). During his pregnancy, which lasts several weeks, a female could produce several clutches of eggs. Therefore, males become a limiting resource for female reproductive success and females compete for males, with males preferring larger, more ornamental females who produce larger clutches (Rosenqvist, 1990; Berglund *et al.*, 2006).

Seasonal changes in which sex competes the most in katydids ...

Sometimes there may be seasonal variation in sexual competition. Darryl Gwynne and Leigh Simmons (1990) have shown how seasonal variation in food availability leads to changes in sex roles in *Kawanaphila* katydids in Australia. When food is scarce, the male's large protein-rich spermatophore is costly to produce and also very valuable to females (Fig. 7.15). Females compete for males and males are choosy, preferring larger females who lay more eggs. However, when pollen-rich grass trees come into flower,

Fig. 7.14 Sex role reversal. (a) A pair of pipefish *Syngnathus typhle*. The male is in front, the female is upside down below him. (b) A pregnant male pipefish with a brood pouch full of developing young. Photos © Anders Berglund.

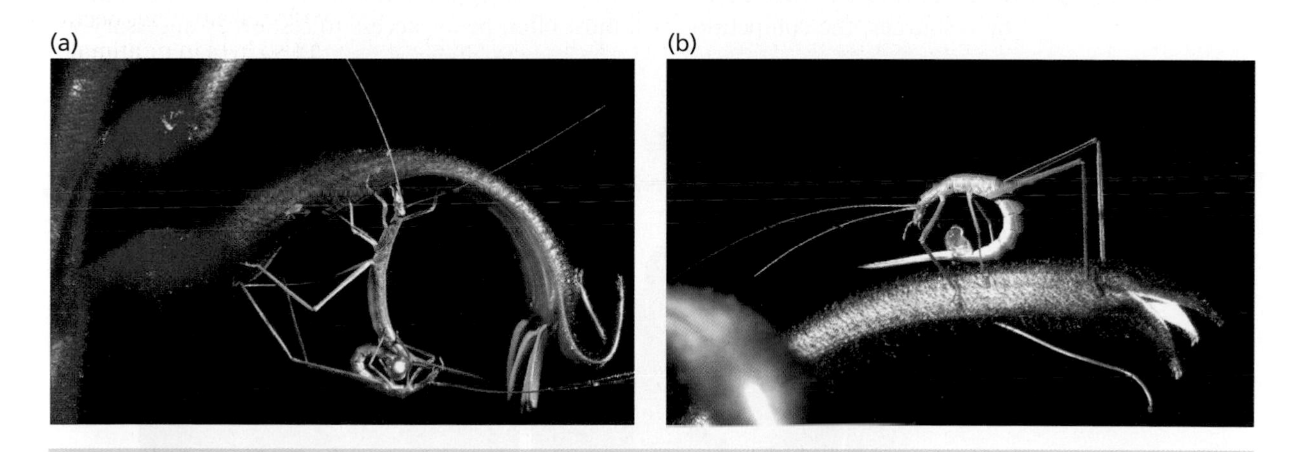

Fig. 7.15 The katydid (bush cricket) *Kawanaphila nartee* from Australia on kangaroo paw flowers. The male produces a large protein-rich spermatophore. (a) A pair together at the end of copulation, with the male (curled up behind the female) depositing his spermatophore. (b) The female bending over to eat the spermatophore. She will use this food to help her form eggs. Photos © Darryl Gwynne.

males can produce spermatophores more rapidly. Access to receptive females now limits male success, and males compete for females, with females rejecting some males.

... and in gobies

Similar seasonal changes occur in two-spotted gobies, *Gobiusculus flavescens*, a common fish along the rocky shores of western Europe. Breeding males defend nest sites in seaweed and empty mussel shells to which they attract females for spawning.

A male can care simultaneously for the clutches of several females. Early in the season, when the sex ratio is 1:1, males compete for females and actively court females at their nest sites. However, during the summer the sex ratio becomes female-biased because of higher male mortality, likely due to the costs of parental care and mating competition. This leads to sex role reversal, with females now competing for access to males and playing the most active role in courtship (Forsgren *et al.*, 2004).

Sex differences in competition

In some cases the two sexes may have different ornamentation which is related to sex differences in competition. The remarkable eclectus parrot, *Eclectus roratus*, of north-eastern Australia has pronounced sexual dimorphism (Fig. 7.16). The bright red and blue females compete for scarce nest hollows, particularly ones less susceptible to flooding, where fledging success is higher. Females are hidden inside these nest hollows during incubation and nestling care, so they do not need to be cryptic. They display below the canopy, where their bright colours contrast with the dark limbs and trunks of the trees. The bright green males compete for access to females with the best nest sites and their colouration reflects a compromise between camouflage from predators and the advantage of conspicuousness (scarlet underwing coverts) for displays (Heinsohn *et al.*, 2005; Heinsohn, 2008).

Eclectus parrots: sex differences in colours and in competition

The eclectus parrots remind us that both sexes are likely to have to compete with members of their own sex. For females, whose reproductive success is most often limited by resources, the competition will most often be for access to resources necessary for

Fig. 7.16 Sexual dimorphism in the eclectus parrot. The male is bright green, with red underwing coverts. The female is bright red and blue. Photos © (a) Lochman Transparencies, (b) © Michael Cermak.

successful reproduction (breeding sites, parental care, social rank) rather than for male sperm. For males, whose reproductive success is most often limited by mates, the competition will usually be for access to females, either directly (through force or charm) or indirectly (by monopolizing resources that females need).

Sperm competition

The process of sexual selection as envisaged by Darwin is only half the story. A century after Darwin first proposed the theory Geoff Parker (1970c) came up with a novel insight. He had been observing dung flies (*Scatophaga stercoraria*) competing for mates around cowpats. He noted firstly that a female often mated with more than one male, and secondly that females stored sperm in organs called spermathecae. He then realized that sexual selection must continue after the act of mating, as the ejaculates from different males compete for fertilizations inside the female tract. He called this process 'sperm competition'. This is now recognized to be a powerful selection pressure on reproductive behaviour throughout the animal kingdom (Birkhead & Moller, 1998; Parker, 2006). Just as with sexual selection before mating, there is the potential for two processes, namely competition between sperm from rival males (analogous to male combat) and female sperm choice (often called 'cryptic female choice', analogous to female mate choice; Table 7.2). And just as with Darwin's theory, the evidence for the first process was readily accepted whereas it is only recently that the importance of female sperm choice has been recognized.

Geoff Parker shows that sexual selection continues after mating

Why do females copulate with more than one male?

As we saw with Bateman's classic experiments (Fig. 7.3), the advantage to a male from copulating with several females is clear – he fathers more offspring. But why should a female copulate with more than one male, especially as in many species (like Bateman's *Drosophila*) a single insemination provides sufficient sperm for a female to fertilize all her ova? We shall consider three hypotheses.

Three hypotheses

Costs of resistance exceed the costs of acquiescence

In some cases it may pay a female to accept an extra mating even though she might gain no benefit. For example, a female dung fly has to go to a cowpat in order to lay her eggs. Here she meets an army of eager males who attempt to grab her (Chapter 5; Fig. 5.3). After one male has mated, another may displace him and mate too. In some cases the

	Two processes	
Sexual selection	**Male–male competition**	**Female choice**
Before copulation (Darwin, 1871)	Between rival males	Of mates
After copulation (Parker, 1970c)	Between rival sperm	Of sperm

Table 7.2 Sexual selection operates both before and after mating

female drowns in the liquid dung as the males struggle for possession (Parker, 1970c). Males are larger than females and in this case there seems to be little scope for mate choice.

Material (or direct) benefits from multiple mating

In some cases, multiple mating increases the number of young that a female can produce. She may simply ensure fertility by acquiring more sperm, or she may gain more resources from males, either food gifts or nutrients in spermatophores which enable her to lay more eggs (Thornhill & Alcock, 1983). Where some males control the best egg laying sites in the form of territories, a female may have to copulate each time she comes to lay, simply to gain access to these sites (e.g. many dragonflies and damselflies). In other cases, mating with multiple males may increase the amount of care a female can gain for her offspring because several males will cooperate to help feed a brood if they all have a chance of paternity; for example, Galapagos hawks *Buteo galapagoensis* (Faaborg *et al.*, 1995) and dunnocks *Prunella modularis* (Davies *et al.*, 1996b). A female may similarly reduce male harassment of herself and her offspring if she mates multiply. In some primates where females live in multimale troups, it pays a female to give each male a sufficient paternity chances that he will not commit infanticide (Hrdy, 1999).

Genetic (or indirect) benefits

Here the female increases the genetic quality of her offspring by mating with more than one male. Good evidence for this hypothesis comes from a surprising source, namely studies of song birds, where social monogamy (one male breeding with one female) is the most common mating system (Lack, 1968). Since 1985, when DNA markers first became available for precise measures of paternity, studies have revealed that socially monogamous song birds engage in frequent extra-pair mating. Typically, 10–40% of the offspring are sired by males who are not the female's social mate (Griffith *et al.*, 2002; Westneat & Stewart, 2003). In song birds, males cannot easily force a mating because the female can fly away to escape. Instead, extra-pair matings occur because females sneak away from their home territories to solicit to these extra-pair males. A comparative study of bird plumage suggests that extra-pair matings are an important component of sexual selection. Owens and Hartley (1998) compared the plumage of males and females across 73 bird species where there were data on rates of extra-pair paternity. They ranked plumage dimorphism on a scale from zero (no difference between males and females) to ten (males much more colourful than females). They found no correlation between the degree of plumage dimorphism and the number of social mates but a strong correlation with the rate of extra-pair paternity (Fig. 7.17). This suggests that males have evolved colourful plumage not to attract their own females but rather to attract other males' females for extra-pair matings!

Extra-pair matings in birds: female choice for male plumage or display traits

Detailed studies of several species have now confirmed that females do indeed seek extra-pair matings with males whose displays or plumage traits are more elaborate than those of their social partners, for example: greater song repertoires, longer tails or brighter plumage (Table 7.3). Females paired to attractive males (as measured by these traits) tend to be more faithful; those paired to males with poorer developed traits are the ones who

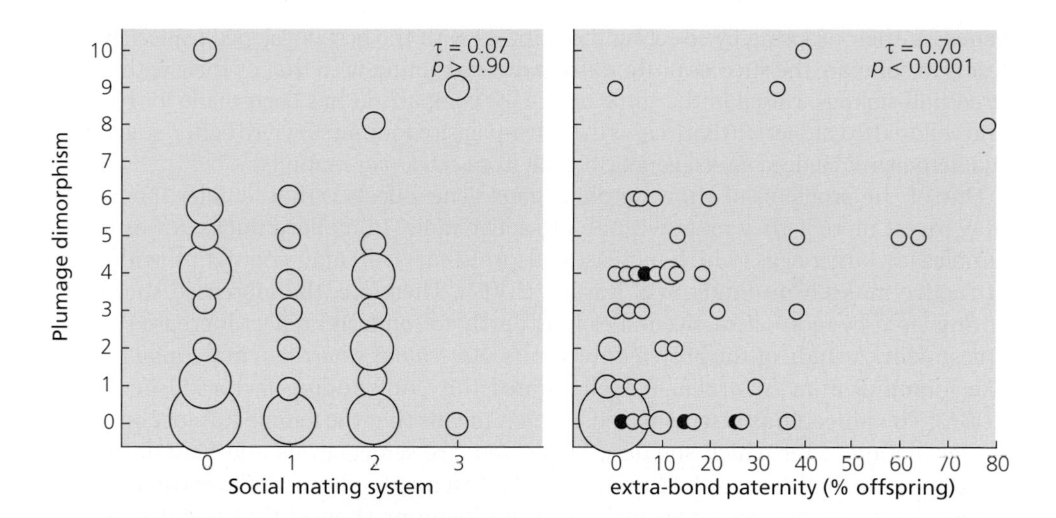

Fig. 7.17 A comparative study of bird plumage dimorphism by Owens and Hartley (1998) showed that sexual dimorphism did not correlate with social mating system (number of mates: 0 – social monogamy, 1 to 3 – increasing levels of male polygyny), but increased significantly with extra-bond paternity (percentage offspring sired by extra-bond males). Plumage dimorphism is scored on a scale from zero (no difference between males and females) to ten (males much brighter than females). Circle sizes reflect sample size.

Species	Attractive male trait	Trait of extra-pair male compared to social mate	Reference
Great reed warbler *Acrocephalus arundinaceus*	Song repertoire	Larger	Hasselquist *et al.* (1996)
Barn swallow *Hirundo rustica*	Tail length	Larger	Saino *et al.* (1997)
Blue tit *Cyanistes (Parus) caeruleus*	Ultraviolet reflectance of crown feathers	Brighter	Hunt *et al.* (1999) Kempenaers *et al.* (1997)
Collared flycatcher *Ficedula albicollis*	White patch on forehead	Larger	Michl *et al.* (2002) Sheldon *et al.* (1997)

Table 7.3 In some song birds, females seek extra-pair matings from males whose display traits are more elaborate than those of their social mates

most actively seek extra-pair matings. In all four examples in Table 7.3, there is experimental evidence that these display traits are attractive to females. Furthermore, males with the most developed traits survive better, suggesting these display traits are viability indicators. Extra-pair males do not provide care, only sperm. Could females be shopping for good

Females increase the genetic quality of their offspring by extra-pair matings

genes for their offspring by selecting those males with the best developed traits? The key test is to compare the success of the extra-pair sired young with that of their within-pair sired half-siblings, raised in the same nest. This comparison has been made for blue tits and collared flycatchers; in both cases the extra-pair sired young survived better, suggesting that females did indeed gain a genetic benefit from extra-pair matings.

One of the problems of showing male 'good gene' effects is that females themselves may invest more if they mate with an attractive male. In mallard ducks, for example, females lay larger eggs (which increases offspring survival) after copulating with more attractive males (Cunningham & Russell, 2000). Therefore, the increased success of young sired by more attractive males may partly (or entirely) reflect increased female investment. A study of the brown antechinus (*Antechinus stuartii*), a marsupial mouse-like mammal from Australia, has eliminated this confounding factor (Fisher *et al.*, 2006). The antechinus is semelparous, which means that the female has just one litter in her lifetime, after which she dies. Males, too, are semelparous and play no part in offspring care. In this case, the female should invest maximally in her litter (there are no future litters to save resources for). Field observations showed that females usually mated with several males during their two-week mating period. Why? In a carefully-controlled experiment, females were caught and mated in the laboratory with a pool of 41 'stud' males. Each female was allowed to mate three times. One group of females (monandrous) mated three times with the same male. A second group (polyandrous) mates three times, but with three different males (once each). When the young were born, they were marked individually with little microchips under the skin. The mothers and their litters were then released back into the wild in nesting boxes. Two months later, the number of surviving young was scored from each litter.

The brown antechinus: multiple mating improves a female's reproductive success

The results were striking. There was no difference between the two female groups in their initial litter size, but polyandrous females produced three times the number of surviving young to the post-weaning stage. A closer look at the success of the monandrous group was revealing; a monandrous female did better if the male she had been paired with was a good competitor in polyandry, in other words had gained high paternity in competition with other females. This suggests that the advantage of polyandry to a female was she was more likely to gain a good genetic sire for her offspring (Fisher *et al.*, 2006).

Constraints on mate choice and extra-pair matings

These results leave us with two puzzles. Firstly, why do females not choose to mate with an ideal male in the first place? In cases where competition for the best males is intense, females may be forced to settle for less than the best social mate and then rely on extra-pair matings to increase the genetic quality of their offspring.

Cuckolded males may reduce parental care

Secondly, what limits extra-pair mating? Two hypotheses have been suggested. One is that in cases where males provide help with parental care, the social mate would reduce his care in response to cuckoldry. There is good evidence for this in collared flycatchers: when a female was removed experimentally for one hour during her fertile period (to simulate the female going off to seek extra-pair matings), then her social mate reduced his effort in chick feeding (compared to control female removals outside the female's fertile period) (Sheldon *et al.*, 1997). Thus a female may have to trade-off benefits from extra-pair fertilizations with loss of help in raising the chicks.

Different sires best for sons versus daughters

A second hypothesis is that different sires are best for sons and daughters. This could arise because of sexually antagonistic genes, which have a beneficial effect in one sex but a harmful effect in the other sex. Imagine, for example, a gene in our hominid ancestors that caused an increase in hip width. This might be good for daughters (easier for them to give birth) but bad for sons (reduces locomotion efficiency). Selection for sex-limited gene expression could result in independent evolution in the two sexes, so enabling each sex to reach its optimal outcome. Nevertheless, selection experiments with *Drosophila* have shown that selection for genes that have beneficial effects in one sex can lead to reduced fitness in the other sex (Chippindale *et al.*, 2001). There is also evidence for sexually antagonistic genetic effects in natural populations. In red deer *Cervus elaphus* on the isle of Rum, Scotland, males with high lifetime reproductive success fathered, on average, daughters with low fitness (Foerster *et al.*, 2007). In crickets, breeding experiments reveal that more attractive males in the wild sire higher quality sons but poorer quality daughters than do less attractive males (Fedorka & Mousseau, 2004). This could impact on mate choice. A female might temper her extra-pair matings because her social mate is a better sire for offspring of one sex while an extra-pair male is a better sire for the other sex.

Sexual conflict

We now come to the third main theme of this chapter. We began with the origin of anisogamy as the outcome of a primeval conflict in which microgametes evolved to exploit the resources of macrogametes. We then showed that the greater parental investment by females before mating often combines with great investment after mating, leading on the one hand to intense competition among males for mates, and on the other to female choice among males for better access to resources or for genetic benefits. This will often result in sexual conflict, both over the act of mating and over sperm use after mating (Table 7.4). Sexual conflict occurs whenever the optimal outcome is different for the male and female. In theory, it should lead to each sex evolving adaptations that bias the outcome towards its own interests, leading to sexually antagonistic coevolution between traits in males and females (Parker, 1979; Chapman *et al.*, 2003). We now consider the evidence for such sexual conflict: firstly over mating itself, and secondly concerning fertilization of ova after mating. In the next chapter we explore sexual conflict concerning parental investment.

Sexual conflict over mating

Sexually antagonistic coevolution

We have seen that a male's reproductive success is often more limited by access to mates than is a female's reproductive success. Thus, for a given encounter it will often pay a male to mate but a female to resist. An extreme manifestation of this conflict is enforced copulation, as exemplified by scorpion flies (*Panorpa* spp.). Male scorpion flies usually acquire a mate by presenting her with a nuptial gift in the form of a special salivary

Table 7.4 Sexual conflict: summary of some male and female adaptations and counter-adaptations

Male traits	Female traits
Enforced copulation.	Resistance.
Intromittant organs which enhance mating success.	Elaborate reproductive tracts which pose obstacle courses for sperm.
Mate guarding, frequent copulation, strategic allocation of sperm.	Seek extra-pair copulations.
Remove or displace sperm of rival males.	Sperm ejection.
Copulatory plugs and anti-aphrodisiacs to discourage matings with other males.	Sperm choice.
Accessory gland proteins to manipulate female.	Chemical defence.

secretion or a dead insect (this is very similar to the *Hylobittacus* described earlier). The female feeds on the gift during copulation and turns the food into eggs. However, sometimes a male enforces copulation: he grasps her with a special abdominal organ (the notal organ) without offering a gift (Thornhill, 1980). Enforced copulation appears to be a case of sexual conflict. The female loses because she obtains no food for her eggs and has to search for food herself, while the male benefits because he avoids the risky business of finding a nuptial gift. Scorpion flies feed on insects in spiders' webs and quite often get caught up in the web themselves, so foraging is certainly risky (65% of adults die this way). Why do not all males enforce copulations? The exact balance of costs and benefits is not known, but it appears that it results in a very low success rate in fertilizing females, so perhaps males adopt this strategy only when they cannot find prey or make enough saliva to attract a female.

Water striders: coevolution of male grasping and female resistance

Studies of water striders (Gerridae) by Goran Arnqvist and Locke Rowe (2002a, 2002b) provide particularly good evidence for how conflicts over mating lead to antagonistic coevolution between the two sexes. These insects can often be seen skating over the surface of ponds and streams in search of food or mates. A male pounces on top of a female and tries to secure a mating by grasping her. However, superfluous matings are costly for females; with a male on her back, the female has reduced mobility, which reduces feeding success and increases predation, so females who have already mated struggle to avoid males. Comparison across species shows correlated evolution of male morphology to increase grasping (elongation of grasping genitalia) and female morphology to resist (elongation of abdominal spines, which help thwart the male) (Fig. 7.18). Thus, adaptations in one sex are matched by counter-adaptations in the other sex. Differences in the opportunity for matings and the costs of armaments may explain why different species are at different equilibrium armament levels.

Mate guarding and frequent copulation for paternity assurance

In other cases the sexual conflict goes in the other direction: it pays a female to gain extra-pair matings but it pays her social mate to prevent her from doing so, to guard his paternity. In many socially monogamous birds, the social male follows the female closely throughout the week or so that she is fertile and chases off any intruding males. For her

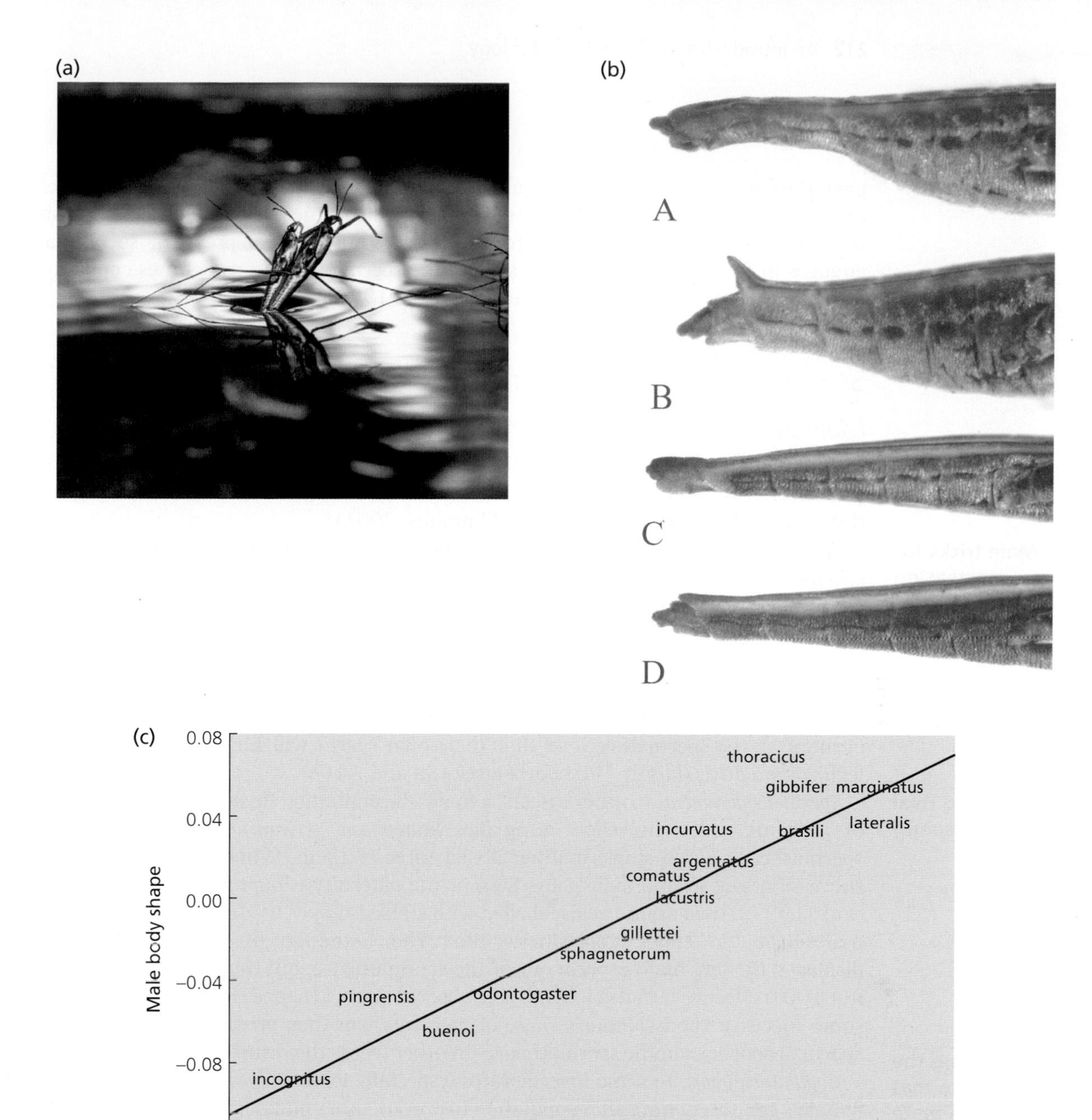

Fig. 7.18 (a) A pair of water striders *Gerris lacustris* engaged in a pre-mating struggle. The female (below) raises up to push the tips of the pair's abdomens onto the water's surface tension, which often causes the male to lose his grip. Photo © Ingela Danielsson and Jens Rydell. (b) In some species, like *Gerris incognitus*, males have evolved grasping genital segments (A) and females have evolved abdominal spines (B), which help the female resist the male's grip. Other species, like *Gerris thoracicus*, are less armed and less dimorphic (C male; D female). (c) When indices of armament levels of male and female across different species of *Gerris* are plotted together, the species tend to fall along a line suggesting sexually antagonistic coevolution, in which increased male armament for grasping is matched by increased anti-grasping defences in females. The most armed species are bottom left (e.g. *incognitus*) and the least armed are top right (e.g. *thoracicus*). From Arnqvist and Rowe (2002b). Reprinted with permission from the Nature Publishing Group.

part, the female may attempt to sneak off to achieve extra-pair matings. In cases where males cannot guard their mates closely because one partner has to defend the nest site while the other goes off to forage (many seabirds and birds of prey) the social male engages in frequent copulations to swamp the sperm of rivals, sometimes copulating hundreds of times per clutch (Birkhead & Møller, 1992).

Sexual conflict after mating

Male adaptations

Males employ a diverse array of tactics to increase their success in sperm competition. The invertebrates provide the most extraordinary examples, and we can only give a taster here (Birkhead & Møller, 1998; Simmons, 2001).

Male tricks for increasing paternity after mating

Sperm removal. In many insects, females store sperm in special sacs called spermathecae. Jonathan Waage (1979) was the first to show that male damselflies and dragonflies may remove sperm deposited by rival males before inserting their own. The penis of male *Calopteryx maculata* has two specialized scoops at the end which are used to scrape out of the female any sperm left by previous males, before he injects new sperm into the female's sperm stores (Figs 7.19c and 7.19d). In other species, males use an inflatable penis with horn-like appendages to pack sperm of previous males into the corners of the spermatheca, so that their own sperm will have first access to the fertilization ducts (last in – first out; Figs 7.19a and 7.19b).

Removing rival sperm

Sperm displacement. In other insects, a male's insemination flushes out inseminations of previous males. In yellow dung flies *Scatophaga stercoraria* females have three spermathecae and a single mating fills all three of them. When two males mate in succession, the second male gains 80% of the paternity (Chapter 3). Leigh Simmons *et al*. (1999a) used amino acids labelled with stable isotopes to identify sperm from two males inside the female's reproductive tract. They found that the second male's sperm displaced the first male's sperm out of the spermathecae, but this flushing action was not 100% effective because it led to some sperm mixing (hence the first male achieved some success). The two male's share of paternity was then predicted by their relative sperm proportions in the spermathecae (in other words the result of a 'fair raffle').

Reducing the chance that females mate again

Copulatory plugs. In some invertebrates (especially insects) the male cements up the female's genital opening after copulation to prevent other males from fertilizing her. The males of *Moniliformes dubius*, a parasitic acanthocephalan worm in the intestine of rats, produces a chastity belt of this kind but, in addition to sealing up the female after copulation, the male sometimes mounts rival males and applies cement to their genital region to prevent them from mating again (Abele & Gilchrist, 1977). No less remarkable are the habits of the hemipteran insect *Xylocoris maculipennis*. In normal copulation of this species the male simply pierces the body wall of the female and injects sperm, which then swim around inside the female until they encounter and fertilize her eggs. As with the acanthocephalan worms, males sometimes engage in homosexual 'copulation'. A male *Xylocoris* may inject his sperm into a rival male. The sperm then swim inside the body to the victim's testes, where they wait to be passed on to a female next time the victim mates (Carayon, 1974).

Anti-aphrodisiacs. Larry Gilbert (1976) noticed that female *Heliconius erato* butterflies have a peculiar odour after they have mated. He showed experimentally that the scent

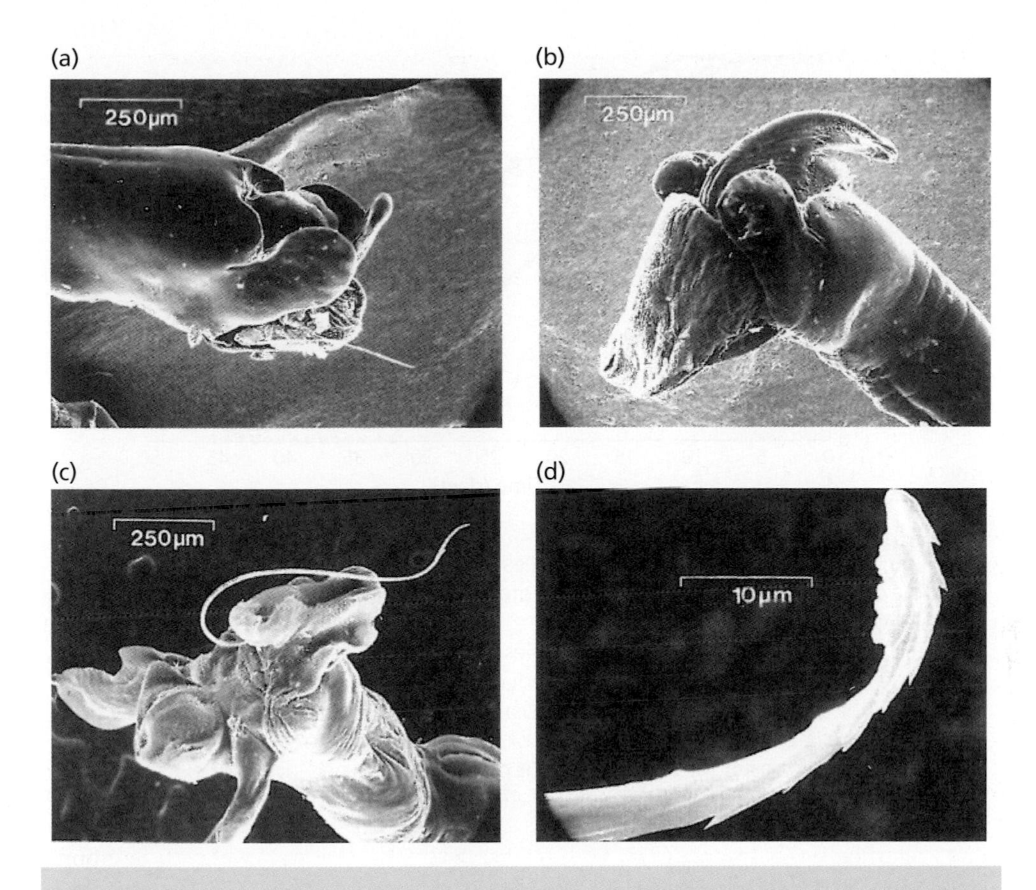

Fig. 7.19 Two sperm displacement mechanisms in Odonata. *Crocethemis erythraea*: (a) uninflated penis; (b) inflated penis. The horn like structure repositions sperm of previous males in the spermatheca. *Orthetrum cancellatum*: (c) the whip-like flagellum is everted during copula; (d) it carries barbs which remove sperm from the narrow ducts of the spermatheca. Siva-Jothy (1984). Photos by Michael Siva-Jothy.

does not originate from the female herself, but is deposited by the male at the end of mating. The scent discourages other males from mating with the female, perhaps because it resembles a scent used by males to repel one another in other contexts.

Sterile sperm as 'cheap fillers'

Sterile sperm. In some invertebrates, males produce two types of sperm: 'eusperm', which have the potential to fertilize the female's ova, and 'parasperm' which are sterile (and may or may not contain a nucleus). In butterflies and moths, parasperm act as 'cheap fillers' of the female sperm storage organs, leading to a delay in female remating (Cook & Wedell, 1999). In *Drosophila pseudoobscura*, parasperm help protect their fertilizing brother eusperm from spermicide inside the female tract (Holman & Snook, 2008).

Chemical manipulation of females

Accessory gland proteins (Acps). In many insects the male's ejaculate contains not only sperm but also a cocktail of proteins that influence female behaviour and physiology. In *Drosophila melanogaster* no fewer than 80 Acps have been identified and their functions

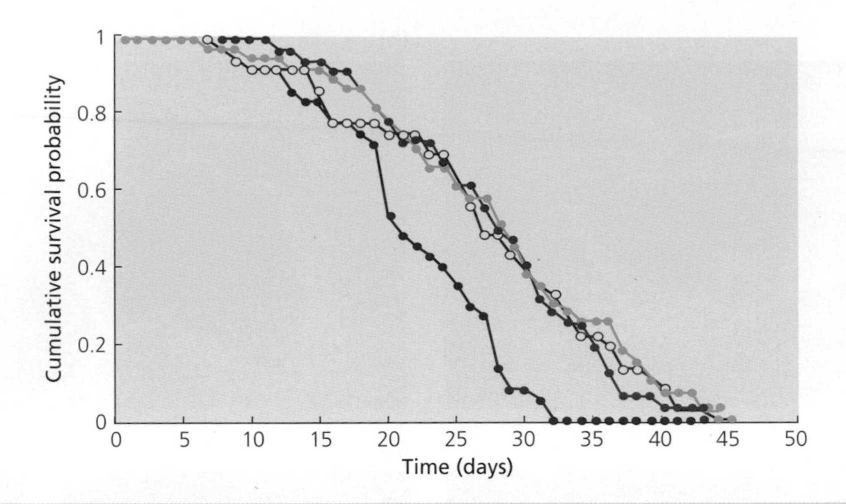

Fig. 7.20 In this experiment, female fruit flies *Drosophila melanogaster* were given varying exposure to male accessory gland proteins (Acps) at mating, while keeping constant other costly aspects of reproduction, such as egg production, non-mating exposure to males and rate of mating. Females exposed to males which produced Acps (dark blue line) died significantly sooner (median lifespan 21 days) than females exposed to three other types of males (median 29 days), namely: males genetically engineered to lack Acps (red line), and two control groups of males (open symbols and pale blue line) which courted females at the normal rate but could not mate because their external genitalia were ablated. From Chapman *et al.* (1995). Reprinted with permission from the Nature Publishing Group.

have been investigated by producing males genetically engineered to lack them or overexpress them. Among their functions are: incapacitating rival male sperm; protecting a male's own sperm from enzymatic attack in the female reproductive tract; increasing a female's egg laying rate; decreasing a female's propensity to remate. All these will help to increase the male's reproductive success, but experiments have shown that these male benefits can be at the expense of female fitness because they reduce a female's longevity (Fig. 7.20). The deleterious side-effects to the female may arise because the Acps enter the female haemolymph through the vaginal wall and then perhaps interfere with essential enzymatic processes inside the female body cavity (Chapman *et al.*, 1995, 2003).

Strategic allocation of sperm. Comparative studies of many taxa (primates, bats, other mammals, birds, frogs, fish and various insects) have shown that testis size relative to body size (a measure of investment in sperm) increases with the degree of female promiscuity (a measure of sperm competition; Wedell *et al.*, 2002). The potential for evolutionary change is revealed by selection experiments with male dung flies; when exposed to increased sperm competition, larger testes and ejaculates evolved within ten generations (Hosken *et al.*, 2001). These results show that investing resources in sperm production is costly and evolves only when this brings a competitive benefit. Within species, too, it is clear that males do not have limitless potential to copulate (Dewsbury, 1982). For example, male adders (*Vipera berus*) lose as much body mass during spermatogenesis (when they are inactive) as during the subsequent phase of searching and competing for females (Olsson *et al.*, 1997).

Males vary sperm allocation in response to sexual competition

It is not surprising, therefore, that males allocate their sperm strategically in response to both sperm competition and female fecundity (Wedell *et al.*, 2002). The study by Tommaso Pizzari *et al.* (2003) of a free-ranging population of feral fowl *Gallus gallus* provides an excellent example (Fig. 7.21). They collected ejaculates by fitting females with a harness and the sperm were then counted. Dominant males increased their sperm investment in a female when there were more male competitors and they allocated more sperm to females with larger combs, which laid larger eggs. Furthermore, a male's sperm investment declined with repeated exposure to the same female but was renewed by the arrival of a novel female. This is known as the 'Coolidge effect', named after President Coolidge – during a visit to a chicken farm with his wife, when Mrs. Coolidge was told that a rooster could copulate dozens of times each day she is reported to have said: 'Tell that to the President'. The President then enquired whether this involved the same hen every time, and was informed that many different hens were involved. 'Tell *that* to Mrs. Coolidge!', he replied.

Fig. 7.21 A pair of feral fowl copulating. The female is fitted with a harness, so sperm can be collected after a mating to measure the ejaculate. Photo © Charlie Cornwallis.

Sperm depletion in males

Sometimes males become sperm depleted and this can reduce female fertility. In blueheaded wrasse (*Thalassoma bifasciatum*), a coral reef fish, dominant males attract the most females but females suffer fertility costs when pairing with these males because they release fewer sperm per mating (fertilization is external; Warner *et al.*, 1995). Sperm limitation of attractive males can lead to competition among females for mating access. In the great snipe (*Gallinago media*) females compete for repeated copulation with the most popular males on a lek, but males reject females with whom they have already mated (Saether *et al.*, 2001).

Female adaptations

When Parker (1970c) first identified sperm competition as an important component of sexual selection, researchers focussed on male adaptations because theory led them to believe that success in post-mating conflicts was more important to a male than to a female. Essentially, males were thought to play a 'numbers game', where increased fertilization success leads to more offspring, whereas females played a 'quality game', where increased control of fertilization improved only offspring quality. However, adaptations will be influenced by costs as well as benefits: surely once sperm are inside the female's body, she will have better control over the outcome? Bill Eberhard (1996), in particular, championed the potential for female control and the possibility of female sperm choice (cryptic female choice). There is now growing evidence for this view.

Cryptic female choice

A dramatic example of female control is provided by feral fowl *Gallus gallus*. Females prefer to copulate with dominant males. However, subordinate males can sometimes force matings despite female resistance. In these cases, the female retaliates immediately after mating by cloacal contractions which eject the subordinate male's sperm (Pizzari & Birkhead, 2000). If more than one male succeeds in inseminating a female, could she then later bias fertilization success by sperm choice? Experiments with *Drosophila* have shown that the relative success of two males in sperm competition depends not only on male genotype but also on female genotype (Clark *et al.*, 1999). This shows that females are not simply passive providers of a reproductive tract for male–male sperm competition, but rather have some control over the outcome.

Females can control sperm storage

Recent experiments by Tom Tregenza, Nina Wedell and their colleagues have provide convincing evidence for female sperm choice after mating. They studied field crickets (*Gryllus bimaculatus*), where females readily mate with more than one male (relatives or non-relatives) both in the field and in the laboratory. In laboratory experiments, females were mated successively with two males: two of the female's siblings, or two non-siblings, or a sibling and a non-sibling (the order was varied). Male crickets transfer a spermatophore to the female during mating and they prepare this in readiness before they encounter a female, so siblings and non-siblings transferred equal numbers of sperm. There was also no difference between treatments in the numbers of eggs laid. However, females mated with two of her siblings had decreased hatching success compared to those mated with two non-siblings (Fig. 7.22), which is the typical outcome of inbreeding depression arising from homozygosity for deleterious recessive alleles. If sperm from the two males mixed inside the female reproductive tract, then in the sibling plus non-sibling treatment we would expect, on average, half of the eggs to be fertilized by the sibling, so these females should have had an intermediate reproductive success. However, they did just as well as those mated to two non-siblings (Fig. 7.22). This suggests that a female can bias fertilization in favour of the ejaculate of unrelated males. By using DNA markers to identify sperm from different males, it was shown that a female preferentially stored sperm from the unrelated males inside her spermathecae, and it was this biased storage that enabled females in the sibling and non-sibling treatment to avoid the costs of inbreeding (Bretman *et al.*, 2009).

Sperm choice in female field crickets

A key question for future research is whether male versus female control of fertilization varies between species and, if so, why? In general, we might expect most female control post-copulation in species where females have least control over which males mate.

Sexual conflict: who wins?

If each sex evolves suites of sexually antagonistic adaptations that bias the outcome towards their own interests, who will win the conflict? Theoretical models suggest that the outcome is often a never-ending evolutionary chase leading to rapid evolutionary change by both parties (Parker, 1979; Chapman *et al.*, 2003). For example, male

(a)

(b)

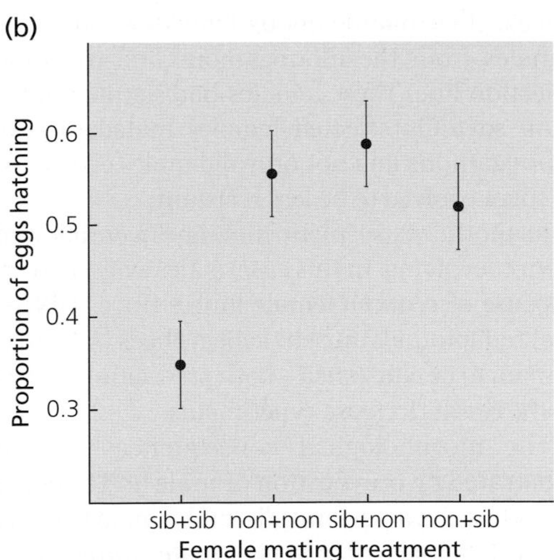

Fig. 7.22 (a) Field cricket, *Gryllus bimaculatus*. Photo © Tom Tregenza. (b) Results of a double mating experiment. Females mated to two of the female's own siblings suffer reduced hatching success due to inbreeding depression. Females mated to a sibling plus a non-sibling (in either order) do just as well as those mated to two non-siblings, suggesting the female can bias fertilization success in favour of the unrelated male. From Tregenza and Wedell (2002).

accessory gland proteins in *Drosophila* have high levels of amino acid polymorphism and differ markedly between species; they are estimated to evolve at twice the rate of non-reproductive tract proteins, which is strongly suggestive of a continuing sexual arms race.

Selection experiments with *Drosophila* reveal antagonistic co-evolution between the sexes

The power of sexual conflict to produce antagonistic coevolution between the sexes has been demonstrated by some wonderfully elegant artificial selection experiments with *Drosophila melanogaster*. Bill Rice and Brett Holland asked the question: are evolving male adaptations kept in check by evolving female counter-adaptations? In one experiment (Holland & Rice, 1999), they had two selection lines, each run for 47 generations in the laboratory, during which they selected for the most successful males and females. In one line, there was intense sexual selection: each vial had three males and one female. In this environment there was strong selection for males who were successful at sperm competition and for females able to cope with male–male competition. In the second line, sexual selection was eliminated altogether by the neat trick of enforced random monogamy. In these vials, one male and one female spent their whole lives together. Here, in the absence of male–male competition, a male was guaranteed paternity of all his female's ova and it obviously paid him to behave in a way which maximized her lifetime success.

The results showed that in the monogamous line males did indeed evolve to be less harmful to females. For example, they had a decreased courtship and mating rate. As a result, female survival and fecundity was greater than in the sexual selection line. Did females evolve too? If sexual conflict leads females to evolve costly defences against males, then monogamous females should have reduced these defences. To test this, females from the monogamous line were mated to males from the intense sexual selection line. These females had significantly lower survival and reproductive success than sexually selected females mated to sexually selected males. Therefore, in the monogamous line not only did males evolve to be less harmful to females, but females, in turn, evolved to be less resistant.

In another experiment, male lines were allowed to evolve but female lines were prevented from coevolving. In this case, males evolved increased success in sperm competition at the expense of reduced female fitness (Rice, 1996). These experiments show that normally male adaptations must be kept in check by female counter-adaptations. The genes involved remain to be elucidated – both pre-mating and post-mating conflict may have contributed to the results in these experiments.

Coevolution of male and female genital morphology in waterfowl

The morphological consequences of post-mating conflicts are dramatically illustrated by coevolution of male and female genital morphology in waterfowl. Most birds have simple genitalia and a male transfers sperm by placing his cloaca over the female's cloaca. However, male waterfowl have a phallus whose length varies between species from 1.5 to 40 cm and is positively correlated with the frequency of forced extra-pair matings (Coker *et al.*, 2002). This suggests that the phallus has evolved to enhance a male's ability to force intromission. However, female reproductive traits have coevolved with male morphology; in species where males have the longest and most elaborate phallus (with spines and grooves), females have the most elaborate vaginal morphology, including dead end sacs and coils, which are likely to reduce the chances of male intromission without female cooperation (Brennan *et al.*, 2007; Fig. 7.23).

Chase-away sexual selection

Sexual conflict may be likened to an evolutionary dance, where each sex tracks the other with costly tricks and counter-tricks. The discovery of antagonistic coevolution between the sexes has led Holland and Rice (1998) to propose a new model of sexual selection. This envisages a process in which males are selected to induce females to mate, either by force or by charm, and females are then selected to resist, leading to 'chase-away' coevolution of male traits to stimulate females and female traits to improve resistance. Recall the 'good genes' and 'Fisher's runaway' models, where females evolve *preferences* for male traits because of genetic *benefits*. The 'chase-away' model is the precise opposite; females evolve *resistance* to male ploys because acquiescence is *costly*. The results of the *Drosophila* selection experiments and the complex coevolution of male and female genital morphology in waterfowl certainly fit the chase-away model. Could this model also explain the exaggeration of some male traits to attract females?

Females evolve resistance to male ploys

Holland and Rice (1998) point to two possible cases. The fish genus *Xiphophorus* includes the swordtails, which have elongated caudal fins, and the platyfishes, which do not. Their closest relatives (genus *Priapella*) lack long tails, so do more distant relatives, so the ancestor of *Xiphophorus* probably lacked the ornament (Fig. 7.24). Female swordtails prefer males with longer tails but, surprisingly, so do

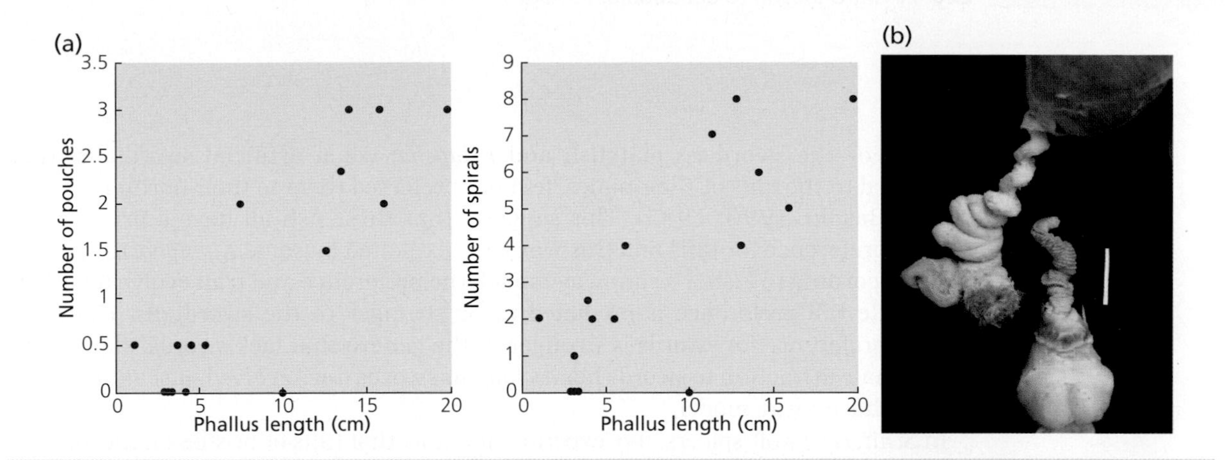

Fig. 7.23 Coevolution of male and female genitalia in waterfowl. (a) In species where the male has a longer phallus, the female has a more elaborate vagina, with more spirals (right) and 'dead end pouches' (left). The vaginal spirals are in the opposite direction to male phallus spirals, suggesting antagonistic rather than mutualistic 'lock and key' coevolution. (b) Mallard duck, *Anas platyrhynchos*, a species with high levels of forced copulations in which the male has a long phallus (bottom right) and the female has a long and elaborate vagina (top left). The white bar is 2 cm. From Brennan *et al*. (2007).

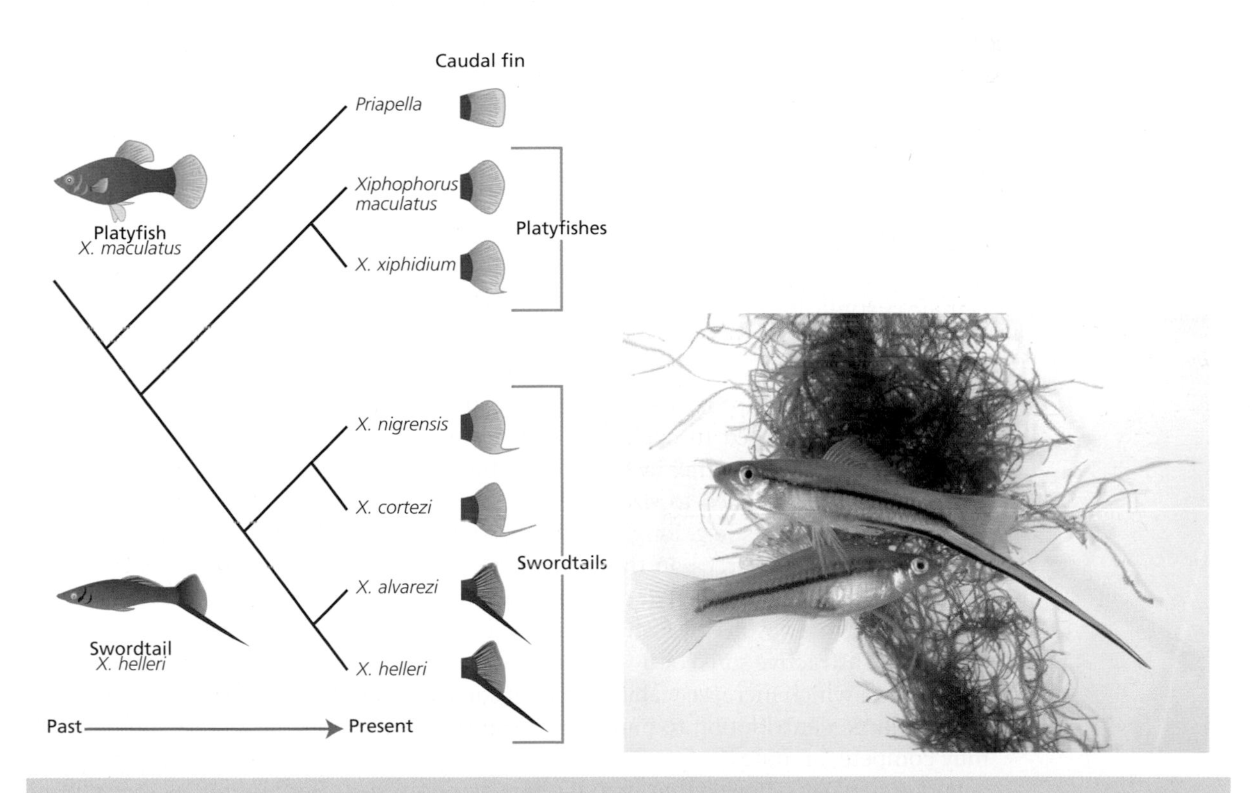

Fig. 7.24 The phylogeny of the swordtails suggests that the ancestor of the genus *Xiphophorus* lacked an elongated tail, with the swordtail evolving in the lineage that diverged from the platyfish. Surprisingly, female platyfish and *Priapella* prefer males of their species with experimentally elongated tails. This suggests that a sensory bias in favour of tails evolved first and that this was then exploited by the swordtail species. (Basolo 1990, 1995). Photo of female and male swordfish *Xiphophorus helleri*. © Alexandra Basolo.

females of the swordless platyfish and *Priapella*: when artificial swordtails were appended to the tails of their males, females preferred them to their normal tail-less males (Basolo, 1990, 1995). This suggests that these fish all have a pre-existing female preference for tails and this was then exploited by some *Xiphiphorus* species. Now, according to Fisher's runaway model, the preference and trait evolve together, so female tail preference is predicted to be stronger in the swordtails. However, female preference for swords is stronger in the genera that lack swords, suggesting that female attraction to swords has *declined* as swords have evolved, just as predicted by the chase-away model.

In *Schizocosa* wolf spiders, too, experiments show that tufts of bristles on the forelegs of males have stronger effects on female sexual receptivity in species which lack the tufts than in species with tufts (McClintock & Uetz, 1996). Once again, this suggests that the evolution of a male trait has been accompanied by female resistance.

Different models of sexual selection may apply to different traits

So, to conclude, we have three models for the evolution of female preferences by sexual selection and there are examples which provide evidence for all three. Future studies need to investigate the relative importance of the three different processes in nature. It is possible that males of some species will have traits that have evolved by all three processes; a male duck, for example, might have bright glossy plumage which signals his good genes for parasite resistance, a long-tail which has been exaggerated by the Fisher runaway process, and an elaborately shaped penis which has evolved to enhance extra-pair matings, but which has provoked the 'chase-away' coevolution of elaborate female genitalia as a counter-defence.

Summary

Females usually invest more in gametes and parental care, so males spend more time in the mating pool and successful males have a greater potential rate of reproduction. This leads to males competing for females and females choosing among males based either on the resources they provide or their genetic quality. Darwin's (1871) theory of sexual selection was proposed to explain the evolution of traits that are of advantage in competition for mates, either by force or by charm. There is good evidence that a male's mating success is related to size and strength (e.g. elephant seals) or to ornaments which attract females (e.g. long tails in widowbirds). Females may choose males that provide the best resources, in the form of nest sites, food or paternal care. However, females are also choosy even in cases where males provide only sperm. Here, elaborate male displays may evolve by Fisher's runaway process (where the benefits of choice are genetically attractive sons) or by honest advertisement of genetic quality (females gain good genes, which increase viability of both sons and daughters). In cases where males make a large contribution to parental investment, males may be choosy and females may compete for mates.

Parker (1970c) showed that sexual selection continues after mating, as sperm from rival males compete for fertilizations inside the female tract (sperm competition). Females may gain both material benefits and genetic benefits from mating with more than one male. This leads to sexual conflict, with male adaptations to improve success in sperm competition and female adaptations involving sperm choice and resistance to male manipulation. The result may be antagonistic coevolution between the sexes, where females evolve resistance to male force or charm.

Further reading

The topics discussed in this chapter are a particularly vigorous field of research, and the literature is vast. There are many excellent books: Malte Andersson's (1994) book reviews *Sexual Selection*; Tim Birkhead and Anders Møller's (1998) edited volume surveys *Sperm Competition*; and Göran Arnqvist and Locke Rowe's (2005) book covers *Sexual Conflict*. Tim Birkhead's book *Promiscuity* (2000) is an excellent popular scientific account of all three topics. For shorter reviews, see Chapman *et al.* (2003) on Sexual Conflict, Wedell *et al.* (2002) on Sperm Competition, Birkhead and Pizzari (2002) on post-copulatory sexual selection, Andersson and Iwasa (1996) on Sexual Selection and Clutton-Brock (2009a) on selection in males and females. Kokko *et al.* (2003) is a fine review of models of mate choice; their key point is that direct and indirect (genetic) benefits often go hand in hand. Bart Kempenaers (2007) reviews mate choice for genetic quality.

Robert Brooks (2000) provides evidence for Fisher's model of sexual selection with some clever experiments on guppies (see also Brooks & Couldridge, 1999). Pizzari and Foster (2008) review the 'social life' of sperm, showing how sperm sometimes cooperate as a team to enhance their fertilization success.

Trevor Price's book *Speciation in Birds* (2008) discusses sexual selection and speciation (Chapters 9–11) and Seddon *et al.* (2008) show that sexual selection has promoted speciation in the antbirds of South America. Kokko and Jennions (2009) review the links between parental investment and sexual selection.

Richard Prum (1997) shows how a phylogenetic analysis of male behaviour and plumage in manakins (a tropical bird family) can test between alternative models of sexual selection.

TOPICS FOR DISCUSSION

1. Tim Clutton-Brock (2009a) suggests we should broaden the definition of sexual selection, from one focusing on competition for mates to one focusing on competition for reproductive opportunities, and we should emphasize the contrasting ways in which selection operates in males versus females. Do you agree?
2. How would you test the hypothesis that females should have greater control over sperm choice in species where they have least control over mate choice?
3. A recent experimental study by Bro-Jørgensen *et al.* (2007) suggests that variation in tail length in barn swallows reflects natural selection, because the aerodynamically optimal tail length varies significantly between males. How might this complicate the interpretation of long tails as a sexually selected ornament?
4. Suggest hypotheses for why male satin bowerbirds prefer blue objects while male spotted bowerbirds prefer white and green objects. How would you test your hypotheses?
5. Why in some species do males compete for females by force while in others they do so by charm? Discuss the factors that might lead to one form of competition rather than the other.

6. Geoffrey Hill (1991) has shown that female house finches prefer to mate with redder males (this colour is related to dietary intake of carotenoids). Redder males provide better parental care (good resources). Could females also be gaining 'good genes' by preferring redder males? What experiments would you do to test this?
7. Same-sex sexual behaviour has been extensively documented in non-human animals. How would you explain this? How would you test your hypotheses? (see the review by Bailey & Zuk, 2009).

CHAPTER 8

Parental Care and Family Conflicts

Photo © Bruce Lyon

In the last chapter, we showed that there is sexual conflict over mating as rival males compete for copulations and females choose between males, and that this conflict continues after mating, as rival sperm compete for fertilizations and females choose sperm. In this chapter, we shall see that conflicts continue further when there is parental care of the eggs or young. We shall explore three interrelated conflicts (Fig. 8.1): conflicts between male and female parents over how much care each should provide; conflicts between siblings over how much care each should demand; and conflicts between parents and offspring over the supply and demand of care. We shall discuss the theory and evidence for each of these conflicts in turn, but before we do so we need to consider the circumstances that favour the evolution of parental care.

Three inter-related conflicts: between male and female parents, between siblings and between parents and offspring

Evolution of parental care

There are remarkable differences in patterns of parental care across the animal kingdom (Table 8.1). In many species there is no parental care (most invertebrates), either because parents are unable to protect their young effectively, or because selection favours the production of a large number of eggs, whose fate is then left to chance. Parental care in invertebrates tends to occur where fewer young are produced and they can be protected from the physical or biotic environment (predators, parasites). Parental care includes the preparation of nests and burrows, the provisioning of eggs with yolk food reserves and the feeding and protection of eggs and young before and after birth (Clutton-Brock, 1991). In some species only the female cares (most mammals), in others only the male (many fish), and in some both sexes care (most birds). To explain these patterns, we need to examine two factors. Firstly, different groups of animals have

Variation in parental care across the animal kingdom

An Introduction to Behavioural Ecology, Fourth Edition. Nicholas B. Davies, John R. Krebs and Stuart A. West.

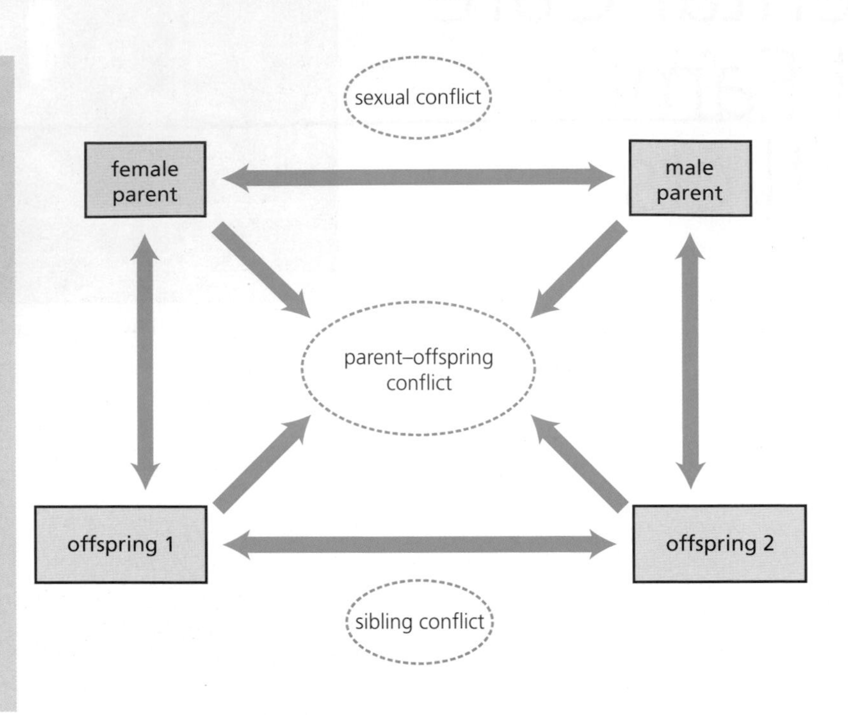

Fig. 8.1 The three types of intra-familial conflict discussed in this chapter. Male and female parents are in conflict over who should invest and how much. Siblings compete over the investment from parents. Offspring have different interests from their parents over supply and demand of investment. From Parker *et al.* (2002).

Table 8.1 Sex roles in parental care (Kokko & Jennions, 2008)

Taxon	Parental care	References
Invertebrates	Uncommon, but when it occurs mainly female only. Biparental care uncommon. Male-only care rarer.	Zeh & Smith (1985); Tallamy (2000).
Fish	Ratio of genera with male-only: biparental: female-only care is 9:3:1.	Reynolds *et al.* (2002).
Amphibians	Female-only care and male-only care equally common. Low frequency of biparental care.	Beck (1998); Summers *et al.* (2006).
Reptiles	Either female alone or both parents.	Reynolds *et al.* (2002).
Birds	In 90% species, biparental care (including 9% where helpers assist), but females often invest more in care. Females care alone in majority of the remaining species. Male-only care is rarer.	Lack (1968); Cockburn (2006).
Mammals	Females care in all species. In 95% species, the female cares alone. In 5% the male helps too. No cases of male-only care.	Clutton-Brock (1991).

different physiological and life history constraints which may predispose one sex to care more. Secondly, ecological conditions and mating opportunities will influence the costs and benefits of care for each sex. We shall illustrate these two themes by contrasting patterns of care in birds, mammals and fish.

Birds

As we saw in the first chapter, reproductive success in birds can be limited by the rate at which food is delivered to the nest. When two parents can feed twice as many young as a single parent, both male and female will increase their reproductive success by staying together. If either sex deserted, they would approximately halve the output of the last brood and would also have to spend time searching for a new mate and nest site before being able to start again. So monogamy and parental care by both male and female are not hard to understand.

Biparental care common in birds

When the constraint of having to have two parents feeding the young is lifted, it is usually the male who deserts and the female who is left caring for the brood. Comparative evidence shows that polygyny often occurs in fruit and seed eaters, probably because these food supplies become so seasonally abundant that one parent can feed the young almost as efficiently as two (e.g. weaver birds, Chapter 2). Why is it the male who deserts? There are two factors which may be important. Firstly, the male has the opportunity to desert before the female. With internal fertilization, she is left literally holding the babies inside her. Secondly, the male can often gain more by desertion than the female because his lifetime reproductive success depends more on his number of matings (Chapter 7).

Mammals

Female only care common in mammals

In mammals, females are even more predisposed to care for the young. The offspring often have a prolonged period of gestation inside the female, during which the male can do little direct care (though he can protect and feed the female). Once the young are born they are fed on milk and only the female lactates. Because of these constraints on the opportunity to care for offspring and also because, with internal fertilization, the male can desert first, it is not surprising that most mammals have parental care by the female alone, with the male deserting to seek further matings.

Monogamy and biparental care occur in a few species where the male contributes to feeding (carnivores) or to carrying the young (e.g. marmosets). Perhaps it is surprising that male lactation has not evolved in these cases (Daly, 1979).

Fish

In the bony fish (teleosts), most families (79%) have no parental care (Gross & Sargent, 1985). In those families which do care for the eggs or young, it is usually done by one parent; biparental care occurs in less than 25% of the families which show care. Compared with the elaborate care of offspring by birds, parental care in fish is a simple affair often consisting of just guarding or fanning eggs. These tasks can usually be done effectively by one parent alone. Which parent will provide care? Table 8.2 shows that

Table 8.2 Distribution of male and female parental care with respect to mode of fertilization in teleost fishes. The table shows number of families; a single family may appear in more than one category, but is not listed under 'no parental care' unless care is completely unknown in the family (Gross and Shine, 1981)

	Internal fertilization	External fertilization
Male parental care	2	61
Female parental care	14	24
No parental care	5	100

Uniparental male or female care common in fish

female care is commonest with internal fertilization (86% female care) and male care with external fertilization (70% male care). The overall predominance of male parental care in fish is related to the prevalence of external fertilization.

Three hypotheses have been proposed to explain why mode of fertilization influences which sex cares (Gross & Shine, 1981).

Three hypotheses for why male care is commonest with external fertilization and female care with internal fertilization

Hypothesis 1: Paternity certainty

Trivers (1972) suggested that reliability of paternity will be affected by mode of fertilization. Since external fertilization occurs at the time of oviposition, reliability of paternity may be greater than with internal fertilization where sperm competition may take place inside the female's reproductive tract (Chapter 7). According to this hypothesis, with internal fertilization a male should be less prepared to provide parental care than the female because he is less certain that the offspring are his. While this idea is sound in theory (Queller, 1997), it is not yet known whether paternity certainty really is greater with external fertilization. In some external fertilizers, for example sunfish *Lepomis*, cuckoldry takes place during oviposition.

Hypothesis 2: Order of gamete release

Dawkins and Carlisle (1976) suggested that, as with birds and mammals, internal fertilization gives the male the chance to desert first and thus leave the female to care. With external fertilization they suggested that the roles may be reversed; sperm are lighter than eggs so perhaps the male must wait until the eggs are laid before he can fertilize them, or else his gametes will float away. Therefore, the female has the opportunity to desert first and swim off while the male is still fertilizing the eggs! This is an ingenious idea but it must be rejected on empirical grounds. In fact, the most common pattern of gamete release in external fertilizers is simultaneous release by male and female. In these cases both sexes have an equal chance to desert, but 36 out of 46 species which have simultaneous gamete release and monoparental care have care provided by the male. Secondly, in some families of fish (Callichythyidae, Belontiidae) the male builds a foam nest and releases sperm before the female lays eggs. In these cases, the 'opportunity for desertion' hypothesis predicts that males can desert first, but nevertheless parental care is provided by the male. Therefore, male parental care remains correlated with external fertilization independently of the order of gamete release or opportunity to desert (Gross & Sargent, 1985).

Hypothesis 3: Association

Male care in fish is associated with male territoriality

Williams (1975) suggested that association with the embryos preadapts a sex for parental care. For example, with internal fertilization the female is most closely associated with the embryo and this may set the stage for the evolution of embryo retention and live birth, followed by care of the young fry. With external fertilization, on the other hand, the eggs are often laid in a male's territory and it is the male who is most closely associated with the embryos. Defence of the territory in order to attract further females becomes, incidentally, defence of the eggs and young, and therefore provides a preadaptation for more elaborate parental care by males. Therefore, male care involves fewer opportunity costs (lost matings) than in other cases, because a male that guards eggs can still attract more mates. In fact, females sometimes prefer males that already have eggs in their nest (Hale & St Mary, 2007). This hypothesis is the best predictor of the data in Table 8.2. Male parental care is more common in territorial species and the prevalence of male parental care with external fertilization results from the fact that male territoriality is particularly common with external fertilization.

Parental investment: a parent's optimum

We now change focus, from broad comparisons across different taxa (birds, mammals, fish) to consider what factors influence the amount of care a parent should devote to its offspring. Robert Trivers (1972) introduced the concept of *parental investment*, which he defined as 'any investment by the parent in an individual offspring that increases the offspring's chance of surviving (and hence reproductive success) at the cost of the parent's ability to invest in other offspring.' Parental investment will include any investment, such as guarding or feeding, that benefits the eggs and young. Lifetime parental investment will be the sum of all the resources a parent can gather in its lifetime and use for offspring care.

Investment trade-offs within broods and between broods

From a parent's point of view, what is the optimal parental investment per offspring? The key point is that this will involve trade-offs, because whereas increased investment in any one young will bring benefits to that offspring, there will be costs to the parent in that it reduces the resources available for other offspring. These trade-offs will operate at two stages. The first was recognized by David Lack (Chapter 1) and involves the trade-off between offspring quantity and quality *within a brood*. If a parent spreads its limited resources thinly among too many offspring, then few of them will survive. On the other hand, if it uses its resources too generously among a small brood, then other parents will produce more surviving young and will outcompete it over the generations. In theory, there will be an optimal brood size to maximize productivity per brood. The second trade-off, first recognized by G.C. Williams (1966b), involves that between investment in *current versus future broods*. To maximize its lifetime success, a parent needs to allocate parental care optimally not only within broods but also between broods, as increased investment in any one brood will reduce a parent's ability to invest in future broods.

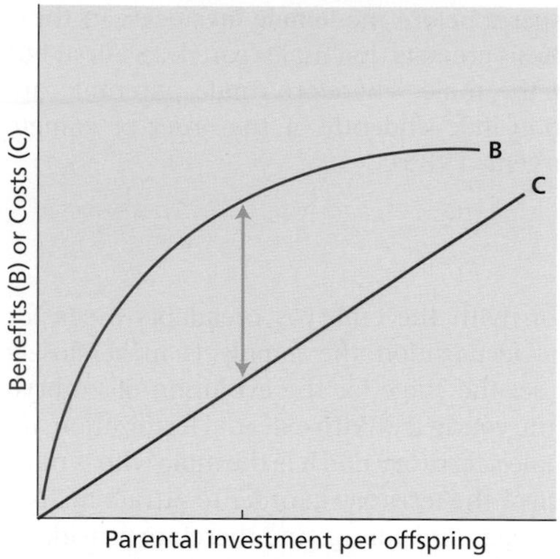

Fig. 8.2 shows the theoretical optimal investment per offspring from a parent's point of view. Increased investment in any one offspring will bring diminishing benefits, as the care provided begins to saturate the offspring's needs. Increased investment will also bring increasing costs in terms of reduced resources available for other offspring, both in current and future broods. There will be an optimum where the benefits minus the costs are at a maximum.

Benefits (B) or Costs (C)

B

C

Parental investment per offspring

Fig. 8.2 The optimal parental investment per offspring from a parent's point of view is where the Benefits minus Costs are at a maximum. Increasing investment brings diminishing benefits as the offspring's needs become saturated, but costs continue to increase because every unit of continued investment deprives other offspring (current and future) of a parent's limited lifetime resources for care.

Evidence for trade-offs: increased investment may reduce adult survival or future fecundity

There is good experimental evidence for this trade-off between the benefits and costs of parental care, but the form of the trade-off varies between species. In some cases increased investment in the current brood reduces the parent's survival. In side-blotched lizards *Uta stansburiana*, gravid females not only have the extra mass of their eggs to carry, but their distended abdomens hinder their leg movements. When some females had half their eggs removed surgically, they had improved locomotary performance (measured on a treadmill) and were more likely to survive to produce another clutch, probably because of reduced predation (Miles *et al.*, 2000). In other cases, increased investment reduces an adult's future fecundity, rather than survival. When male common gobies, *Pomatoschistus microps*, were induced to invest more in fanning the eggs in their nest (by reducing levels of dissolved oxygen in their aquaria) they lost more mass and were more likely to abandon their next clutch (Jones & Reynolds, 1999). When male and female burying beetles *Nicrophorus vespilloides* were induced to care for a large brood of larvae in their first breeding attempt, they subsequently produced fewer larvae from future broods than those that cared for a small brood the first time they bred (Ward *et al.*, 2009). Similarly, when collared flycatchers *Ficedula albicollis* were induced to increase their feeding rates to their current brood (by increasing brood size), both males and females survived as well as control birds but they had reduced fecundity the following year (Gustafsson & Sutherland, 1988).

Obviously, increased reproductive effort may involve greater exposure to predators or compromise the breeder's ability to maintain its own body condition, and so reduce a breeder's survival. But why would increased current reproduction sometimes reduce only future fecundity? One possibility is that resources allocated to reproduction are at the expense of resources available for the immune system, so increased reproductive effort reduces the breeder's physiological condition (Sheldon & Verhulst, 1996; Norris & Evans, 2000).

Varying care in relation to costs and benefits

Do parents vary their care in relation to these costs and benefits? Studies have investigated this question both by comparing the behaviour of related species which face different selection pressures, and by experiments in which costs and benefits have been manipulated for particular species.

Comparing North and South American passerine birds

Cameron Ghalambor and Thomas Martin (2001) compared parental risk-taking in response to predators by passerine birds breeding in North and South America. Species in temperate North America tend to have large clutch sizes (typically four to six eggs) and low adult survival to the next breeding season (around 50% or less). By contrast, species in tropical South America tend to have small clutches (typically two to three eggs) and high adult survival (around 75% or more). The selective forces leading to these different life histories are likely to be complex, including a greater flush of food during northern temperate breeding seasons (which permits larger clutches) but harsher climatic conditions in the non-breeding seasons (which reduces adult survival). Ghalambor and Martin predicted that these life history differences should lead to different parental risk-taking responses. Their previous work had shown that increased nest visits by parents led to increased risks of predation to both parents and young, by attracting the attention of predators. Therefore, they measured parental responses in terms of how much parents reduced their nest visits to a brood in the presence of predators of the adult parents versus predators of the nestlings.

Parent birds modulate risk-taking in relation to value of current versus future broods

Five pairs of species were tested, matched for phylogeny and ecology, including a North and South American species of each of the following families: flycatcher, thrush, wren, bunting and warbler. Ghalambor and Martin predicted that South American species should respond more strongly than their North American counterparts to a predator of adult birds (a hawk) because South American parents have greater expectations of future reproduction. Conversely, North American parents should respond more strongly to a nest predator (a jay) because their current brood is more valuable than is a brood for a South American parent. The data supported both these predictions (Fig. 8.3). Therefore, parents modulate their risk-taking in response to both the costs and benefit of parental investment.

Flexible parental response to current brood demands

Prospects of future reproduction can also influence a parent's response to increased demands from its current brood. Rose Thorogood and colleagues (2011) showed this in a neat field experiment with a nectar-feeding passerine bird from New Zealand, the hihi

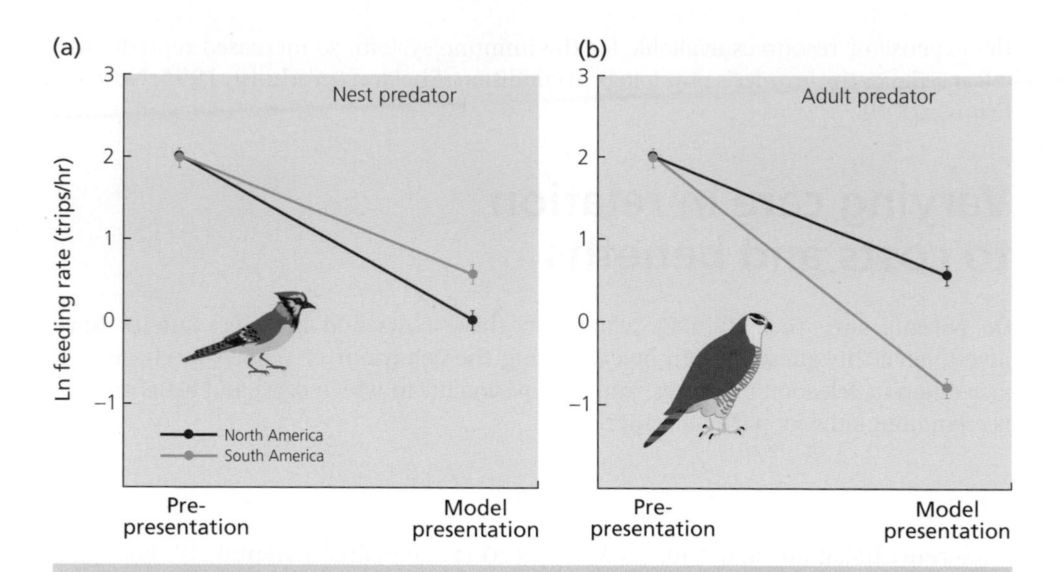

Fig. 8.3 Responses of North American and South American songbird parents to presentations of: (a) a nest predator (experimental playback of calls of a jay) and (b) a predator of adults (presentation of taxidermic mount of a hawk). Responses measured as reduction in feeding visits to a brood of nestlings. South American parents value their own lives (future broods) more (stronger response to adult predator) while North American parents value their current brood more (stronger response to nest predator). From Ghalambor and Martin (2001). Reprinted with permission from AAAS.

Response to current brood demands varies with parents' future breeding prospects

Notiomystis cincta. On territories where broods were fed experimentally with extra carotenoids in a sugar solution, their mouths became redder and this enhanced begging display led to increased provisioning by the parents, probably because redder mouths signalled healthier offspring, worth more investment. However, on other territories, where the adults were also provided with carotenoid-rich sugar feeders, this increased the chance that they had a second brood that season. The pairs that had second broods did not respond to the enhanced begging signals of their current brood. Therefore, parents strategically varied their sensitivity to their current brood's demands in relation to their future prospects of breeding that season.

Flexible care in St Peter's fish

Galilee St Peter's fish, *Sarotherodon galilaeus*, is a mouth brooding cichlid found in rivers and lakes throughout Africa and Asia minor. Parental care may be provided by either sex alone or by both parents. This flexibility within a species provides an ideal opportunity to test whether males and females vary their care in relation to the costs and benefits. Mating is monogamous; pairs dig a shallow depression in the substrate together, then the female lays batches of 20–40 eggs into the depression and the male glides over them, fertilizing each batch in turn until the clutch is complete. Then either the female, or the male, or both parents pick up the eggs in their mouths, where they protect the eggs and young fry for about two weeks. The pair bond dissolves after mouth brooding begins, even in cases where both sexes provide care.

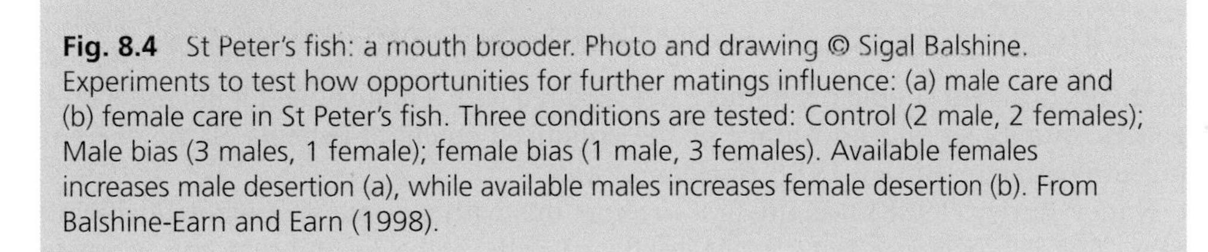

Fig. 8.4 St Peter's fish: a mouth brooder. Photo and drawing © Sigal Balshine. Experiments to test how opportunities for further matings influence: (a) male care and (b) female care in St Peter's fish. Three conditions are tested: Control (2 male, 2 females); Male bias (3 males, 1 female); female bias (1 male, 3 females). Available females increases male desertion (a), while available males increases female desertion (b). From Balshine-Earn and Earn (1998).

Opportunities of further matings influence parental investment

Sigal Balshine-Earn and David Earn (1998) performed experiments in tanks. Firstly, they varied the opportunities for further matings. When the sex ratio was female-biased, males increased their likelihood of desertion. Conversely, when the sex ratio was male-biased, female desertion increased (Fig. 8.4). Therefore, both males and females deserted their offspring more frequently when the costs of care were higher (in terms of lost mating opportunities). Next, they varied the benefits of biparental care by pairing same-sized females with males that were larger than the female, the same size or smaller. Smaller males have smaller buccal cavities so are able to carry fewer eggs and fry. Therefore, biparental care involving a small male brings fewer additional benefits compared to uniparental care. There were no differences in clutch size between the three treatments, but smaller males were more likely to desert. Therefore, males are more likely to desert when the benefits they gain from care are reduced. Overall, uniparental care (either by male or female) was more likely when clutch size was small, again suggesting that patterns of care were influenced by benefits as well as costs.

Filial cannibalism

The act of eating one's own offspring (filial cannibalism) seems bizarre at first. It is particularly common in fish and was long regarded as abnormal behaviour. However, Sievert Rohwer (1978) suggested that it might sometimes be adaptive for a parent to use its offspring as an extra source of food, eating part of a brood to improve parental care of the current brood, or eating the whole brood to cut parental losses and improve future reproductive success.

It sometimes pays a parent to eat all or part of its brood

Andrea Manica (2002, 2004) has shown that male scissor-tailed sergeant fish *Abudefduf sexfasciatus* vary filial cannibalism in response to experimental changes in the costs and benefits of care. Males defend territories on coral reefs. They alternate between a two to three day mating phase, when they become golden in colour and display to attract egg laying females to their territories, and a four to five day parental phase, when they lose their gold colour, cease displays and guard their eggs until they hatch. When males had their clutches reduced by 75% on the first day of the parental phase they were more likely to cannibalize the remaining eggs and revert to the mating phase than control males, who had the same disturbance but without egg reduction. Egg reduction on the third day of the parental phase led to no increase in cannibalism, most likely because parental care was then less costly (fewer days to hatching). In another experiment, the provision of supplemental food (conspecific eggs or crabmeat) led to a reduction in partial filial cannibalism compared to control males, suggesting that males sometimes ate part of their brood to fuel their energy requirements for parental care.

In other fish, increased cuckoldry during spawning (which reduces the benefit of care) also led guarding males to increased cannibalism (Gray *et al.*, 2007) or reduced care (Neff, 2003).

Varying investment in response to mate attractiveness

In theory, a parent should invest more when paired with a mate of better phenotypic or genetic quality, to take advantage of the enhanced potential benefits of the current breeding attempt (Burley, 1986; Sheldon, 2000).

Nancy Burley (1988) was the first to show this with classic experiments involving zebra finches, *Taeniopygia guttata*. Males have bright red beaks and they can be made more attractive to females experimentally, by giving them red leg bands. When paired to these attractive males, females increased their effort in chick-feeding, and raised more young, compared to when given less attractive males with blue or green leg bands.

Females may invest more when paired to an attractive male

Similarly, experiments have shown that female mallard ducks lay larger (better provisioned) eggs when paired with more attractive males (Cunningham & Russell, 2000), and female peacocks lay more eggs after copulating with males with more elaborate tails (Petrie & Williams, 1993).

Sexual conflict

The examples we have discussed so far reveal the first of the conflicts we shall encounter in this chapter, namely sexual conflict. It is useful to consider this in two stages: conflict over who should care, and conflict over how much care to provide.

		Female	
Male		**Care**	**Desert**
Care	Female gets	wP_2	WP_1
	Male gets	wP_2	WP_1
Desert	Female gets	wP_1	WP_0
	Male gets	$wP_1 (1 + p)$	$WP_0 (1 + p)$

Table 8.3 An ESS model of parental investment (Maynard Smith, 1977). Each sex has the possibility of caring or deserting. The matrix gives the reproductive success for males and females (see text for details)

Who should care?

Each parent faces the decision of whether to care or desert. As we have seen, the costs and benefits of these two options will depend on constraints (e.g. female mammals lactate, males do not), and ecological factors which will influence both offspring survival (intensity of predation, availability of food) and the opportunities for further matings. However, a key influence on these costs and benefits will be the behaviour of the other parent. If the female cares, it may pay the male to desert; but if the female deserts, it may pay the male to care. In St Peter's fish, for example, each parent might do best if the other did the caring and it was then free to desert to seek further matings. John Maynard Smith (1977) was the first to propose a model which explored the outcomes of joint decisions by both parents. There is a flaw in his original model, but it is instructive to explain this, and to explore both the fundamental insights and limitations of the model.

The best strategy for one parent depends on the strategy adopted by the other parent

The model introduced a game theoretic approach to parental conflicts. It assumed that each parent decides independently whether to care or desert (these 'blind bids' are a simplifying and often unrealistic assumption – see later). The model looks for a pair of strategies, say I_m for males and I_f for females, such that it would not pay a male to diverge from strategy I_m so long as females adopt I_f, and it would not pay females to diverge from I_f as long as males adopt I_m. In other words, we seek the evolutionarily stable strategies (ESS) for male and female (Chapter 5).

An ESS model for parental investment

Let P_0, P_1 and P_2 be the probabilities of survival of eggs which are not cared for, cared for by one parent and cared for by two parents, respectively; $P_2 > P_1 > P_0$. A male who deserts has a chance p of mating again. A female who deserts lays W eggs and one who cares lays w eggs; $W > w$ (caring females have fewer resources for eggs).

The pay-off matrix for this game is illustrated in Table 8.3. There are four possible ESSs:

ESS 1: female deserts and male deserts. This requires $WP_0 > wP_1$, or the female will care, and $P_0 (1 + p) > P_1$, or the male will care.

ESS 2: female deserts and male cares. This requires $WP_1 > wP_2$, or the female will care, and $P_1 > P_0 (1 + p)$, or the male will desert.

ESS 3: female cares and male deserts. This requires $wP_1 > WP_0$, or the female will desert, and $P_1(1+p) > P_2$, or the male will care.

ESS 4: female cares and male cares. This requires $wP_2 > WP_1$, or the female will desert, and $P_2 > P_1(1+p)$ or the male will desert.

For given values of the parameters in the model, ESS 1 and ESS 4 can be alternative possibilities, as can ESS 2 and ESS 3. For example, ESS 2 is most likely if the female can lay many more eggs if she does not invest in caring ($W >> w$) and if one parent is much better than none ($P_1 >> P_0$) but two parents are not much better than one ($P_2 \approx P_1$). This situation probably applies to many fish, as discussed above, where the female tends to desert and the male often cares. However, ESS 3 is an alternative possibility, especially if a male who deserts has a much better chance of mating again; this may apply to some species of birds and mammals. If two parents can raise twice as many young as one ($P_2 >> P_1$), or if the chance that a deserting parent will remate is small, then ESS 4 is the likely outcome, as in many species of passerine birds.

Correcting a flaw in the model

If you haven't spotted the flaw in the pay-off matrix in Table 8.3, then you are in good company because it remained unnoticed for 25 years (Kokko & Jennions, 2003). The problem, pointed out by Michael Wade and Stephen Shuster (2002) is that deserting males gain extra offspring by mating with females who 'appear from nowhere', in the sense that these females do not appear in the calculations of female fitness. Males, therefore, have more total paternity than is possible from the number of offspring produced by females. Wade and Shuster refined the model to show exactly where these extra offspring might come from. For example, deserting males might steal some paternity in care-giving male broods, or they might mate with females who have deserted care-givers. These considerations complicate the pay-off matrix. Nevertheless, the fundamental insight provided by Maynard Smith's model remains: the key to understanding decisions by one sex is the decisions made by the other parent.

While Maynard Smith's model still provides a useful framework for thinking about the evolution of parental care, the assumption that parents make independent decisions (blind bids) over caring and deserting is unlikely to apply to most cases in nature. For example, independent decision making would sometimes lead both parents to desert in St Peter's fish, but at least one parent always cares (Balshine-Earn, 1995). More realistic models need to incorporate sequential decision making and opportunities for deserting first.

How much care?

There is also sexual conflict over how much care to provide. For example, a female scissor-tailed sergeant fish would do best if the territorial male guarded all her eggs safely through to hatching, but the male might do best to cannibalize some, or even all, of her brood to maximize his success over all the clutches he will fertilize in his lifetime. Sexual conflict will also occur during biparental care. We may not see obvious conflicts as a pair of birds works hard to provision a hungry brood, but a simple experiment reveals the underlying conflict. If either parent is removed temporarily, then the other parent often increases its work rate. This shows that each parent has the capacity to work harder. How then do cooperating parents come to an agreement over how hard each should work? The problem is that of cheating; a partner may be tempted to do less than its fair share of work by relying on the compensatory reactions of the other partner.

Conflicts over how much care each member of a pair should provide

This conflict over parental investment was first modelled as an evolutionary game, in which each partner independently plays a fixed effort (a 'blind bid') and the optimal effort for each parent is then resolved over evolutionary time (Chase, 1980; Houston & Davies, 1985). At the ESS, each parent will invest a fixed level of effort that maximizes its own fitness, given the effort invested by its mate. Consider a pair. The male will have a 'best response', in terms of parental effort, to a given effort put in by the female. Likewise, the female will have a 'best response' to any given effort by the male. If brood productivity is an increasing but decelerating function of total parental effort, and the costs of increased effort for a parent are non-decelerating (as in Fig. 8.2), then it can be shown that the 'best responses' for each parent will involve incomplete compensation. That is to say, if one parent reduces its effort, the other will increase its effort but not sufficiently to compensate for the loss (Figs 8.5a, 8.5b). Such incomplete compensation leads to stable biparental care, as explained in Fig. 8.5c.

If conditions led to a partner fully compensating, or even over-compensating for a reduction in effort by the other partner, then biparental care would be unstable and the ESS is for uniparental care (as explained in Fig. 8.5d).

This basic framework has been extended to incorporate behavioural negotiation between the parents, where each parent may adjust its own effort in response to that of its partner. Here, it is the 'response rules' that evolve rather than the effort levels. The mathematics now become complex and the evolutionary outcome of the negotiation game differ in detail from the 'blind bid' analysis. Nevertheless, the model still predicts incomplete compensation, this time on a behavioural time scale (McNamara *et al.*, 1999).

In theory, incomplete compensation stabilizes biparental care

The key theoretical prediction, therefore, is that in cases of biparental care, parents should respond to reduced partner effort with incomplete compensation. A partner who cheats, by reducing its effort, will suffer reduced fitness because its young will get less well fed. This has been tested in numerous experiments with birds (e.g. starlings *Sturnus vulgaris*), where the effort of one partner has been reduced by various techniques, including temporary removals, feather cutting, tail weighting and testosterone implants. Overall, the mean response of the other partner was indeed to increase its effort, but with partial compensation (Harrison *et al.*, 2009). Nevertheless, in some studies the unmanipulated partner showed no response, or full compensation. And a careful study by Camilla Hinde (2006) produced a surprising result. She tricked parent great tits, *Parus major*, into increasing parental effort by augmenting the begging calls of their brood with playback of extra calls through a little loudspeaker placed next to the nest (Fig. 8.6). When one parent only (either male or female) was exposed to the playback, it increased its provisioning rate (the expected response to an apparently more hungry brood). However, the other parent also increased its provisioning (Table 8.4), even though it did not experience increased chick begging (the playback had no effect on the chicks themselves). The unmanipulated adult must, therefore, have responded directly to its partner, increasing its own effort in response to the partner's increase.

An experiment with great tits: partners match each other's response

This matching increase in response is not what our model predicts. How might we explain it? The most likely explanation is that the model is too simple. When one parent changes its level of investment in parental care, this may influence the behaviour of the other parent for two reasons. The first, addressed by the model, is that a change in

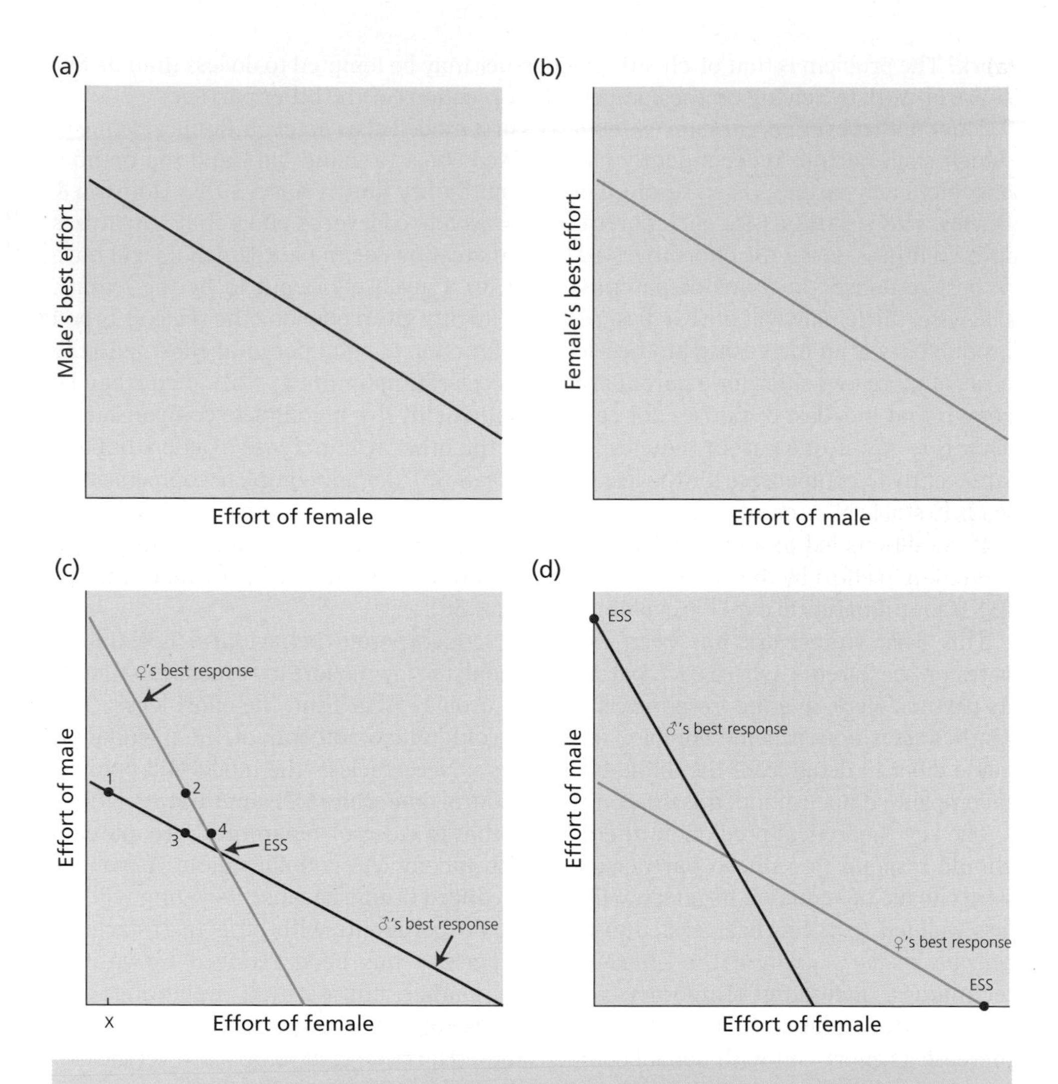

Fig. 8.5 Male-female conflict over how much care to provide (Houston & Davies 1985). (a) A male's best response to the female's parental effort; (b) a female's best response to the male's parental effort. These shallow slopes involve incomplete compensation, so if the partner reduces its effort, the other increases but not sufficiently to compensate for the loss. (c) Plotting both responses together shows that these lead to stable biparental care. Imagine, for example, that the female plays effort *x*. The male's best response is one. The female then replies with two, the male with three, the female with four, and so on, reactions proceeding by smaller and smaller amounts until the intersection which is the ESS. (d) If reactions have slopes steeper than one (over-compensation) the intersection is unstable; responses proceed by larger and larger amounts until one parent ends up doing all the work. The reader is invited to start with any female effort and then follow the male's best response, the female's best reply, and so on. The ESS is for uniparental care by either male or female. Which parent it is depends on the starting point of the game.

Fig. 8.6 Camilla Hinde's (2006) experiment with great tits. (a) A great tit brood. A speaker is hidden inside the nest, so begging calls of the brood can be augmented by playback. Photo © Simon Evans. (b) Male parent great tit. Photo © Joe Tobias.

Table 8.4 Parental response to playback of extra calls (Hinde, 2006) Each parent responded by increasing its provisioning rate to the brood. The partner who did not experience the playback also increased its effort.

Treatment	Feeds/hour to the brood		
	Female	Male	Total
Control (no playback)	15	18	33
Playback to female parent	22	25	47
Playback to male parent	19	27	46

partner effort directly affects the benefit of additional investment by its mate. If investment yields decelerating returns (as assumed in the model) then greater effort by one parent should lead to a decline in the marginal benefit of care and so favour a compensatory reduction in investment by its mate. However, the model ignores a second, indirect way in which a change in partner effort might influence its mate's behaviour, namely by conveying information about brood needs. If one partner has better information about how hungry the chicks are, then the other parent may use the partner's work rate as a cue to how hard it should work. This might lead to matching responses if increased effort by one parent signalled to the other parent that the brood's needs had increased (Johnstone & Hinde, 2006).

Future work needs to investigate the relative importance of compensation and cueing in parental responses. The main message from this example is that there can be fruitful exchanges between theory and empirical studies; when a model fails to predict observed behaviour this can often provide insights into aspects of natural history that the theory has neglected.

Sibling rivalry and parent–offspring conflict: theory

Robert Trivers's theory

Everyone is familiar with human family conflicts, if not from personal experience then from literature and art. So it is perhaps surprising that scientists were slow to recognize the importance of conflict in the evolution of interactions in animal families. In 1974, Robert Trivers published a paper which has changed our perception of family life.

Intrabrood conflict: each offspring should demand more than its fair share from the parent's point of view

Let's begin by returning to Fig. 8.2, where we considered a parent's view of how investment should be allocated to a particular offspring. Trivers's insight was that the optimal investment would be different from an offspring's point of view. It is useful to distinguish two sources of conflict. The first is *intrabrood* conflict. Consider a species with sexual reproduction and a brood of two offspring (Fig. 8.7a). A parent is related genetically to each offspring by 0.5 (half an offspring's genes come from the other parent). Therefore, there is no genetic reason for a parent to favour either offspring

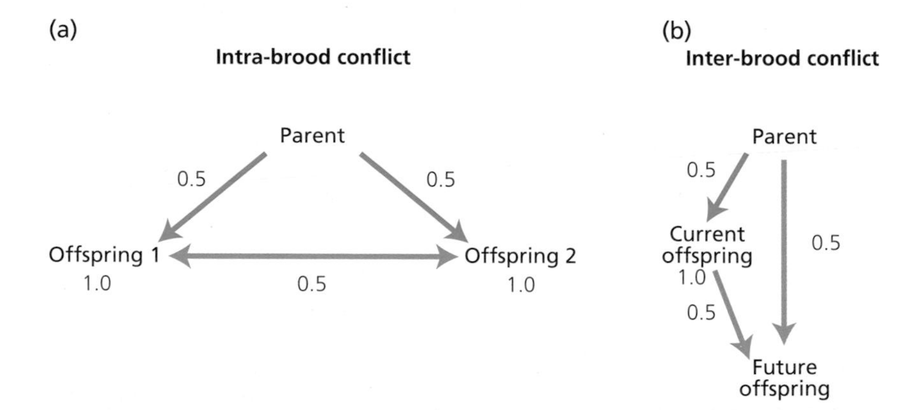

Fig. 8.7 (a) *Intrabrood conflict.* A parent with two offspring in a brood; the parent is equally related to both offspring (r = 0.5) but each offspring is more related to itself (r = 1) than to its sibling (r = 0.5, if full sibling). (b) *Interbrood conflict.* A parent with one offspring per brood. Again it is assumed that the offspring in the next brood is a full-sibling (same father and mother). The current offspring values itself (r = 1) more than its future full sibling (r = 0.5), whereas the parent is equally related to both offspring (r = 0.5).

(there may be practical reasons, for example if one is smaller, but we ignore these for the moment). However, from either offspring's point of view, it should be more interested in its own welfare (it is genetically related to itself by a fraction, one) than its sibling (if its sibling is a full sibling, then relatedness is 0.5; Chapter 11). Therefore, each offspring should try to grab more than its fair share of parental investment.

Interbrood conflict: current broods should demand more, at the expense of future broods

There will also be *interbrood conflict* (Fig. 8.7b). Imagine a parent which has just one offspring at a time (e.g. a seal). A time will come where it will pay a parent to terminate care of this offspring and save further investment for the next one (the B–C maximum in Fig. 8.2). However, the current offspring will benefit by continuing to demand care because, once again, it is genetically more interested in its own welfare than that of its future sibling.

We can illustrate parent–offspring conflict graphically (Fig. 8.8). The benefit and cost curves from a parent's point of view remain the same as in Fig. 8.2; increased investment in any one offspring brings diminishing benefits, yet costs continue to increase because every unit of investment decreases resources available for other (current and future) offspring. We now add benefit and cost curves from an offspring's point of view. For an offspring, the benefit curve will be twice the parental benefit curve (an offspring is twice as related to itself compared to the parent's relatedness to it). However, the offspring will also experience costs from increased investment because this deprives its siblings, in

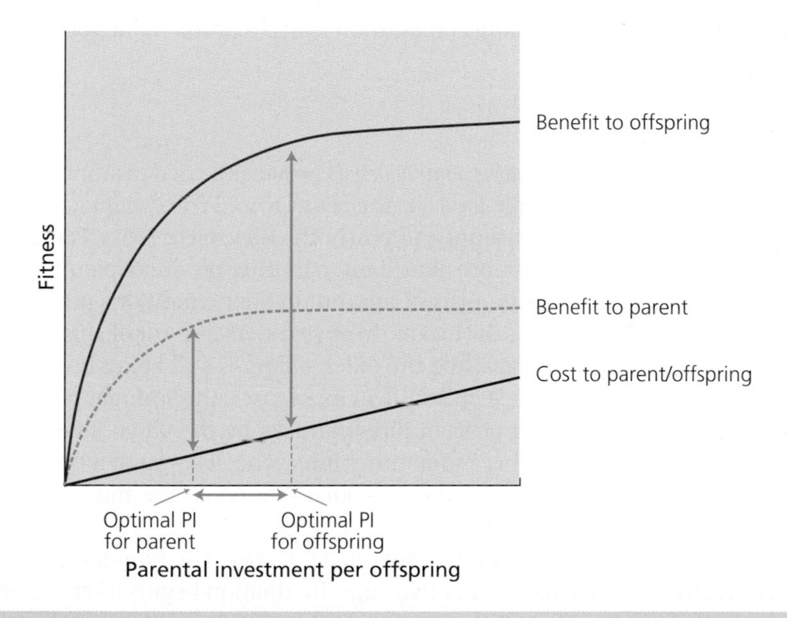

Fig. 8.8 Trivers's (1974) theory of parent–offspring conflict. The benefits and costs from the parent's point of view are the same as for Fig. 8.2. However, an offspring will value its own life (r = 1) twice as much as it is valued by its parent (r = 0.5), so the benefit curve for the offspring is twice that for the parent. If siblings are full siblings (r = 0.5) then the cost curve for the offspring is the same as that for the parent (see text). The optimal parental investment from an offspring's point of view is greater than the parental optimum. From Lazarus and Inglis (1986). With Permission from Elsevier.

whom it has a genetic stake. If these are full siblings, then the cost curve will be the same as that for the parent because relatedness is 0.5 in both cases; a unit of extra investment costs the parent investment in future offspring of relatedness 0.5, and costs the offspring investment in future siblings, also of relatedness 0.5. Fig. 8.8 shows that the optimal parental investment from an offspring's point of view is greater than that from the parent's point of view. Between these optima there will be conflict, where it pays the offspring to continue to demand care but it pays the parent to resist offspring demands. Beyond the offspring optimum, both parent and offspring agree that investment should cease.

Sibling rivalry leads to parent–offspring conflict

The key point from Trivers's theory is that it is sibling rivalry which leads to parent–offspring conflict. If a parent was designed to have just one offspring in its lifetime, there would be no conflict.

Sibling rivalry: evidence

We now consider the evidence, firstly for sibling rivalry and then for parent–offspring conflict. There is abundant evidence that siblings compete for parental resources. This often arises because food availability in the environment is unpredictable. It then pays mothers to produce an optimistic brood size, in the hope that conditions will be good. If food turns out to be scarce, sibling competition leads to brood reduction.

Facultative siblicide

The Galapagos fur seal *Arctocephalus galapagoensis* provides a good example of *interbrood conflict* arising from unpredictable food. In the seas around the Galapagos Islands, the availability of fish varies with seasonal and yearly changes in currents. Female fur seals have one pup at a time. When fish are abundant, a mother produces plenty of milk and can wean her pup when it is 18 months of age, but in poor conditions pups grow more slowly and suckling can continue for two to three years. As a result of this variation, up to 23% of pups per year are born while the older sibling is still being nursed. The two pups then compete for the mother's milk and, in most cases, the younger pup dies within a month, either from starvation or from direct attacks by the older sibling, who may grab it and toss it in the air. Mothers sometimes intervene, leading to a fatal tug of war as the older sibling pulls one end of the newborn pup while the mother attempts to retrieve it (Trillmich & Wolf, 2008).

Sibling rivalry in fur seals and boobies

Unpredictable fishing can also lead to *intrabrood conflict*. The blue-footed booby *Sula nebouxii* is a tropical seabird which lays two eggs. Incubation begins after the first egg is laid, so the first chick has about four days' growth before its younger sibling hatches. This size advantage means the older chick can reach up higher to intercept the regurgitated fish from its parent's bill, and only after it is satiated does the younger chick get fed. If food is abundant, then both chicks can take their turn. However, when food is scarce the younger chick rarely gets fed and it starves to death within the first two to three weeks. The key predictor of the younger chick's survival chances is the weight of its elder sibling. When the elder sibling is 20–25% below its expected weight, it attacks the younger sibling by pecking it. The younger sibling then cowers, becomes reluctant to beg and starves to death. During an experiment conducted in a year when food was abundant, both chicks were deprived of food for a day in some nests. The elder chick then increased its attacks on the younger sibling but these attacks declined once feeding was resumed. Thus siblicide is facultative and depends on the hunger of the elder chick. Remarkably, the parents never intervene to protect the younger chick (Drummond & Chavelas, 1989).

Many other studies have documented increasing sibling rivalry within broods as food becomes scarcer, with older siblings hastening the demise of their younger siblings directly, by wounding or killing them, or indirectly, by dominating parental feeds (Mock & Parker, 1997).

Obligate siblicide

Second eggs as an insurance

In some birds of prey, pelicans and boobies the mother lays two eggs, yet the older sibling *always* kills the younger sibling. This raises the question of why the mother lays two eggs rather than just one. The obvious hypothesis is that in these cases the second egg is an insurance in case the first egg fails to hatch. In support of this, in Nazca boobies *Sula granti* the second egg often produces a surviving offspring after the first egg's failure (Anderson, 1990). Furthermore, in both this species (Clifford & Anderson, 2001) and American white pelicans *Pelecanus erythrorhynchos* (Cash & Evans, 1986), experimental removal of the second egg leads to reduced reproductive success.

Sibling relatedness influences rivalry

In theory, an offspring's demand for parental resources should depend not only on the benefits it gains but also on the costs from depriving siblings of resources (Fig. 8.8). As relatedness to other offspring declines, they become less valuable (genetically), so the costs of depriving them of parental resources will decline (a shallower slope of the cost curve in Fig. 8.8). Therefore, an offspring's demands for parental resources should increase.

Two studies have tested this prediction by comparing the begging displays of various species of passerine birds. Here, offspring attempt to gain parental resources through charm rather than by force. (There's an obvious analogy with the two means by which males compete for females, Chapter 7.) Passerine nestlings are naked and poorly developed at hatching, and they beg by calling loudly and by displaying their brightly coloured gapes to their parents. Jim Briskie and colleagues (1994) found that nestlings begged more vigorously (as measured by the loudness of their begging calls) in species with higher levels of extra-pair paternity in the brood (where fellow nestlings were more likely to be half-siblings, and hence of lower relatedness; Fig. 8.9a).

Increased offspring selfishness when brood mates are of lower relatedness

In some species, a nestling's mouth becomes redder with increased hunger and parents use this as a signal to direct their food to the most needy chicks in the brood (Kilner, 1997). Rebecca Kilner (1999) predicted that if decreased relatedness to other offspring in the brood led to increased offspring selfishness, then offspring should have redder mouths in species with higher rates of extra-pair paternity. The data supported this prediction, but only for species breeding in well-lit nests, suggesting that signalling environment as well as sibling conflict influenced nestling displays (Fig. 8.9b).

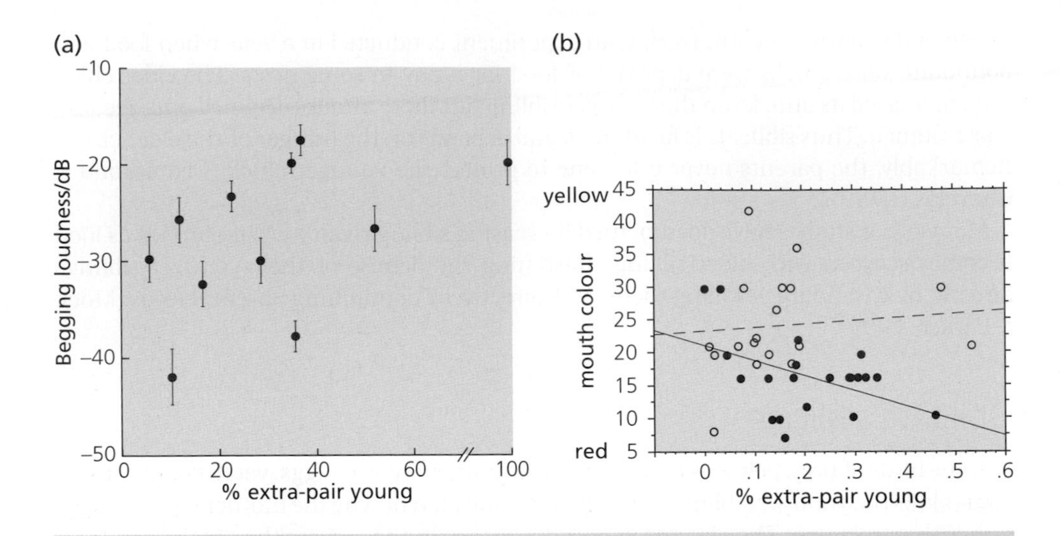

Fig. 8.9 Sibling conflict within broods of nestling birds increases as relatedness declines. (a) Nestlings beg more loudly in species where there is higher extra-pair parentage (i.e. lower average relatedness between siblings). This significant relationship still holds when controlling statistically for phylogeny, brood size and body mass. The species with 100% extra-pair parentage is the brown-headed cowbird, a brood parasite unrelated to the host young (Briskie *et al.* 1994). (b) Nestlings also have redder mouths in species with higher extra-pair parentage, but only in species nesting in open nests (solid symbols; solid line), not in those nesting in dark nests (open symbols: dashed line) (Kilner, 1999).

In precocial birds, the newly-hatched chicks have downy feathers and can run or (in waterbirds) swim soon after hatching. In the rail family (Rallidae), 36 of the 97 species studied had ornamented chicks, in the form of brightly coloured bills, fleshy patches or plumes. A phylogenetic analysis revealed that chick ornamentation has evolved multiple times within this family and is associated with increased sibling competition, as measured by larger brood sizes and mating systems involving multiple parentage (hence lower relatedness between siblings; Krebs & Putland, 2004).

So far, parental feeding preferences have been studied in just one species of this family, the American coot *Fulica americana*, in which newly-hatched chicks have long, bright orange tips to their black body feathers (Fig. 8.10a). Bruce Lyon and colleagues (1994) performed an elegant experiment in central British Columbia, Canada. When whole broods of chicks had their orange plumes trimmed, so they became black in colour, they were fed by their parents just as well, and grew just as well as whole broods left with their normal plumes intact (Fig. 8.10b). This shows that these plumes were entirely ornamental and did not influence chick viability directly, for example by improving warmth. However, when broods were manipulated so that half of the chicks had their orange plumes intact and half were trimmed, then parents showed a clear preference for feeding the ornamented chicks and the black chicks grew less well (Fig. 8.10b). Therefore, parental preference is relative, a key element in the evolution of exaggerated traits (Chapter 7 gives analogous examples where female mate choice selects for ornaments in males).

Chick ornaments charm parents

Further studies are needed to investigate why parents prefer ornamented chicks; ornaments may signal chick age or quality or may involve sensory exploitation of parental

(a)

(b)

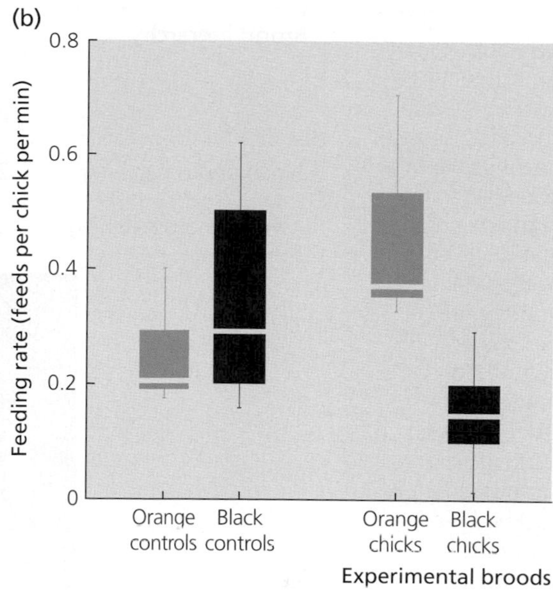

Fig. 8.10 (a) American coot chicks have orange-tipped ornamental plumes. Photo © Bruce Lyon. (b) Parental provisioning to broods where all the chicks have had their orange plumes trimmed (black controls) is no different than to normal broods (orange controls). However, in experimental broods where half the chicks have been trimmed, the black chicks are fed less than their orange sibs. (Lyon *et al.* 1994).

preferences in other contexts. Nevertheless, it is clear that in all these examples increased sibling rivalry from lower relatedness leads to increased signalling for parental resources.

Parent–offspring conflict: evidence

Behavioural squabbles

While the evidence for sibling rivalry is compelling, it has proved more challenging to provide convincing evidence for parent–offspring conflict (Kilner & Hinde, 2008). Simply observing behavioural squabbles does not necessarily demonstrate an evolutionary conflict. Robert Trivers (1974) suggested that weaning tantrums might reflect an offspring's attempts to manipulate parents into continued investment beyond the parental optimum, but equally they may be part of behavioural development with no underlying conflict (Bateson, 1994). A parent might gain information about offspring needs when observing an older offspring attack a younger sibling, and even brief interventions on behalf of the younger offspring, as observed in fur seals, may be the way the mother assesses their relative fitness. In fact, all the examples of sibling rivalry we have discussed so far might reflect the outcome of a parental strategy to optimize investment in relation to variable resources.

David Lack, who pioneered the idea of optimal clutch size (Chapter 1), proposed that brood reduction was a parent adaptation (Lack, 1947). He suggested that parents

Table 8.5 Robert Magrath's (1989) experiment with blackbirds, to test the influence of a brood hierarchy on the parent's reproductive success. All broods were of four chicks (Synchronous, same size; Asynchronous, different size). Photo of a female blackbird feeding a worm to her chicks © W.B.Carr.

Brood hierarchy	**Mean no. young surviving to two weeks after fledging (n = no. broods)**	
	Good food supply	**Poor food supply**
Synchronous hatching	2.9 (n = 8)	1.3 (n = 21)
Asynchronous hatching	2.3 (n = 13)	2.1 (n = 25)

profited by beginning incubation before their clutch was complete, because the resulting asynchronous hatching produced a brood hierarchy which led to efficient brood reduction if food was scarce. By contrast, Lack supposed that synchronous hatching would produce chicks of similar size; with no clear dominance hierarchy among the brood a parent would waste resources on producing many weedy offspring, all with poor survival prospects, rather than a few, healthy survivors.

Brood hierarchies facilitate brood reduction

Robert Magrath (1989) has provided experimental evidence in favour of Lack's view that parents influence the degree of sibling rivalry to optimize their own reproductive success. He studied blackbirds *Turdus merula*, a member of the thrush family which feeds its chicks largely on earthworms, whose availability varies unpredictably with rainfall. Blackbirds normally have asynchronous hatching and the younger members of the brood often starve to death during dry spells, when worms are too deep in the soil for the blackbirds to reach. Magrath created experimental broods of four chicks by swapping newly-hatched young between nests. There were two treatments: synchronous broods had chicks of similar size, while asynchronous broods had a size hierarchy of the same magnitude as natural broods. Conditions were rather dry during the experiment, so Magrath was able to compare productivity of the two brood types under these poor conditions and in good conditions, where he provided some pairs experimentally with extra food.

As predicted by Lack's hypothesis, under poor conditions asynchronous broods produced more surviving young than did synchronous broods (Table 8.5) because, with a clear brood hierarchy, the smallest chicks quickly died, leaving parents with a reduced brood which they could raise effectively. In the synchronous treatment, nestlings took longer to die and more died. Under good conditions, however, synchronous broods tended to do best (Table 8.5). Further work is needed to make Lack's hypothesis quantitative rather than qualitative.

Sex ratio conflict

If sibling rivalry might sometimes benefit parents, how then are we to test for parent–offspring conflict? The ideal test would be to calculate the optimal level of parental investment for both parent and offspring, and to show that these differ (Fig. 8.8). This has been demonstrated just once, in a brilliant paper by Robert Trivers and Hope Hare (1976) which revealed parent–offspring conflicts over offspring sex ratio in social hymenoptera.

Parent–offspring conflict over sex ratios in social hymenoptera

This is explained in detail in Chapter 13; just a brief summary is provided here. Sex in hymenoptera is determined by haplodiploidy. Fertilized eggs produce daughters (who are diploid) while unfertilized eggs produce sons (who are haploid). In a colony, the queen is equally related to her sons and daughters (r = 0.5), so in theory she should prefer a 1:1 sex ratio in her reproductive offspring. However, the workers will prefer a different sex ratio in their mother's reproductive offspring. If the queen is singly-mated, then the workers are three times as related to their reproductive sisters (r = 0.75) as to their brothers (r = 0.25). Therefore, the optimal sex allocation from the workers' perspective is 3:1 in favour of reproductive sisters (Trivers & Hare, 1976).

Who wins the conflict? Workers can bias sex allocation in their favour by selectively destroying male eggs, and sometimes their 3:1 optimum is the outcome. In other cases, however, sex ratios are closer to the queen's optimum and sometimes the outcome is a compromise intermediate sex ratio, so either party can win depending on circumstances (Ratnieks *et al.*, 2006).

Conflicts during pregnancy

If parental and offspring optima cannot be measured precisely, parent–offspring conflict can, nevertheless, be shown by demonstrating that traits have antagonistic lifetime fitness consequences for parents and offspring. Recall that we used the same argument to demonstrate male–female mating conflicts (Chapter 7) where, for example, a male trait, such as grasping genitalia, was counteracted by a female trait, such as spines to resist male grasping. Similar antagonistic evolution has been demonstrated during pregnancy in mice, where genes in offspring function to gain extra resources from the mother, while genes in the mother function to resist offspring demand (Haig & Graham, 1991; Haig, 2000).

Genomic imprinting may evolve from parent–offspring conflict

To understand this conflict, we first need to introduce the concept of 'genomic imprinting'. Most genes are expressed in the same way, whether they are inherited from the mother or the father. However, imprinted genes behave differently depending on the parent they come from. David Haig and his colleagues suggested that genomic imprinting evolves in the context of parent–offspring conflict. In many species (like mice), a female mates with several different males during her lifetime. A maternally-derived gene in a current offspring is thus more likely to have copies in future offspring (because the mother remains the same) than a paternally-derived gene (because different offspring can have different fathers). Therefore, paternal genes in offspring are predicted to demand more maternal resources than are maternal genes in the same offspring. Genomic imprinting would be favoured under these conditions, enabling genes to play conditional strategies depending on whether they were derived from the mother or father (Moore & Haig, 1991).

Antagonistic genes in mice: a tug-of-war between offspring and mother

Two antagonistic genes in mice support this idea. *Insulin-like growth factor 2* (*Igf2*) is paternally-imprinted (expressed only when inherited from the father). It encodes IGF-II, an insulin-like polypeptide that plays a role in extracting resources from the mother during pregnancy. When expression of this paternal allele is experimentally inactivated, offspring are 60% their normal weight at birth, whereas inactivation of the maternal allele has no effect on birth weight. Counteracting the effects of *Igf2* is a maternally-imprinted gene, the *insulin-like growth factor 2 receptor* (*Igf2r*). This encodes a receptor that degrades the product of *Igf2*, thus reducing the resource transfer from mother to offspring. When expression of the maternal allele is inactivated, offspring are 20% larger than normal at birth, whereas inactivation of the paternal allele has no effect on birth weight (Haig, 1997).

Thus, there is a tug of war between offspring and mother, with the paternally-imprinted *Igf2* gene functioning to extract extra resources from the mother and the maternally-imprinted *Igf2r* gene functioning to resist extra investment. It is not yet clear whether other examples of genomic imprinting have evolved in response to parent–offspring conflict (Haig, 2004).

Conflict resolution

Charles Godfray (1995) has made a useful distinction between two types of model of parent–offspring conflict. So far, we have considered a 'battleground' model that defines the zone of conflict (Fig. 8.8). Our last two examples provide evidence for this battleground; tugs of war between parent and offspring over sex ratios in social hymenopteran societies, and between parent and offspring genomes over maternal resources during mouse pregnancy. However, there are also 'resolution' models that try to predict how the conflict might be resolved. If evolution leads to a continuing arms race, then the battleground may still be evident. However, if a stable resolution has been reached, then the original difference in parent and offspring optima may be concealed by the two parties having now become coadapted to each other's strategies.

'Battleground' models define the conflict, 'resolution' models predict the outcome

Nestling begging displays provide a good example of how the conflict might lead to a stable resolution. In theory, an offspring should increase its demand with need. However, if it pays offspring to demand more than the parental optimum, parents should require an honest, unfakable demonstration of need, otherwise they will be tricked into providing too much investment. In theory, an evolutionarily stable resolution to this conflict can be achieved if begging nestlings suffer a fitness cost from soliciting care (Godfray, 1991, 1995). This might explain why nestling begging is so exuberant, with loud calls, stretching and colourful gapes. Just as a female demands an honest signal from a male in mate choice (Chapter 7), so a parent might demand an honest signal from offspring in investment choice. In both cases, the cost of the signal ensures its honesty (Grafen, 1990a).

Costly begging by offspring may resolve parent–offspring conflict

Experiments by Rebecca Kilner with canaries *Serinus canaria* support this view, that costly begging might resolve parent–offspring conflict. Nestlings begged more vigorously when they were more hungry, and parents provided more food as begging signals increased (Kilner, 1995). Furthermore, unrewarded begging was costly because it retarded chick growth. In an experiment, pairs of siblings were hand-fed with the same amount of food, but one member of the pair had to beg for just ten seconds before it was fed, while the other had to beg for 60 s (both times were within the natural range for begging bouts). The sibling that begged for longer had lower mass gain (which reduces survival to independence), demonstrating that increased begging is costly to chick fitness and thus restrains chick selfishness (Kilner, 2001).

According to resolution models, therefore, offspring demand and parental provisioning become coadapted, so that the underlying conflict is now obscured. Mathias Kölliker and colleagues (2000) were the first to reveal this coadaptation by clever cross-fostering experiments with great tits *Parus major*. This enabled them to measure offspring demand and parental generosity independently. They swapped newly-hatched young between nests, so that parents raised a mixed brood of foster-young, half of which came from one foreign nest and half from another foreign nest. When the chicks were ten days old, each had its begging measured in the laboratory at two levels of hunger; after 60 min and 150 min of food deprivation. This gave a measure of a nestling's begging intensity, or demand for food, in response to increased hunger. Parental responses to chick begging signals were recorded in the field by measuring their increase in provisioning in response to playbacks of high versus low intensity begging calls.

Coadaptation of offspring demand and parental provisioning

The results showed that a nestling's demand varied with the nest of origin. In other words, nestlings from the same original brood, but reared in different foster nests, tended to have similar demands. Furthermore – and this was the most exciting discovery – a nestling's demand was related to its genetic mother's generosity; nestlings with more generous mothers demanded more, while those with less generous mothers demanded less (Fig. 8.11a).

Experiments with great tits and burying beetles: generous mothers have more demanding offspring

Cross-fostering experiments with burying beetles *Nicrophorus vespilloides* have shown the same positive correlation between offspring demand and parental provisioning (Lock *et al.*, 2004). These beetles lay their eggs in the soil near a vertebrate carcass. When the larvae hatch, they crawl to the carcass and beg (rather like nestling birds). Their parents feed them by regurgitating a digested soup of carrion into the larva's mouth. Laboratory experiments revealed that a mother's level of care, when provisioning a foster brood, was positively correlated with the begging levels of her offspring, when they were reared by a foster mother (Fig. 8.11b).

In theory, such covariation between offspring demand and parental supply could result from genetic variation, with mothers genetically pre-disposed to be more generous having offspring genetically pre-disposed to demand more resources (Kölliker *et al.*, 2005). However, it could also result from a so-called 'maternal effect', where individual mothers have the capacity to match their offspring's demands to the local resources available for care. This may sound far-fetched but there are many cases where offspring phenotype varies as the result of maternal control. For example, towards the onset of winter vole mothers give birth to offspring with longer fur, and in ponds with more predators small crustaceans give birth to offspring who have more spines for protection

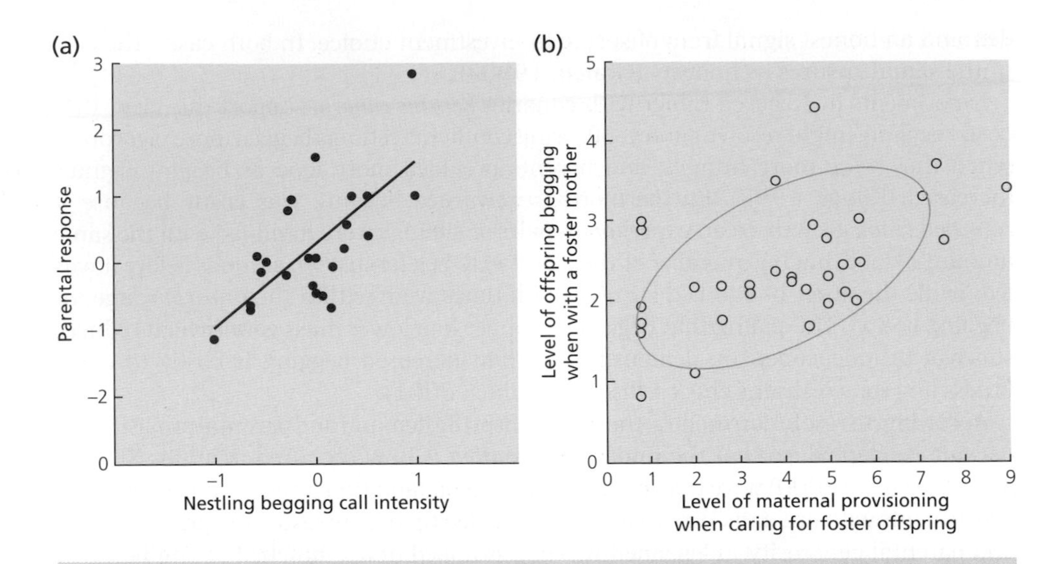

Fig. 8.11 Cross-fostering experiments reveal that within families parental supply is coadapted to offspring demand. (a) Great tits: each point refers to a different brood. Offspring begging intensity (measured in a foster-parent's nest) is correlated with its genetic mother's generosity (measured as increased provisioning response to begging playbacks). From Kölliker *et al.* (2000). (b) Burying beetles: each point again refers to a different brood. More generous mothers (response to foster offspring) have offspring which beg more strongly (response measured when raised by a foster mother). From Lock *et al.* (2004).

(a review is given in Kilner & Hinde, 2008). These maternal effects arise because mothers influence how genes are expressed in their offspring.

Maternal effects can match offspring demand to maternal capacity

Maternal effects can also influence offspring begging behaviour. In canaries, an experimental increase in yolk testosterone leads to more vigorous nestling begging at hatching, suggesting that mothers could vary their nestlings' demand through varying maternal hormones in the egg (Schwabl, 1996). Camilla Hinde and Rebecca Kilner tested this by varying the food available to female canaries before and during egg laying, and then cross-fostering broods so offspring begging and parental provisioning could be measured independently. They found that an increase in food quality led to increased maternal androgens and increased provisioning effort by the mother, and also to an increase in nestling androgens and nestling begging intensity (Hinde *et al.*, 2009). Thus, nestling demand was matched to maternal provisioning capacity, most likely by means of maternal hormones in the egg.

An experiment with canaries

The discovery of maternal control of offspring begging behaviour then enabled an elegant test of the consequences of parent–offspring conflict. By cross-fostering nestlings, parent canaries were exposed to foster broods that begged more, less or appropriately for the parents' capacity for parental care (Hinde *et al.*, 2010). The results showed that foster young did best when their begging levels matched those that the parents expected from their own brood (Fig. 8.12). Foster broods that begged less than the parents expected under-demanded and grew less well, but broods that over-demanded suffered too, because although their foster parents continued to feed them, this did not compensate for the high levels of energy expended during begging. In canaries, therefore, coadaptation leads to a resolution of parent–offspring conflict where offspring will pay a price if they attempt to demand more resources than parents plan to provide.

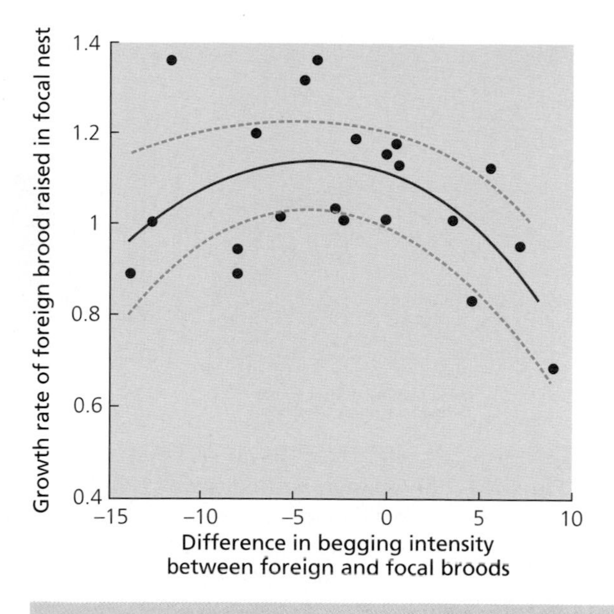

Fig. 8.12 Cross-fostering experiments with canaries. The growth rate of a foreign brood is greatest when its begging levels match those that the parents expected from their own (focal) brood. From Hinde *et al.* 2010. Reprinted with permission from AAAS.

Brood parasites

Some species of birds, fish and insects are brood parasites; they lay their eggs in the nests of other (host) species and trick the hosts into providing all the parental care. In theory, parasitic offspring should behave exceptionally selfishly because they are unrelated to the host parents and host offspring. As predicted, parasite offspring provide what Darwin (1859) called some of the most 'strange and odious instincts' in nature. Their extreme selfishness also provides new insights into the evolution of parent–offspring interactions.

Tricking the host species into raising the brood parasite's offspring

In some species of parasitic cuckoos (e.g. the common cuckoo, *Cuculus canorus*), the female cuckoo lays one egg per host nest and soon after hatching the young cuckoo chick ejects all the host eggs (and any newly-hatched host young) by balancing them on its back, one by one, and heaving them over the rim of the nest (Chapter 4; Fig. 4.19). Young parasitic honeyguides, *Indicator spp*, dispose of the host young by stabbing them to death using sharp-bill hooks. The dead host nestlings are then either trampled into the nest lining or removed by the host parents (Spottiswoode & Koorevaar, 2012). Caterpillars of large blue butterflies (genus *Maculinea*) secrete cuticular hydrocarbons that mimic those made by *Myrmica* ants. This tricks the worker ants into taking the caterpillar into their nest where, depending on the species of butterfly, it is either a predator (devouring ant larvae or pupae) or a cuckoo (begging like ant larvae to claim regurgitated food from the worker ants; Thomas & Settele, 2004). In one parasitic large blue butterfly, *Maculinea rebeli*, the caterpillar gets treated like royalty by mimicking the sounds of queen ants. This induces the worker ants to kill and feed their own brood to the parasite if food is scarce (Barbero *et al.*, 2009).

Sometimes it may pay a parasite chick to tolerate host young

In the last section, we saw how parents might avoid over-exploitation by their own young. This suggests that parasite offspring might not always get an easy ride when they attempt to extract extra resources from the host parents. A comparison between the begging behaviour of parasitic cowbirds and cuckoos reveals some of the trade-offs faced by a young parasite. The brown-headed cowbird, *Molothrus ater*, is a parasitic bird, widespread throughout North America. In contrast to many cuckoos and honeyguides, the young cowbird tolerates the company of the host young. Why? One hypothesis is that the collective begging of a brood evokes a higher total level of provisioning, which the

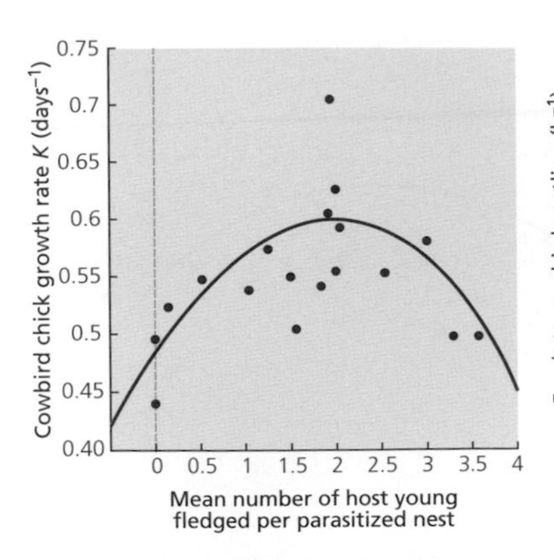

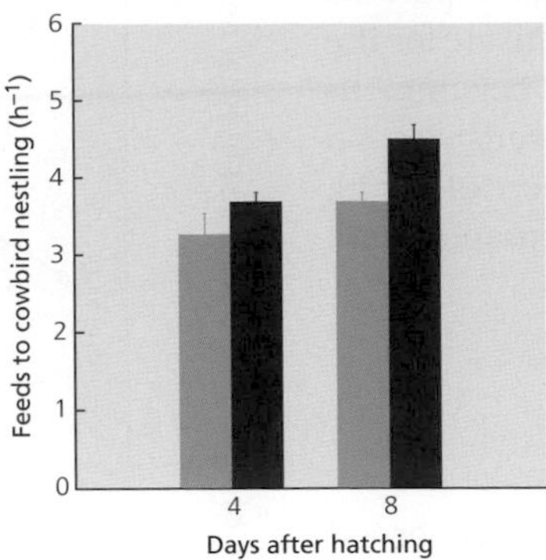

Fig. 8.13 (a) A brown-headed cowbird chick grows best when it shares the nest with two host young. Each point refers to a different host species. The curve is the fitted polynomial regression. (b) An experiment in which a cowbird chick is raised by eastern phoebe host parents, either on its own (light blue bars) or together with two host young (dark blue bars). The cowbird grows best when it is with host young. From Kilner *et al.* (2004). Reprinted with permission from AAAS. The photograph shows an eastern phoebe with two of its own chicks (yellow gapes) and a brown-headed cowbird chick. Photo © Marie Read.

cowbird chick can then exploit by grabbing more than its fair share in competition with the host young. Comparisons of the growth rate of a cowbird chick when raised in the nests of various host species showed that the cowbird did best when it had the company of two host young (Fig. 8.13a). Experiments with cowbirds raised in the nests of Eastern phoebes, *Sayornis phoebe*, also revealed that the cowbird grew better when it shared the nest with two host young than when it was reared alone (Fig. 8.13b). Therefore, the cowbird uses the host young to help it procure resources from the host parents. The optimal number of host young is two; presumably a larger number would take too much of the additional food collectively solicited by the brood (Kilner *et al.*, 2004).

How, then, does a common cuckoo chick manage on its own (see back to Fig. 4.19d)? By ejecting all the host young from the nest, it benefits by removing the competition, but then faces the cost of having to do all the work in soliciting food. The common cuckoo's trick is a remarkable rapid begging call, which sounds like many hungry host young (Fig. 8.14a). The way this works is subtle (Kilner *et al.*, 1999). Experiments with reed warbler *Acrocephalus scirpaceus* hosts show that when parents provision a brood of their own young, they respond both to the total gape area on view (a visual cue) and the begging rate of the brood (a vocal cue). The visual cue gives the parents a rough guide of how much food to bring, because it is related to chick number (more chicks, more gapes) and chick age (older chicks, larger gapes). The vocal cue enables parents to fine-tune their provisioning in relation to chick hunger (hungrier chicks beg more rapidly). So if host parents are given more chicks, or older chicks, they work harder and if the brood's begging calls are augmented by playback of extra calls, they work harder still.

Vocal trickery by common cuckoo chicks

How does the common cuckoo chick exploit this system? The cuckoo needs as much food as a brood of four reed warblers. Its problem is that although it is larger than any one reed warbler chick, its gape area cannot match that of a whole brood of reed warblers. To compensate for this deficient visual component of the begging display, the

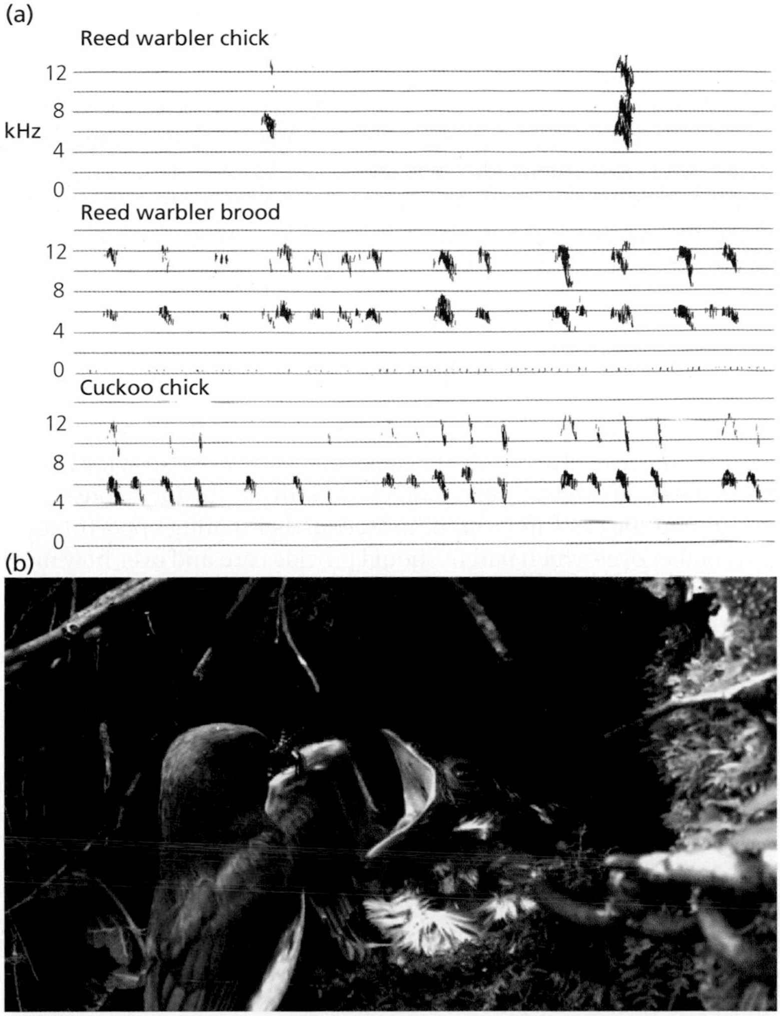

(a)

(b)

Fig. 8.14 Vocal and visual trickery by cuckoo chicks. (a) A common cuckoo chick's vocal trickery in a reed warbler nest. The sonograms, each 2.5 s long, show the begging calls of six day-old chicks recorded in the laboratory one hour after they had been fed to satiation. The cuckoo's begging calls are much more rapid than a single reed warbler chick and at a week of age are more like those of a whole brood of hungry host chicks. From Davies *et al.* (1998). (b) A Horsfield's hawk-cuckoo exposing a false gape – a yellow wing patch – next to its own yellow gape. The host is a blue and white flycatcher *Cyanoptila cyanomelana*. Photo courtesy of Keita Tanaka.

cuckoo has to boost the vocal component by producing extraordinarily rapid calls. Thus, the cuckoo succeeds by tuning into the host's provisioning rule (Kilner *et al.*, 1999).

And visual trickery by Horsfield's hawk-cuckoo chicks

In Japan, Horsfield's hawk-cuckoo *Cuculus fugax* has an equivalent trick, but it manipulates the visual component of the begging display. When it begs for food it exposes yellow wing patches which are the same colour as its yellow gape (Fig. 8.14b).

Two components of begging signals: stimulating provisioning to the brood and competing for the food brought

These false gapes spur the hosts into collecting more food; hosts sometimes try to place food into a patch instead of the gape, and experimental darkening of the patches with dyes reduces provisioning (Tanaka & Ueda, 2005).

These tricks of brood parasites suggest that begging displays are likely to involve two components; a cooperative component in which increased signalling by the brood leads to more food brought by the parents, and a competitive component, where nestlings compete for the food once it has been brought (Johnstone, 2004). Future experiments should investigate how the displays of individual chicks (e.g. larger, smaller) contribute to these two components.

Summary

In some species there is no parental care (most invertebrates), in some only the female cares (most mammals), in others only the male (many fish), and in some both sexes care (most birds). These differences reflect differences in life history constraints, in the benefits of care and the costs in terms of missed further mating opportunities. There is often sexual conflict over which parent should provide care and over how much care to provide. Game theory models predict the evolutionarily stable strategies of the two sexes, but current models often do not incorporate all the complexities of nature, such as sequential desertion by the sexes and how parents can learn from each other about brood needs. Robert Trivers (1974) showed that individual offspring are selected to demand more care than is optimal from a parent's point of view. There is good evidence for both interbrood conflict (e.g. Galapagos fur seals) and intrabrood conflict (begging nestlings in a brood). As predicted, conflict increases as sibling relatedness declines. The resultant conflict between parents and offspring is evident as a 'battleground', for example tugs of war over optimal sex ratios in hymenopteran societies and over maternal investment during pregnancy in mice. The conflict can also lead to an evolutionary resolution, for example costly begging displays in nestling birds, where parent supply and offspring demand become coadapted (e.g. canaries). Brood parasites reveal extreme examples of offspring selfishness; nevertheless, the parasitic offspring have to 'tune in' to their host's provisioning systems.

Further reading

Clutton-Brock (1991) reviews the evolution of parental care. Klug and Bonsall (2007) consider the life history characteristics likely to lead to the evolution of parental care, offspring desertion or offspring consumption. David Queller (1997) provides an elegant model to show how lower probability of parentage for males tends to make males less likely to provide care than females. Mock and Parker (1997) provide a superb review of sibling rivalry, both theory and evidence. Royle *et al.* (2002) discuss begging and honest signalling. Kilner and Hinde (2008) review parent–offspring conflict and suggest that the outcome is influenced by the information each party has concerning supply and demand for resources. Grodzinski and Lotem (2007) show, by means of hand-feeding broods of nestlings, that allocating food in response to begging intensity leads to better nestling growth than allocating the same amount of food at random among the brood. This provides evidence that it pays parents to respond to offspring begging signals. Emlen (1995) reviews theoretical predictions for conflicts in families. The books by Hrdy (1999), Mock (2004) and Forbes (2005) are wide-ranging reviews of animal family evolution. Houston *et al.* (2005) discuss models of sexual conflict over parental care.

TOPICS FOR DISCUSSION

1. Why don't male mammals lactate?
2. In penduline tits *Remiz pendulinus* sometimes the male deserts, leaving the female to care, sometimes the female deserts, leaving the male to care, and sometimes both parents desert, in which case the clutch fails (Persson & Öhrström, 1989; Szentirmai *et al.*, 2007). Discuss how these various outcomes might reflect sexual conflict.
3. Discuss alternative hypotheses for weaning tantrums. How would you test these?
4. Why under good feeding conditions might synchronous broods produce most surviving offspring (Table 8.5)?
5. Would you ever expect parent blue-footed boobies to intervene to protect their younger chick from siblicide by the elder chick?
6. In experiments with nestling tree swallows, *Tachycineta bicolor*, Marty Leonard and Andrew Horn (2001) found that nestlings increased their rate of calling in the presence of a sibling. They suggest that conspicuous and costly begging displays may evolve simply through selection for effective individual signal transmission in the face of sibling competition for parental attention. What evidence would distinguish this hypothesis from that discussed in this chapter, namely that conspicuous begging has evolved to enforce honesty in signals of offspring need?
7. Discuss alternative hypotheses for the evolution of ornamentation in offspring. How would you test these? Would you expect offspring ornaments to be less extravagant than male ornaments?
8. Compare the methods used to study sexual conflict over mating (Chapter 7) and parent–offspring conflict over parental investment (this chapter). Could researchers on these two topics learn from each other?

CHAPTER 9
Mating Systems

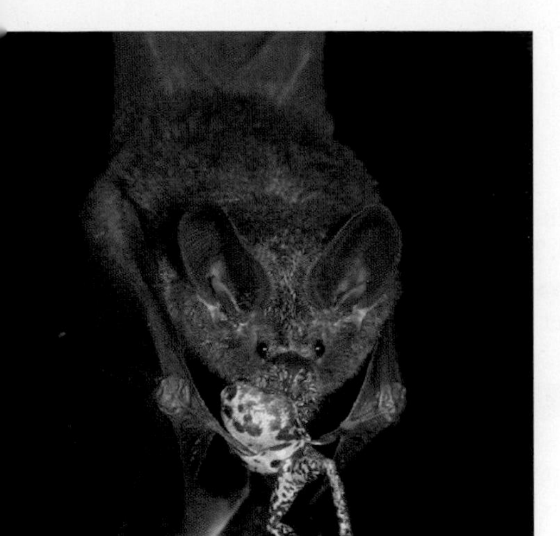

Photo © Alexander T. Baugh

Our main theme in the last two chapters has been conflict; conflict over mating and fertilization (Chapter 7) and conflict over parental care (Chapter 8). We now explore the outcomes of these conflicts in terms of mating systems, which refer to how mates are acquired, how many mates and associated patterns of parental care. At first sight, the variety in nature appears bewildering (Table 9.1). How can we explain this diversity?

Two factors influencing mating system diversity

Beginning with an influential paper by Stephen Emlen and Lewis Oring (1977), a fruitful approach has been to view mating systems as outcomes of the behaviour of individuals competing to maximize their reproductive success. Different mating systems might emerge depending on two factors: (i) *male and female dispersion in space and time* (which will influence how easy it is for either sex to gain access to mates); (ii) *patterns of desertion by either sex* (which will depend on the costs and benefits of parental care).

Mating systems with no male parental care

We begin with cases where the male does not provide parental care. In theory, the mating system should result from a two-step process (Fig. 9.1). Firstly, female reproductive success will be limited most by access to resources (Chapter 7), so female distribution should depend primarily on resource dispersion (e.g. food, breeding sites), modified by the costs and benefits of associating with other individuals (e.g. through influences on predation and resource competition; Chapter 6). On the other hand, male reproductive success will be limited more by access to females (Chapter 7), so the second step in the sequence is that males should then distribute themselves in relation to female dispersion. Males could

An Introduction to Behavioural Ecology, Fourth Edition. Nicholas B. Davies, John R. Krebs and Stuart A. West.

Table 9.1 A classification of mating systems

Mating system	Who mates with whom?
Monogamy	One male restricts his matings to one female, and she to him, either for one breeding season or longer. Both partners may forgo other mating opportunities by choice, or one partner may enforce monogamy by keeping other potential mates at bay. Often both parents care for the eggs and young.
Polygyny	One male mates with several females in a breeding season by defending them directly (a harem or female-defense polygyny); or by defending resources that the females require (resource-defence polygyny); or by attracting females to a display site, sometimes where many males aggregate together (leks); or by the male roaming in search of widely dispersed females (scramble competition polygyny). Often the female provides most or all of the parental care.
Polyandry	One female mates with several males in a breeding season by defending them simultaneously or in succession. Often the male provides most or all of the parental care.
Promiscuity	Both male and female have multiple partners during a breeding season.
Polygamy	A general term for when an individual of either sex has more than one mate.

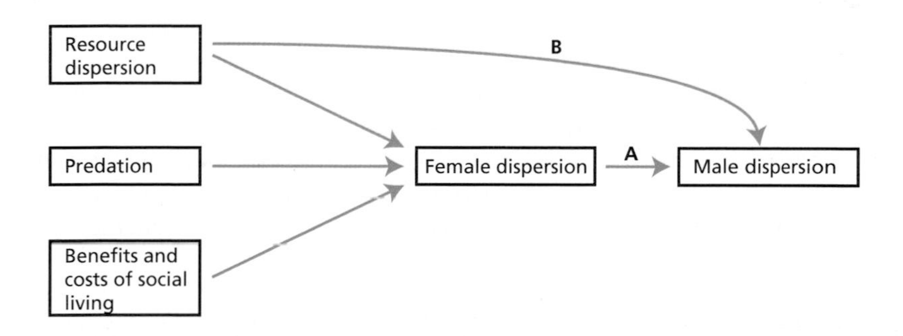

Fig. 9.1 The two-step process influencing mating systems in cases where males do not provide parental care. Because female reproductive success tends to be limited by resources, whereas male reproductive success tends to be limited by access to females, female dispersion is expected to depend primarily on resource dispersion (modified by predation and benefits and costs of social living), while male dispersion is expected to depend primarily on female dispersion. Males may compete for females directly (A) or indirectly (B), by anticipating how resources influence female dispersion and competing for resource-rich sites.

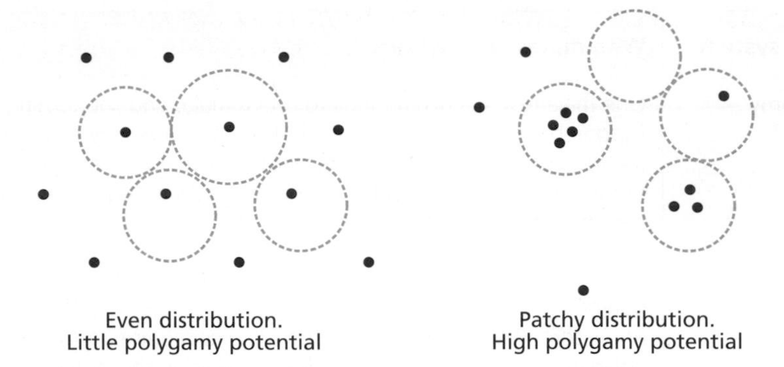

Fig. 9.2 The influence of the spatial distribution of resources (food, nest sites) or mates on the ability of individuals to monopolize more than others. Dots are resources or mates and circles are defended areas. With a patchy distribution of resources or mates there is greater potential for some individuals to 'grab more than their fair share'.

compete for females either directly (A in Fig. 9.1) or indirectly (B) by anticipating how resources will influence female dispersion and competing for resource rich sites.

Female dispersion is influenced by resources, male dispersion is influenced by females

The economics of female defence or resource defence by males will depend on their distribution both in space and in time. When mates or resources are more patchily distributed in space there will be greater opportunities for polygyny (Fig. 9.2). The key factor for determining the temporal distribution of mates is the 'operational sex ratio' (Emlen & Oring, 1977; Chapter 7), which is the ratio of receptive females to sexually active males at any one time. If all the females bred in synchrony, then with a real sex ratio of 1:1 in the population, the operational sex ratio at breeding would also be 1:1 and there would be little opportunity for a male to mate with more than one female because by the time he had mated once all the other females would have finished breeding. This applies, for example, to common toads (*Bufo bufo*) which are 'explosive breeders': all the females spawn within a few days, so a male has time to mate with one, or at most two, females before the breeding season is ended. Bullfrogs, *Rana catesbeiana*, by contrast, are 'prolonged breeders' with females arriving at the pond over several weeks. Males which can defend the best spawning sites may mate with up to six females in a season (Wells, 1977).

Mating systems will be influenced by spatial and temporal distribution of mates

We now examine experimental and comparative evidence for the scheme in Fig. 9.1.

Experimental evidence: voles and wrasse

Grey-sided voles, Clethrionomys rufocanus

Rolf Anker Ims (1987) showed that female dispersion was influenced by food; when food was provided in abundance at particular sites female ranges became smaller and overlapped in the resource-rich areas. The males also homed in on these sites. Was the change in male dispersion due to males following females or because they, independently, followed changes in resource distribution? To test this, Ims (1988) introduced a small population of voles on to a little wooded island in south-east Norway. In one experiment, females were kept individually in small cages and their positions were moved each day to simulate movement about a home range. When females were spaced out, free-ranging males (tracked by radiotelemetry) became dispersed, overlapping their ranges with the female ranges. When females were clumped, by placing cages close together, the males aggregated on the female clumps. By contrast, when males were then kept in individual cages the dispersion of free-ranging females was not affected by experimental changes in male dispersion. This study shows that the causal links are from resources to female dispersion and then from female dispersion to male dispersion, as in Fig. 9.1.

Blue-headed wrasse, Thalassoma bifasciatum

Robert Warner (1987, 1990) studied these coral reef fish in the Caribbean. Females spawned in favourite sites on the downcurrent edges of a reef where the pelagic eggs were swept quickly into the open sea, so avoiding predation from other reef fish. Individual females returned almost every day to particular sites where they laid a few eggs. Males competed to defend territories at these preferred sites, with the largest males defending the best sites and so gaining most mates.

To assess the roles of the two sexes in determining spawning sites, Warner removed either all the breeding males or all the females from local isolated populations and replaced them with fish from other populations. When males were replaced, most of the spawning sites remained the same as before. By contrast, when females were replaced there were marked changes in sites used, even though some males initially continued to display at and defend the original sites. This neat experiment shows that females choose the spawning sites and males simply compete to defend sites which females prefer.

In both voles and wrasse, females follow resources, males follow females

Comparative evidence: mammalian mating systems

The influence of variation in resource and female dispersion on mating systems is well illustrated by mammals, where the economics of female monopolization by males is influenced by three main factors: female group size, female range size and the seasonality of breeding (Fig. 9.3). The following comparative survey is based on the review by Clutton-Brock (1989).

Females solitary: range defensible by male

In over 60% of mammalian species females are solitary and a male defends a territory which overlaps one or more female ranges. If female ranges are small relative to the area which a male can defend then the male can be polygynous. If female ranges are larger, then the male may only be able to defend one female, hence monogamy (e.g. most rodents and nocturnal prosimians; Kleiman, 1977). Usually the male simply mates with the female then leaves her to care for the offspring alone. More rarely (3% of mammalian species) the male may help defend the young against predators (e.g. klipspringer, *Oreotragus oreotragus*) or carry them (e.g. siamangs, marmosets, tamarins) or help to feed them (e.g. jackals, wild dogs). Obligate monogamy occurs where female ranges are small enough for a male to defend but where a male is unable to defend a large enough area to have more than one female (Rutberg, 1983). In these cases a male may then maximize his reproductive success by providing parental care. Species with obligate monogamy tend to have large litter sizes. For example, this mating system is common in canids, which have large litters, and rare in

Fig. 9.3 Diversity of mammalian mating systems, illustrated by ungulates. (a) The dik-dik *Madoqua kirki* is monogamous: a male defends one female, probably because female ranges are too large for a male to defend more than one mate. Courtesy Oxford Scientific Films. Photo by Zig Leszczynski. (b) Male impala *Aepyceros melampus* defend herds of females temporarily during oestrus. Here a male is preventing a group of three females from leaving his territory. Courtesy Peter Jarman, photo by Martha Jarman. (c) Male Uganda kob *Kobus kob thomasi* defend tiny territories (15–30 m diameter) on leks and display to attract females. The male in the centre of the photo is mating with a female who has visited his territory. Photo by James Deutsch. (d) In the buffalo *Syncerus caffer*, several males associate with a large group of females and compete for matings in the multimale group. Courtesy Oxford Scientific Films. Photo by G.I. Barnard.

felids, which have smaller litters. Marmosets, which have male parental care, produce twins whereas most monkeys produce only single offspring. If a male marmoset dies then the female often deserts the young so male assistance with parental care seems to be important.

In mammals, different mating systems arise from variation in female home range, group size and movements

Females solitary: range not defensible by male

Where females wander more widely, then males may rove over wide ranges, associating with females temporarily while they are in oestrus. This occurs in moose, *Alces alces* and orang-utans, *Pongo pygmaeus*; in the latter species the females move over large ranges following the fruiting seasons of different species of plants (Mackinnon, 1974).

Females social: range defensible by male

Where females occur in small groups in a small range, then a single male may be able to defend them as a permanent harem within his territory (e.g. black and white colobus *Colobus guereza*, Hanuman langurs *Presbytis entellus*). When a new male takes over the territory, he often kills the young offspring fathered by the previous male, thus bringing the female into oestrus sooner and hastening the day he has a chance to sire his own young (Hrdy, 1977). Where females occur in larger groups several males (often relatives) may defend the territory together (e.g. red colobus *Colobus badius*, chimpanzees, lions). Joint defence by several males may increase the length of tenure of a harem and may also be necessary for economic defence of large groups of females wandering over a large range (Bygott *et al.*, 1979).

Females social: range not defensible by male

Sometimes groups of females wander over ranges which are uneconomic for one or more males to defend. The ways in which males compete for females then depend on how predictable female group movements are in time and space.

Sometimes males wait for females

(a) *Daily female movements predictable*. Sometimes the group of females wanders over a large range but uses regular routes to particular water holes or rich sources of food. In these places the males may defend small territories, much smaller than the females' range, and attempt to mate with them as they pass through (e.g. topi *Damaliscus lunatus korrigum*, Grevy's zebra *Equus grevii*). Such defence of mating territories may occur where more direct competition between males, such as fights for harems, would be costly because males are unable to build up the food stores necessary for them to engage in intense male–male interactions (Owen-Smith, 1977).

Sometimes males follow females

(b) *Daily female movements not predictable*. Here males tend to follow the females, rather than waiting for the females to come to them. Where females live in small groups males may rove and associate with individuals in oestrus (e.g. mountain sheep *Ovis canadensis*, elephants *Loxodonta africana*). Where female groups are larger, the males may attempt to defend harems. Harems may be seasonal or permanent.

Seasonal harems. If all the females come into oestrus at a particular season, then it may pay a male to put on energy reserves to enable him to have a burst of energy expenditure on harem defence. For example, male red deer (*Cervus elaphus*) stags compete to defend harems during the one month in which all females come into oestrus. A male's reproductive success depends on his harem size and the length of time for which he can defend the harem and this, in turn, depends on his body size and fighting ability. After the mating season the males are reduced to very poor body condition and are literally 'rutted out'! (Clutton-Brock *et al.*, 1982).

Harem defence

As another example, female Northern elephant seals (*Mirounga angustirostris*) haul up on beaches to drop their pups and mate again for the production of next year's offspring. Because the females are grouped, due to the localized nature of the breeding grounds, they are a defendable resource and the males fight with each other to monopolize them. The largest and strongest males win the biggest harems and in any one year all the matings are performed by just a few males (Le Boeuf, 1972, 1974; Cox & Le Boeuf, 1977). To be a harem master is so exhausting that a male usually only manages to be top ranking for a year or two before he dies. In the process of defending his harem against other males he sometimes tramples on his females' new-born pups. Although this is obviously not in the females' interests, these pups were probably not sired by the male himself because he is unlikely to have been a harem master the previous year. From the male's point of view, therefore, there is little cost in damaging or even killing the pups; his main concern is to protect his paternity.

Permanent harems. Where females do not all come into oestrus at a particular time, males may defend permanent harems for the whole duration of their reproductive life (e.g. hamadryas baboons (*Papio hamadryas*) and gelada baboons (*Theropithecus gelada*), Dunbar, 1984; Burchell's zebra (*Equus burchelli*), Rubenstein 1986). Often several groups (male plus harem) go around together forming a large 'super-group'. Where females go around in still larger groups, several males may associate with large groups of females and compete with each other for matings. Such 'multimale' groups occur in buffalo, *Syncerus caffer*, and olive baboons, *Papio anubis* (Altmann, 1974).

Leks and choruses

Leks are aggregations of males on small mating territories

In the examples discussed so far, males compete for females directly (female defence) or indirectly by defending resources to which females are attracted (resource defence). In some cases, by contrast, males aggregate into groups and each male defends a tiny mating territory containing no resources at all – often the territory is no more than a bare patch of ground just a few metres across. The males put a great deal of effort into defending their territories and advertise themselves to females with elaborate visual, acoustic or olfactory displays. In these mating systems, known as leks, females often visit several males before copulating and appear to be very selective in their choice of mate. Mating success is strongly skewed, with the majority of matings performed by a small proportion of males on the lek (Fig. 9.4).

Leks have been reported for seven species of mammals – the walrus, hammer-headed bat and five ungulates – and some 35 species of birds, including three shorebirds, six grouse, four hummingbirds, two cotingas, eight manakins, eight birds of paradise, the kakapo and great bustard (Oring, 1982). This breeding system is, therefore, not common. Similar systems occur in some frogs (Wells, 1977) and insects (Thornhill & Alcock, 1983), where females visit male groups, choose a mate and then lay eggs away from the display site.

Leks may occur when neither females nor resources can be defended economically

It has been suggested that leks occur when males are unable to defend economically either the females themselves or the resources they require (Bradbury, 1977; Emlen & Oring, 1977). This may arise where females exploit widely dispersed resources and so have large, undefendable ranges, or because high population density, and thus high rates of interference between males, precludes economic female or resource defence. Thus, in both antelope and grouse, the lekking species are those with the largest female home ranges (Bradbury *et al.*, 1986; Clutton-Brock, 1989) and in Uganda kob, topi and fallow deer, males lek at high population density but defend resource-based territories or harems at low density, where defence of females is presumably more economic.

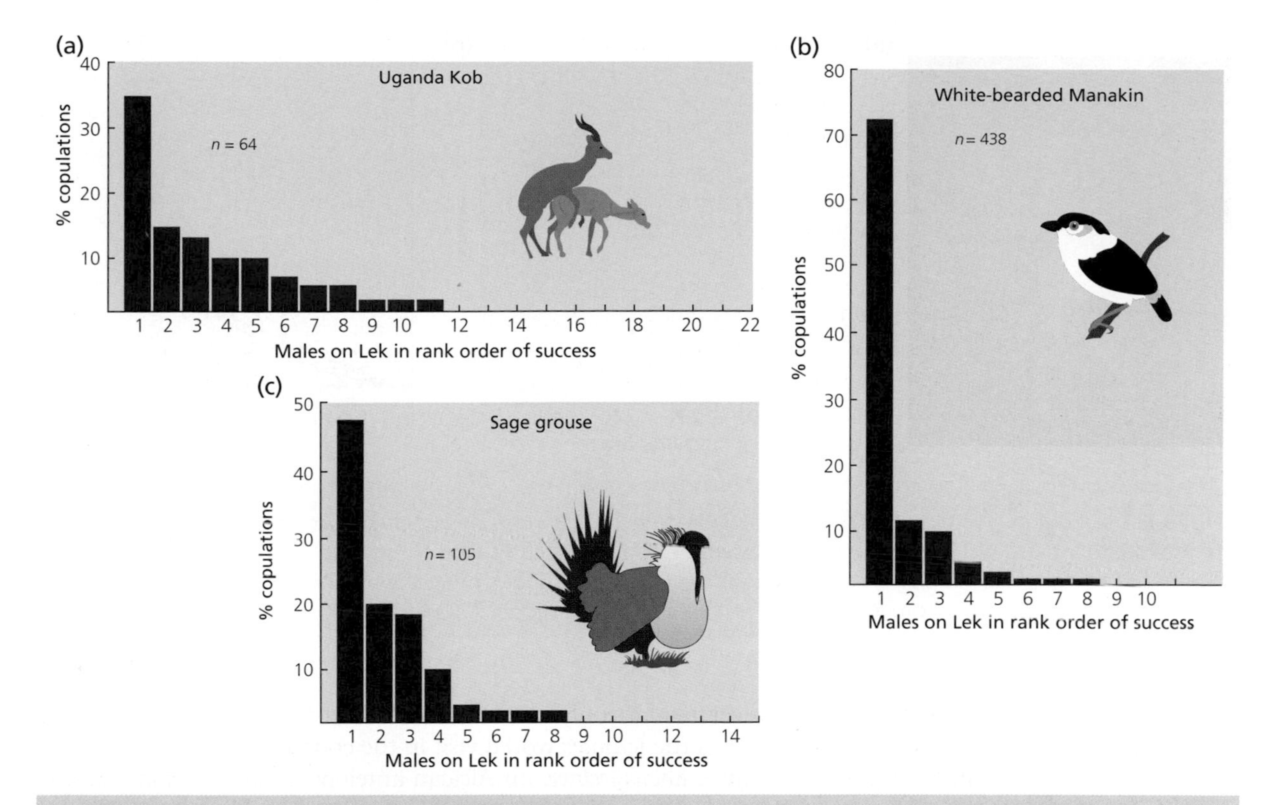

Fig. 9.4 On leks, most of the copulations are performed by just a few of the males. (a) Uganda kob *Adenota kob thomasi*. From Floody and Arnold (1975). (b) White-bearded manakin *Manacus manacus trinitatis*. From Lill (1974). (c) Sage grouse *Centrocercus urophasianus*. From Wiley (1973).

Why do the males aggregate into leks? Five main hypotheses have been proposed (Bradbury & Gibson, 1983).

Males aggregate on 'hotspots'

Leks may occur where females are abundant

Male aggregations may be explained by the familiar scheme in Fig. 9.1, with males settling in areas where female encounter rate is particularly high (hotspots). In sage grouse, leks are located in areas where females travel between their wintering and nesting ranges. Furthermore, the numbers of males on a lek is related to the number of females nesting within a 2 km radius, suggestive of 'hotspot' settlement by males (Gibson, 1996). In the lekking sandfly *Lutzomyia longipalpis*, too, males aggregate on hotspots, namely vertebrate hosts, which females visit to obtain a blood meal. Several hundred males may occur in a lek, each defending a tiny territory (about 2 cm radius) where they jostle and fight for space. Experiments in which the numbers and distribution of chicken hosts were varied, using cages, showed that males quickly aggregated on new hotspots when host distribution was changed, with larger leks where there were most hosts (Jones & Quinnell, 2002).

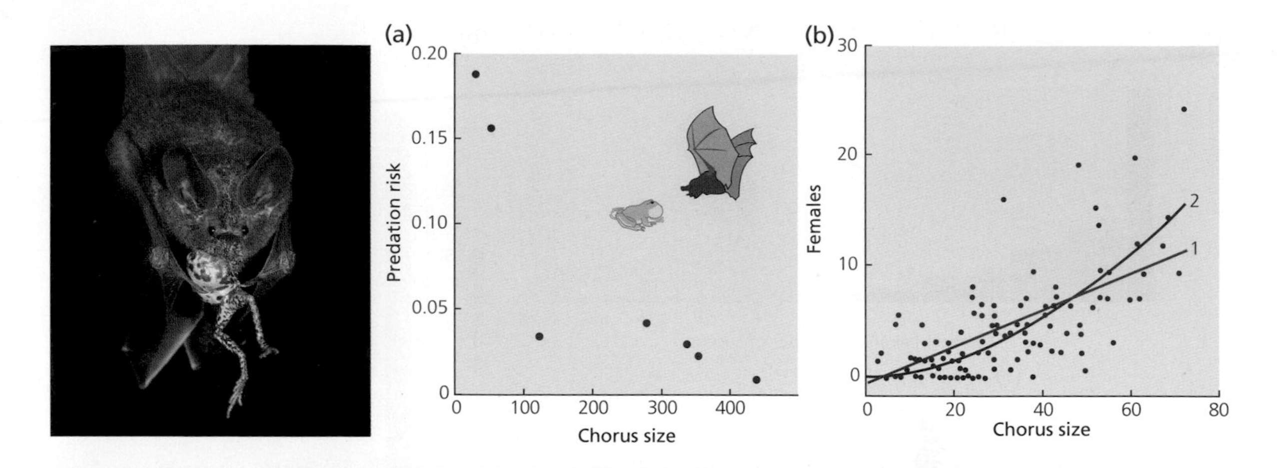

Fig. 9.5 Male frogs, *Physalaemus pustulosus*, aggregate into choruses. In larger choruses individuals are safer from predatory bats (a). The number of females attracted also increases with chorus size (b). The curve (2) gives a better fit to the observed points than a straight line (1), which suggests that the number of females per male increases with chorus size. From Ryan *et al.* (1981). Photo © Alexander T. Baugh.

In many cases, however, females only visit lek sites in order to mate, so the males are not simply settling in areas the females would visit in the course of their normal daily routine (e.g. Kafue lechwe, *Kobus lechwe*, an African antelope; Balmford *et al.*, 1993). Furthermore, males are often aggregated far more closely than would be expected from settlement 'on top' of the female pattern of dispersion.

Avoiding predation

Males aggregate to reduce predation

In the neotropical frog *Physalaemus pustulosus*, calling males suffer heavy predation by bats, *Trachops* spp., which home in on the male calls. Calling males are safer in larger choruses because of the dilution effect (Fig. 9.5a; Chapter 6). However, this is unlikely to be a general explanation of lekking; in many bird leks predation pressure seems to be extremely low.

Synergistic displays

Males aggregate to increase female attraction

Males may gain from 'stimulus pooling'; by displaying together they may provide a greater attraction for females and draw in mates from a larger distance. To explain the male aggregation the increase in female attraction would have to be marked, so that pay-offs per individual male increased with lek size. Figure 9.5b suggests this is true for the frog *Physalaemus*, but the relationship may merely reflect the fact that larger choruses form in areas where females are more abundant anyway. An experimental approach is needed to test whether larger choruses *cause* greater female attraction. Todd Shelly (2001) varied lek size experimentally in two species of tephritid flies, where males aggregate on leaves and emit pheromones or acoustic signals to attract females. He placed varying numbers of males in small pots covered with mesh and then released hundreds of females nearby to see how many were attracted to each lek. In one species, there was no variation with lek size in the number of females attracted per male. In the other species, larger leks (18–36 males) did attract more females per male than smaller leks (six males), but curiously leks in this species tend to be small, not large.

Males aggregate around attractive 'hotshot' males

Poorer signallers may parasitize better signallers

Average success per male may not be what we really need to measure because individuals may vary in their ability to signal. If some males had particularly effective displays ('hotshots'), it could pay poorer signallers to cluster around them to parasitize their attractiveness (Beehler & Foster, 1988).

This process certainly seems to explain male aggregation on a small scale, such as that involving calling and satellite male toads (Chapter 5), but two sources of evidence suggest that it cannot explain the larger scale aggregations of leks. Firstly, when the most successful males are removed from a lek, their territories are quickly taken-over by other males (sharp-tailed grouse, Rippin & Boag, 1974; white-bearded manakins, Lill, 1974). This suggests that there is something about the site which influences female preference. The 'hotshot model' predicts that the next most preferred male would remain on his territory with the male aggregation rearranging around him, rather than for replacement to occur on particular sites. Secondly, in an experiment on fallow deer (*Dama dama*) leks, Clutton-Brock *et al.* (1989) covered the territories of the most successful males with black polythene, so forcing them to change site. Even though these males set up new territories several hundred metres away, they remained favoured by females. In this case, therefore, females were apparently choosing particular males rather than particular sites. The hotshot model predicts that the other males should have followed the movements of these attractive males to set up a lek at the new site. However, most remained on their old territories. Thus, the hotshot model may not explain male aggregation on leks even when females are choosing males rather than particular sites.

Females prefer male aggregations because these facilitate mate choice

Do females choose males directly or particular sites on a lek?

Although lek mating systems are rare, they have attracted considerable attention because of the fact that females seem to exercise careful choice before they mate, with particular males gaining most of the matings. Because males of lekking species do not provide any parental care, the only possible gains from such choice are either a safe mating (safety from male harassment or predation) or genetic benefits (Chapter 7). It is still not clear whether one or both of these benefits is important. For example, female black grouse (*Lyrurus tetrix*) prefer the males with the most vigorous displays and these chosen males also have higher survival (Alatalo *et al.*, 1991). One possibility, therefore, is that leks provide a testing ground where males reveal their health and viability through the vigour of their displays; if there is any heritability of male viability then females will gain good genes for their offspring by mating with the most vigorous males. Thus, female choice may lead to male aggregations either because choice of a particular site enables females to gain matings with the most vigorous male (the one able to win that site), or because preference for aggregations facilitates comparison among males.

Female choice may bring genetic benefits or simply a safe mating

In conclusion, the factors leading to lek mating systems are likely to be diverse. All five hypotheses may be important, with different explanations applying to different species or at different spatial scales.

Mating systems with male parental care

Where males provide parental care the males themselves become a resource which may influence female dispersion, so the simple scheme in Fig. 9.1 no longer applies. As we saw at the beginning of the chapter, male parental care is particularly common in birds, so we shall mainly use examples from birds to illustrate the ideas in this section.

Monogamy

Obligate monogamy: fidelity and divorce

David Lack (1968) suggested that monogamy is the predominant mating system in birds (90% of species) because 'each male and each female will, on average, leave most descendants if they share in raising a brood'. This hypothesis certainly explains obligate monogamy in many seabirds and birds of prey, where male and female share incubation or where males feed females on the nest, and where both sexes are essential for chick-feeding. In these species, the death or removal of one partner leads to complete breeding failure.

Sometimes it pays both male and female to breed as a monogamous pair

In some of these 'obligate monogamous species' males and females form life-long pair bonds and their annual reproductive success increases with the duration of the pair bond. Is this because pairs function better the more that a male and female get to know each other? Or does it result simply from improved individual experience with age? A long-term study of oystercatchers *Haematopus ostralegus* on the island of Schiermonnikoog, in the Dutch Wadden Sea, revealed that pair bond duration influenced reproductive success independently of male and female age and other confounding factors (e.g. territory quality). This shorebird is socially and genetically monogamous and pairs often return to the same territories to breed year after year. Newly-formed pairs have low reproductive success, but success increases with duration of the pair bond up to 5–7 years (Fig 9.6). With more experience together, pairs breed

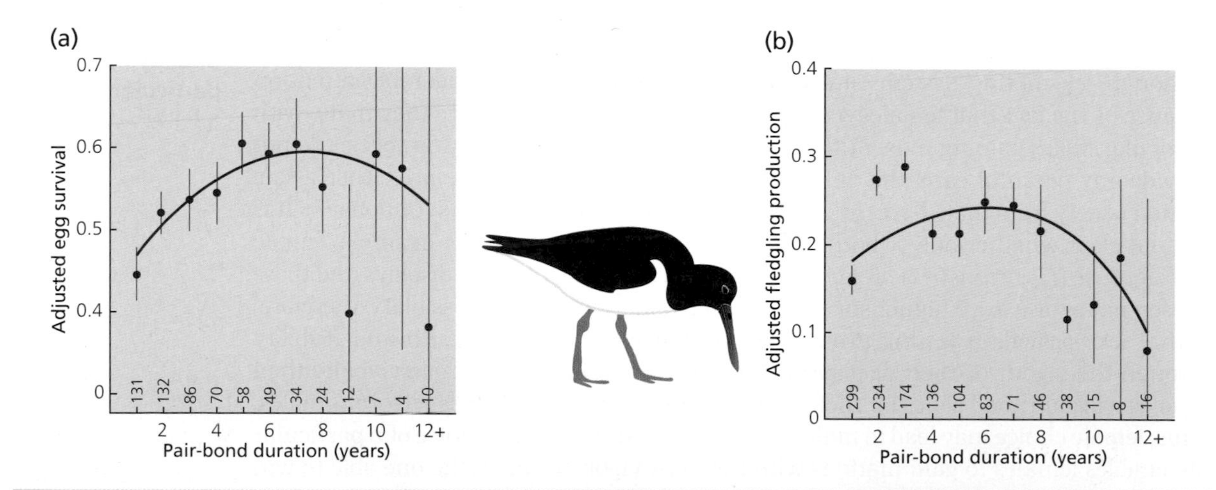

Fig. 9.6 Effect of pair-bond duration in oystercatchers on: (a) egg survival and (b) annual fledgling production. These 'adjusted' measures control statistically for other effects, such as male and female age, individual identity and territory quality. Sample sizes are shown above the x axes. From van de Pol *et al.* (2006).

earlier and are better at caring for their eggs and chicks, perhaps through improved behavioural coordination. The decline in success in later years most likely reflects reproductive senescence. Experiments confirmed that there was a causal link between pair-bond duration and reproductive performance; when either male or female partner was removed, reproductive success was lower with the new partner but it increased over the subsequent four years (van de Pol *et al.*, 2006).

These results reveal an initial cost of breeding with a new partner. Nevertheless, in this oystercatcher population there was an annual divorce rate of 8%. Why? It is important to distinguish two causes of divorce and to consider the consequences for each individual concerned (Fig. 9.7, Table 9.2). Oystercatchers that deserted their mates gained increased survival and reproductive success because they often left for better breeding territories (ones nearer good feeding grounds). By contrast, those that were forced to change mate because they were usurped by competitors tended to suffer reduced fitness because they ended up in poorer territories. Therefore, initiators of divorce gained while victims lost (Heg *et al.*, 2003).

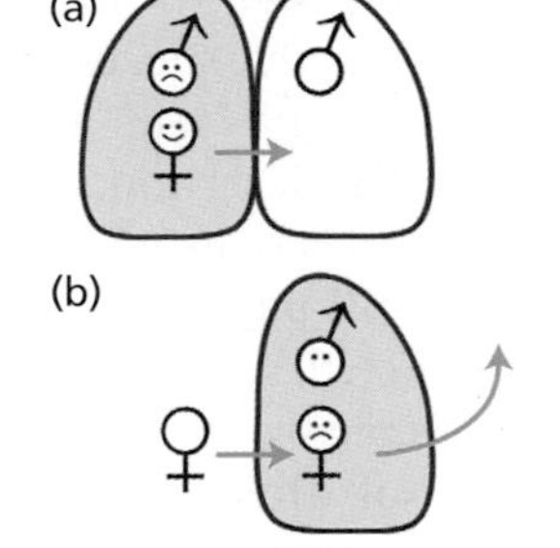

Fig. 9.7 Two causes of divorce: (a) desertion; (b) usurpation. This cartoon refers to a territorial species, such as the oystercatcher, where the male remains on his territory and the female moves to another territory. The territory where the divorce occurs is shaded. From Ens *et al.* (1996).

Initiators and victims of divorce in oystercatchers

Constrained to be monogamous

However, Lack's hypothesis does not explain the monogamous mating systems of many songbirds, where both sexes also commonly rear the young together. If males are removed during the nestling period then females are usually able to raise at least some of the young to independence. For example, in song sparrows (*Melospiza melodia*) male removal caused success to decrease to 51% of that of pair-fed broods, and in seaside sparrows (*Ammodramus maritimus*) and dark-eyed

Participant	Role	Change in territory
(a) **Desertion**		
Victim of desertion	Abandoned by mate	Remains
Deserter	Abandons mate	Moves to another
(b) **Usurpation**		
Victim of usurpation	Forced to abandon mate	Forced to leave
Bystander	Loses old mate, gains new mate	Remains

Table 9.2 The fitness consequences of divorce for each participant in desertion and usurpation in Fig. 9.7 (Heg *et al.*, 2003)

juncos (*Junco hyemalis*) the figures were 66 and 38%, respectively (Smith *et al.*, 1982; Greenlaw & Post, 1985; Wolf *et al.*, 1990). For some species, removals suggest that male help is more important when food is scarce (Lyon *et al.*, 1987; Bart & Tornes, 1989). These experiments show that male help can clearly increase reproductive success, but it is not essential. If male desertion reduces productivity to a fraction $1/x$ of a pair-fed brood, then provided a male can gain more than x females, desertion will be the more profitable option from his point of view. Even if success is reduced to a half or less, and a male can gain only two females, polygyny will still pay provided the male helps to provision at least one of the broods. As predicted, male songbirds readily desert to gain extra females if given the chance, for example by removal of a neighbouring male, and they often help to provision either one of their female's broods full-time, or several females' broods part-time. Occasional polygyny has been reported in 39% of 122 well-studied European songbirds.

Monogamy in birds because of limited opportunities for polygamy

These experiments suggest that the predominance of monogamy in many birds arises not, as Lack proposed, because each sex has the greatest success with monogamy but because of the limited opportunities for polygyny. The two most obvious constraints are: (i) strong competition among males may make it difficult for a male to gain a second female; and (ii) females are likely to suffer in polygyny through the loss of male help and, as predicted, females are often aggressive to other females, which may decrease the chance that their partners are able to gain a second mate. This latter constraint is particularly well illustrated by burying beetles *Nicrophorus* where, just like birds, males help to feed offspring (Chapter 8). Females lay eggs on vertebrate carcasses and both parents provision the larvae on regurgitated carrion. On small carcasses, the male cooperates to feed the offspring of his one female. However, on large carcasses the male adopts a 'headstand' posture and exposes his last abdominal segment to emit pheromones in an attempt to attract a second female. A second female increases the male's reproductive success from the carcass, but it is costly to the first female because, with reduced male care, she produces fewer offspring herself (Eggert & Müller, 1992). It is not surprising, therefore, that the first female interferes with her male's attempts to gain polygyny; when he displays, she mounts him, pushes him over or bites his abdomen with her mandibles (Eggert & Sakaluk, 1995).

Burying beetles: females oppose their male's attempts to gain polygyny

In the last chapter, we saw that social monogamy does not necessarily imply genetic monogamy because extra-pair matings are rife in many species (Box 9.1). In some species of birds males protect their paternity by following the female closely during her fertile period ('mate guarding', e.g. magpies and swallows). In other species this is not possible because one partner has to defend the nest site while the other goes off to forage (many seabirds and birds of prey). Here males engage in frequent copulations to swamp the sperm of rivals, sometimes copulating several hundred times per clutch, clearly far more than necessary simply to make sure that the eggs are fertilized (Birkhead & Møller, 1992). Even so, despite these paternity guards the frequency of extra-pair paternity can be very high (25–35%) in some species (Westneat & Stewart, 2003). For example, in the red-winged blackbird (*Agelaius phoeniceus*) Lisle Gibbs and his colleagues (1990) found that extra-pair fertilizations accounted for on average 21% of a male's reproductive success (Fig. 9.8).

Social monogamy in birds does not mean genetic monogamy: extra-pair matings are often common

BOX 9.1 USING DNA PROFILES TO ASSIGN PARENTAGE

In 1985, Alec Jeffreys and colleagues discovered that there was enormous genetic variability which could be used to assign paternity and maternity with great precision. Since then, methods for DNA profiling have been refined and they have revolutionized field studies by linking behaviour to individual reproductive success. An individual's DNA is isolated from a tissue sample (e.g. blood) or non-invasively, for example by analysis of epithelial cells on dung (these cells are sloughed off as faeces pass through the intestine). Microsatellites, or Simple Sequence Repeats, are stretches of DNA that contain short nucleotide repeats of two to six base pairs. The microsatellite below contains a two base pair (dinucleotide) repeat of the bases G and T.

AGATTTTAAAGTCGTGTGTGTGTGTGTGTGTGTGTGTGTGTGTGTGTGTGTGTGTGATGACAA
GTTGGTG

This particular allele has 19 repeats of the bases GT. Microsatellites have several important features which make them ideal for parentage assignment and kinship analysis.

(1) They are highly polymorphic. Microsatellite polymorphism occurs when homologous microsatellite loci differ in the number of repeats between individuals (often ten or more alleles per locus). The repetitive nature of microsatellites is thought to lead to an increased rate of slipped strand mispairing (slippage) during DNA replication in comparison to other neutral regions of DNA, leading to a change in length of the microsatellite (e.g. from 19 to 20 dinucleotide repeats).

(2) Microsatellites are inherited in a Mendelian fashion; every individual contains two copies of each microsatellite locus, one allele inherited from the individual's mother and the other from its father. Comparing the alleles present in offspring to those present in putative parents allows us to assign the most likely parents.

Microsatellites are amplified using PCR (polymerase chain reaction) and then either labelled with fluorescent dye and visualized using a DNA sequencer or visualized by gel electrophoresis as dark bands, as shown in the example below. This example is from a study by Hazel Nichols *et al.* (2010) of banded mongooses *Mungus mungo* in Queen Elizabeth National Park, Uganda. In the microsatellite locus shown in Fig. B9.1.1, male A is heterozygous (two bands, representing two different 'lengths', or alleles, of the microsatellite), male B is homozygous for a third allele and the female is homozygous for one of male A's alleles (homozygous means two copies of the same allele, hence just one band on the gel). There are seven pups. Pup five matches the mother and male B. The rest of the litter match the mother and male A. By using a number of different microsatellite loci (fourteen in this study), parenthood can be assigned with great precision. In the banded mongoose study, the oldest three males in each group fathered on average 85% of the group's pups. The older males achieved this high reproductive success by choosing to focus their mate-guarding effort on the oldest, most successful females.

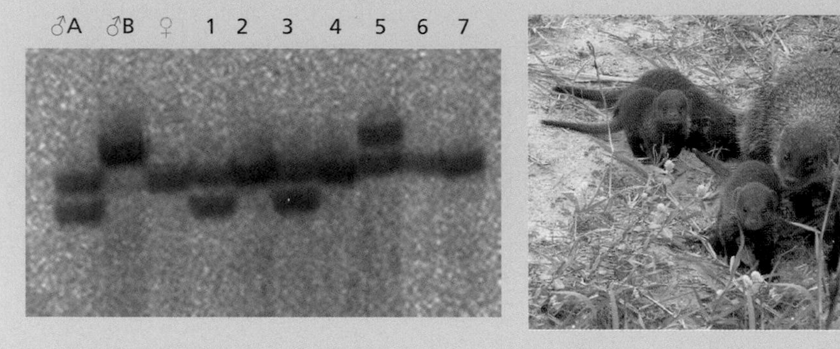

Fig. B9.1.1 Using DNA microsatellites to assign paternity in banded mongooses. Photos © Hazel Nichols.

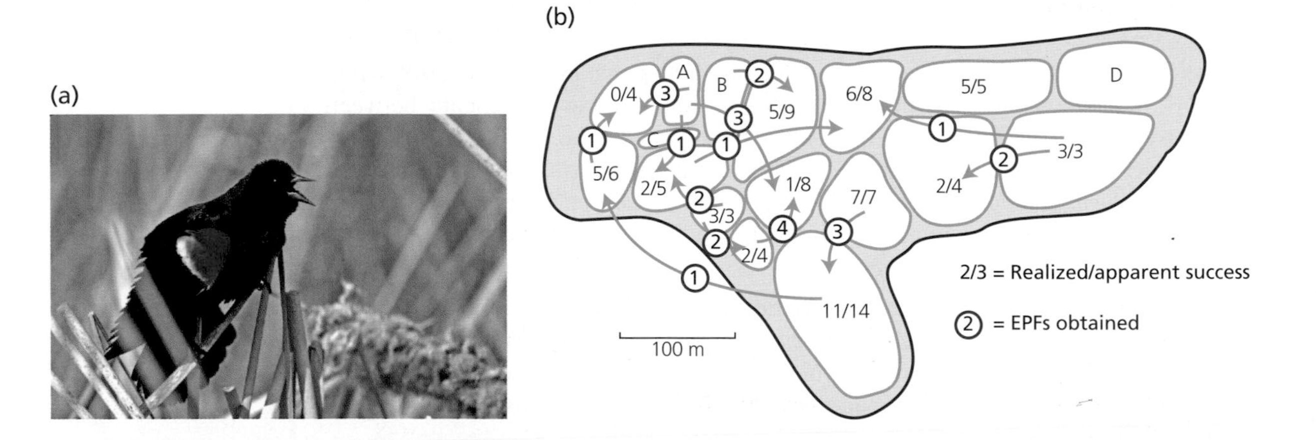

Fig. 9.8 (a) A male red-winged blackbird displaying his red epaulets. Photo © Bruce Lyon. (b) Reproductive success of male red-winged blackbirds on a marsh in Ontario, Canada, assessed by DNA markers. The fractions in each male territory show the number of chicks sired by the resident male over the total chicks raised. Arrows refer to extra-pair fertilizations (EPF's): the origin of the arrow shows the identity of the cuckolding male; the arrowhead indicates the territory in which he fertilized chicks; the number in the circle indicates the number of extra-pair chicks he sired. The map shows that most, but not all, cuckolders were near neighbours. From Gibbs *et al.* (1990). Reprinted with permission from AAAS.

Polygyny

If monogamy in birds often occurs because males are unable to gain another female, rather than because it pays males to remain faithful to one mate, what permits regular polygyny in some species? Polygyny (one male with several females) in birds usually arises through males monopolizing females indirectly, by controlling scarce resources such as food or nest sites. Where these are patchily distributed, males able to defend the best patches can gain the most mates (Fig. 9.2). It is useful to distinguish various ways in which such 'resource defence polygyny' can arise, bearing in mind that we need to consider the costs and benefits for each sex separately (Searcy & Yasukawa, 1989).

Resource defence polygyny in birds: males controlling the best resources get more mates

No cost of polygyny to females

In some species the males contribute very little to parental care and so females suffer little, if any, cost from mating polygynously. For example, female yellow-headed blackbirds (*Xanthocephalus xanthocephalus*) build their nests in marshes and feed in fields away from the breeding site. In one study there were apparently no costs or benefits from settling near other females and they settled more or less at random in the marsh (Lightbody & Weatherhead, 1988). In yellow-rumped caciques (*Cacicus cela*) females also do not suffer from sharing a male but they benefit from nesting close together in safe sites and by cooperative nest defence against avian predators (Robinson, 1986). In both species females may be largely indifferent to the mating system that emerges, which is determined simply by a male's ability to monopolize mates. If a small number of males is able to control the area with the most nesting females then high degrees of polygyny may occur.

In some species, females may experience no polygyny costs

Cost of polygyny to females

In many species, however, females will suffer costs from polygyny through having to share either the resources a male controls (food, nest sites) or his contribution to parental care. Females may be forced to accept these costs if a fraction of the males control all the suitable breeding habitat, their choice being 'accept polygyny' versus 'forego breeding'. For example, in Leonard and Picman's (1987) study of marsh wrens, *Cistothorus palustris*, females settled with mated males only after all the bachelor males had paired. For these later settling females there was no choice but to accept the costs of polygyny.

In other species, females may be forced to accept polygyny costs

In other cases, however, most of the males may be able to gain breeding territories. If there is variation in male territory quality then a female's choice may be 'settle on a good territory with an already-mated male, that is choose polygyny' versus 'settle on a poor territory with an unmated male, that is choose monogamy'. Jared Verner and Mary Willson (1966) suggested that females may choose the polygyny option if the costs of sharing a male's help with parental care were outweighed by the benefits of gaining access to good resources, such as food or nest sites. Gordon Orians (1969) presented this idea in a graphical model, known as the 'polygyny threshold model' (Fig. 9.9).

In others, females may choose polygyny because the costs are outweighed by the benefits

In many species males with the best territories are indeed the ones to attract the most females, just as the model predicts. However, showing that females are making the best choice among the breeding options available is difficult unless a great deal is known about the costs of sharing and the choices available to them (Fig. 9.10). We consider two species which have provided excellent systems for testing the model.

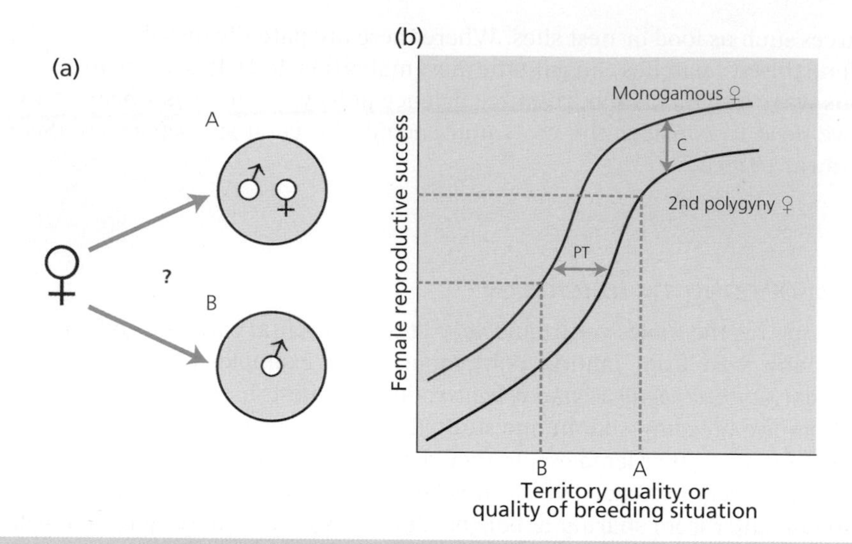

Fig. 9.9 The polygyny threshold model. (a) A female has the choice of settling with an unmated male on a poor quality territory B, or with an already-mated male on a good quality territory A. (b) Female reproductive success increases with territory quality. There is a cost C of sharing with another female, so the curve for the second female in polygyny lies below that for a monogamous female. Provided the difference in territory quality exceeds PT (the polygyny threshold), a female does better by choosing to settle with an already-mated male on territory A rather than with an unmated male on territory B. Modified from Orians (1969).

Great reed warblers (Acrocephalus arundinaceus)

This species breeds in reed beds on lake edges in Europe and Asia, weaving its nest around the reed stems. In a study in Sweden, Bensch and Hasselquist (1992) captured some newly arrived females in spring and fitted them with radio transmitters. They then released them onto a study area where male territories had been mapped to see how they sampled territories before pairing. Most females paired up within 24 hours having visited the territories of from three to eleven different males, sometimes going back to pair with a male they had previously sampled. Some females selected already-mated males despite the conspicuous presence of another female on the territory, and even though they had previously sampled the territories of unmated males. These observations show that females sample and choose male territories in exactly the way envisaged in the polygyny threshold model.

Evidence for the polygyny threshold model in great reed warblers

Do females make the best choice available? This is a more difficult question to answer. In another study of the same species in Lake Biwa, Japan, Ezaki (1990) found that from 30 to 80% of the males were polygynous each year, some attracting up to four females to their territories, while other males were monogamous or remained unpaired. The polygynous males were those who claimed territories containing the best nest sites, namely dense reeds where predation was lowest. Females who settled polygynously as second females did not seem to suffer from their choice because they did at least as well as simultaneously nesting monogamous females on poor territories. Thus the

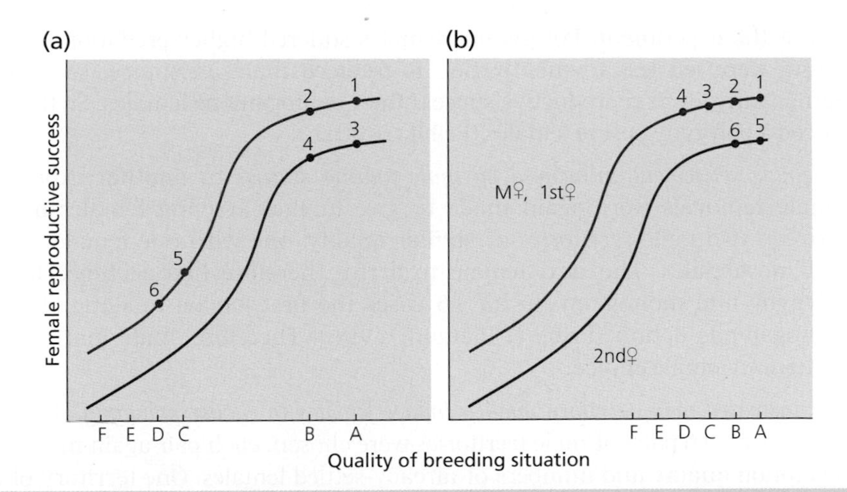

Fig. 9.10 Female settlement patterns predicted by the polygyny threshold model for two distributions (a and b) of male territory quality (territories A to F). It is assumed that the first female does not suffer from the arrival of a second female, so the top line represents the reproductive success of both monogamous (M) females and the first females in polygyny, while the bottom line refers to second females in polygyny. The sequential settlement patterns of six females (1–6) are shown for the six male territories, assuming that females settle where their expected reproductive success is greatest. In both cases two males become polygynous (A and B), two monogamous (C and D) and two remain unmated (E and F). However, settlement patterns and the reproductive success of monogamous versus polygynous females vary depending on the choices available. After Altmann *et al.* (1977) and Davies (1989).

differences in male territory quality seem sufficient for females to cross the polygyny threshold (Fig. 9.9).

Red-winged blackbirds (Agelaius phoeniceus)

This North American species also breeds in marshes, where females build nests in emergent vegetation, such as cattails *Typha*. Males defend territories and attempt to attract females by singing and displaying their striking red epaulets (Fig. 9.8). Stanislav Pribil, Jaroslav Picman and William Searcy studied a population in Ontario and tested four key predictions of the polygyny threshold model by some clever field experiments. At the start of the breeding season they selected pairs of adjacent male territories such that each member of a pair was matched in terms of territory quality (emergent vegetation) and number of newly-arrived females. Each territory of a matched pair was then given, at random, a different experimental treatment.

An experimental test of the polygyny threshold model

(i) *Is there a cost of polygyny?* In one experiment, females were removed from 40 pairs of male territories, so that one territory in each pair had two females and the other had one female. The removed females were kept in aviaries and then released at the end of the experiment. Polygynous females suffered higher predation and their young were fed less frequently due to reduced male assistance, so they had significantly lower reproductive success than monogamous females. So there was indeed a polygyny cost to females (Pribil, 2000).

(ii) *Is female settlement influenced by male mating status?* In another experiment, female removals were again made to give further arriving females a choice between two male territories of similar quality, one with one female and one with no females. The next female to arrive therefore had a choice between polygyny and monogamy. In all 16 cases the first female to settle chose the monogamous option (Pribil & Picman, 1996). Therefore, male mating status influenced female choice.

(iii) *Can increased male territory quality induce females to choose polygyny?* In a third experiment, 16 pairs of male territories were chosen, each pair again matched for vegetation quality and numbers of already-settled females. One territory of each pair was allowed to retain one female (any further females were removed) and the quality of this territory was improved by adding wooden nesting platforms of cattail shoots, supported underneath by chicken wire and placed over open water. Females preferred to build nests over water because this reduced predation by raccoons and weasels, so these territories were given additional good nesting sites. The other territory of each pair had all the females removed and was also provided with the same additional cattails, but this time the platforms were placed over dry land, less suitable for nesting. On these territories, cattails over water were also trimmed.

Thus, newly-arriving females now had a choice between joining an already-paired male on an improved quality territory versus an unpaired male on a reduced quality territory. There were 16 pairs of territories in this experiment. In two pairs new females settled on both territories on the same day. In the remaining 14 pairs, only one new female settled; in 12 cases she chose polygyny on the good territory and in only two did she choose monogamy on the poor territory. Therefore, females can be induced to choose polygyny, provided the male's territory is of sufficient quality (Pribil & Searcy, 2001).

In red-winged blackbirds, females benefit by choosing polygyny on a good territory rather than monogamy on a poor territory

(iv) *Is female choice for polygyny on a good territory adaptive?* Pribil and Searcy (2001) then calculated that the benefit of nesting over water was 1.02 more young raised, compared to nesting over land. The cost of polygyny to a female was 0.62 fewer young raised, compared to monogamy. Therefore, female choice for polygyny on a good territory was adaptive; the benefits exceeded the costs.

These results provide strong support for the polygyny threshold model, but the model may not apply to all populations of red-winged blackbirds. In another study in Pennsylvania, females did not suffer costs from polygyny (Searcy, 1988). Thus, the costs and benefits of female choice are likely to vary geographically depending on local predation pressures and food supply. Further studies are needed to examine how females make their choices in the face of varying selective pressures.

Sexual conflict and polygamy

The assumptions of the polygyny threshold model are like those of the 'ideal free distribution', which were discussed in Chapter 5. The different resource patches available are male territories of varying quality and females are assumed to be 'free' to settle where they choose. Under 'ideal' conditions, they are expected to settle where their reproductive success is greatest. However, we saw that ideal free conditions often do not hold in nature because dominant individuals attempt to grab more than their fair share of resources, so the assumption that polygyny arises from ideal free female settlement may often be unrealistic. For example, if the first female suffers from the arrival of a second female then it will pay the first female to try to prevent her from settling. Males, too, may attempt to change the mating system in ways which are detrimental to female success. The following two case studies provide good examples of such sexual conflict.

The pied flycatcher (Ficedula hypoleuca)

Rauno Alatalo, Arne Lundberg and their colleagues have studied this bird in the woodlands around Uppsala, southern Sweden. Males defend nest sites, holes in trees or nest boxes, and sing to attract a female. Once a male has attracted one female and she has laid her eggs, he then goes to another nest hole and tries to attract a second female (Fig. 9.11). Males do not simply advertise from the next nearest nest site but go, on average, 200 m away and even up to 3.5 km from their first site! About 10–15% of the males succeed in gaining a second female. They then desert her and go back to help their first female with chick-feeding. Compared to monogamous females the first female suffers little, if at all, from polygyny because she usually gains the male's full-time help, but the second female, who is left to raise her brood on her own, suffers reduced success, raising on average only 60% of the number of young she would have gained in monogamy (Alatalo *et al.*, 1981).

Why, then, do females ever settle polygynously? Three hypotheses have been proposed.

In pied flycatchers, second females suffer, so why do they settle polygynously? Three hypotheses

(i) *The 'sexy son' hypothesis.* Weatherhead and Robertson (1979) have suggested that although second females produce fewer offspring, this may be offset if they have sons who inherit their father's ability to be polygynous. The female loses out in the first generation, but then makes up for this in the second generation when her sexy sons sire lots of grandchildren compared to the sons of monogamous females. According to this hypothesis (which derives from Fisher's sexual selection argument, Chapter 7), secondary females are still making the correct choice when they settle polygynously, but offspring quality is another factor which must be taken into account in the y axis of the polygyny threshold model (Fig. 9.9).

For the pied flycatcher, calculations show that the heritability of male mating status would have to be 0.85 for 'sexiness' of sons to offset the loss in offspring numbers. The heritability is not known because young birds disperse and breed away from their natal area. However, the probability that an individual male is polygynous in successive years is only 0.29, which must give an upper limit to the heritability value. Therefore, we can reject this hypothesis.

Fig. 9.11 Once a male pied flycatcher has attracted one female, he flies off to another nest site some distance away and tries to attract another. Secondary females suffer because they get little or no help from the male in chick rearing. However, females probably are unable to assess whether the male they pair with has another female because of the large distance between a male's two nest sites.

(ii) *Deception.* Alatalo *et al.* (1981) proposed that second females are deceived into settling polygynously because the male's habit of setting up nesting territories several hundred metres apart (polyterritoriality) prevents females from distinguishing mated from unmated males. By the time the second female has laid her clutch and the male has deserted her to go back to his first female, it is too late in the season for it to be profitable to start another clutch, so she has to make the best of a bad job and rear her offspring alone.

(iii) *Unmated males hard to find.* An alternative hypothesis is that second females are not deceived but they choose polygyny as their best option simply because unmated males are hard to find. According to this view, male polyterritoriality is not to aid female deception but to decrease the chance that aggression from the first female will prevent the second female from settling (Stenmark *et al.*, 1988; Dale *et al.*, 1990).

An experimental test of deception by males

Testing between these last two hypotheses requires detailed observations on how females sample males and territories, and measurements of the profitability of their alternative options. Alatalo *et al.* (1990) performed a clever experiment to test between them. By erecting nest boxes in careful sequence, they arranged for neighbouring boxes, less than 100 m apart, to be occupied by an unmated male and a mated male, whose first female was incubating a clutch in another box 100–300 m away. Boxes were put up at randomly chosen sites, so there was no difference in territory quality between mated and unmated males. In this situation females could clearly sample both males (some were seen to do so) and the songs of both could be heard from either nest site. In 20 such paired choices, nine females settled with the unmated male and 11 with the mated male – clearly no difference. Furthermore, the females who chose the mated males raised significantly fewer young than those who later chose the unmated males they had rejected. This result supports the deception hypothesis; females did not discriminate between mated and unmated males even when they had a simultaneous choice between them, and even though it would have paid them to make a choice.

The dunnock (Prunella modularis)

Conflicts of interest have led to a very variable mating system in another European songbird, the dunnock (Fig. 9.12), including simple pairs (monogamy), a male with two females (polygyny) and a female with two (unrelated) males (polyandry).

A female has least success in polygyny, where she has to share a male's help with parental care. She has greater success with monogamy, where she gains a male's full-time help with chick feeding. And she has greatest success of all with polyandry because if she copulates with two (unrelated) males she then gains both their help and three adults provisioning a brood increases the number of young that survive (Fig. 9.12a). In polyandry, shared copulations often lead to mixed paternity in the brood. Observations and experiments (temporary removal of a male for parts of the female's fertile period) revealed that a male will help to feed the brood only if he gained a share of the matings. Furthermore, he increased his help in proportion to his mating share, which predicts his paternity share. A female maximized the total care she gained from her two males if she gave equal mating shares to each male (Davies *et al.*, 1992).

From a male's point of view, however, reproductive success is least in polyandry (the system where a female does best) because although more young are raised through extra male help, the increased production of a trio-fed brood does not compensate a male for shared paternity. In fact, a male does best in polygyny (the system where a female does least well) because despite the cost that each female suffers, the total output of two females with part-time help exceeds that from monogamy (Fig. 9.12a).

In dunnocks, males prefer polygyny, females prefer polyandry

These conflicts of reproductive interests make good sense of male and female conflicts in behaviour. In polygyny, the dominant female attempts to drive the other female away to claim the male to herself, while the male tries to keep between his squabbling females so that both remain with him. On the other hand, females encourage copulations from subordinate males in the hope that they will remain and provide parental care, while dominant males attempt to guard the female and drive subordinate males away to claim full paternity for themselves. The variable mating system can, therefore, be viewed as the different outcomes of these conflicts of interest. Sometimes the conflict reaches a 'stalemate', in which two males end up sharing two females (polygynandry; Fig. 9.12b). Here the dominant male is unable to drive the other male off to claim both females for himself (polygyny) and the dominant female is unable to drive the other female off to claim both males for herself (polyandry).

Polygynandry as a 'stalemate' outcome to the conflict

A variable mating system reflects the different outcomes of sexual conflict

So the key question to ask is this: under what circumstances can particular individuals gain their best mating option despite the conflicting interests of others?

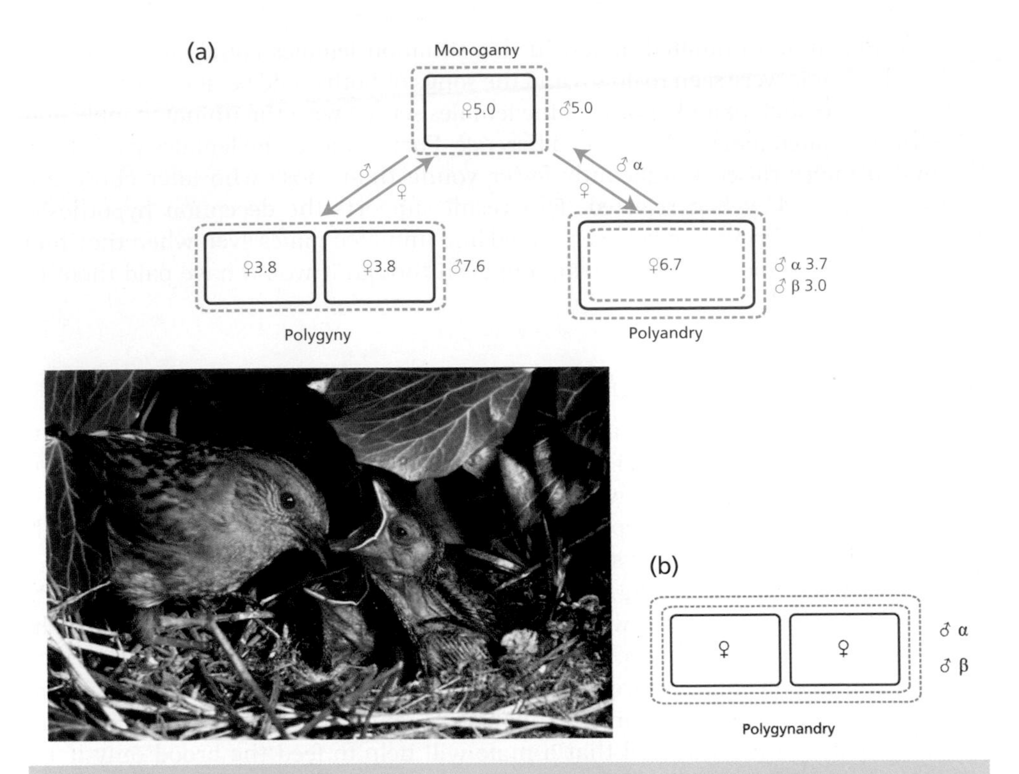

Fig. 9.12 Male dunnock feeding a brood of chicks. Photo © W. B. Carr. (a) Sexual conflict in dunnocks. Female territories (solid lines) are exclusive and may be defended by one or two unrelated males (dashed lines). The numbers refer to the number of young raised per season by males and females in the different mating combinations (maternity and paternity measured by DNA fingerprinting; Burke *et al.* (1989)). Arrows indicate the directions in which dominant (alpha) male and female behaviour encourage changes in the mating system. A male does best with polygyny; the cost of polygyny to females is shared male care. A female does best with polyandry; the cost of polyandry to males is shared paternity. (b) Polygynandry as a stalemate to the conflict: the alpha male is unable to drive the beta male off to claim polygyny, and neither female can evict the other to claim polyandry. From Davies (1989, 1992).

Various factors influence the conflict outcome. Firstly, differences in individual competitive ability are important. For example, young males are more likely to be subordinate males and older, experienced, males are more likely to defend larger territories which can encompass two female territories (polygyny or polygynandry). Secondly, the adult sex ratio influences mating systems; after harsher winters the breeding sex ratio is more male-biased (females are subordinate at feeding sites and more likely to die), so there is more polyandry. Thirdly, territory characteristics can influence the conflict outcome; on territories with denser vegetation, a female can more easily escape dominant male guarding and so promote mixed paternity (Davies, 1992).

Polyandry threshold

In dunnocks, polyandry is advantageous to the female but not to the alpha male. Indeed, it occurs despite the best attempts of the alpha male to prevent it! In theory, however, there may be conditions when the increased production of offspring from cooperation by a team of males does offset the costs to an individual male of paternity sharing. Therefore, there would be a 'polyandry threshold' (analogous to the 'polygyny threshold' discussed earlier), at which males would do better by agreeing to share a female rather than attempting to go it alone in monogamy (Gowaty, 1981). Two situations in which the benefits of cooperation among males are likely to outweigh the cost of sharing paternity are: a scarcity of food (so more than two individuals are needed to raise offspring effectively) and intense competition for territories or females (so a team of males is more effective in gaining reproductive opportunities).

Male lions benefit from polyandry ...

A good example of this last scenario is provided by lions *Panthera leo*, where larger coalitions of males are both more likely to gain control of a pride of females, and to maintain longer tenure of a pride, with the result that individual male lifetime fitness increases with male coalition size. In lions, male teams often consist of relatives (brothers, half brothers), so there is kin selection for cooperation (Chapter 11). But it pays even unrelated males to cooperate in small teams rather than to attempt to take-over a pride on their own (Packer *et al.*, 1991).

... and so might male Galapagos hawks

In some populations of Galapagos hawks, *Buteo galapagoensis*, groups of unrelated males (usually two to three, but up to eight) cooperate to defend a breeding territory. They share matings with the single breeding female and share paternity of the offspring (Faaborg *et al.*, 1995). In remarkable contrast to the dunnocks, males do not squabble over matings and they share paternity roughly equally (DeLay *et al.*, 1996). Two likely factors promoting male cooperation are the larger size of the female (which may enable her to control copulation shares) and the increased number of years of tenure of a breeding territory when it is defended by a larger male team. This means that a male's short-term paternity sharing costs each season could be compensated by breeding over many more years, leading to increased lifetime reproductive success. Such compensation cannot occur in shorter-lived dunnocks, where males breed for just one or two seasons.

Female desertion and sex role reversal

In most birds, if one sex deserts it is usually the male, because he has the opportunity to desert first. Usually, he also has more to gain from increasing his number of mates because he can potentially fertilize eggs at a faster rate than a female can lay them (Chapter 7). Thus, polygyny is far commoner than polyandry. Some studies, however, suggest sexual conflict over opportunities to desert first (Chapter 8). In the Florida snail kite, *Rostrhamus sociabilis*, either sex may desert leaving the other to care for the brood. Which sex deserts depends on who has the greatest chance of gaining another mate, which varies depending on the operational sex ratio. Desertion is also more frequent when food is abundant, so that the remaining partner is better able to raise the young unaided (Beissinger & Snyder, 1987).

In shorebirds (Charadrii), most species are monogamous with biparental care. However, in some species the normal sex roles are reversed, with the females being

Fig. 9.13 Sex role reversal in birds. This male African jacana performs all the parental duties. Females are larger than males and compete for males by defending large territories. Photo © Tony Heald/naturepl.com.

larger and more brightly coloured and females competing for males to incubate their clutches for them. In phalaropes (*Phalaropus* spp.) a female defends one male, lays a clutch for him to incubate and then goes off to find a second male (sequential polyandry; Reynolds, 1987). In spotted sandpipers (*Actitis macularia*) and jacanas (Fig. 9.13) a female competes to defend a large territory in which she may have several males simultaneously incubating her clutches or caring for her chicks (resource defence polyandry).

Some shorebirds show sex role reversal: females compete for males

Why should some shorebirds show such sex role reversal? A likely explanation is that shorebirds are characterized by a small clutch size, which never normally exceeds four eggs. The four eggs are large and fit snugly together and experiments suggest that they represent an incubation limit: adding an extra egg reduces hatching success. If shorebirds are indeed 'stuck' with a maximum clutch of four, selection may particularly favour female desertion because, with a fixed clutch size, the only way females can increase their reproductive output if conditions become more favourable is to lay more clutches. In the spotted sandpiper, productivity on the breeding grounds can be so high that the female becomes rather like an egg factory, laying up to five clutches in 40 days, a total of 20 eggs, which represents four times her own body weight. Her reproductive success is no longer limited by her ability to form reserves for the eggs but rather by the number of males she can find to incubate them. This has led to the evolution of sex role reversal with females being 25% larger than males and females competing to gain as many mates as they can (Lank *et al.*, 1985).

In jacanas, too, the female is larger than the male and her territory may overlap the smaller territories of up to four males (Butchart *et al.*, 1999a; Emlen & Wrege, 2004).

Females compete to takeover neighbouring male territories and sometimes kill other female's chicks in order to make extra males available to receive their clutches (Emlen *et al.*, 1989). Such female infanticide is the sex role reversed analogy of male infanticide in lions (Chapter 1), where males attempt to make more fertile females available.

While polyandry benefits a female jacana, it brings potential costs to a male because other males in the female's harem (co-mates) may fertilize the eggs laid in his territory. Indeed, a female often copulates with several males while she lays a clutch for one of them. In the bronze-winged jacana *Metopidius indicus* co-mates compete for their female's attention by yelling for copulations (Butchart *et al.*, 1999b). A study of the wattled jacana *Jacana jacana*, in Panama revealed that monandrous males (one female with one male) gained full paternity of their clutches while in polyandry 41% of broods had chicks sired by other males in the harem (17% of all chicks were sired by co-mates; Emlen *et al.*, 1998). Co-mates are, therefore, competitors both for receipt of clutches and for copulations.

Female jacanas compete for male care

A hierarchical approach to mating system diversity

We have discussed three broad themes that might influence mating systems:

(i) *Life history constraints*. For example, male mammals (which do not lactate) might often be less able to make a worthwhile contribution to the feeding of offspring than male birds.

(ii) *Ecological factors*. For example, the dispersion of food and nest sites will influence the dispersion of potential mates and, hence, their economic defendability.

(iii) *Social conflicts*. Individuals often behave to limit the choices of their mates through coercion or deception.

How well do these three themes explain the diversity of mating systems in Table 9.1?

Ian Owens and Peter Bennett (1997) have suggested that we need all three themes but that each will be most relevant at a different taxonomic level of analysis (Table 9.3). Consider bird mating systems. If we want to explain differences between higher taxonomic levels, for example different orders or families, then differences in life history constraints are likely to be important (theme (i) above). For example, the offspring of pheasants (family Phasianidae) are precocial and can run about and feed themselves soon after hatching. Therefore, it is not too costly for a male to desert and seek further mates because the female can care for a brood alone. By contrast, young hawks (family Accipitridae) are born naked and helpless and need the care of both parents. Initially the female guards the nest and keeps the young warm while the male does all the hunting for the family.

Differences between orders or families reflect life history differences

If we want to explain differences in mating systems at lower taxonomic levels, for example differences between species within a genus, then here differences in ecology are likely to be most important (theme (ii) above). Thus, weaverbird species that exploit seeds (large patches of easily-found food) are more likely to be polygynous than

Table 9.3 A hierarchical approach to mating system diversity (Owens & Bennett, 1997)

Taxonomic comparison	Main factors leading to different mating systems
Between orders or families (e.g. pheasants versus hawks)	Differences in life history constraints (e.g. precocial versus altricial young)
Between species within a genus	Differences in ecology (e.g. food, breeding sites)
Between individuals within a population of one species	Social conflicts

Differences between related species reflect differences in ecology

weaverbird species that exploit insects (harder to exploit, more likely to need both parents to raise a brood; Chapter 2).

Differences within species or populations reflect social conflicts

Finally, differences within populations of a species are likely to be explained by social conflicts (theme (iii) above), for example differences in the success with which individuals can exploit their mates. A male dunnock or burying beetle that can gain a second mate despite the attempts of his first female to prevent this will profit, though his first female will suffer. Different outcomes may occur depending on which individuals can gain their preferences, despite the conflicting preferences of others.

Unravelling the interactions between these three processes is a challenge for future studies.

Summary

When males do not provide parental care, the mating system often emerges as the outcome of a two-step process: (i) females distribute in relation to resources and (ii) males distribute in relation to female dispersion, either competing for females directly (female defence) or competing for resource rich sites (resource defence). The economics of defence by males depends on both the spatial and temporal distribution of females/resources. These themes are illustrated by experiments with voles and wrasse, and comparative studies of mammalian mating systems.

In some species, males aggregate on leks. Five hypotheses for these male aggregations are: settlement on female 'hotspots'; reducing predation; increasing female attraction; parasitizing 'hotshot' males; and facilitating female choice.

Where males provide care (most birds), males become a resource for females and the mating system often depend on patterns of desertion by either sex. Monogamy may be obligate because both male and female need to cooperate for any success, but often individuals are constrained to be monogamous, either by their partner's behaviour or competition for mates. Social monogamy often does not lead to genetic monogamy because of extra-pair matings. Females may choose polygyny on a good quality male territory rather than monogamy on a poor one (polygyny threshold model, e.g. red-winged blackbirds).

There are often conflicts between the sexes over the mating system that maximizes an individual's success (burying beetles, pied flycatcher, dunnock), which may lead to variable mating systems within a population. In some shorebirds there is sex role reversal with females competing for males. A hierarchical analysis suggests that different factors explain mating system diversity at different taxonomic levels.

Further reading

Bennett and Owens (2002) review mating systems in birds. Höglund and Alatalo (1995) review lek mating systems. David McDonald's study of long-tailed manakins shows that unrelated males cooperate in displays on leks. The alpha male gains all the matings while the beta male cooperates to maintain the attractiveness of the site for when he becomes alpha in later years (McDonald & Potts, 1994; McDonald, 2010). Andersson (2005) and Berglund and Rosenqvist (2003) consider the theories and evidence for the evolution of sex role reversal and polyandry. Setchell and Kappeler (2003) review sexual conflict and mating systems in primates.

TOPICS FOR DISCUSSION

1. Why are most mammals polygynous whilst most birds are monogamous?
2. Why might female red-winged blackbirds suffer polygyny costs in some populations but not in others? How would this variation influence mate choice and mating systems?
3. Compare the evolution of sex role reversal in shorebirds and pipefish (read Andersson, 2005, and Berglund and Rosenqvist, 2003).
4. How could the comparative survey of mammalian mating systems discussed in this chapter be enhanced by mapping mating systems onto a phylogeny?
5. Compare the techniques needed to study differences in mating systems at the various taxonomic levels in Table 9.3.

Photo © David Shuker and Stuart West

CHAPTER 10 Sex Allocation

All sexually reproducing organisms must decide how to allocate resources to male and female reproduction. This encompasses a range of related questions across the different types of breeding systems. In dioecious species, where individuals are either male or female for their entire lifetime, such as birds and mammals, the problem is whether to produce male or female offspring? In sequential hermaphrodites, or sex changers, where individuals function as one sex early in their life, and then switch to the other, such as many reef fish, the problem is what sex to be first, and when to change sex? In species where the offspring sex is determined by the environment (environmental sex determination), such as some shrimps and fish, what cue should determine sex, and how?

Sex allocation is the allocation of resources to male versus female reproduction in sexual species

It used to be the conventional wisdom that whether interesting things happened with sex allocation depended upon the method of sex determination (Box 10.1). In species with genetic sex determination, such as birds or mammals, it was assumed that sex determination was random, and could not be controlled by parents. This idea had been supported by selection experiments on domestic animals such as chickens (where a female biased sex ratio would bring huge economic benefits), which had failed to shift offspring sex ratios from equal numbers of males and females. In contrast, sex determination mechanisms such as haplodiploidy and environmental sex determination allowed greater control over offspring sex, so interesting patterns of sex allocation could occur, with biases towards sons or daughters. For example, in haplodiploid species, such as ants, bees and wasps, a female can adjust the sex of her offspring, in response to local conditions, by whether or not she fertilizes an egg (Box 10.1).

However, over the last 40 years, this picture has changed dramatically, with the discovery that a huge variety of organisms, even birds and mammals, are manipulating the sex of their offspring in ways that increase their fitness. Indeed, research on sex allocation has provided one of the most productive and successful areas of behavioural ecology, illustrating several general points about how natural selection shapes

An Introduction to Behavioural Ecology, Fourth Edition. Nicholas B. Davies, John R. Krebs and Stuart A. West.

BOX 10.1 SEX DETERMINATION

The sex of an individual can be determined by sex chromosomes, the environment, or even change during its lifespan (Bull, 1983).

In species with genetic (chromosomal) sex determination, such as birds or mammals, sex is determined by whether individuals have two of the same kind of sex chromosome (homogametic) or two distinct sex chromosomes (heterogametic). In mammals, females are the homogametic sex (XX) and males the heterogametic sex (XY). Consequently, all ova are X and sex is determined by whether an ovum is fertilized by an X or a Y sperm. In birds, females are the heterogametic sex (ZW) and males the homogametic sex (ZZ). Consequently, all sperm are Z and sex is determined by whether the ovum is Z or W. To control the sex of her offspring, a female would, therefore, need to bias either acceptance of X or Y sperm (mammals), or the production of W or Z ova (birds).

Fig. B10.1.1 Temperature dependent sex determination in reptiles. In many reptiles sex is determined by the temperature during development. For example, in (a) the box turtle (*Terrapene ornate*) and (b) the green turtle (*Chelonia mydas*), males are produced at cool incubation temperatures and females at warm incubation temperatures. In other species, such as (c) the Australian freshwater crocodile (*Crocodylus johnstoni*), the opposite pattern occurs, with males produced at relatively high temperatures. Finally, one sex may be preferentially produced at extreme temperatures (both hot and cold), such as (d) the frill-necked dragon (*Chlamydosaurus kingii*), where both sexes are produced at intermediate temperatures, but only females at extreme temperatures. Photo (a) © Fred Janzen; (b) © Annette Broderick; (c) and (d) © Ruchira Somaweera.

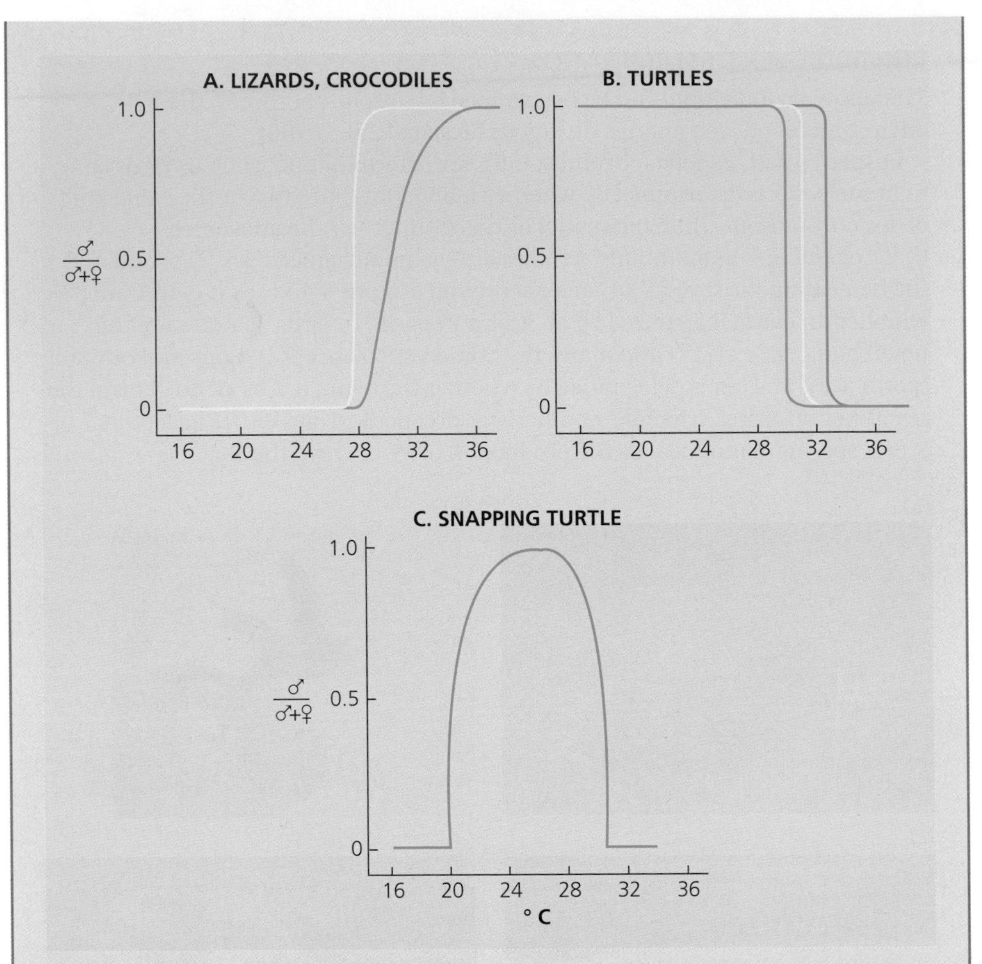

Fig. B10.1.2 Example sex ratio responses to incubation temperature in reptiles with temperature-dependent sex determination. Different lines represent different species. From Bull (1980).

In haplodiploid species, such as ants, bees, wasps and some beetles, sex is determined by whether or not an egg is fertilized. Fertilized eggs are diploid and develop as females, whereas unfertilized eggs are haploid and develop as males. This allows a female to control the sex of her offspring, by whether or not she fertilizes an egg. In some wasp species, an observer can even determine the sex of an egg as it is laid, because there is a visible pause, while the female fertilizes it.

In species with environmental sex determination, some feature of the environment such as temperature or day length determines sex. This occurs in a range of species including some turtles, crocodiles, shrimps and worms. For example, in turtles, males are produced at relatively low temperatures and females at relatively high temperatures. In contrast, the opposite pattern occurs with many lizards and crocodiles, with males being produced at relatively high temperatures. (Fig. B10.1.2).

behaviour. To understand interesting patterns of sex allocation, it is first necessary to explain why equal investment in males and females is the null model.

Fisher's theory of equal investment

If one male can fertilize the eggs of dozens of females why not produce a sex ratio of, say, one male for every 20 females? With this ratio the reproductive success of the population would be higher than with a 1:1 ratio since there would be more eggs around to fertilize. Yet in nature the ratio is often very close to 1:1, even when males do nothing but fertilize the female. As we saw in Chapter 1, the adaptive value of traits should not be viewed as being 'for the good of the population', but 'for the good of the individual' or, more precisely, 'for the good of the gene' that controls that trait. Darwin struggled with why a 1:1 sex ratio should be favoured, but a clear answer was provided by R.A. Fisher (1930).

Suppose a population contained 20 females for every male. Every male has 20 times the expected reproductive success of a female (because there are on average 20 mates per male) and, therefore, a parent whose children are exclusively sons can expect to have almost 20 times the number of grandchildren produced by a parent with mainly female offspring. A female-biased sex ratio is, therefore, not evolutionarily stable because a gene which causes parents to bias the sex ratio of their offspring towards males would rapidly spread, and the sex ratio will gradually shift towards a greater proportion of males than the initial 1 in 20. But now imagine the converse. If males are 20 times as common as females a parent producing only daughters will be at an advantage. Since one sperm fertilizes each egg, only one in every 20 males can contribute genes to any individual offspring; therefore, females have 20 times the average reproductive success of a male. So a male-biased sex ratio is not stable either. The conclusion is that the rarer sex always has an advantage, and parents which concentrate on producing offspring of the rare sex will, therefore, be favoured by selection. Only when the sex ratio is exactly 1:1 will the expected success of a male and a female be equal and the population sex ratio stable. Even a tiny bias favours the rarer sex: in a population of 51 females and 49 males where each female has one child, an average male has 51/49 children. This *average* value is the same whether one male does most of the fathering or whether fatherhood is spread equally among the males.

All else being equal, a 1:1 sex ratio will be favoured by natural selection

One way to test Fisher's theory is by perturbing the sex ratio away from 1:1 and then examining whether it evolves back towards this point. Alexandra Basolo (1994) did this, taking advantage of the unusual sex determination in the southern platyfish *Xiphorus maculatus*. In this species, sex is determined by a single locus with three sex alleles, with three female (WX, WY, XX) and two male (YY, XY) genotypes. She showed that if the relative frequency of the different alleles is varied, to set up populations with biased sex ratios, then selection favours the rarer sex, as predicted, and quickly moves the sex ratio back to a 1:1 ratio (Fig. 10.1).

If the sex ratio is perturbed from 1:1, it will evolve back to this point

The above argument, that the sex ratio should be 1:1, can be refined by re-phrasing it in terms of resources invested. In the above discussion, we have implicitly assumed that sons and daughters are equally costly to produce. However, suppose sons are twice as costly as daughters to produce because, for example, they are twice as big and need twice as much food during development. When the sex ratio is 1:1 a son has the same average number of children as a daughter. But since sons are twice as costly to make they are a bad investment for a parent: each of its grandchildren produced by a son is twice as

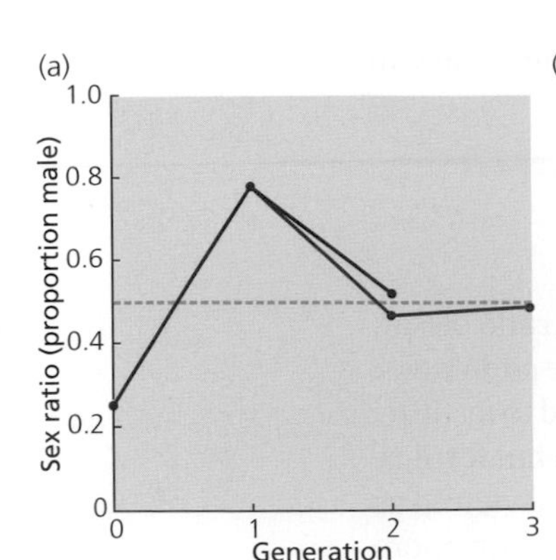

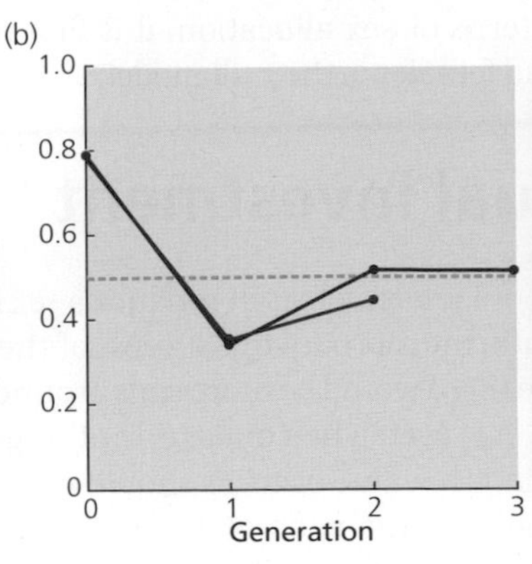

Fig. 10.1 Sex ratio evolution in the southern platyfish. When the sex ratio is perturbed away from equal numbers of males and females, it quickly evolves back to this point. The perturbation is towards a female (a) or a male (b) biased sex ratio. Different lines are different replicates. From Basalo (1994). Reprinted with permission from the University of Chicago Press.

More precisely, equal investment is favoured in the two sexes

costly as one produced by a daughter. It would, therefore, pay the parents to concentrate on making daughters. As the sex ratio swings towards a female bias, the expected reproductive success of a son goes up until at a ratio of two females to every male an average son produces twice the number of children produced by an average daughter. At this point sons and daughters give exactly the same return per unit investment; a son costs twice as much to make but yields twice the return. This means that when sons and daughters cost different amounts to make, the stable strategy in evolution is for the parent to *invest* equally in the two sexes and not to produce equal numbers. An example to illustrate this point is Bob Metcalf's (1980) study of the sex ratio of two species of wasp: *Polistes metricus* and *P. variatus*. In the former females are smaller than males, while in the latter they are similar in size. As predicted, the population sex ratio is biased in *P. metricus* and not in *P. variatus*. In both species the investment ratio is 1:1.

Sex allocation when relatives interact

Fisher's theory assumes that related individuals do not interact, either cooperatively or competitively. If such interactions occur, then individuals can be favoured to bias their offspring sex ratios, to reduce competition between relatives, or increase cooperation (Box 10.2).

Local resource competition

Anne Clark (1978) found that the African bushbaby, *Galago crassicaudatus*, has a male-biased investment ratio among its offspring. She pointed out that this could be explained by the species' life history. As with most mammals, female *Galago* do not disperse as far as males, and often end up competing both with their mother and with each other for rich sources of food, such as gum and fruit trees in the mother's home range. This local

BOX 10.2 SEX RATIOS WHEN RELATIVES INTERACT

The fitness of a female will be equal to her fitness through daughters plus her fitness through sons, which will be given by the equation:

fitness = (number of daughters produced multiplied by the fitness of each daughter) + (number of sons produced multiplied by the fitness of each son)

Consider the case of where daughters compete for resources, and so the mean fitness of each daughter decreases with the number of daughters produced. Such local resource competition would favour a male-biased sex ratio, to lessen competition between daughters and, hence, increase the fitness of each daughter produced (Fig. B10.2.1a). As the sex ratio becomes more male biased, it will decrease the average number of mates obtained by each male, and hence decrease the fitness return of producing sons. The evolutionarily stable (ES) sex ratio would be that at which these two forces cancels out.

In contrast, if daughters cooperate, then the average fitness of daughters increases with the number of daughters produced. Such local resource enhancement would favour a female-biased sex ratio, to increase cooperation between daughters and, hence, increase the fitness of each daughter produced (Fig. B10.2.1b). As the sex ratio becomes more female biased, it will increase the average number of mates obtained by each male, and hence increase the fitness return of producing sons. The ES sex ratio would be that at which these two forces cancels out.

(a) Local resource competition

(b) Local resource enhancement

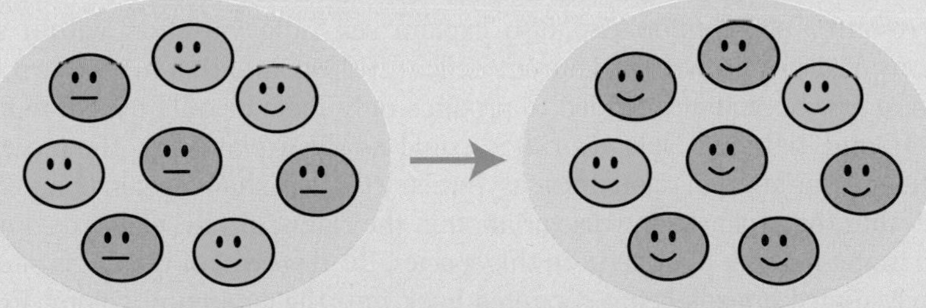

Fig. B10.2.1 Selection for biased sex ratios when relatives interact. (a) If sisters compete for resources, then a male-biased sex ratio is favoured to reduce competition between sisters. (b) If sisters cooperate with each other, then a female-biased sex ratio is favoured to facilitate cooperation. Males are blue, females are pink. From West (2009).

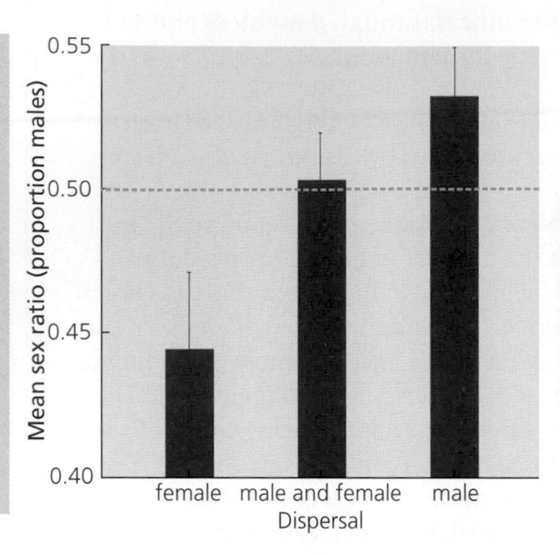

Fig. 10.2 The sex ratio at birth in primate species where either females, males and females or males are the dispersing sex. The sex ratio is biased towards the dispersing sex. From Silk and Brown (2008). The photo shows chimpanzee by Joan Silk.

If relatives of one sex compete for resources, then the sex ratio may be biased towards the other sex

resource competition (LRC) among females reduces their value as offspring: in the extreme case only one daughter might be able to survive on the food available near home, and so investment in other daughters would be wasted. The extent to which such adaptive adjustment of the offspring sex ratio occurs in primates has proved extremely contentious, because the data is so variable, and because it has been argued that chromosomal (genetic) sex determination acts as a constraint that prevents control of offspring sex ratios in taxa such as birds and mammals.

In a comprehensive survey of data from 102 primate species, Joan Silk and Gillian Brown (2008) showed that there was support for this prediction, but with relatively weak effects, showing only a slight bias to approximately 53% males in species where they are dispersing sex, and 55% females in species where they are the dispersing sex (Fig. 10.2). Much greater biases are seen in species where LRC is likely to be more extreme. For example, in the army ant *Eciton burchelli*, new colonies are only formed by the division of older colonies into two swarms, one of which is headed by the old queen and the other headed by one of her daughters. The resultant competition between sisters to head the swarm can explain why colonies produce six females and 3000 males, giving a sex ratio of about 99.8% males amongst the reproductives (Franks & Holldobler, 1987).

Local resource competition can also explain sex ratio variation within species or populations. William Brown and Laurent Keller (2000) found that in the narrow-headed ant *Formica exsecta*, colonies tended to produce only male or only female reproductives (Fig. 10.3), and that such 'split sex ratios' could not be explained by the more common explanation of variation in relatedness asymmetry that we shall meet in chapter 13. They suggested that the explanation was variation in the extent of LRC across colonies, due to variation in the number of queens. In this species, the dispersal of queens is often limited, with newly mated queens being recruited back into their parental colony, from which they may eventually disperse with workers to initiate new colonies nearby. In colonies where there are more queens, or lower resource availability, then there will be greater LRC, so the relative benefit of producing new queens will be reduced. Brown, Keller and colleagues tested this prediction by examining how the sex ratio of reproductives in a colony varied queen number and a number of ecological variables. In support of their hypothesis, they found that colonies produced a higher proportion of males when they contained a greater number of queens (Fig. 10.3), and when there was a lower availability

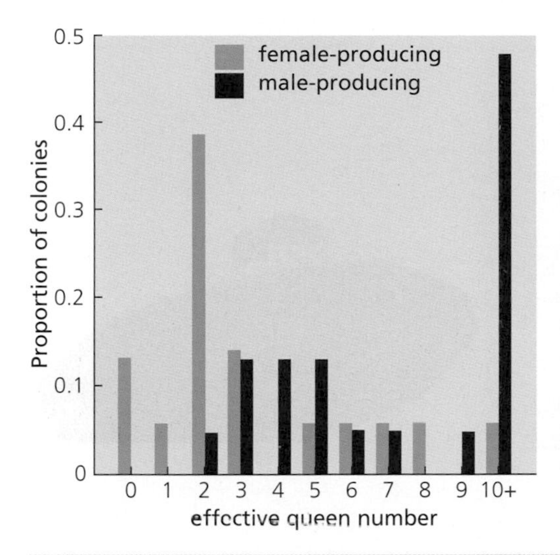

Fig. 10.3 The distribution of number of queens in a colony for colonies that produce either female (light blue) or male (dark blue) reproductives, in the narrow-headed ant *Formica exsecta*. Colonies with relatively few queens produced females, whereas colonies with many queens produced males. From Brown and Keller (2000). Photo by Rolf Kümmerli.

of resources, such as honeydew-bearing aphids on conifer trees. Furthermore, they then went on to test these ideas experimentally, showing that both the removal of queens and increasing food resources (by placing tuna and honey at the nest) led to colonies producing a higher proportion of female reproductives (Kümmerli *et al.*, 2005; Brown & Keller, 2006).

If the extent of competition between related females varies, then females will be produced in the environments where competition is lowest

Local mate competition

Bill Hamilton (1967) noticed that a number of insect and mite species which had extremely female-biased sex ratios, also tended to have a life history where brothers compete with each other for mates, many of whom are their sisters. He explained this as a special case of LRC that is termed 'local mate competition' (LMC). The exact reason why LMC should favour a female-biased sex ratio was the subject of much controversy during the 1970s and 1980s, but it is now accepted that it occurs for two reasons (Taylor, 1981). Firstly, suppose, for example, that two sons have only one chance to mate and that they compete for the same female. Only one of them can be successful in mating, so from their mother's point of view one of them is 'wasted'. This is an extreme example, but it illustrates the general point that when sons compete for mates their value to their mother is reduced. The mother should, therefore, bias her ratio of investment towards daughters. Secondly, if sons are able to mate with their sisters (inbreeding), then a female-biased sex ratio has the additional bonus of providing more mates (daughters) for sons, and so the higher proportion of daughters a mother produces the greater value of each son produced.

If brothers compete for mates, the sex ratio should be female biased

The exact degree of bias predicted by Hamilton's theory depends on the degree of local mate competition. Consider the extreme case of complete inbreeding, where a mother 'knows' that all her daughters will be fertilized by her sons. The best sex ratio in this instance is to produce just enough sons to fertilize the daughters, since any other

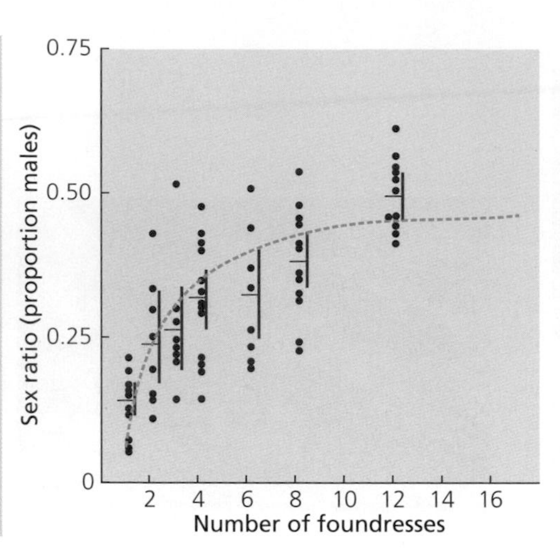

Fig. 10.4 Sex ratio adjustment in the parasitoid wasp *Nasonia vitripennis*. A less female-biased sex ratio is produced when larger numbers of females lay eggs in a patch. From Werren (1983). Photo by Michael Clark.

males will be wasted. The crucial difference between this and the earlier argument for a 1:1 sex ratio is that here the ratio of males to females in the rest of the population does not matter. A female-biased ratio within a brood will not give other parents a chance to benefit by concentrating on sons. Hamilton noticed that in many species where there was a high likelihood of inbreeding there was a tendency to produce just one or a small number of males in each brood. An example which supports this prediction is the viviparous mite, *Acarophenox*, which has a brood of one son and up to 20 daughters. The male mates with his sisters inside the mother and dies before he is born.

As more females lay eggs on a patch, a less female-biased sex ratio is favoured.

Hamilton also suggested that if individuals could assess the likely degree of LMC that their offspring would experience, then they should adjust their offspring sex ratio accordingly. Specifically, if N mothers lay eggs on a patch, and mating then occurs between these offspring, before only the females disperse, then the evolutionarily stable (ES) sex ratio (proportion sons) is $(N - 1)/2N$. This predicts that the sex ratio should vary from 0.5 for large N (Fisher's case) to increasingly female biased as N is reduced. In the extreme, if $N = 1$, then a sex ratio of zero is predicted, which is interpreted to mean that the mother should produce just enough sons to fertilize her daughters.

Parsitoid wasps adjust the sex of their offspring, depending upon how many females are laying eggs in that patch

Jack Werren (1983) tested this prediction with the parasitoid wasp, *Nasonia vitripennis*, which lays its eggs inside the pupae of flies such as *Sarcophaga bullata*. In this species, the winged females mate with the wingless males, either in, on or near the host pupae in which they developed, before only the females disperse. Consequently, if only one female parasitizes the pupa or pupae in a patch, her daughters are all fertilized by her sons and as predicted the sex ratio of her clutch of eggs is highly biased towards females. Only 8.7% of the brood is male. If more females lay eggs on the patch, then the extent of LMC is reduced and they should lay a less female-biased sex ratio. Werren found this pattern in laboratory experiments, with individuals producing a less female-biased sex ratio when more mothers laid eggs on a patch (Fig. 10.4).

Since then, recent advances in molecular methods (microsatellite markers) have been exploited to examine whether the same pattern also occurs in natural field populations, examining wasps that emerge from fly pupae found in birds nests. Max Burton-Chellew and colleagues (2008) genotyped all the wasps emerging from a number of nests (the offspring), reconstructed the genotypes of their mothers and hence determined how many mothers

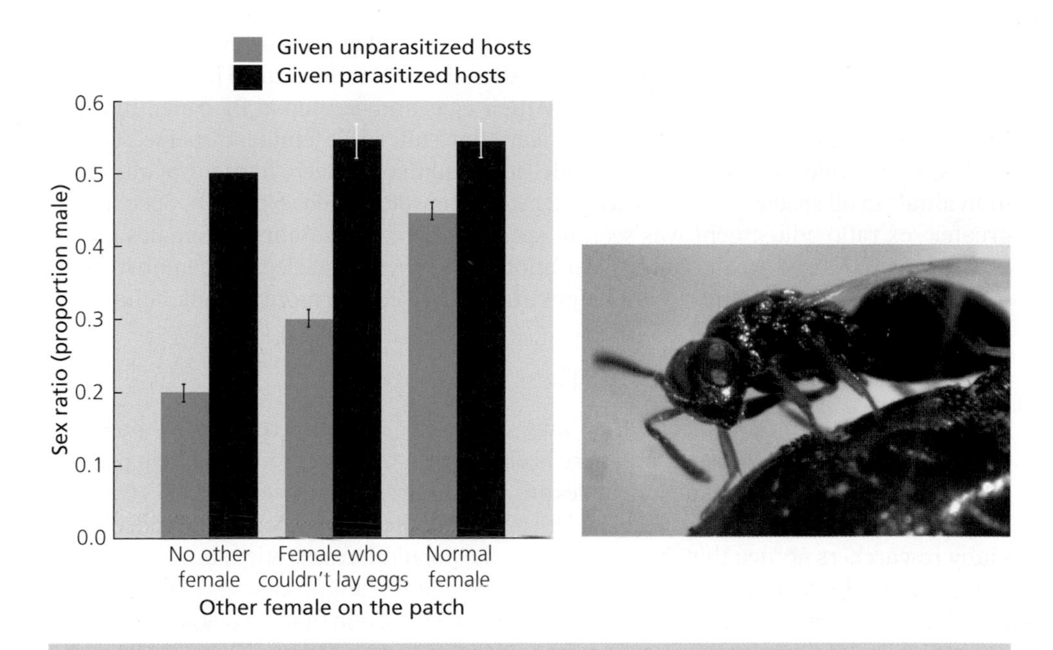

Fig. 10.5 The mechanism of sex ratio adjustment in the parasitoid wasp *N. vitripennis*. Females produced a less female-biased sex ratio in response to the presence of previously parasitized hosts and the presence of other females (photograph shows the red eye mutant used to follow the behaviour of individuals). From Shuker and West (2004). Photo © David Shuker and Stuart West.

had laid eggs in each nest, which they then correlated with the sex ratio of those emerging wasps. They found a very similar pattern to Werren, with the offspring sex ratio being less female biased in host pupae where the larger numbers of females had laid eggs.

How perfectly should we expect animals to behave? Research on LMC has proved extremely useful for addressing this question, because theory is able to make relatively clear predictions about how individuals should behave. One factor is the ability of individuals to process relevant information about the environment. Within the context of LMC, a key factor is how do individuals assess the number of females laying eggs on a patch, and hence the extent of LMC. Shuker and West (2004) investigated this in *N. vitripennis*, examining the relative importance of direct cues from the presence of other females, and indirect cues from eggs laid by other females. They were able to separate these factors by using females that could not lay eggs because they had cut their ovipositors off, and they followed the behaviour of individuals by using eye colour mutants. Their results showed that the shift in offspring sex ratios with increasing number of females on a patch is primarily caused by the presence of eggs laid by other females, and to a lesser extent by the presence of other females (Fig. 10.5). Consequently, we should expect variation in behaviour between females depending upon when they arrive on a patch, relative to other females, and over time, as more eggs are laid. This shows how the answers to 'how' (proximate) questions can help us to better understand the answers to 'why' (ultimate) questions.

Females can assess the number of other females laying eggs on a patch by directly observing them, or through indirect cues of their presence, such as recently laid eggs

Animals can be expected to show more variable behaviour in more variable environments

Another factor which may affect how perfectly we should expect animals to behave is the extent to which the environment varies. Allen Herre (1987) studied LMC in thirteen species of pollinating fig wasps, which lay their eggs and develop in fig fruits. In these species, the males are wingless and do not leave the fruit, while females disperse, so their life history provides a close fit to that modelled by Hamilton. Herre found that although individuals in all species adjusted their sex ratios, they did not do so equally. Specifically, greater sex ratio adjustment was seen in species where the number of females laying eggs in a fruit showed greater natural variation. This provides an elegant, demonstration of the general point that a more variable environment selects for more variable behaviour.

Local resource enhancement

Relatives may not only compete, they may cooperate. In many cooperative breeding vertebrates, offspring of one sex are more likely to remain in the group and help parents rear further offspring. For example, females are more likely to help in the Seychelles warbler, whereas males are more likely to help in African wild dogs and red-cockaded woodpeckers. Many researchers argued that, in such species, the population sex ratios should be biased towards the helping sex (Box 10.1). However, Ido Pen and Franjo Weissing (2000) showed that this prediction would only hold in the simplest of cases, and that it is possible to predict sex ratio biases in any direction, depending upon life history details. This illustrates the advantage of producing mathematical models, which force the underlying assumptions to be made explicit, and shows that verbal arguments can sometimes be misleading. Instead, Pen and Weising showed that the clear prediction that could be made was that groups with relatively few helpers should produce the helping sex, and groups that already have helpers should produce the other sex. This prediction has been supported in a number of species. For example, in African wild dogs, groups with relatively few helpers produced male-biased litters with 63% sons, whereas groups with relatively many helpers produced female-biased litters with 64% daughters (Creel *et al.* 1998).

Groups without helpers should overproduce the helping sex

The most striking example of sex ratio adjustment in a bird or mammal is provided by the Seychelles warbler, where both LRE and LRC occur. In this species, breeding pairs remain together on the same territory for up to nine years, producing one offspring a year, with female offspring often staying behind on their natal territory to help, through a range of behaviours, including territory defence, nest building, incubation and feeding of young. Jan Komdeur and colleagues showed that the advantage of helping depends strongly on territory quality. On high-quality territories, where there is a high density of insect prey, having a helper is advantageous (local resource enhancement). In contrast, on low-quality territories, where there is a low density of insect prey, the increased competition for food means that having a helper is disadvantageous (local resource competition). Amazingly, Komdeur *et al.* found that females adjust their offspring sex ratios with precision in response to this, producing 90% females on high-quality territories, and 80% sons on low-quality territories (Fig. 10.6).

The benefit of producing the helping sex can be reduced or removed on low quality territories

In addition to their observational data, Komdeur *et al.* carried out an experiment in which warblers were translocated to a new island, where they took high-quality territories. Pairs moved from low-quality to high-quality territories switched from producing 90% males to 85% females, while those moved from high-quality to high-quality territories continued to produce 80% females. Furthermore, on high-quality territories, while having one or two helpers is advantageous, the increased competition

Fig. 10.6 Sex ratio adjustment in the Seychelles warbler. The offspring sex ratios (proportion male) produced on different quality territories in (a) 1993, (b) 1994 and (c) 1995. Mothers produced daughters on high-quality territories and sons on low-quality territories. (d) The data from 1995 are also shown distinguishing between nests that had either one (solid circles) or more than one (open circles) helper already at the nest. When mothers already had more than one helper, they produce sons, irrespective of territory quality. From Komdeur *et al.* (1997). Reprinted with permission from the Nature Publishing Group. Photo © Martijn Hammers.

for food and breeding means that having more than two helpers is disadvantageous. Consistent with this, on high-quality territories, pairs switched from producing 85% females when no or one helper was already present to producing 93% males when two or more helpers were already present. This was also shown experimentally – when one helper was removed from high-quality territories with two helpers, the breeding pairs switched from producing 100% males to 83% females.

At least some birds can adjust their offspring sex ratio with remarkable precision

Komdeur's results had a huge impact when they were first published, crushing the conventional wisdom that vertebrates such as birds could not manipulate their offspring sex ratios. In addition, more recent work has shown that the pattern cannot just be the result of differential mortality. Females of this species normally produce one egg, but when moved to vacant high-quality territories females produced two eggs, one day apart.

Komdeur *et al.* (2002) showed that just as extreme sex ratio biases occur in the second egg as in the first. Given that this egg is laid only one day after the first egg, the sex ratio bias could not result from post-ovulation mechanisms, such as sex-selective re-absorption of ova or dumping eggs of the wrong sex. Consequently, this suggests that the control of offspring sex involves some pre-ovulation control, such as biasing segregation towards the required sex chromosome (Z or W; see Box 10.1). More generally, this stresses that we should always test whether patterns of sex ratio bias can be explained by the alternate explanation of differential mortality, such as individuals of one sex requiring more resources, so being more likely to die during development, especially when conditions are relatively harsh.

In species where helpers provide less benefit, there is reduced selection for sex ratio adjustment

There are also a number of species where the offspring sex ratios do not vary with the number of helpers in a group, such as acorn woodpeckers and superb fairy wrens. How can we explain this variation across species? Ashleigh Griffin and colleagues suggested that weaker sex ratio adjustment might be seen in species where the benefits of adjusting sex ratios were lower. If helpers provide little actual benefit, then there is little benefit in groups without helpers preferentially producing the helping sex. They found support for this hypothesis this with a meta-analysis (Box 10.3) across 11 species, with lower levels of sex ratio adjustment in species where the actual benefit provided by helpers was lower (Fig. 10.7; Griffin *et al.*, 2005). In the species where helpers provide negligible benefits, sex ratio adjustment is not observed.

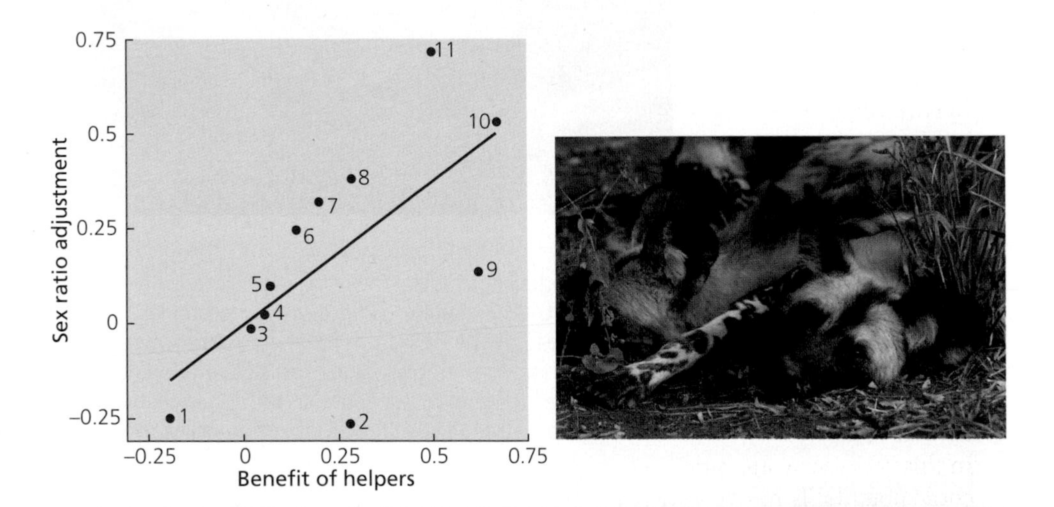

Fig. 10.7 The correlation between the extent to which sex ratios are adjusted and the benefit provided by the presence of helpers. A more positive extent of sex ratio adjustment signifies a greater tendency to produce offspring of the sex that helps more, in patches where there is a lack of helpers. Across species, the significant positive correlation indicates that sex ratio adjustment is greater in species where the presence of helpers leads to greater fitness benefits. The data points represent (1) laughing kookaburra, (2) sociable weaver, (3) Harris's hawk, (4) acorn woodpecker, (5) green wood-hoopoe, (6) western bluebird, (7) alpine marmot, (8) redcockaded woodpecker, (9) bell miner, (10) Seychelles warbler and (11) African wild dog. Griffin et al. (2005). Reprinted with permission of the University of Chicago Press. Photo © Andrew Young.

BOX 10.3 META-ANALYSIS

In most of the comparative studies described in this book, we are asking whether two variables are correlated across species. For example, in Chapter 2 we asked whether sexual dimorphism is correlated with mating system across primates, and found that it is (Fig. 2.6). However, in some cases, the relevant question is not whether variables are correlated, but rather whether consistent patterns occur across species. Meta-analysis provides a method for this, and is becoming an increasingly important tool in all the areas discussed in this book (Arnqvist & Wooster, 1995).

To illustrate the usefulness of meta-analysis, imagine that we are interested in whether female mammals are able to adjust their offspring sex ratio in response to their maternal condition, as originally suggested by Trivers and Willard (1973), or whether genetic sex determination constrains them from doing this. Imagine that data had been collected on eight species. Two studies showed a significantly positive correlation between maternal condition and offspring sex ratio, with better condition females being more likely to produce sons, as predicted by Trivers and Willard. One study showed a significantly negative correlation between maternal condition and offspring sex ratio, with better condition females being less likely to produce sons, in the opposite direction to that predicted by Trivers and Willard. The remaining five studies showed no significant correlation between maternal condition and sex ratio.

One possible conclusion from this data would be that Trivers and Hare hypothesis is not supported, and that the positive results could just have arisen by chance. However, this conclusion throws away data on the direction of the non-significant results and relies on the implicit assumption that all the studies were equally good, with the same sample size. This can lead to incorrect conclusions. For example, suppose that the two positive results were from large studies of 150 individuals, while all the others were from small studies of 10 individuals. In this case, we would want the larger studies (which were both positive) to carry more weight towards the final conclusion, because there is a greater probability that they show the 'correct' result. Furthermore, suppose that all the non-significant results were in the positive direction, so that we actually had seven studies showing the predicted direction, and only one not. In this case, we would have a consistent pattern and the non-significant results could potentially be explained by a low sample size. Overall, this closer examination of sample size and directionality would have changed our conclusion, and given support to the Trivers and Willard hypothesis.

Meta-analysis was developed to solve these problems of sample size and directionally. Firstly, instead of just counting whether a study is significant or not, it uses a standard measure of effect size, the correlation coefficient r (where r^2 is the proportion of variance in the data explained – here it would be the proportion of the variance in the sex ratio explained by maternal condition). Secondly, it gives more weight to studies with larger sample sizes. A meta-analysis of the real ungulate data shows that although there is much variation, there is consistent support for the Trivers and Willard hypothesis (Sheldon & West, 2004). As well as asking whether consistent patterns occur, meta-analysis provides a method for examining whether variation across species in strength of a relationship can be explained. For example, could variation across species in whether the Trivers and Willard hypothesis is supported be explained because it doesn't always apply (for an analogous analysis see Fig. 10.7)? Furthermore, methods have recently been developed which allow such 'comparative meta-analyses' to be carried out in ways that allow for the fact that, as discussed in chapter 2, species are not independent data points (Hadfield & Nakagawa, 2010).

Sex allocation in variable environments

Maternal condition

Females in better condition can be selected to preferentially produce sons

Robert Trivers and Dan Willard (1973) suggested that individuals should adjust the sex of their offspring in response to environmental conditions. They envisaged a mammal population in which three assumptions held: (1) females in better condition have more resources for reproduction and produce better quality offspring; (2) higher quality offspring become higher quality adults; and (3) sons gain a greater fitness benefit from being higher quality adults. They assumed that this last assumption would hold when competition for mates among males is intense, with the highest quality mates gaining a disproportionate share of the matings, as is the case in many polygynous mammals (Chapter 7). The consequence of these assumptions is that if offspring fitness is plotted against maternal quality, then fitness increases more rapidly for sons than it does for daughters (Fig. 10.8). In this case, Trivers and Willard argued that relatively low quality mothers would be selected to produce daughters, and relatively high quality mothers to produce sons.

Red deer adjust their offspring sex ratios in response to maternal condition, with mothers in better condition being more likely to produce sons

Tim Clutton-Brock and colleagues (Clutton Brock *et al.*, 1984) tested this prediction with red deer, where female condition is determined by their rank in a dominance hierarchy. They found that females shifted their offspring sex ratio in the predicted direction, from 47% males in low-ranking females to 61% males in high-ranking females. Furthermore, they also showed that the assumptions made by Trivers and Willard's held: (1) higher ranking females produced heavier (better quality) offspring; (2) heavier offspring became bigger and higher quality adults; and (3) sons gained a greater benefit from increased adult body size. Increased quality is more important for males because red deer are polygynous, with males fighting to defend harems of females during the rutting season, leading to a much higher mating success for larger, better condition males. Overall, they were able to examine the combined consequences of these effects, showing that a mother's rank had a significantly greater effect on the lifetime reproductive success of her sons than that of her daughters (Fig. 10.9).

Mate attractiveness

The application of the Trivers and Willard hypothesis to ungulates has proved controversial, with mixed empirical support. However, much clearer support has come from other areas. Ben Sheldon and colleagues showed that the same logic can explain sex ratio variation in response to the quality or attractiveness of a mate. In blue tits, males have an ultraviolet patch on the top of their head, which appears to act as a reliable signal of quality (Chapters 7 and 14). Sheldon *et al.* (1999) showed that females mated to males with a brighter UV patch laid a higher proportion of sons (Fig. 10.10). They argued that females should adjust their offspring sex ratio in this way, because higher quality mates would lead to higher quality offspring, either through passing on good genes or through higher quality paternal care, and that sons would benefit more from being higher quality. The idea here is very similar to the classic Trivers and Willard (1973) argument shown in Fig. 10.8, except that mate quality replaces maternal condition as the factor on the x axis that influences offspring fitness. Sheldon *et al.* also confirmed their result experimentally, showing that when the UV signal was blocked out with sunblock, this led to a higher proportion of female offspring. Whilst these results were initially controversial, both the observational and experimental patterns have since been replicated in three different European populations.

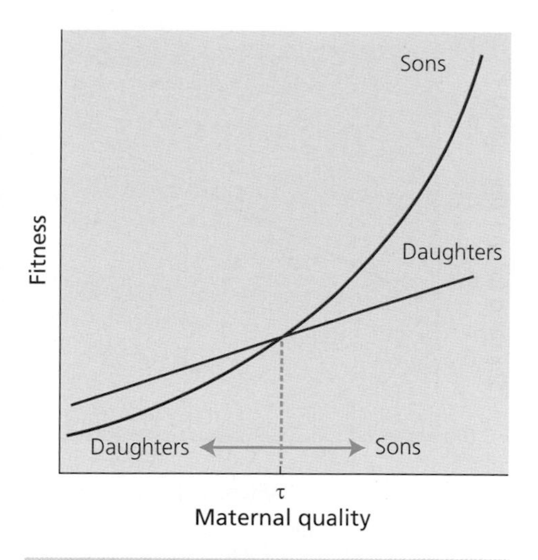

Fig. 10.8 The Trivers and Willard hypothesis. The relative fitness of sons increases more rapidly with maternal quality than that of daughters. Consequently, females in relatively good condition (>τ) would do best by producing sons and females in relatively poor condition (<τ) would do best by producing daughters. From Trivers and Willard (1973). Reprinted with permission from AAAS.

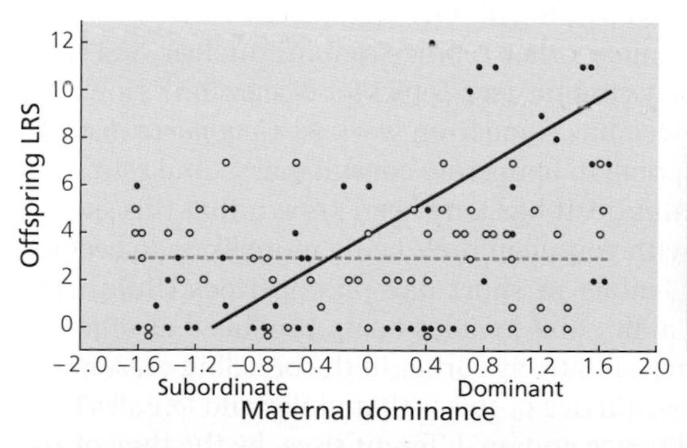

Fig. 10.9 In red deer, the lifetime reproductive success (LRS) of sons (filled circles and solid line) increases more rapidly with their mother's social rank than daughters (open circles and dashed line). From Clutton-Brock *et al.* (1984). Photo © Alison Morris.

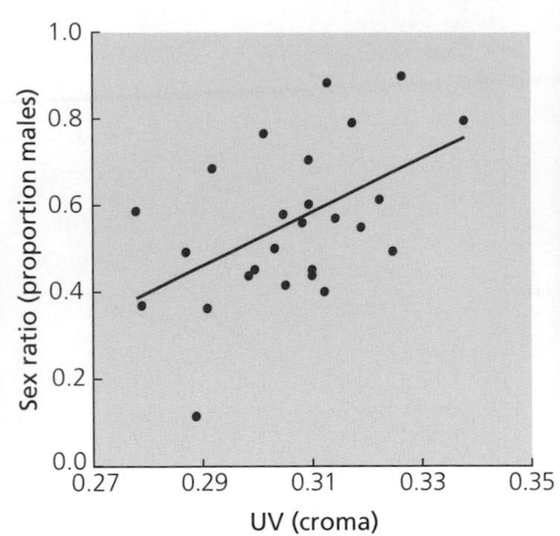

Fig. 10.10 Female blue tits which mate males with a brighter UV patch (crown) produced a higher proportion of male offspring. From Griffith *et al.* (2003). Photo © Joseph Tobias.

Environmental sex determination

Females who mated with attractive males can be selected to preferentially produce sons

In some animal and plant species, the sex of an offspring is determined by the environment in which embryonic development occurs. This is termed *environmental sex determination*, or ESD. For example, in many turtles males are produced at relatively low temperatures and females at relatively high temperatures (Box 10.1). Ric Charnov and Jim Bull (1977) suggested that an analogous form of the Trivers and Willard hypothesis could explain the occurrence and pattern of ESD. Their idea was that the environment in which an individual developed could have different fitness consequences for males and females. For example, suppose that developing in a 'good' environment carried much greater fitness benefit to males, with a poor environment producing just below average females or exceptionally bad males and a good environment producing females just above average or exceptionally good males. In this case, selection would favour development as males in the good environment and as females in the poor environment. The idea here is again very similar to the classic Trivers and Willard (1973) argument shown in Fig. 10.8, except that some environmental factor replaces maternal condition as the factor on the x axis that influences offspring fitness.

If one sex gains a greater benefit from developing under certain conditions, then this can lead to sex being determined by the environment

The reasons for ESD in turtles and many other reptiles remain unclear and have even been referred to as an 'evolutionary enigma' (see Topics for discussion). However, clear support for Charnov and Bull's idea has come from work on *Gammarus duebeni*, a brackish water shrimp that is widespread in temperate coastal marsh and estuarine habitats on both sides of the North Atlantic. It has long been known that this species has ESD in response to temperature, with young animals being more likely to become male in long day photoperiods and female in short day photoperiods (Bulnheim, 1967). This pattern of ESD leads to males and females being produced at different times of the year—males being produced relatively earlier in the breeding season and females relatively later. The major consequence of this is that males and females have different lengths of time to grow, and hence end up different sizes, by the time of the next breeding season. Specifically, earlier born young are predominantly males, giving them longer to grow, so males are bigger than females during the breeding season. This pattern of ESD is selected for if males gain a greater fitness advantage

The time of year can be an important environmental factor, because it is a cue of how long there is to grow before the next breeding season

from size than females. Jennie McCabe and Ali Dunn (1997) tested this hypothesis by examining the consequences of size under field conditions. They found that whilst larger females produce more eggs, larger males are more successful at obtaining mates and obtain larger mates, which therefore produce more eggs (Fig. 10.11). When all of these factors are combined, they found that size had much greater fitness consequences for males.

Sex change

If being old and big provides a greater benefit to one sex, then sex change can be favoured

Sex change, where individuals mature as one sex, and later change to the other, occurs in a variety of fish, invertebrates and plants (Fig. 10.12). Ghiselin (1969) argued that sex change would be favoured if the fitness of an individual varies with age or size, and the relationship is different for males and females. In this case, natural selection favours individuals who mature as the sex whose fitness increases more slowly with age (first sex) and then change to the other sex (second sex) when older. The idea here is very similar to the classic Trivers and Willard (1973) argument shown in Fig. 10.8, except that age or size replaces maternal condition as the factor on the x axis that influences offspring fitness. A special feature of sex change, discussed in Box 10.4, is that it is one of the few applications of the Trivers and Willard hypothesis where useful predictions can also be made about the population sex ratio.

When old large males can monopolize the mating with females, then individuals may be selected to mature as females, and then change sex later in life to males

In many coral reef fish, female first (protogynous) sex change occurs. Robert Warner and colleagues have argued that this occur in species where the mating system leads to male mating success being monopolized by the oldest, largest individuals. For example, in the bluehead wrasse *Thalassoma bifasciatum*, males set up territories where the females come to spawn (Warner *et al.*, 1975). Females choose to mate with the largest terminal phase males. This leads to a huge size advantage, with large males spawning more than 40 times a day and small ones less than twice a day (Warner *et al.*, 1975). Furthermore, individuals are able to time their sex change with remarkable precision in response to the social conditions. If the largest males on a reef are removed, the next largest individuals (females) will change sex and become brightly coloured males. The cues involved in stimulating such social sex change are the matter of debate, and could be behavioural, visual or chemical.

When males are not able to monopolize matings, then sex change can be favoured in the other direction from male to female

Sex change can also be favoured in the other direction, from male to female (protandry), if male size has little effect on breeding success. In this case, an individual may then reproduce best as a male when small because it is able to spawn with some of the large, most fecund females. An example of a fish that changes from male to female is the anemonefish, or clownfish *Amphiprion akallopisos*, which lives on coral reefs in the Indian Ocean. It lives in close symbiosis with sea anemones and because there is usually only enough space for two fish to inhabit the same anemone this species lives in pairs. In effect, the habitat forces them to be monogamous. The reproductive success of a pair is limited more by the female's ability to produce eggs than by the male's ability to produce sperm, so each individual does better if the larger one is female. Like the wrasses, sex change is socially controlled. If the female is removed, the male is then joined by a smaller individual, so he changes sex and lays the eggs while the newcomer functions as a male (Fricke & Fricke, 1977).

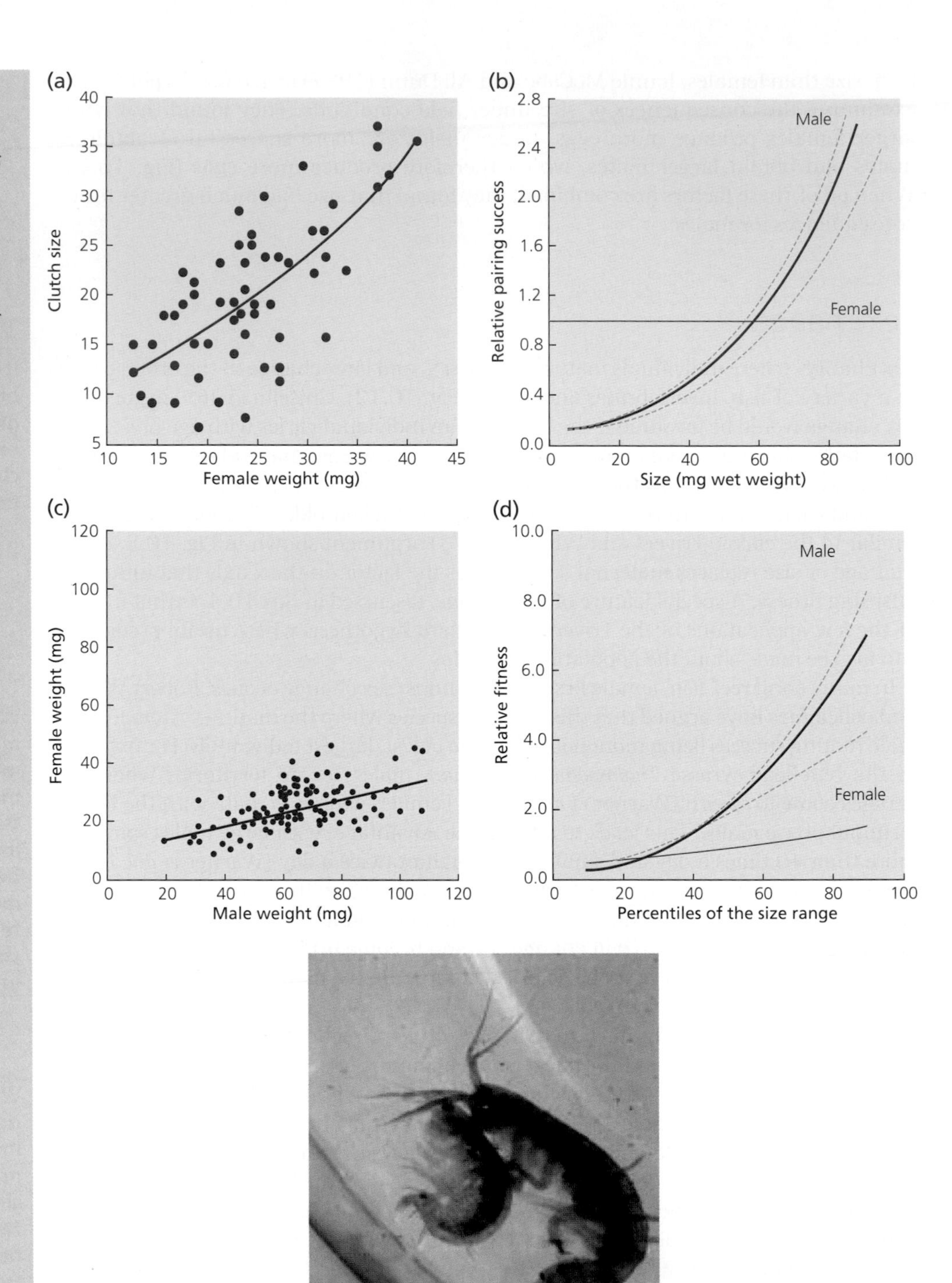

Fig. 10.11 Size and fitness in the shrimp *Gammarus duebeni*. (a) Larger females laid more eggs. (b) Larger males were more likely to mate. (c) Larger males mated with larger females. (d) When all of the consequences of size are summed, the relative fitness of males increases more rapidly with size than that of females (photograph of a mating pair, where male is the larger individual). From McCabe and Dunn (1997). Photo © Alison Dunn.

Fig. 10.12 Sex changers. Sex change may be from female to male, as in (a) the bluehead wrasse (terminal male); or male to female, as in (b) the Clownfish *Amphiprion percula*; (c) the common slipper limpet (*Crepidula fornicata*; photograph of a mating stack, where the largest individuals at the bottom are female and the smaller individuals at the top are male; and (d) the Pandalid shrimp. Photo (a) © Kenneth Clifton; (b) © Peter Buston; (c) © Rachel Collin; (d) © David Shale/naturepl.com

BOX 10.4 POPULATION SEX RATIOS, SEX CHANGE AND GONADS

As the examples in this chapter show, sex allocation theory is often able to make clear predictions about when individuals should adjust their offspring sex ratio conditionally in response to environmental conditions. In contrast, when such facultative sex ratio adjustment occurs, theory has been much less successful in predicting and explaining variation in the overall population or breeding sex ratio. The reason for this is that the population sex ratio is often predicted to depend upon biological details that are rarely known, such as the details of male and female life histories, and whether other behaviours, such as clutch size, are also adjusted conditionally (Frank 1987, 1990). Indeed, a lack of an appreciation of this problem is one of the most common misunderstandings in the field of sex allocation. Ric Charnov and Jim Bull (1989) showed that the major exception to this problem is in sex changing animals, where the sex ratio should be biased towards the sex that individuals mature as, termed the *first sex*.

Consider the case of a protogynous species where individuals mature as females and then change sex to males when older (bigger). In this case, the relative fitness of males increases faster with age than it does for females. Males and females must make an equal genetic contribution to the next generation, because all offspring have two parents. Consequently, it must be true that:

$$Nm\ Wm = Nf\ Wf \qquad \text{(B10.4.1)}$$

where Nm and Nf are the number of mature males and females, and Wm and Wf represent the mean fitness of males and females. Given that the fitness of a male at the point of sex change will be equal to that of a female, and that male

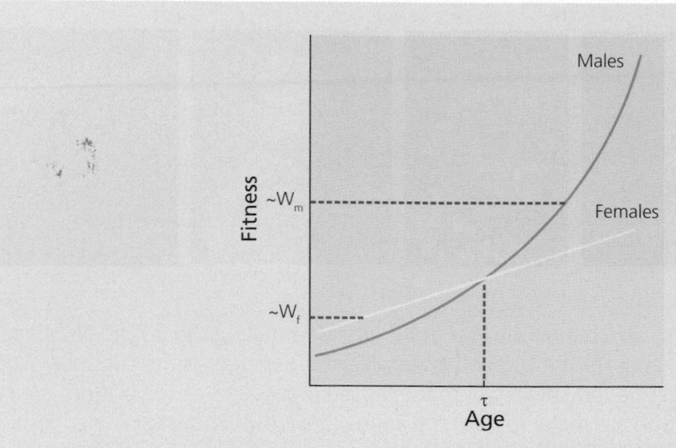

Fig. B10.4.1 Sex change from female to male. The relative fitness of males increases more rapidly with age than that of females. Consequently, individuals are selected to mature as females and then change sex to males at age τ. Note the similarity to figure 10.8.

fitness increases faster with age (that is why sex change in this direction is favoured!), this means that $Wm > Wf$ (Fig. B10.4.1), so, in order for Equation B10.4.1 to hold, then Nm < Nf. This means that there will be more females than males and, hence, a female-biased sex ratio. The converse prediction for protandrous (male first) species can equally be made, showing that a male-biased sex ratio is predicted.

Support for this prediction has been provided with data from 121 sex changing animal species, spanning a wide range of taxa, including fish, crustaceans, mollusks, echinoderms and annelid worms (Fig. B10.4.2). The average population sex ratio of female first (protogynous) species was significantly female biased, while that of male first (protandrous) species was significantly male biased. Furthermore, the difference between these two groups was significant when considering species as independent data points or phylogenetically controlled independent comparisons.

More recently, a study of 116 fish species by Philip Molloy and colleagues (2007) has shown that the relative gonad size of males correlates with the occurrence of sex change. Specifically, males in female first (protogynous) species had relatively smaller testes than males of species where sex change did not occur (Fig. B10.4.3). This covariation across strategies can be explained because the mating system influences both sex change and sperm competition. In species where males can monopolize females, this selects for female first sex change, but also means that male–male competition during spawning will be reduced, and hence the benefits from larger testes are reduced (Chapter 7). In contrast, in species where males cannot monopolize females, there will not be selection for sex change, and male–male sperm competition will be maximized during spawning, leading to a greater benefit from increased testes size.

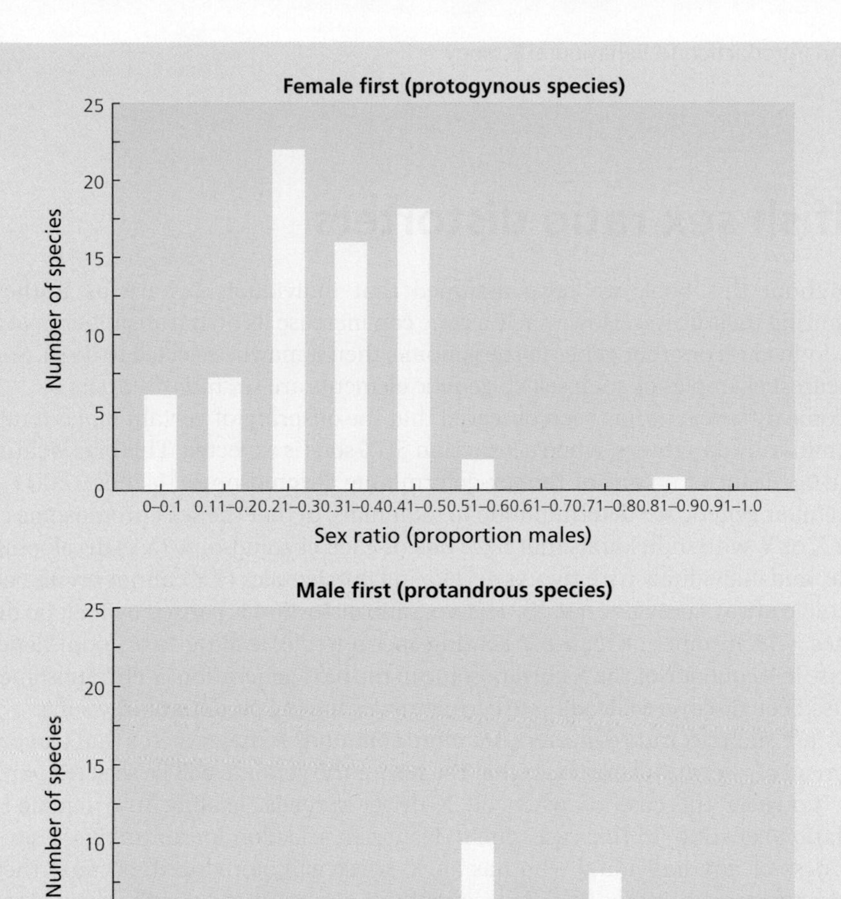

Fig. B10.4.2 The distribution of population sex ratios in sex changing species that are either (a) female (protogynous) or (b) male (protandrous) first. The population sex ratio tends to be biased towards the first sex. From Allsop and West (2004).

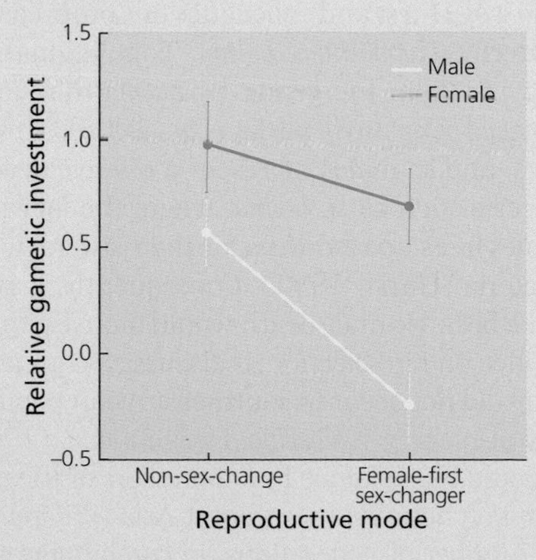

Fig. B10.4.3 The correlation between gametic investment and the occurrence of sex change. Species with female first sex change had relatively similar size female gonads, but smaller testes, than species where sex change did not occur. From Molloy *et al*. (2007).

Selfish sex ratio distorters

Throughout this book we have assumed that individuals behave as if they are maximizing their fitness. However, if a gene can increase its own transmission, at a cost to the individual or other genes in the genome, then it may be selected to do so. Some of the clearest examples of such selfish genetic elements are *sex ratio distorters*.

Genes can be selected to distort the sex ratio away from that which is optimal for the individual, if it increases their transmission to the next generation

In some fly species it has been observed that the offspring of certain males tend to be predominantly daughters, when a Fisherian 50% sons is expected. This bias is caused by a sex ratio distorter on one of the sex-determining chromosomes (Jaenike, 2001). Flies have similar genetic sex determination to mammals, in that the sex chromosome can be either X or Y, with individuals that have one of each chromosome (XY) developing into a male, and individuals with two Xs developing into females (YY cannot occur, because the female always provides one X). The sex ratio distortion is caused by meiotic driving X genes, which somehow cause Y bearing sperm to die, leading to an non-Mendelian increase in frequency of the X chromosome in the next generation. Y chromosome drive has also been discovered, leading to certain males having predominantly sons.

Why are such sex ratio distorters not more common? A major factor that can prevent the spread of sex ratio distorters is that the rest of the genome will be selected to repress them. Consider the case of when an X driver spreads, leading to a female-biased population sex ratio. In this case, due to Fisherian selection for an unbiased sex ratio, the fitness of any individual who has an X driver will be reduced, because they will produce the more common sex. Consequently, if a mutant at any other locus is able to suppress the driver, and restore a more normal sex ratio to the population, then it will be favoured. Egbert Leigh (1971) pointed out that this leads to the rest of the genome becoming united as a 'parliament of genes' to suppress sex ratio distorters. A huge variety of suppressors have been found that return the sex ratio to more equal values, acting against both X and Y drivers (Burt & Trivers, 2006). Indeed, crosses between fly species that produce unbiased sex ratios often show that they contain sex ratio distorters, just that these are hidden because they also contain suppressors.

The rest of the genome is selected to suppress the action of sex ratio distorters

An amazing example of just how quickly suppressors of sex ratio distorters can spread is provide by the work of Greg Hurst and colleagues on south-east Asian and Polynesian populations of the butterfly *Hypolimnas bolina*. Populations of this butterfly from Polynesia contain a bacteria from the genus *Wolbachia* that causes males to die, and hence causes a female-biased sex ratio (Charlat *et al*., 2005). This bacteria is only passed maternally, through eggs, and so males represent a relative dead end for it. Such male killing is favoured in species such as *H. bolina*, where the larvae develop gregariously, because the death of males frees up resources for their sisters, who are likely to contain genetically identical bacteria (Hurst, 1991). Consequently, a gene in the bacteria for male killing would spread, because male death would increase the transmission of other copies of that gene. This idea of 'kin selection' is discussed in greater detail in Chapter 11.

Maternally transmitted bacteria can be selected to kill males that carry them

In contrast, male killing did not occur in south-east Asian populations of this butterfly, despite the fact that they also carry *Wolbachia*. Hornett *et al*. (2006) tested whether this variation across species could be explained by suppressors of the male killer. They did this by taking butterflies from Polynesia and south-east Asia (Philippines and Thailand), and then mating them for a number of generations, in combinations that produced genetic backgrounds from butterflies from each area, with *Wolbachia* from each area. They found that the genetic background of the host butterfly played the key role – *Wolbachia* from both

Polynesia and south-east Asia caused male killing when in Polynesian genetic backgrounds, but not in south-east Asian genetic backgrounds. The data from these crossing experiments suggested that this variation is due to a single gene in the south-east Asian populations that suppresses the male killing by *Wolbachia*. Data collected over recent years suggest that this suppressor has spread at a rapid rate through south-east Asia and is starting to do so in Polynesia, as would be predicted from theory. For example, on the Polynesian island of Upolu, the suppressor seems to have spread from a prevalence of about 0% in 2001 to about 100% in 2005–2006, only eight to ten generations later (Charlat *et al.*, 2007).

Genes in the host butterfly are selected to suppress male killing, and then spread extremely rapidly

Summary

There are two major hypotheses for why sex allocation should be shifted away from everyone investing equally in the two sexes. Firstly, if relatives interact, then biased sex ratios can be favoured to reduce competition between relatives, or increase cooperation between relatives. Secondly, if the sexes benefit differently from environmental variation, then sex allocation can be adjusted conditionally to take advantage of this. This can be done by adjusting the offspring sex ratio, environmental sex determination or sex change. In Chapter 13 we discuss examples of where there can be conflict over sex allocation in social species.

Much work on sex allocation has focused on hymenopteran insects, the ants, bees and wasps. This is because their haplodiploid genetics allows a clear mechanism for controlling offspring sex, by whether an egg is fertilized with sperm – males develop from unfertilized eggs (haploid) and females develop from fertilized eggs (diploid). However, there is rapidly growing evidence for conditional adjustment of offspring sex ratios in vertebrate taxa such as birds and mammals, where it was previously assumed that chromosomal sex determination would have acted as a constraint that would prevent this. The mechanism by which offspring sex ratios are adjusted in such species remains a major puzzle.

Further reading

Hamilton's (1967) paper was pivotal for the field of behavioural ecology. It showed that simple mathematical models could be used to make predictions for how animals should behave that could be easily tested, by either comparing across species or by looking at how individuals vary their behaviour under different conditions. Whilst this approach is taken for granted today because it forms the daily bread of behavioural ecology research programmes, it should be remembered just how astounding this was at the time, to suggest that a few lines of simple maths could make testable predictions about how organisms should behave.

Charnov's (1982) book unified the field of sex allocation from a conceptual perspective. West's (2009) book provides an overview of the theoretical and empirical literature, as well as a discussion of how research on sex allocation can shed light on broader issues and questions in behavioural ecology. Hardy's (2002) book provides a thorough introduction to the practical methods that are required to study sex allocation, from how to work with different organisms to data analysis. Munday et *al.* (2006) review the diversity of sex-change strategies in animals. Burt and Trivers (2006) review the natural history all forms of genetic conflict, including sex ratio distorters.

Badyaev *et al.*'s (2001) data on recently introduced populations of the house finch provide a stunning example of how selection can rapidly lead to different patterns of sex ratio adjustment in different populations of the same species. Charnov and Hannah (2002) use data from an impressive 30 years of commercial catches of Pandalid shrimp, to show that individuals change their sex-changing strategy in response to the local age distribution of shrimp. Nothing is known about the physiological or social mechanisms that shrimp could use to assess this. Burley (1981) was the first to suggest that mate attractiveness should influence selection on the sex ratio. She obtained the remarkable result that, in the zebra finch (*Poephila guttata*), the colour of a plastic leg band placed on males influenced their attractiveness, and that females adjusted their sex ratio accordingly.

TOPICS FOR DISCUSSION

1. In honey bees, new colonies are formed by colony fission, where the worker force splits into two *swarms*, one of which is headed by a new queen that was produced in that nest, and the other by the old queen. Would you expect the sex ratio of reproductives to be biased and, if so, why?
2. Why is there often a female-biased sex ratio in the sexual stage of malaria and related blood parasites (see Read *et al.*, 1995; Reece *et al.*, 2008)?
3. In the parasitoid wasp *N. vitripennis*, the shift in offspring sex ratios with increasing number of females on a patch is primarily caused by the presence of eggs laid by other females, and, to a lesser extent, by the presence of other females (Fig. 10.5). Why would this have been favoured? How would you test your hypothesis?
4. Sex allocation is often argued to be one of the most successful areas of behavioural ecology. Discuss why the fit between theory and data could be expected to be especially close in this area.
5. Discuss the extent to which the Trivers and Willard hypothesis is supported in ungulates (Hewison & Gaillard, 1999; Cameron, 2004; Sheldon & West, 2004).
6. Joan Roughgarden (2004) has argued that traditional Darwinian evolutionary theory has problems explaining mating systems such as sex change. Do you agree?
7. Discuss whether temperature-dependent sex determination in reptiles is an 'evolutionary enigma' (see Shine, 1999; Janzen & Phillips, 2006; Warner & Shine, 2008).
8. Discuss how meta-analysis can be used to test whether there has been selective reporting of results (Palmer, 1999; Simmons *et al.*, 1999b).

CHAPTER 11

Social Behaviours: Altruism to Spite

Photo © David Pfennig

So far throughout this book we have championed the view that natural selection designs individuals to behave in their own selfish interests and not for the good of their species or for the good of the group in which they live. For example, observed clutch size, foraging behaviour and mating patterns are what would be expected if selection optimized behaviour and life history strategies to maximize an individual's reproductive success.

However, it will be obvious to any naturalist that animals do not behave selfishly all the time. Often individuals apparently cooperate with others. Several lions may cooperate to hunt prey; in many species of birds and mammals individuals give alarm calls to warn others of the approach of a predator; in many cooperative breeding species, such as meerkats, an individual may help others to produce offspring rather than have young itself. In some cases, cooperation provides an immediate or delayed benefit to the survival or reproduction of the actor, that outweighs the cost of performing the behaviour. For example, one individual may help another because that individual will then help it back. In this case, cooperation is mutually beneficial and can be explained by selfish interests, as is discussed in further detail in the next chapter.

Mutual benefit: benefit to others, benefit to actor

In other cases, and more troubling for evolutionary theory, cooperative behaviours provide no benefit to the actor and are altruistic. Specifically, a behaviour is defined as altruistic if it is costly to the personal reproduction of the actor who performs that behaviour, but beneficial to another individual or individuals (Box 11.1). For example, in the social insects, the ants, bees, wasps and termites, the workers are often sterile, giving up any opportunity to reproduce, in order to help raise the offspring of the queen. If natural selection favours individuals that behave in their own selfish interests, how can we explain such altruistic helping behaviours?

Altruism: benefit to others, cost to actor

To understand how natural selection can lead to altruistic, and even spiteful behaviours, we have to return to the fundamental issue of exactly how natural selection works at the genetic level.

BOX 11.1 CLASSIFYING SOCIAL BEHAVIOURS

Bill Hamilton (1964) showed how social behaviours can be defined according to their fitness consequences for the actor and recipient. A behaviour is social if it has fitness consequences for both the individual that performs that behaviour (the actor) and another individual (the recipient). Social behaviours are classified according to whether the consequences they entail for the actor and recipient are beneficial (increase fitness) or costly (decrease fitness) (Table B11.1.1). A behaviour which is beneficial to the actor and costly to the recipient (+/–) is selfish, a behaviour which is beneficial to both the actor and the recipient (+/+) is mutually beneficial, a behaviour which is costly to the actor and beneficial to the recipient (–/+) is altruistic, and a behaviour which is costly to both the actor and the recipient (–/–) is spiteful (Hamilton, 1964). Selfish and mutually beneficial behaviours can be explained from the perspective of individuals maximizing their reproductive success. Altruistic and spiteful behaviours can only be explained by also taking account of the indirect consequences of the behaviour.

Table B11.1.1 Classification of social behaviours

Effect on actor	**Effect on recipient**	
	+	–
+	Mutually beneficial	Selfish
–	Altruistic	Spiteful

Kin selection and inclusive fitness

The most familiar example of an individual giving aid to another is, of course, parental care. We are not surprised to see a parent bird hard at work feeding its offspring because natural selection favours individuals who maximize their genetic contribution to future generations. The young will have copies of their parent's genes, so parental care is one way in which parents can increase their genetic contribution to the next generation.

The coefficient of relatedness measures the genetic similarity of two individuals relative to the population at large

We can quantify the probability that a copy of a particular gene in a parent is present in one of its offspring. In diploid species, when an egg and a sperm fuse to form a zygote, each parent contributes exactly 50% of its genes to the offspring. Therefore, the probability that a parent and an offspring will share a copy of a particular gene because the offspring obtained it from that parent is 0.5. This likelihood that individuals share a gene that is identical by descent (inherited from the same ancestor) provides a measure of the *coefficient of relatedness*, often denoted by r.

Now offspring are not the only relatives to share copies of the same genes identical by descent. Again we can calculate the probability that a copy of a gene in one

BOX 11.2 CALCULATION OF r, THE COEFFICIENT OF RELATEDNESS

r is the probability that a gene in one individual is an identical copy, by descent, of a gene in another individual.

General method

Draw a diagram with the individuals concerned and their common ancestors, indicating the generation links by arrows. At each generation link there is a meiosis, so a 0.5 probability that a copy of a particular gene will get passed on. For L generation links the probability is $(0.5)^L$. To calculate r, sum this value for all possible pathways between the two individuals:

$$r = \sum(0.5)^L.$$

Specific examples

These diagrams show calculations of r between two individuals represented by solid circles. Other relatives are indicated by open circles. In all cases, outbreeding is assumed. The solid lines are the generation links used in the calculations; the dotted lines are the other links in the pedigrees.

(a) Parent and offspring

$r = 1(0.5)^1$
$= 0.5$

(b) Grandparent and grandchild

$r = 1(0.5)^2$
$= 0.25$

(c) Full siblings (brother, sister)

$r = 2(0.5)^2$
$= 0.5$

(Identical genes by descent can be inherited by two pathways, either mother or father)

(d) Half-siblings

$r = 1(0.5)^2$
$= 0.25$

(Identical genes by descent can only be inherited from one parent)

(e) Cousins

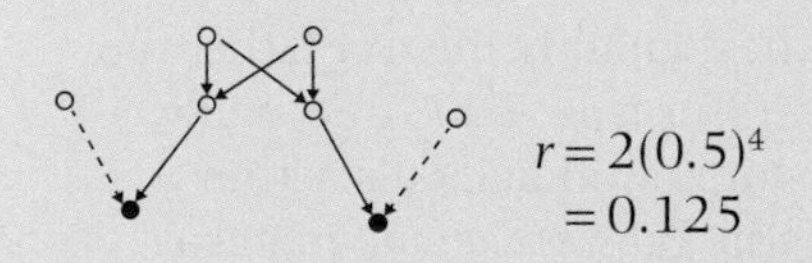

$r = 2(0.5)^4$
$= 0.125$

individual will be present, by virtue of descent from a common ancestor, in a brother, sister, cousin and so on. For brothers and sisters r is 0.5, for grandchildren it is 0.25 and for cousins it is 0.125 (Box 11.2). Bill Hamilton (1963, 1964) realized the important implication of this for the evolution of altruism, showing that just as gene proliferation can occur through parental care so it can through care for siblings, cousins or other relatives. Table 11.1 shows various values of r for descendant kin and non-descendant kin, and Box 11.3 describes how neutral molecular markers can be used to measure relatedness in natural populations.

Table 11.1 Coefficients of relatedness (r) for descendant and non-descendant kin, calculated as the probability that a gene in one individual is an identical copy, by descent, of a gene in another individual (assuming outbreeding).

r	**Descendant kin**	**Non-descendant kin**
0.5	Offspring	Full siblings
0.25	Grandchildren	Half-siblings
		Nephews and nieces
0.125	Great-grandchildren	Cousins

BOX 11.3 MEASURING RELATEDNESS WITH MOLECULAR MARKERS

The simplest way to estimate relatedness between individuals is if pedigree relationships are known. In this case, relatedness values can be assigned such as $r = 0.5$ between a pair of full siblings, $r = 0.25$ for a pair of half-siblings and so on. However, detailed pedigrees would be impossible to assign for many natural populations. The reasons for this are that parents, especially fathers, often cannot be assigned with complete confidence, and that we usually lack detailed breeding histories of populations over past generations. One way around this problem is to measure relatedness by the extent to which individuals share common alleles in molecular markers such as microsatellites (microsatellites are explained in Box 9.1). Queller and Goodnight (1989) provided a method of estimating relatedness with genetic data, which can be implemented with an easy to use computer program. This method measures the genetic similarity between two individuals, relative to that between random individuals in the population as a whole (Grafen, 1985). We shall return to this statistical definition of relatedness in Box 11.5.

Direct molecular methods for measuring relatedness offer a number of advantages. Firstly, they allow a detailed picture to be built up of the relatedness structure within social groups. For example, in meerkats it has been shown that natal subordinates tend to be half-siblings of the juveniles that they help raise, but that immigrant male subordinates can also be related to the juveniles that they help raise, because they can be related to the dominant male or father the offspring of subordinate females (Fig. B11.3a; Griffin *et al.*, 2003). Secondly, they have overturned previous assumptions about how individuals within groups are related. For example, in the social wasp, *Polistes dominulus*, it was shown that about 35% of helpers at the nest are unrelated to the dominant queen, so are not helping a close relative reproduce, as is usually assumed in the social insects (Fig. B11.3b; Queller *et al.*, 2000). Thirdly, they allow relatedness to be assessed in cases where it would not even be possible with pedigrees. For example, it has been shown that when individuals of the slime mould *Dictyostellium discoideum* come together to form fruiting bodies in the wild, the mean relatedness is extremely high, $r = 0.98$, suggesting that most fruiting bodies are composed of clone-mates (Fig. B11.3c; Gilbert *et al.*, 2007).

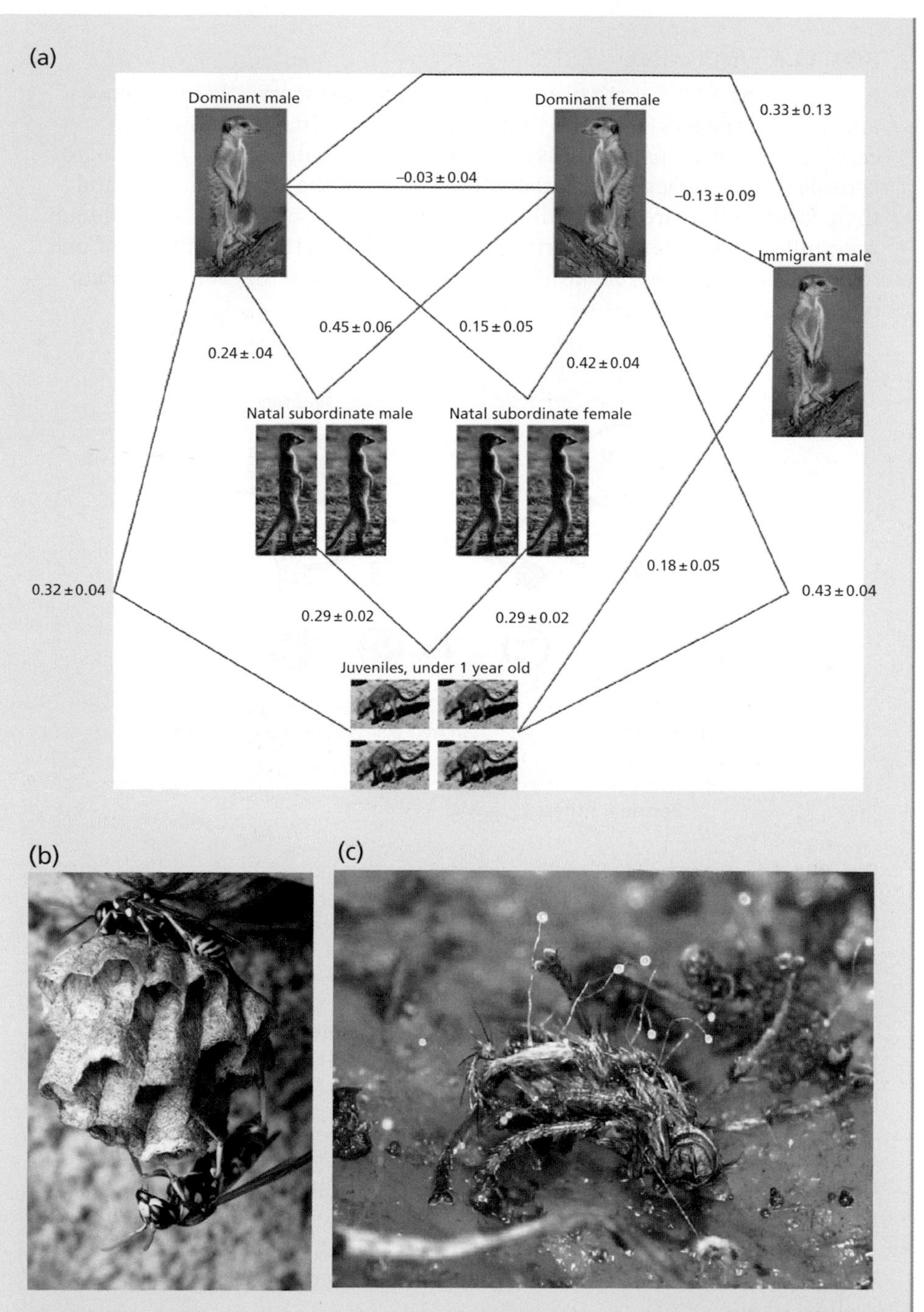

Fig. B11.3 (a) The genetic relatedness (± s.e.) of different members of a meerkat group. From Griffin *et al.* (2003). (b) The social wasp, *P. dominulus.* Photo © Alex Wild. (c) Fruting bodies of the slime mould *D. discoideum*, sprouting from a dead fly and white-tailed deer (*Odocoileus virginianus*) faeces. Photo © Owen Gilbert.

BOX 11.4 INCLUSIVE FITNESS

Inclusive fitness has a long history of being defined incorrectly in textbooks and scientific papers (Grafen, 1982; Queller, 1996). In particular, when considering the inclusive fitness consequences of a behaviour, it is crucial to consider only the *change* in reproductive success of the focal individual and their relatives that are due to this behaviour. Otherwise the problem of double accounting arises, where offspring can be assigned to the direct fitness of one individual and the indirect fitness of another individual (or even more than one other individual!).

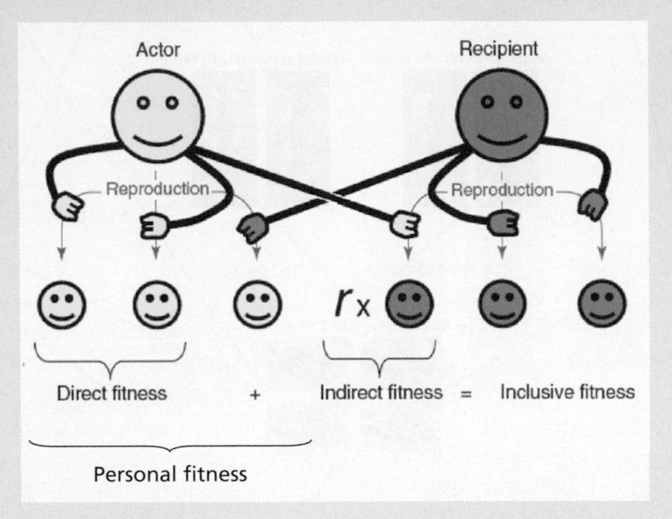

Fig. B11.4.1 Inclusive fitness is the sum of direct and indirect fitness. A key feature of inclusive fitness is that, as defined, it describes the component of reproductive success which an actor can influence, and therefore which they could be appearing to maximize. From West *et al.* (2007b).

Figure B11.4.1 provides an illustration highlighting how two individuals behave in ways that influence each other's reproductive success. The impact of the actor's behaviour (yellow hands) on its reproductive success (yellow offspring) is the direct fitness effect. The impact of the actor's behaviour (yellow hands) on the reproductive success of social partners (blue offspring), weighted by the relatedness of the actor to the recipient, is the indirect fitness effect. The key point here is that inclusive fitness does not include all of the reproductive success of the actor (yellow offspring) or the actor's relatives (blue offspring), only those which are due to the behaviour of the actor (yellow hands).

In practise though, it will often be impossible or extremely hard to calculate inclusive fitness. The easier and more useful way to test inclusive fitness theory is with Hamilton's rule (Grafen, 1991), as is illustrated in the main text.

The main point is that there is nothing particularly special about offspring as kin; if we saw a bird helping to feed a younger brother or sister this could also be favoured by selection as a means of passing on copies of genes to future generations. To put this more generally, a gene can increase its transmission to the next generation either by increasing the reproductive success of the individual in which it is in, or by increasing the reproductive success of other individuals who carry copies of that gene. These different routes for passing a copy of a gene to the next generation are termed direct and indirect respectively.

A gene's eye view: direct and indirect transmission

Hamilton showed that when these indirect effects are taken into account, natural selection on genes will lead to individuals behaving in a way that maximizes their *inclusive fitness* rather than their own or direct reproductive success. Inclusive fitness is defined as the sum of direct and indirect fitness, where *direct fitness* is defined as the component of fitness gained from producing offspring, and *indirect fitness* is defined as the component of fitness gained from aiding related individual, both descendant and non-descendant (Box 11.4). Maynard Smith (1964) coined the term *kin selection* to describe the process by which characteristics are favoured due to their effects on relatives.

Inclusive fitness is the sum of direct and indirect fitness

Kin selection: effects on relatives

Hamilton's rule

The conditions under which an altruistic act will spread due to kin selection are as follows (Hamilton, 1963, 1964). Imagine an interaction between an altruist (or actor) and a recipient in which the costs and benefits of the interaction can be assessed in terms of survival chances of the actor and recipient. The actor could be helping the recipient in any way, such as giving an alarm call, or feeding it. If the actor suffers cost C (through, for example, making itself more visible to a predator, or losing a food item) and the recipient gains a benefit B as a result of the altruistic act, then the gene causing the actor to behave altruistically will increase in frequency if:

$$\frac{B}{C} > \frac{1}{r}, \text{ or alternatively, if } rB - C > 0$$

where r is the coefficient of relatedness of the actor to the recipient. This pleasingly simple result is known as 'Hamilton's rule' (Charnov, 1977). Put into words, altruistic cooperation can be favoured if the benefits to the recipient (B), weighted by the genetic relatedness of the recipient to the actor (r), outweigh the costs to the actor (C).

Hamilton's rule predicts when altruistic acts will be favoured by selection

An intuitive understanding of Hamilton's rule can be seen as follows. As an extreme example of altruism imagine a gene that programs an individual to die in order to save the lives of relatives. One copy of the gene will be lost from the population in the death of the altruist, but the gene will still increase in frequency in the gene pool if, on average, the altruistic act saves the lives of more than two brothers or sisters ($r = 0.5$), more than four nieces or nephews ($r = 0.25$) or more than eight cousins ($r = 0.125$). Having made these calculations on the back of an envelope in a pub one evening, J.B.S. Haldane announced that he would be prepared to lay down his life for the sake of two brothers or eight cousins!

It is often useful to measure the costs and benefits in terms of offspring lost and gained, in which case we use the following form of Hamilton's rule:

$$\frac{B}{C} > \frac{r_{\text{donor to own offspring}}}{r_{\text{donor to recipient's offspring}}}$$

Two examples will help to make this clear. Imagine an individual has a choice between rearing its own offspring and helping its mother to produce offspring. The individual's own offspring and its mother's offspring, assuming they are full siblings, both have $r = 0.5$, so the expression above becomes $B/C > 1$. Therefore, helping will be favoured by kin selection if by your help your mother produces more extra offspring than you have 'sacrificed' through providing help (i.e. through forgoing the chance to produce your own offspring). If the individual was faced with the alternative of rearing its own offspring or helping its sister to produce offspring, then the expression becomes $B/C > 2$ (0.5/0.25). In this case helping would evolve only if it resulted in two or more extra offspring produced by the sister for every one offspring lost by the donor. The above examples stress that kin selection isn't just about genetical relatedness (r), it is also about the ecological factors that determine the cost (C) and benefit (B) of behaviours.

Examples of altruism between relatives

Extreme altruism: suicide and sterility in the social insects

The social insects provide good examples of extreme altruism. Worker bees have barbed stings and attack predators which approach their nests. In the act of stinging the predator the barbs of the sting become embedded in the victim and the worker bee dies as a result. The evolution of such suicidal behaviour poses a problem until we discover that the beneficiaries of the altruistic act are, in fact, close relatives of the worker. Workers are altruistic in another way because they rarely reproduce themselves, but instead help others in the nest to produce offspring. Darwin regarded this observation as potentially fatal to his theory of natural selection. How can such altruism evolve? The theory of kin selection immediately suggests a possible answer to this problem: the sterile workers usually help their mother (the queen) to produce offspring, and so she is the one who passes the altruistic genes onto future generations (Chapter 13).

Not all acts of altruism are as extreme as suicide or sterility. We now consider two examples where the costs to the altruist are smaller, but where it is still likely that kin selection has been a major force in driving the evolution of the behaviour.

Cooperation and alarm calls in ground squirrels and prairie dogs

Less extreme altruism – alarm calls

Paul Sherman made an extensive study of Belding's ground squirrel, *Urocitellus beldingi*, a diurnal social rodent which inhabits the subalpine meadows of the far western United States (see photographs in Figs. 11.5 and 11.6). This species hibernates during the winter, emerging above ground in May. Soon after emergence the females become sexually receptive and mate. After mating the males wander off and leave the females to rear their young alone. A female establishes a territory surrounding its nest burrow and produces a single litter of three to six young per year. The pups first emerge above ground at the time of weaning, when they are three to four weeks old. Soon after this the juvenile males disperse while the juvenile females tend to remain near their natal area. This means that males seldom, if ever, interact with close relatives whereas females spend their whole lives surrounded by close female kin.

Sherman found that closely related females (mother and daughters, sisters) seldom fought for nest burrows and seldom chased each other from their territories. Indeed, they cooperated to defend each other's young against infanticidal conspecifics. Of all young born, 8% were dragged from their burrows and killed by other ground squirrels. The killers were not close relatives of the victims and were either young males wandering in search of an easy meal or immigrant adult females searching for new nesting burrows. These females attempted to take over occupied burrows, killing any young they found as a way of clearing the territory of potential competitors (Sherman, 1981a, 1981b). Such cooperation among close relatives, in contrast to the conflict among unrelated individuals, is exactly what would be predicted from the theory of kin selection.

Unrelated ground squirrels kill each other's offspring, relatives do not

Individuals also gave alarm calls whenever a predator, such as a coyote or weasel, approached. Callers probably suffered a cost from giving the alarm because they were more likely to be attacked by the predator, perhaps because the calls made them more conspicuous. Others, however, benefited from the early warning and were more likely to escape. Sherman (1977) found that females were much more likely to give alarm calls than males and, furthermore, females with close relatives nearby were more likely to give calls than females without. Although in most cases the beneficiaries of the alarm calling were likely to be offspring, individuals also gave alarm calls when only parents or non-descendant relatives were nearby. For example, young females who had yet to produce their own young gave alarm calls to warn their mother and sisters of the approach of a predator. In another species of ground squirrel, *Spermophilus tereticaudus*, males were much more likely to give alarm calls before leaving their mother's home area, when they had relatives nearby, but when they dispersed and left their close relatives they were more likely to remain silent when a predator came along (Dunford, 1977).

Alarms are given when relatives are nearby

These data on alarm calls show clearly that individuals are more likely to incur the cost of calling when relatives are nearby to gain a benefit. Warning offspring (i.e. parental care) and warning sisters are, of course, just different ways of increasing gene propagation to future generations. However, it is still interesting to ask whether alarm calling could evolve mainly because of benefit to non-descendant relatives. The best evidence for this comes from a long-term study by John Hoogland (1983, 1995) of another colonial rodent, the black-tailed prairie dog, *Cynomys ludovicianus*.

Black-tailed prairie dogs live in social groups called coteries, typically one adult male with three to four adult females and their offspring. Young females remain in their natal coterie all their lives while young males disperse in their second year. All the females and yearling males within a coterie are, therefore, usually close genetic relatives. Hoogland studied alarm calling responses by presenting a stuffed specimen of a natural predator, the badger *Taxidea taxus* (Fig. 11.1). This enabled him to get more data than could be obtained by waiting for natural predator attacks and also allowed him to control for proximity of the predator to the prairie dogs. Figure 11.1 summarizes the results of over 700 experiments. The data show that individuals gave alarm calls just as frequently when there were only non-descendant kin in their home coterie as when there were

Prairie dog alarms are mainly given in the presence of offspring or other relatives...

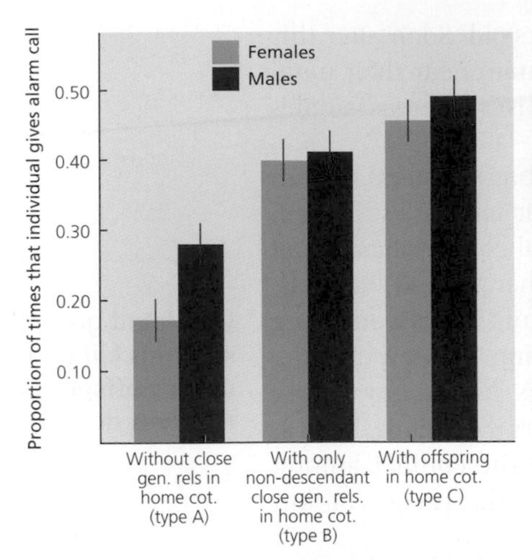

Fig. 11.1 Alarm calling by black-tailed prairie dogs to a stuffed badger. For both males (dark blue histograms) and females (light blue), there are significant differences between type A and type B individuals and also between type A and type C. There was, however, no significant difference for either sex between type B and type C. Data are means ± 1 SE. From Hoogland (1983). With permission of Elsevier. Photographs show a female making an alarm call, and a stuffed badger that is pulled across ground to simulate the presence of a predator. Both photos © Elaine Miller Bond.

... but also when there are no relatives nearby

offspring present. Factors other than the warning of relatives must also be involved, however, because immigrants who had no relatives nearby nevertheless sometimes called (Fig. 11.1). There may sometimes be direct benefits to the caller itself in giving the alarm, for example signalling to the predator 'I've seen you', which may reduce the likelihood of attack because the predator then does not have the advantage of surprise. Another possibility is that it pays to warn others of the approach of a predator, even if they are not close relatives, because if a neighbour was caught successfully then the predator may be more likely to return to hunt in the same area again. Alarm calling, may, therefore reduce the likelihood of future attacks by the same predator.

These studies of cooperation among relatives in ground squirrels and prairie dogs are consistent with the kin selection model but they do not constitute a quantitative test of Hamilton's rule. Indeed, it would be difficult to assess the costs and benefits of alarm calls in terms of offspring lost and gained. We now turn to an example where these values could be estimated.

Cooperative courtship in wild turkeys

A quantitative test of Hamilton's rule

In the wild turkey, *Meleagris gallopavo*, pairs of same aged males sometimes form coalitions to court females, and then defend those females against other males (Fig. 11.2). In these groups, one of the males appears to be dominant, obtaining all the

Wild turkeys: brothers form display partnerships to attract females

matings with females, whilst the other male is a subordinate who appears to obtain no mates. Given this, why do the subordinates remain in the coalition? Alan Krakauer (2005) examined whether this cooperative courtship could be explained as altruistic cooperation between kin. In order to do this, he estimated the three parameters of Hamilton's rule: r, B and C (Table 11.2). Firstly, he used microsatellite markers to estimate relatedness, and found that the mean relatedness between males in coalitions was $r=0.42$, and not significantly different from that expected if they were full siblings ($r=0.5$).

Krakauer then went on to estimate the costs to the subordinate and benefits to the dominant of this cooperative courtship. To estimate the benefit, B, he compared the reproductive success of dominant males and non-cooperating solitary males. Dominant paired males mated with more females and, on average, produced 6.1 more offspring than solitary males. To estimate the cost, C, he compared the reproductive success of subordinate males and non-cooperating solitary males. While solitary males

Fig. 11.2 A pair of male wild turkeys, *Meleagris gallopavo*, that have formed a coalition to court females. Photo by Maslowski/National Wild Turkey Federation.

Variable	Description	Calculation	Value
r	Coefficient of relatedness	Mean pairwise relatedness of subordinates to their dominant display partner	0.42
B	Benefit to dominant (number of extra young produced due to subordinate male's help)	(Mean number of offspring produced per dominant male) – (Mean number of offspring produced per solitary male)	6.1
C	Cost to subordinate (number of own young sacrificed in order to help dominant male)	(Mean number of offspring produced per solitary male) – (Mean number of offspring produced per subordinate male)	0.9
$rB-C$	Is Hamilton's rule satisfied?	$(0.42 \times 6.1) - 0.9$	+1.7 (> 0)

Table 11.2 Parameterizing Hamilton's rule with wild turkeys (Krakauer, 2005). B and C are measured in units of offspring per male.

produced an average of 0.9 offspring each, the subordinates were never observed to father any offspring. Therefore, by helping the dominant to reproduce, the subordinate 'sacrificed' 0.9 of its own offspring in order to increase the dominant's reproductive success by 6.1 offspring.

Plugging these values into Hamilton's rule gives $(0.42 \times 6.1) - 0.9 = 1.7$, which is > 0, and so Hamilton's rule is met. Consequently, the cost of helping by the subordinate is outweighed by the indirect benefits they obtain by helping their kin (brothers) obtain more mates. Indeed, the minimum relatedness required to meet Hamilton's rule is $r = 0.15$ (i.e. the value of r at which $r \times 6.1 - 0.9 = 0$), and so even half-brothers ($r = 0.25$) would be expected to cooperate. It should be noted that the data are observational, so these calculations do not allow for complications such as individuals of different quality pursuing different strategies. For example, are subordinates low-quality individuals that were incapable of breeding independently? Ideally, removal experiments are needed to justify the assumption that, in the absence of its partner, both the dominant's and the subordinate's success becomes equal to that of a solitary male.

Cooperative courtship is explained by the benefit of helping a relative obtain more mates ...

Nevertheless, a number of alternative explanations can be ruled out, based upon direct benefits to the subordinates, which may be important in other species (Krakauer, 2005). Firstly, there is the possibility that subordinates help to gain a 'share' of the reproductive success, as appears to occur in the ruff (*Philomachus pugnax*; Lank *et al.*, 2002). This can be ruled out in the wild turkey because subordinates produced no offspring. Secondly, subordinates may help to gain an increased likelihood of future inheritance of a display perch, as appears to occur in *Chiroxiphia* manakins (McDonald & Potts, 1994). In the wild turkey, coalitions form before adulthood and only change through death. Consequently, if the dominant dies, the subordinate just becomes a solitary male.

... and not by immediate or delayed direct fitness benefits to the subordinate

How do individuals recognize kin?

Hamilton's explanation for altruism requires a sufficiently high relatedness between interacting individuals. One way to obtain this is kin discrimination, whereby an individual assesses its relatedness to other individuals and adjusts behaviour accordingly. There is now a rapidly growing body of evidence that individuals can indeed do this, and even distinguish close kin from distant kin. How do they achieve this?

Greenbeards

An entertaining but theoretically unlikely possibility was proposed by Bill Hamilton (1964) and coined the 'green beard effect' by Richard Dawkins (1976). The idea is that there may be 'recognition alleles' which express their effects phenotypically, so enabling bearers to recognize these alleles in others, and also causing bearers to behave altruistically towards others with the recognizable phenotypic effect (Fig. 11.3). Such a gene would, therefore, need to do three things: signal itself, recognize the signal in others and direct cooperation towards those where it detects the signal. Such a gene would, for example, confer on its owner both a green beard and a tendency to be nice to others with a green beard.

Greenbeard genes cause altruistic behaviours to be directed at other individuals who carry that gene

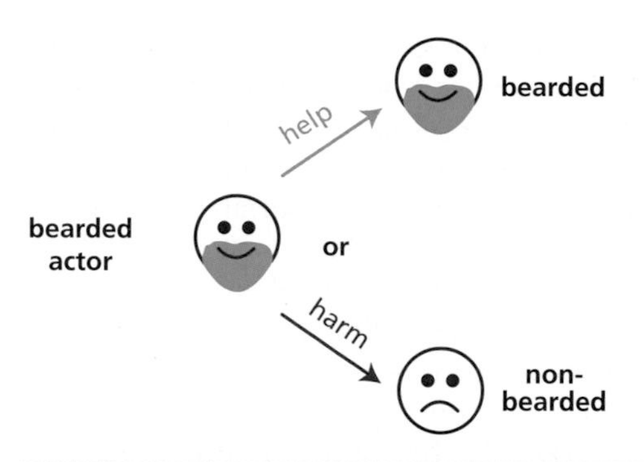

Fig. 11.3 Greenbeard genes cause the actor to either provide help to other individuals who have the beard or harm individuals who do not. Photograph of the fire ant, *Solenopsis invicta,* shows workers executing a queen who does not carry the *b* allele. From Gardner and West (2010). Photo © Ken Ross.

Laurent Keller and Ken Ross (1998) discovered an example of a greenbeard gene in the fire ant, *Solenopsis invicta*. In this species, nests contain multiple queens, with new queens being recruited into nests after they have mated. The green beard gene is the *Gp*-9 locus. Workers with the *b* allele at the *GP*-9 locus use odour to determine whether prospective queens also carry this allele – and decapitate them if they do not. This benefits the queens with the *b* allele, who are recruited and do not have to compete for resources in the nest with females who do not carry the *b* allele.

Fire ants have a greenbeard gene, causing workers to eliminate queens that do not have this gene

However, such greenbeards are unlikely to be common or generally important. One problem is that they would need to be complex, doing three things – signal, recognition and direct cooperation. It is hard enough to imagine a gene that completely encodes for one behaviour, let alone three! Another problem is that they could be easily invaded by 'falsebeards' that displayed the beard, without also performing the altruistic behaviour. Consistent with this, only a very small number of greenbeards have been discovered. Most examples of greenbeards are in microbes, possibly because they have a relatively simpler link between genotype and phenotype, which could prevent a decoupling of beard and social trait, and hence make it harder for falsebeards to arise.

Greenbeard genes are unlikely to be common, because they can be outcompeted by falsebeards

Direct genetic kin discrimination and armpits

Given the problem with directly recognizing a copy of it itself, the next best thing is to use kinship as a probabilistic means of sharing genes (Table 11.1). One way to recognize kin would be on the basis of direct genetic cues that can be recognized phenotypically. For example, if some aspect of scent was genetically determined, then more closely related individuals would smell more similar. In this case, kin discrimination could occur via rules such as 'be altruistic to neighbours who smell similar to you'. Richard Dawkins rather evocatively termed such kin discrimination the 'armpit effect'.

Recognizing kin directly on the basis of genetic similarity

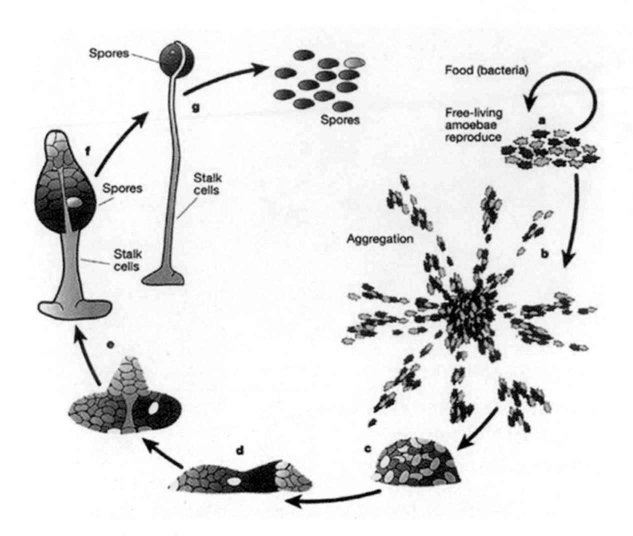

Fig. 11.4 The life cycle of a slime mould. With permission of Dr. Mary Wu and Dr. Richard Kessin.

One of the best examples of kin discrimination based on genetic cues comes from a group of organisms not usually thought of from a social behaviour perspective – amoebae. Joan Strassmann, Dave Queller and colleagues investigated the social behaviour of several species of social amoebae, or slime moulds, in the group *Dictyostelium*. Individuals of these species live in the soil, where they feed upon bacteria (Fig. 114a). However, when starved of food, the cells aggregate in thousands to form a multicellular, motile 'slug' (Fig. 11.4d). Slugs migrate to the soil surface, where they transform into a fruiting body composed of a stalk structure holding aloft a ball of spores (Fig. 11.4g). The stalk cells are non-viable, dying without reproduction after aiding the dispersal of the spores. Consequently, becoming a stalk cell is an example of extreme altruism, analogous to the sterile worker caste in social insects.

Extreme altruism: slime mould stalk cells

Mehdiabadi *et al.* (2006) tested whether kin discrimination occurred in *Dictyostelium purpureum*, by mixing the cells of two unrelated lineages on agar plates, one of which was labelled with a fluorescent dye (Mehdiabadi *et al.*, 2006). When they examined the fruiting bodies that had formed on the plates, they found that cells preferentially formed slugs and fruiting bodies with members of their own lineage, to who they are clonally ($r=1$) related. Overall, this discrimination led to an average relatedness in fruiting bodies of 0.8, as opposed to the expected value of 0.5 (expected because $r=0$ to half the cells and $r=1$ to the other half). Thanks to this high relatedness, kin selection offers a potential solution to why some cells sacrifice themselves and become stalk cells – they are helping their relatives disperse. Further work has since suggested that the underlying genetic cue may come from highly variable *lag* genes, which are involved in signalling and adhesion between cells (Benabentos *et al.*, 2009).

Slime mould cells preferentially form fruiting bodies with relatives

Environmental cues for kin discrimination

Simple rules for recognizing kin

The alternative mechanism for recognizing kin is that individuals use a simple rule, for example 'treat anyone in my home as kin'. Parent birds, for example, may ignore their own young if they are placed just outside the rim of their nests, yet will readily accept a strange chick placed inside their nest. This can lead to odd results; for example a reed warbler, *Acrocephalus scirpaceus*, may mob an adult cuckoo, *Cuculus canorus*, which approaches its nest and then, a minute later, return to the hard work of feeding a baby cuckoo inside its nest! Usually, however, this simple rule will lead individuals to care for their own offspring.

Another mechanism for recognizing kin is to learn that those you grow up with are kin. Konrad Lorenz gave the name 'imprinting' to the phenomenon observed in young geese of following the first conspicuous moving object they see after hatching. Usually this will be their mother and so result in them following someone who will keep them warm and protect them. In an experimental situation, however, young birds have been imprinted on humans and even flashing lights. Experiments by Holmes and Sherman

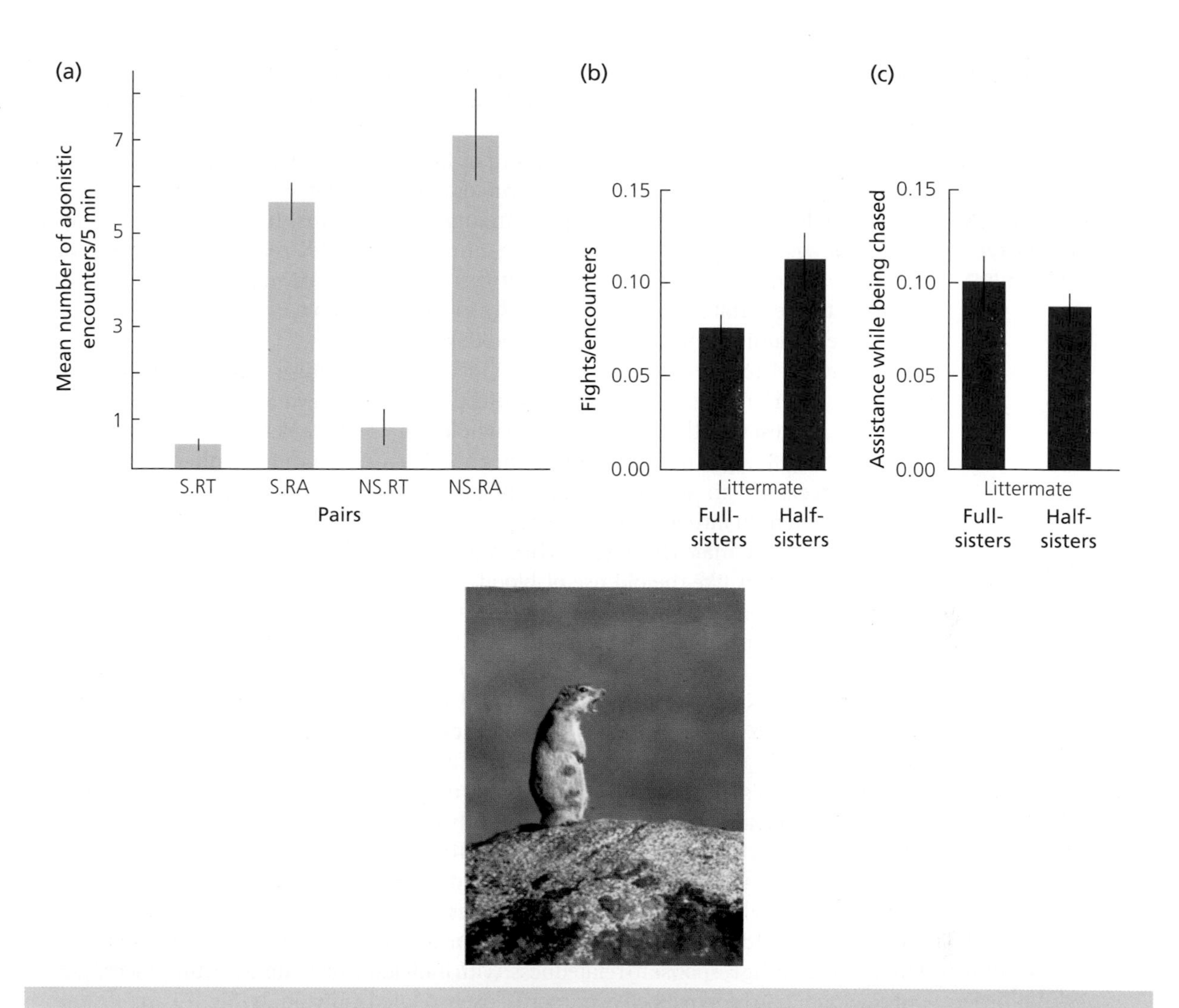

Fig. 11.5 Kin recognition in Belding's ground squirrels. (a) Laboratory experiments: mean number (± 1 SE) of agonistic encounters between pairs of yearling Belding's ground squirrels in arena tests. Non-siblings reared together (NS.RT) are no more aggressive than siblings reared together (S.RT). However, non-siblings reared apart (NS.RA) are more aggressive than siblings reared apart (S.RA). (b) and (c) Field observations: aggression and cooperation among yearling females which were full or half-sisters (genetic relatedness determined by blood proteins). Full sisters are less aggressive to one another (b), and assist each other more (c). From Holmes and Sherman (1982). Photo of a calling female © George D. Lepp.

(1982) show that sibling recognition in ground squirrels (*U. beldingi*) is also in part based on association in the natal nest. They captured pregnant females and used their pups to create four kinds of experimental rearing groups: siblings reared by one mother (their own or a foster mother), siblings reared apart by different mothers, non-siblings reared as a single litter and non-siblings reared apart. When they were older, animals from the four groups were placed in pairs in arenas and their interactions were observed. Holmes and Sherman found that regardless of true genetic relatedness animals that were reared together rarely fought. Figure 11.5 shows that unrelated individuals reared

Female ground squirrels recognize kin partly by learning ...

... and partly by 'phenotype matching'

together were no more aggressive to one another than true siblings reared together. This suggests, therefore, that individuals learn who are their kin from association in early life.

However, it was also found that among animals reared apart, genetic siblings were less aggressive to one another in an arena test than unrelated individuals (Fig. 11.5a). Interestingly, this effect only occurred among females; thus, true sisters reared apart were less aggressive to each other than unrelated females reared apart but genetic relatedness did not affect aggression between male–male or male–female pairs. Only females, the sex that behaves altruistically in the field, show evidence of being able to recognize unfamiliar, but genetically related, kin.

Of course, sisters reared apart may still learn to 'recognize' each other because of prenatal experience in their mother's uterus. Field observations by Paul Sherman, however, suggest that this may not be the whole story. Female Belding's ground squirrels mate with up to eight different males (mean 3.3 males) on the one afternoon they are sexually receptive in spring. Analysis of polymorphic blood proteins collected from mothers, potential fathers and their offspring, showed that 78% of the litters were sired by more than one male (Hanken & Sherman, 1981; this method of assessing probable paternity is rather like the old use of blood groups to settle paternity disputes in court cases). The exciting discovery was that littermates who were full sisters (same mother, same father) were less aggressive to one another and more cooperative than half-sisters (same mother, different father, i.e. half-sisters arise because of multiple matings). For example, when establishing nest burrows and defending territories full sisters fought and chased less often when they encountered each other than did half-sisters (Figs. 11.5b and 11.5c).

Odours from oral and dorsal glands act as a cue of relatedness.

Littermates all share the same nest burrow and the same uterus, whether they are full sisters or half-sisters, so some mechanism other than this common experience must be involved. More recent experiments by Jill Mateo have shown that odours from oral and dorsal glands play a mechanistic role in kin discrimination. She collected the scents of different animals by rubbing a plastic cube over mouth corners or dorsal glands, and then tested how individuals adjusted their behaviour when given cubes from different animals. She found a strong response to relatedness, with individuals spending less time investigating cubes which had been rubbed over more closely related individuals (Fig. 11.6).

In conclusion, a female Belding's ground squirrel seems to categorize others in two ways. Firstly, she recognizes and cooperates with individuals she shared a burrow with as opposed to those she did not; the former will be full siblings or half-siblings. Secondly, she may be particularly cooperative with nest mates who are like herself phenotypically, and hence more likely to be full siblings rather than half-siblings.

Kin selection doesn't need kin discrimination

Behavioural problems can be solved by either fixed or conditional strategies

In the above examples we have examined how individuals vary their behaviour depending upon their relatedness to other individuals. Consequently, we have been examining whether individuals show kin discrimination (a conditional strategy). However, kin selection can also be important without kin discrimination. If kin discrimination were

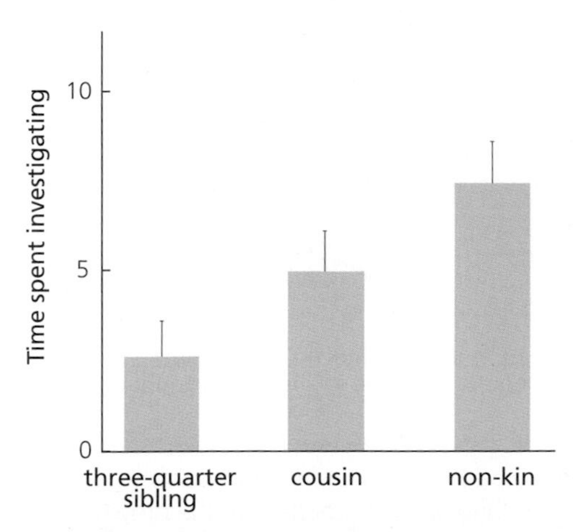

Fig. 11.6 Kin discrimination and odour in Belding's grounds squirells. Individuals spent more time investigating cubes which had been rubbed over closer relatives. From Mateo (2002). With permission on the Royal Society. Photograph of a group of pups at the mouth of a burrow. Photo © George D. Lepp.

not possible, then individuals would be expected to show fixed strategies, where natural selection fine-tunes behaviour in response to the average relatedness between interacting individuals over evolutionary time. In this case, natural selection would favour less selfish (or more cooperative) behaviour in populations or species where relatedness is, on average, higher. Bill Hamilton (1964) pointed out that limited dispersal (population viscosity) could lead to a high relatedness between interacting individuals, and that this would favour indiscriminate altruism (without kin discrimination), because the altruism would be directed towards individuals who are likely to be close relatives.

Limited dispersal provides a mechanism to generate high relatedness between interacting individuals, without kin discrimination

Cooperative iron scavenging in bacteria

Ashleigh Griffin and colleagues tested how dispersal rates could influence relatedness, and hence selection for cooperation, by examining the iron scavenging behaviour of the bacteria *Pseudomonas aeruginosa*. Iron is a major limiting factor for bacterial growth because most iron in the environment is in the insoluble Fe(III) form, and, in the context of bacterial parasites, is actively withheld by hosts. Many species of bacteria address this problem by producing iron scavenging siderophore molecules, which are released from the cell and then bind with iron, allowing it be taken up into cells.

The production of factors such as siderophores leads to the problem of cooperation, because cells who avoided the cost of producing them ('free riders') could still exploit the benefits of those produced by others (Fig. 11.7). This was confirmed experimentally, by showing that when a normal siderophore producing strain is grown in a mixture with a free riding mutant that does not produce siderophores, the mutant increases in frequency. These experiments were facilitated by the handy fact that cooperating and free riding bacteria can be readily distinguished by eye, because siderophores are green, and so colours the colonies of cooperative bacteria, whereas the free riders are white. Kin selection is likely to be a key explanation for cooperation between bacterial cells,

The production of iron scavenging molecules by bacteria is a cooperative behaviour

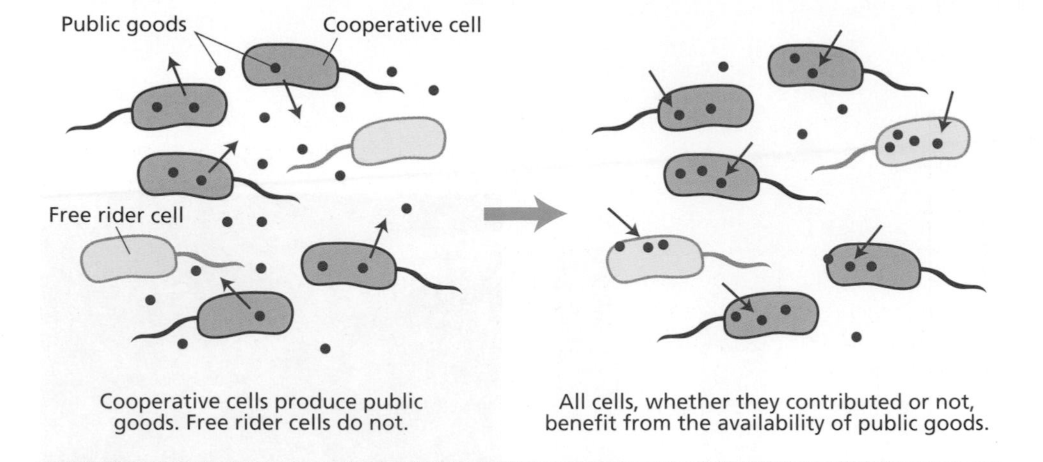

Fig. 11.7 Public goods in bacteria. Bacteria produce a number of factors that they excrete out into the local environment, and which then provide a benefit to the growth or movement of the local population. The factors are open to the problem of cooperation, because 'free riders' who did not produce them will still be able to benefit from the production of these factors by others. Iron scavenging siderophore molecules are an example. Other examples include factors which: digest proteins or sugars; break down host tissues; provide structure for growth; kill or repel competitors or predators; aid movement; modulate immune responses; or inactivate antibiotics.

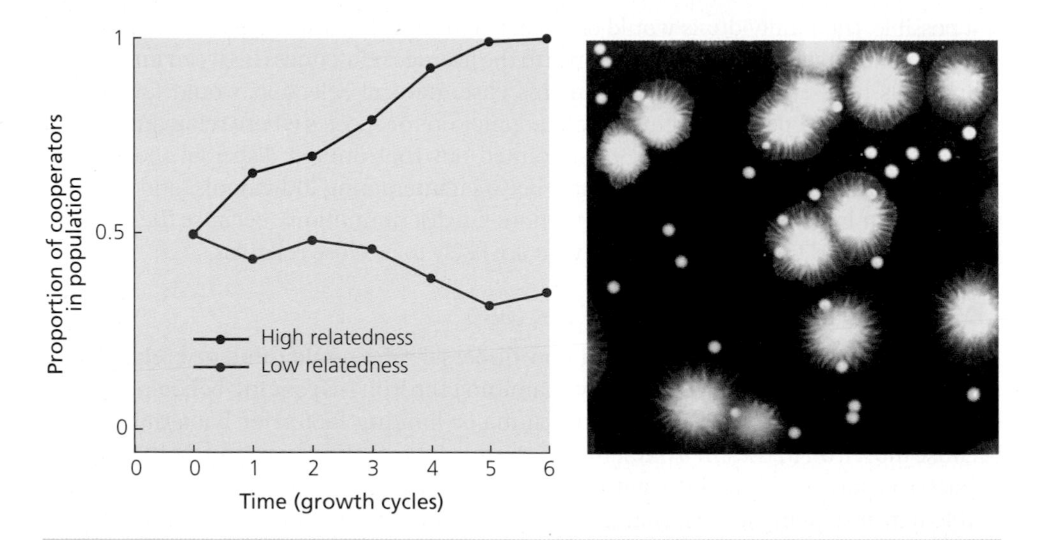

Fig. 11.8 Relatedness and cooperation in the bacterium *P. aeruginosa*. Populations were set up which contained a mixture of cooperative siderophore producers and uncooperative free riders, which did not produce siderophores. When these populations were maintained such that interacting bacteria were highly related, the cooperative siderophore producing cells spread to fixation. In contrast, when they were maintained such that interacting bacteria were less related, the free riders were able to remain in the population. Griffin *et al*. (2004). Reprinted with permission from the Nature Publishing Group. Photograph shows green siderophore producers and white free riders growing (differentially) on an agar plate where iron is limiting. Photo © Adin Ross-Gillespie.

because the clonal growth and limited movement of bacteria mean that they will often be surrounded by and interacting with clone mates, to who they are related by $r=1$.

Griffin *et al.* (2004) tested whether the pattern of dispersal, and hence relatedness mattered with an experimental evolution approach. They initiated populations with a mixture of the cooperator and the free rider, and then maintained them under conditions that would lead to either a high or low average relatedness. They did this by splitting the population into subpopulations, and then initiating each subpopulation, in each round of growth, with either a single bacterial clone (relatively high relatedness) or two bacterial clones (lower relatedness). As predicted by kin selection theory, the wild-type cooperator did much better when relatedness was higher (Fig. 11.8).

Cooperation is favoured when interacting cells are more highly related

Selfish restraint and kin selection

Kin selection is not just about explaining altruistic helping behaviours. As Hamilton (1964) pointed out, the general principle is that with regard to social behaviours (Box 11.1) individuals should value any positive or negative consequences for recipients, according to the coefficient of relatedness (r) between them. Consider a selfish behaviour that provides a benefit to the individual performing the behaviour, but is costly to the recipient of the behaviour. Hamilton (1964) predicted that individuals should show greater selfish restraint when interacting with closer relatives. This is because harming a relative will decrease the ability of that relative to transmit shared genes, and hence incur an indirect fitness cost. More specifically, if we take a genes eye view, then any direct benefit of a selfish behaviour, which increases the chance of that copy of the gene being transmitted to the next generation, must be weighed up against any indirect cost of reducing the likelihood that other copies of the gene are not transmitted to the next generation (e.g. when relatives compete for resources).

Selfish: benefit to actor, cost to others

Individuals should behave less selfishly when interacting with closer relatives

Cannibalism in salamanders

One of the ultimate selfish behaviours is cannibalism, where individuals eat other individuals of the same species. It does not require fancy experiments to realize that this is beneficial for the actor (food, reduced competition) and costly for the recipient (death, eaten). David Pfennig and colleagues investigated the cannibalistic behaviour of Arizonan tiger salamanders. In this species, larvae occur in two morphs, a 'typical' morph that feeds mostly on invertebrate prey, and a larger physically distinctive 'cannibal' morph that has specialized oral structures to facilitate ingestion of conspecifics. This cannibal morph occurs more frequently when the salamander larvae are at higher densities and food is more limited. Pfennig and Collins (1993) tested whether the development into cannibal morphs was also influenced by whether larvae were interacting with relatives or non-relatives. Inclusive fitness theory predicts that cannibalism is less likely to be favoured when interacting with relatives, because eating a relative would incur an indirect fitness cost. To test this, Pfennig and Collins reared larvae in groups of 16, with the groups consisting of either siblings or a mixture of siblings and non-siblings. As predicted, they found that larvae were significantly less likely to develop into the cannibal morph when in a group with only siblings (Fig. 11.9a). Furthermore, when housed with a mixture of siblings and non-siblings, cannibals preferentially consumed non-siblings (Fig. 11.9b; Pfennig *et al.*, 1994).

Interactions with siblings can inhibit cannibalism ...

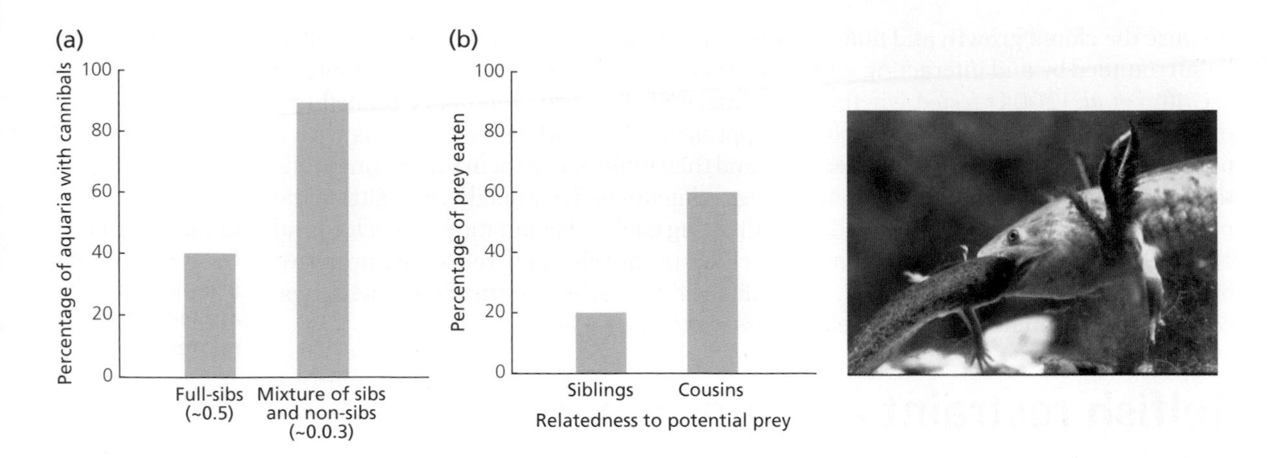

Fig. 11.9 Cannibalism in Arizona tiger salamanders. (a) Larvae are more likely to develop as the cannibal morph when reared with non-relatives. (b) Cannibal morphs were more likely to eat individuals to who they were less closely related. From Pfennig and Collins (1993). Reprinted with permission from the Nature Publishing Group. Photograph of a cannibal morph eating a typical morph. Photo © David Pfennig.

In a series of elegant experiments, Pfennig *et al.* (1999) then went on to test a number of alternative explanations for this observation. They rejected alternate hypotheses that did not rely on the indirect costs of eating relatives. For example, avoiding eating relatives might be a disease avoidance strategy, if it is easier to acquire parasites by eating closer relatives. This might be the case if parasites are adapted to infect specific genotypes, in which case the parasites that have been able to infect an individual, are also likely to be successful at infecting their relatives, who have similar genotypes. However, this hypothesis was rejected because cannibals were not more likely to become infected with bacteria or viruses by eating kin (in fact the opposite occurred), cannibals did not avoid diseased prey, and cannibalism was not less common in patches with higher rates of disease.

In contrast, Pfennig *et al.* (1999) found clear support for the kin selection hypothesis, that individuals are avoiding the indirect costs of eating relatives. He carried out a quantitative test of Hamilton's rule by estimating the costs and benefits of eating siblings. Hamilton's rule predicts that cannibals should be altruistic (avoid eating siblings) if $rB-C>0$, in other words if $B/C>1/r$. For full siblings, $r=0.5$, so the eating of full siblings should be avoided if $B/C>2$.

Pfennig estimated the benefit (B) and cost (C) of not eating relatives, by taking advantage of the fact that there was variation across individuals in the extent to which they avoid eating kin. Some individuals (discriminators) show a strong tendency to avoid eating siblings, whilst others (non-discriminators) show no tendency to avoid eating kin. Pfennig placed either a discriminator or a non-discriminator cannibal larvae in a container with 24 typical larvae, six of which were siblings and the remainder non-siblings. He found a large indirect benefit (B) of kin discrimination, with approximately four siblings surviving in the presence of a discriminator cannibal and two in the presence of a non-discriminatory cannibal, giving $B\approx 4-2=2$. He also found no significant cost of kin discrimination, with non-discriminators growing at the same rate, and reaching maturity at the same age, as discriminators, suggesting $C\approx 0$. This suggests that $B/C>2$, in which case kin discrimination over cannibalism would be explained by kin selection.

... because there is a large indirect fitness benefit to not eating siblings

It is not known why some larvae are discriminators and some not. Perhaps under some conditions, such as harsher competition for resources, or when there is a lack of non-relatives, it could pay to eat kin.

Spite

Bill Hamilton (1970) pointed out that Hamilton's rule, whilst explaining altruism, also has a more sinister interpretation, because it shows how natural selection can favour spiteful harming behaviours, which are costly to both the actor and recipient (Box 11.1). In terms of Hamilton's rule, this means that C is positive (the behaviour is costly to the actor), B is negative (the behaviour is costly or harmful to the recipient). Consequently, for $rB - C > 0$ to occur, a negative relatedness (r) is required. Negative relatedness may seem like a bizarre concept, but it simply means that the recipient of a particular behaviour is less related to the actor than an average member of the population (Box 11.5).

Spite: cost to others, cost to actor

For spite to evolve, a negative relatedness is required

To consider how spite can be favoured, it is useful to take a gene's eye view and consider a gene that will lead to a costly harming behaviour being directed at individuals who do not have this gene. If this harming of individuals who do not carry the spiteful gene frees up resources or reduces competition in a way that benefits other carriers of the spiteful gene, then this gene can spread. One way of conceptualizing this is that spite is favoured when it can be directed at non-relatives, and that this harming of the non-relatives frees up resources (or reduces competition) for relatives of the actor. Spite can, therefore, also be thought of as altruism towards a secondary recipient or recipients – harming an individual can be favoured if this provides a benefit to a closer relatives (Fig. 11.10). The key thing here is relative relatedness, as what is required is that the actor is more closely related to the secondary recipient who benefits from the reduced competition than it is to the individual that it actually harms (primary recipient).

Spite as secondary altruism

It has long been assumed that spite did not occur in the natural world, because it would be hard to obtain situations where harming an individual is the most efficient way of helping another individual. While several possible examples of spite had been given, these were all much more easily explained as selfish behaviours which provide a benefit to the actor in the long term (Table 11.3). For example, birds can engage in aggressive encounters with other individuals over territories. While this aggression may appear costly to the actor, this is only in the short term, and can be explained through the longer-term direct benefits of reducing competition for resources. Consequently, such behaviours are selfish not spiteful. However, it has recently been discovered that spite does occur in the real world.

Many behaviours that were thought to be spiteful actually provide a direct benefit in the long term, and so are selfish

Murderous soldiers in polyembryonic parsitoid wasps

Mike Strand and colleagues studied the parasitoid wasp *Copidosoma floridanum*, which lays its eggs into the eggs of moths. The wasp larvae then develop within the moth caterpillar, consuming it from the inside as it grows. It is often the case that a female will lay two eggs in a host, one male and one female. Each of these eggs divides asexually to produce thousands of larvae, leading to the situation where a caterpillar will contain a

BOX 11.5 RELATEDNESS REVISITED

In this and following chapters we discuss relatedness coefficients in terms of genes shared identically by descent. However, it should be realized that this is just a very useful approximation, which greatly simplifies discussion. More formally, the coefficient of relatedness, *r*, is defined statistically, as a measure of the genetic similarity between social partners, relative to the rest of the population (Hamilton, 1970; Grafen, 1985).

The statistic *r* can be positive or negative like any statistical correlation, but will have a mean value within a population of zero. This is shown graphically in Fig. B11.5.1, where the shaded area represents the proportion of the actors genes that are shared with three individuals (A, B and C) and the population as a whole. For clarity, we have made the actor share ½ of its genes with the population, but this proportion could take any value between zero and one. Individual A carries the actors genes at a higher frequency than the population, so they are positively related ($r>0$). Individual B carries the actors genes at a lower frequency than the population, so they are negatively related ($r<0$). Finally, individual C carries the actors genes at the same frequency as the population, and so they are zero related ($r=0$). It follows that the average relatedness in the population is zero.

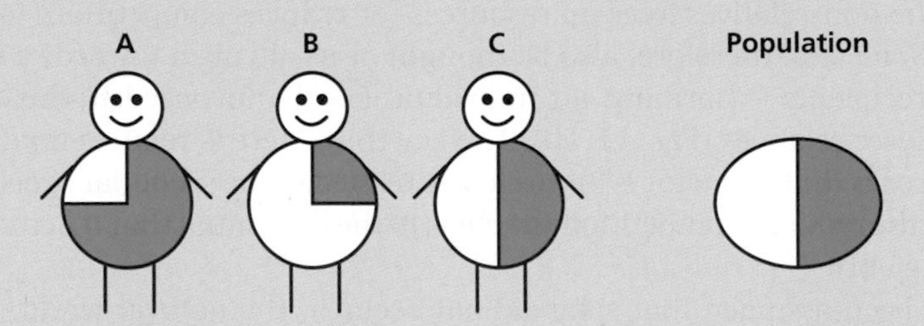

Fig. B11.5.1 The geometric view of relatedness. The shaded area shows the proportion of the actors genes shared with three potential recipients (A, B and C) and the population as a whole. As described in the text, the actor is positively related to A, negatively related to B and zero related to C.

In polyembryonic wasps, some larvae develop as soldiers that attack other larvae

large number of genetically identical females and a large number of genetically clonal males, who are the brothers of the females. While most of these wasp larvae develop normally, a number develop precociously to form a sterile soldier caste that attacks other larvae within the host. Gardner *et al.* (2007) suggested that this is spiteful behaviour aimed at relatively unrelated individuals (brothers; $r = 0.25$), to free up resources for closer relatives (clonal sisters; $r = 1$; see Chapter 13 for the calculation of relatedness coefficients in species such as wasps which have haplodiploid genetics). However, to demonstrate that this really represents spite, the following conditions are required.

(1) *The behaviour really is costly to the actor, and doesn't have some long-term direct benefit.* Developing as a soldier is clearly costly because all soldiers are sterile and do not complete development to adults, paying the ultimate sacrifice of never being able to reproduce themselves.

(2) *Harming behaviours are directed towards relatively unrelated individuals.* Giron *et al.* (2004) tested whether kin discrimination occurred, with harming being preferentially directed at non-relatives. They varied relatedness by introducing either clonal sisters (r = 1.0), brothers (r = 0.25) or unrelated larvae (r=0) into a host caterpillar containing a developing female brood of

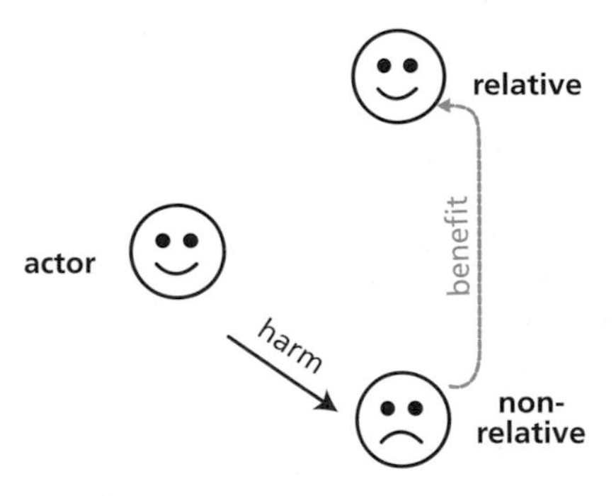

Fig. 11.10 Spiteful harming of another individual is favoured when this provides a benefit to a secondary recipient to whom the actor is more closely related.

Developing as a soldier is costly, because soldiers are sterile

Organism	Trait	Why selfish and not spiteful
Birds	Protection of territories	Reduced competition for resources
Herring gull (*Larus argentatus*) and western gull (*L. occidentalis*)	Siblicide at neighbouring nests	Reduced competition for resources
Macaques (several species)	Harassing infant and juvenile daughters of others	Reduced competition for own offspring
Macaques (stumptail, *Macaca arctoides*)	Sexual interference	Increased reproductive success in long term
Mammals	Infanticide	Reduced competition for own offspring
Mountain sheep (*Ovis Canadensis*)	Harassing injured male	Increases chance the harassed male dies, so reduces competition for mates in next breeding season
Sticklebacks	Egg cannibalism	Reduced competition for own offspring

Table 11.3
Not spite: example behaviours that have been suggested as spiteful, but are explained much more easily as selfish behaviours that provide a direct benefit to the actor (West & Gardner, 2010).

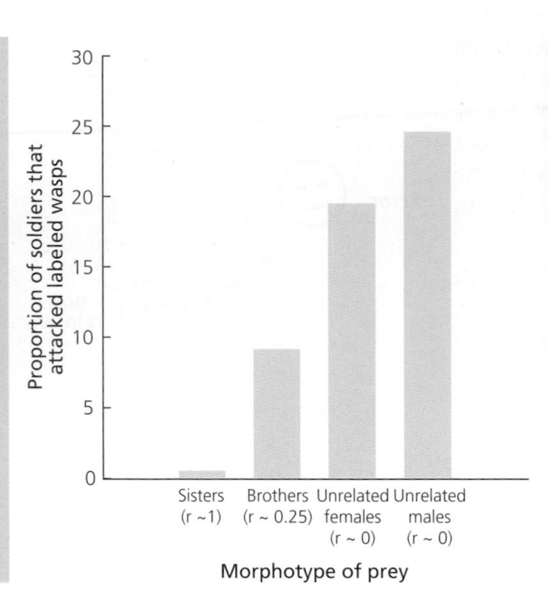

Fig. 11.11 Attack rates and relatedness in the polyembryonic wasp *C. floridanum*. Giron *et al*. (2004). Photograph of a female parasitizing a *Trichoplusia ni* egg. Photo © Paul Ode.

Soldiers preferentially attack larvae to which they are relatively less related

wasp larvae. Giron *et al.* labelled the introduced larvae with a fluorescent tracer and assessed the rate at which they were attacked, measuring how many of the resident soldiers ingested labelled larval tissue. As predicted they found that the soldiers were more likely to attack larvae to which they are less related (Fig. 11.11).

The membrane surrounding each larvae is the cue for relatedness

In a follow up study, Giron and Strand (2004) showed that the cue used for kin discrimination is the extraembryonic membrane surrounding each larva during its development in the caterpillar host. In a series of extremely elegant experiments, they showed that attack rates correlated negatively with kinship when the membrane was present, but not when the membrane was removed. In addition, by transplanting membranes between larvae they were able to fool the soldiers, whose attack rates correlated negatively with their relatedness to the membrane donor but not with the larva encased inside.

The soldiers are closely related to the larvae that benefit from reduced competition than they are to those killed

(3) *There is strong competition for local resources, and considerable variation in relatedness, such that substantially closer relatives can benefit from the harming of others*. The limited resources provided by a moth caterpillar only allow a fraction of the wasp larvae to complete development and emerge as adults. This leads to intense competition for resources among the larvae within the host, which can be reduced by the killing behaviour of the soldiers. However, it is also key that the relatedness of those that benefit from this killing is sufficiently different from those killed. This occurs with these wasps because, by killing brothers, the soldiers free up resources for clonal sisters, to whom they are more closely related.

Chemical warfare in bacteria

Another example of spiteful behaviour is provided by the antagonistic interactions between different strains of bacteria. Many bacteria produce anti-microbial compounds that are termed bacteriocins, which are lethal to members of the same species unless they carry an immunity gene which encodes a factor that deactivates the bacteriocin.

Andy Gardner and colleagues (2004) argued that the production and action of such bacteriocins fulfils all three of the conditions required for spite. Firstly, the production of bacteriocins incurs a large direct cost, so cannot be explained as a selfish behaviour. Indeed, in some species, cell death is required to release the bacteriocins into the environment, and so it represents another sterile trait. Secondly, the negative consequences of bacteriocins are directed towards relatively unrelated cells. The reason for this is that there is genetic linkage between the bacteriocin gene and the immunity gene for that bacteriocin, such that close relatives will have the capacity to both produce and be immune to a particular bacteriocin, or neither. Consequently, the bacteriocin will only harm non-relatives. Thirdly, the benefits of killing non-relatives go to close relatives. This is because the clonal growth of bacteria means that that when multiple clones grow and compete in an area cells can be near both close relatives (clone mates, $r = 1$) and non-relatives ($r = 0$).

Bacteria produce chemicals that kill non-relatives, to reduce the competition experienced by relatives

Summary

Altruism is acting to increase the number of offspring that another individual produces at a cost to one's own chances of survival or reproduction. Extreme examples of altruism include the sterile workers of social insects and the stalk cells of slime moulds. Hamilton showed that altruism could be explained by kin selection. The idea here is that an individual can increase its genetic representation in future generations by helping close relatives, who share copies of its genes. The conditions for the spread of an altruistic behaviour are given by Hamilton's rule, $rB\text{-}C>0$. Examples of traits that may be explained by kin selection include alarm calling in ground squirrels and cooperative courtship in wild turkeys.

Kin selection requires a sufficiently high relatedness between interacting individuals. One way to obtain this is kin discrimination, where individuals are able to assess relatedness and then preferentially help closer relatives. An extreme form of genetic discrimination is greenbeard genes, which lead to both recognition and helping, and hence the gene only helping other individuals that have that specific gene. However, whilst examples exist, such as in the fire ant, this mechanism is unlikely to be of general importance. The more common form of discrimination is to use kinship as an indicator of shared genes. Such kin discrimination can involve genetic cues of kinship (e.g. slime moulds), environmental cues of kinship (e.g. long-tailed tits, chapter 12) or both (e.g. Belding's ground squirrels).

The alternative way to obtain a high relatedness is limited dispersal. Limited dispersal would keep relatives together, and hence indiscriminate altruism could be favoured towards neighbours because they would be likely to be relatives. An example of this is the production of iron scavenging siderophore molecules in bacteria.

Kin selection theory also predicts that individuals should behave less selfishly when interacting with closer relatives. This can explain why salamanders are less likely to cannibalize closer relatives.

Finally, kin selection theory has a darker side, predicting that spiteful behaviours can evolve if they are preferentially directed at non-relatives. Spite appears to be rare, but examples include the sterile soldiers in polyembryonic parasitoid wasps and chemical warfare (bacteriocins) in bacteria.

Further reading

Hamilton's original papers on inclusive fitness theory have been collected together into a single volume (Hamilton, 1996), which also provides illuminating autobiographical notes for each of the papers. Dawkins (1976) provides a very readable popularization of the gene's eye view, whilst Dawkins (1979) gives a lucid discussion of twelve misunderstandings of kin selection, many of which are still made today. Grafen (1991) provides a more technical review of inclusive fitness theory and how it can be tested. Grafen (1985) is the classic text on the concept of relatedness. We discuss a number of altruistic and mutually beneficial behaviours in more detail in the following two chapters. The confusion that may arise from redefining terms such as altruism is reviewed in West *et al.* (2007a).

There is sometimes an overemphasis – both conceptually and empirically – on the importance of the relatedness term (r) in Hamilton's rule and a corresponding neglect of the benefit (B) and cost (C) terms. To some extent, this is the case because genetic similarity can be measured more easily than components of fitness (Box 11.2). However, focusing too strongly on r can lead to misunderstanding and confusion, because variation in B and C is equally important. An excellent example of where variation in C has clear consequences for when individuals cooperate is provided by Field *et al.*'s (2006) study of queuing for reproductive dominance in social groups of the hairy-faced hover wasp. Gorrell *et al.* (2010) use Hamilton's rule to show that indirect fitness benefits explain adoption in asocial red squirrels.

Further examples of greenbeard genes include: the *csa* gene in the slime mould *Dictyostelium discoideum*, which causes individuals to adhere to each other in aggregation streams, and cooperatively form fruiting bodies, whilst excluding non-carriers of the gene (Queller *et al.*, 2003); and the *FLO1* gene in the yeast *Saccharomyces cerevisiae*, which causes individuals to adhere to each other in groups that are better defended from stressful environments (Smukalla *et al.*, 2008). The evolutionary dynamics of greenbeards and other known biological examples are reviewed in Gardner and West (2010).

One way in which individuals could behave less selfishly when interacting with relatives, is by exploiting a resource more prudently and efficiently (analogous to the discussion of producing rather than scrounging in Chapter 5). Frank (1996) provides an overview of how this could explain variation in the damage that parasites cause to their hosts, termed parasite virulence. Specifically, if the parasites infecting a host are highly related, then they have a common interest that favours prudent exploitation of the host over time (and hence lower virulence), to maximize the total amount of resources that can be acquired. Empirical support for the prediction that a lower relatedness between the parasites infecting a host favours a higher virulence comes from Herre's (1993) comparative study on fig wasp nematodes and Boots and Mealors (2007) experimental study in a virus of moths.

Inglis *et al.* (2009) provide an example of how more specific predictions for spiteful traits can be made and tested experimentally. They show how the advantage of spiteful traits, such as bacteriocin production, will vary depending upon aspects of the population structure, including the proportion of individuals in a patch that are clonemates.

TOPICS FOR DISCUSSION

1. All altruism is genetically selfish. Discuss.
2. Does it matter how precisely we use terms such as altruism?
3. Are humans especially altruistic (Fehr & Gachter, 2002)?
4. Discuss whether alarm calls in meerkats are mutually beneficial or altruistic (Clutton-Brock *et al.*, 1999b).
5. Are greenbeards in conflict with the rest of the genome?
6. Discuss how cross-fostering experiments can be used to study the ways in which animals discriminate kin (Mateo & Holmes, 2004; Holmes & Mateo, 2006; Todrank & Heth, 2006).
7. Does kin discrimination occur in ant colonies with multiple queens (Hannonen & Sundstrom, 2003; Holzer *et al.*, 2006; van Zweden *et al.*, 2010)?
8. Will limited dispersal always favour altruism (Taylor 1992; Queller 1994)?
9. When is it OK to bite off your brothers head (West *et al.*, 2001)?
10. Given that spite can also be thought of as 'secondary altruism', is there any use in distinguishing spite from altruism?
11. Should we expect spite to be common, and where would be the best places to look for novel examples?
12. Are greenbeards likely to lead to the complex adaptations that we observe in animals?

Photo © Andrew Young

CHAPTER 12

Cooperation

In the last chapter we explained how altruistic cooperation could be favoured between related individuals by kin selection. However, cooperation can also happen between non-relatives. For example, in many cooperative breeding vertebrates, such as meerkats, dwarf mongooses and superb fairy wrens, some helpers are unrelated to the offspring that they are helping to rear. An even more extreme example is provided by mutualistic cooperation between species, such as when cleaner fish remove parasites from their clients, or when rhizobia bacteria fix nitrogen from the atmosphere and provide it to their legume plant hosts. Consequently, it is clear that cooperation is not just about kin selection.

In this chapter we distinguish four different hypotheses for the evolution of cooperative behaviours. This first of these is kin selection, which can explain altruistic cooperation between relatives. The other three hypotheses all rely on cooperation providing some direct benefit to the cooperator: by-product benefits, reciprocity and enforcement. In these cases, cooperation ends up not being altruistic, and is instead mutually beneficial (Table B11.1.1). We shall see that the ways in which cooperation can provide a direct benefit can be complex, involving delayed benefits that only accrue in the long term or active enforcement mechanisms.

What is cooperation?

A behaviour is cooperative if it provides a benefit to another individual (recipient) and has been selected for (at least partially) because of its beneficial effect on the recipient. The latter clause is added to exclude behaviours which merely provide a one-way by-product to others (West *et al.*, 2007a). For example, when an elephant produces dung, this is beneficial to a dung beetle that comes along and uses that dung but it is not

An Introduction to Behavioural Ecology, Fourth Edition. Nicholas B. Davies, John R. Krebs and Stuart A. West.

Fig. 12.1 Cooperation. (a) Cells of the algae *Volvox carteri weismannia* form cooperative spherical multicellular groups, which contain up to 8000 small somatic cells arranged at the periphery and a handful of much larger reproductive (germ) cells. This distinction between somatic and reproductive cells is analogous to that between workers and reproductives in the eusocial insects. Photo © Matthew Herron. (b) Banded mongooses (*Mungos mungo*) live in cooperative mixed sex groups of about 7–50 individuals across a large part of East, Southeast and South-Central Africa. Photo © Andrew Young. (c) An upside-down jellyfish (*Cassiopea xamachana*) infected with its algal symbiont (*Symbiodinium microadriatum*). The algae (orange in the photograph) provide the jellyfish with photosynthates in exchange for nitrogen and inorganic nutrients. Photo © Joel Sachs. (d) In social insects, such as this ant species *Camponotus hurculeans*, some individuals give up the chance to breed independently and instead raise the offspring of others. Photo © David Nash.

useful to think of this as cooperation, as the elephant produces dung for purely selfish reasons (emptying waste). The production of dung would only qualify as cooperation if a higher level of dung production had been favoured, because of the benefits to dung beetles. The definition of cooperation therefore includes all altruistic (–/+) and some mutually beneficial (+/+) behaviours (Box 11.1).

A behaviour is cooperative if it benefits another individual and has been selected for because of that benefit

Cooperation can take many different forms in different organisms (Fig. 12.1). In cooperative breeding vertebrates, such as meerkats or Florida scrub jays, individuals often live in groups that include the dominant pair, which do most of the breeding, and the subordinates who help care for the young (Hatchwell, 2009;

In cooperative breeders, the subordinates in the group help rear the offspring of the dominant individuals

Clutton-Brock, 2009c). The subordinates can be either individuals who have not dispersed from their natal group, or immigrants who have joined a group. In birds, the subordinates are often termed 'helpers at the nest'. Cooperative behaviours in vertebrates include feeding and protecting the young or other members of a group. Similar forms of help occur in some insect species, reaching their pinnacle in the eusocial insects (e.g. ants, bees, wasps and termites) where the subordinates become sterile workers and give up all chances of breeding independently.

Bacteria cooperate by producing public goods

Cooperation can be found in organisms that are not usually thought of from a social perspective, such as bacteria. The most common form of cooperation in microorganisms is the production of factors that are released from cells, and which provide a benefit to the local group of cells, such as acquiring nutrients for growth, making them analogous to what economists call 'public goods' (Fig. 11.7). This can even lead to bacteria living in cooperative 'slime cities', such as dental plaque and the scum around sink plugholes.

Cooperation also occurs between species. In such mutualisms, the most common form of cooperation is for one or both partners to provide a service or resource to the other. For example: in cleaner fish mutualisms, the cleaners remove parasites from their parasites; or in the legume–rhizobia interaction, the rhizobia bacteria provide nitrogen to their legumous host plants, whilst the plant provides carbon to the bacteria.

Free riding and the problem of cooperation

The problem of cooperation is why should an individual carry out a behaviour that benefits another individual?

The problem of cooperation is that cooperative behaviour can be exploited by free riders, which gain the benefits of others cooperating whilst avoiding the cost of cooperating themselves. This is famously illustrated by the Prisoner's dilemma model, which was originally developed to help us think about human behaviour, but provides a useful model to illustrate the problems of achieving cooperation in animal societies (Axelrod & Hamilton, 1981).

Imagine that two individuals are imprisoned and accused of having performed some crime together. The two prisoners are held separately and attempts are made to induce each one to implicate the other. If neither one does, both are set free. This is the cooperative strategy. In order to tempt one or both to confess (defect), each is told that a confession implicating the other will result in their release and a small reward. If both confess, each one is imprisoned. But if one individual implicates the other, and not vice versa, then the implicated partner receives a harsher sentence than if each had implicated the other. The pay-off matrix for this game is given in Table 12.1 with some illustrative numerical values. From a biological perspective, these values would represent the gain in fitness from the interaction (e.g. number of offspring gained).

In the Prisoner's dilemma game both individuals would benefit from mutual cooperation but both are tempted to cheat

Imagine player A finds another individual B who always cooperates. If A cooperates too it gets a reward of three, whereas if it defects it gets five. Therefore, if B cooperates, it pays A to defect. Now imagine player A discovers that B always defects. If A cooperates it gains nothing (the sucker's pay-off) whereas if it defects it gets one. Therefore, if B defects, it pays A to defect. The conclusion is that irrespective of the other player's choice, it pays to defect even though with both players defecting they get less (one) than they would have got if they had both cooperated (three). Hence the dilemma!

		Player B	
		Cooperate	**Defect**
Player A	**Cooperate**	$R = 3$ Reward for mutual cooperation	$S = 0$ Sucker's pay-off
	Defect	$T = 5$ Temptation to defect	$P = 1$ Punishment for mutual defection

Table 12.1 The Prisoner's dilemma game (Axelrod & Hamilton, 1981); the pay-off to player A is shown with illustrative numerical values

In other words, cooperation is not an evolutionarily stable strategy (ESS) because in a population of cooperators a mutant that defected would spread. Defect, however, is an ESS; in a population of 'all defect' a mutant cooperator does not gain an advantage. Any population with a mixture of heritable strategies will, therefore, evolve to 'all defect'. More generally the conditions for this conclusion to hold in the matrix in Table 12.1 are:

$$T > R > P > S \text{ and } R > \frac{(S+T)}{2},$$

which define the Prisoner's dilemma game. The problem is that while an individual can benefit from mutual cooperation, it can do even better by exploiting the cooperative efforts of others.

It is important to realize that the Prisoner's dilemma game is just an illustration of the problem of cooperation, and not a solution. If we want to find a solution to the problem of cooperation, we need to find situations where the lifetime pay-offs for cooperation or defection are not as given in Table 12.1, and hence where the Prisoner's dilemma does not hold.

Solving the problem of cooperation

How, then, can we account for the evolution of cooperative behaviour? As discussed in Chapter 11, one possible solution to this problem is that cooperation can be favoured by kin selection when it is directed towards relatives, and therefore provides indirect fitness benefits. However, as discussed above, we also need to be able to explain cooperation between non-relatives – in this case cooperation must provide some direct fitness benefit to the cooperator. The distinction here between direct and indirect fitness is whether a gene maximizes its transmission to the next generation by increasing either the fitness of the individual it is in (direct fitness) or another individual with a copy of the same gene (indirect fitness). Within these two broad categories, the explanations for cooperation can be divided further in a number of ways (Fig. 12.2; Sachs *et al.*, 2004, Lehmann & Keller, 2006; West *et al.*, 2007c).

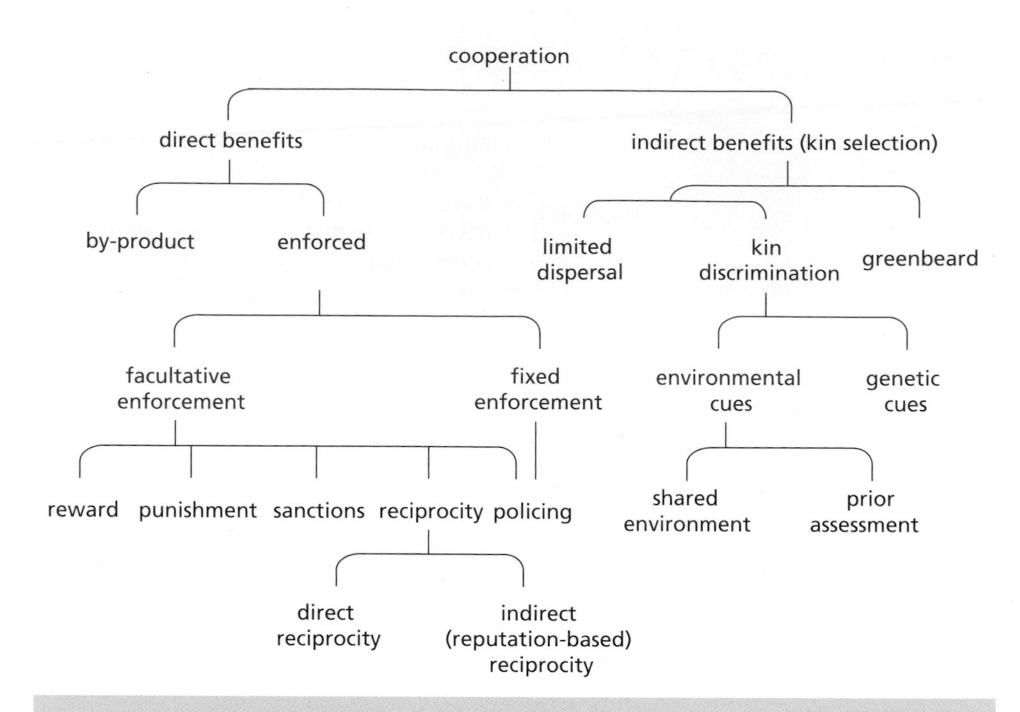

Fig. 12.2 A classification of the explanations for cooperation. Direct benefits explain mutually beneficial cooperation, whereas indirect benefits explain altruistic cooperation. Note that the mechanisms under direct and indirect can be classified in a number of ways, and that if interactions are between relatives, then mechanisms which lead to direct benefits can also lead to indirect benefits. The special case of greenbeards is not considered in this chapter, because they are likely to be of limited importance and have already been discussed in Chapter 11. From West *et al*. (2007b). Reprinted with permission of Elsevier.

Table 12.2 Four hypotheses for cooperation; kin selection can explain altruistic cooperation, while the other three can explain mutually beneficial cooperation

Explanation	Description
Kin selection	Helping individuals that share copies of the same gene
By-product benefits	Cooperation arises as a by-product of an otherwise selfish act
Reciprocity	Helping another individual because that individual will then help them back
Enforcement	Rewarding cooperation and/or punishing free riding

Cooperation can be explained by either direct or indirect benefits

In this chapter we consider how cooperation can be explained in four different ways: kin selection, by-product benefits, reciprocity and enforcement (Table 12.2). The first of these (kin selection) relies on indirect benefits, whereas the other three are different ways in which cooperation can provide direct benefits, and hence be mutually beneficial, rather than altruistic.

Kin selection

In Chapter 11, we explained how altruistic cooperation can be favoured between relatives. This idea is formalized by Hamilton's rule, which states that altruism will be favoured if $rB - C > 0$, where B is the benefit to the recipient, C is the cost to the actor, and r is the genetic relatedness of the recipient to the actor. In the context of the Prisoner's dilemma, cooperation can be favoured when individuals A and B are related, because the pay-off to player B (weighted by r) will then matter to the inclusive fitness of the player A.

Kin selection explains altruistic cooperation between relatives

Kin discrimination in long-tailed tits

One of the best examples of kin selection favouring cooperation comes from an extensive study by Ben Hatchwell and colleagues on the helping behaviour of long-tailed tits, *Aegithalos caudatus*, a small bird which breeds in most of Europe and Asia. In the non-breeding season, long-tailed tits live in fluid winter flocks containing on average 16 individuals, including overlapping generations of kin from one or more families, and also unrelated male and female immigrants (Hatchwell & Sharp, 2006). Each flock occupies a large non-exclusive range. In early spring, monogamous pairs form, with each pair occupying part of the range of their winter flock. All birds start the season by attempting to breed independently and there are no helpers associated with nests at this stage. However, many nests fail due to predation. Whilst failed breeders may make a second attempt at nesting, some instead go and help feed the chicks at other nests, leading to larger chicks with a substantially increased likelihood of surviving to the following year (Fig. 12.3a; Hatchwell *et al.*, 2004).

Helping provides clear benefits to chick survival in long-tailed tits

Using observational data from 52 helpers, Andy Russell and Ben Hatchwell (2001) found that 79% of helpers were closely related to one or both of the breeders that they were helping. They then went on to test a role of kin discrimination with a choice experiment, where potential helpers were approximately equidistant between two breeding nests, one belonging to close relatives and one to non-relatives. The results showed clear kin discrimination – in 16 out of 17 cases (94%), individuals chose to preferentially go and help at the nest of their relatives, rather than their non-relatives (Fig. 12.3b). Furthermore, an analysis of a long-term data set on the same population suggests that indirect fitness benefits are the major reason for the helping behaviour of long-tailed tits (MacColl & Hatchwell, 2004). A role for direct fitness benefits can be ruled out because individuals who help only rarely go on to breed in their later life, due to a high mortality rate between seasons and a low rate of successful breeding. This means that individuals tend to either gain direct fitness from breeding, or indirect fitness from helping, but rarely both.

Long-tailed tits preferentially go and help at the nests of close relatives

Individuals gain fitness through breeding or helping, but rarely both

As discussed in Chapter 11, kin discrimination can occur through direct genetic cues of relatedness, or more indirect environmental cues, such as prior association or shared environment. In long-tailed tits, a major cue for kin discrimination appears to be the 'churr', a contact call given frequently by both sexes for short-range communication during behaviours such as nest building or aggressive interactions (Sharp *et al.*, 2005). The churr calls of relatives are more similar and individuals are more attracted to nests where the recorded churr calls of relatives are played. Hatchwell and colleagues went on to investigate the relative importance of genetic and environmental influences on the development of the churr call with a cross-fostering experiment, where eggs where switched between nests. They found that foster siblings (unrelated individuals reared in the same nest) developed relatively similar calls that were just as similar as true siblings

Long-tailed tits distinguish kin from non-kin by an environmental cue learned as chicks – the churr call

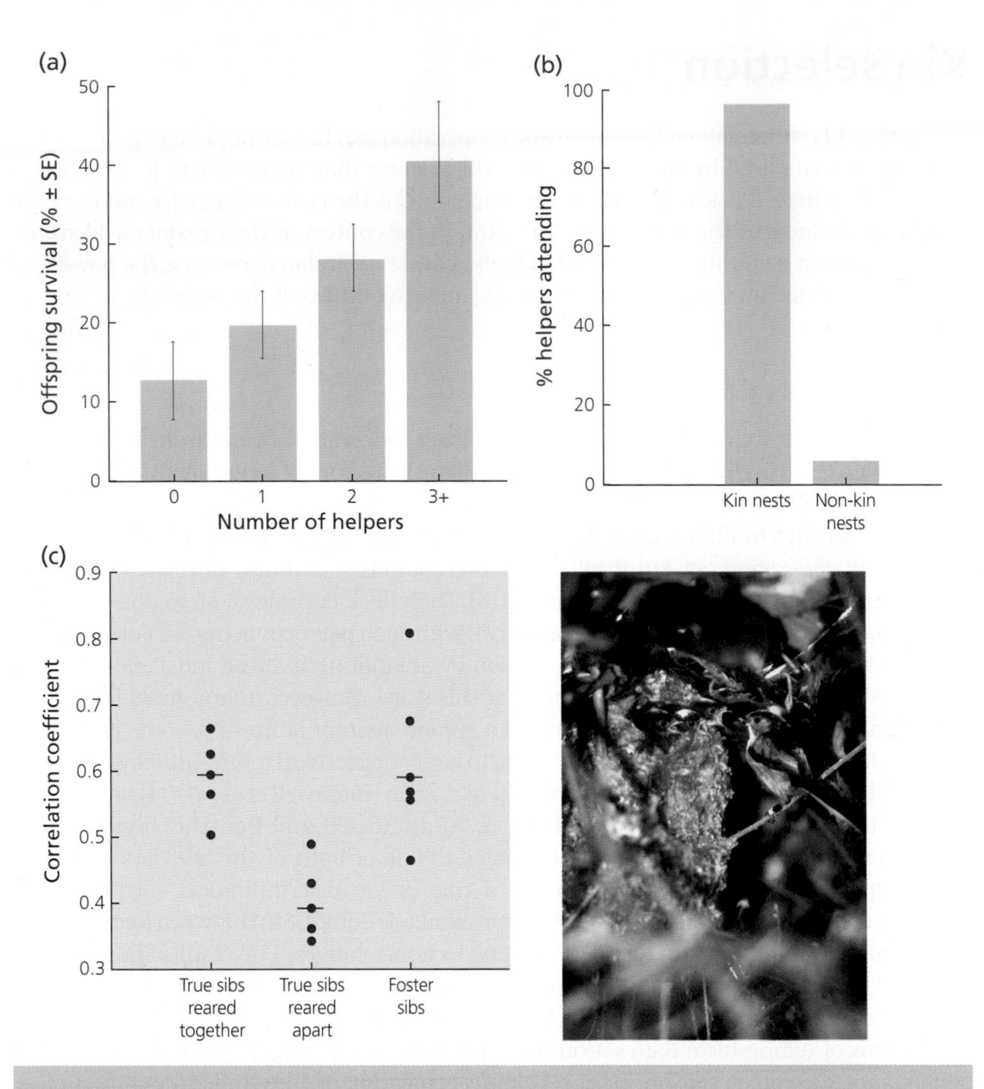

Fig. 12.3 Helping and kin discrimination in long-tailed tits. (a) The presence of helpers leads to a roughly linear increase in the rate at which offspring survive until next year and are recruited into the population. From Hatchwell *et al.* (2004). (b) Individuals that failed to breed, preferentially went and helped at nests of relatives. From Russell and Hatchwell (2001). (c) The churr calls of individuals were similar to those that they had been raised with, rather than those to who they were genetically related. From Sharp *et al.* (2005). Reprinted with permission from the Nature Publishing Group. Photograph of a helper feeding at a nest. Photo © Andrew Maccoll.

(siblings reared together) and significantly more similar than that of siblings who had been reared apart in different nests (Fig. 12.3c). Observations on these birds in later years showed that when individuals failed to breed successfully, and became helpers, they chose who to help on the basis of who they had been reared with, rather than genetic relatedness *per se*. Overall, these result show clear support for kin discrimination via an environmental cue driving altruistic cooperation.

Hidden benefits

In long-tailed tits, the benefits of cooperation are clear, with the presence of helpers leading to a significant increase in chick survival (Fig. 12.3a). However, in other species it can be harder to determine the benefits of cooperation, because they can be hidden and/or delayed. A nice example of this comes from a study by Andy Russell, Becky Kilner and colleagues on the superb fairy-wren, *Malurus cyaneus*, a small cooperative breeding passerine bird found in south-eastern Australia. Observational data showed that the presence of helpers did not lead to an increase in chick mass (Fig. 12.4a), so there was no obvious fitness benefit of helping to the chicks being raised (Russell *et al.*, 2007).

In the superb fairy-wren, the presence of helpers doesn't lead to an increase in chick size ...

Russell *et al.* went on to test whether this could be explained by the mother of the offspring reducing her reproductive effort when she had helpers. Consistent with this, it was found that when females had helpers they laid 5.3% smaller eggs with lower nutritional content (14% smaller yolk; Fig. 12.4b). This suggests that the benefit to chicks from helpers is exactly compensated for by the reduced investment into eggs by mothers. Russell *et al.* separated these effects of egg provisioning and rates of helping with a combination of multivariate statistical analysis and a cross-fostering experiment that moved eggs between nests with different numbers of helpers. This showed that in nests where the eggs had come from nests with the same number of helpers, and hence were the same size, the presence of helpers led to a significant increase in chick size (Fig. 12.4c).

... because mothers with helpers lay smaller eggs

This raises the question of why do mothers with helpers reduce their investment into eggs, such that this cancels the benefit of helpers to their offspring? One possibility is that the presence of helpers leads to greater competition for food, so the mothers have less resources to allocate to eggs. Another possibility is that females with helpers invest less in reproduction, in order to save resources for future breeding opportunities. The latter explanation was supported by the observation from a 16-year data set, showing that the presence of helpers led to a 11% increase in the probability that mothers survived to breed in the next year (Fig. 12.4d). Overall, this suggests that the benefits from helping accrue in the long term, by increasing the chance that breeders survive to breed again in future years. It would be extremely interesting to know if individuals are more likely to help in groups where they are more closely related to the breeding female.

Helping in superb fairy-wrens provides a delayed benefit to the breeders in the group

By-product benefit

In some cases, cooperation can provide a benefit as a by-product or automatic consequence of an otherwise 'self-interested' act. The idea here is that cooperation is always the best option from an individual or selfish perspective, but that this also provides a benefit to others.

This can be illustrated with a cooperative hunting game (Table 12.3), which Ken Binmore (2007a) has referred to as the Prisoners' delight. Imagine that two players have the opportunity to take part in a potentially cooperative endeavour, such as hunting for prey, and that taking part costs one unit of energy. If the hunt is successful, all food is shared between two players irrespective of whether they took part or not. If only one player takes part, the hunt is a mild success, such that the food returned

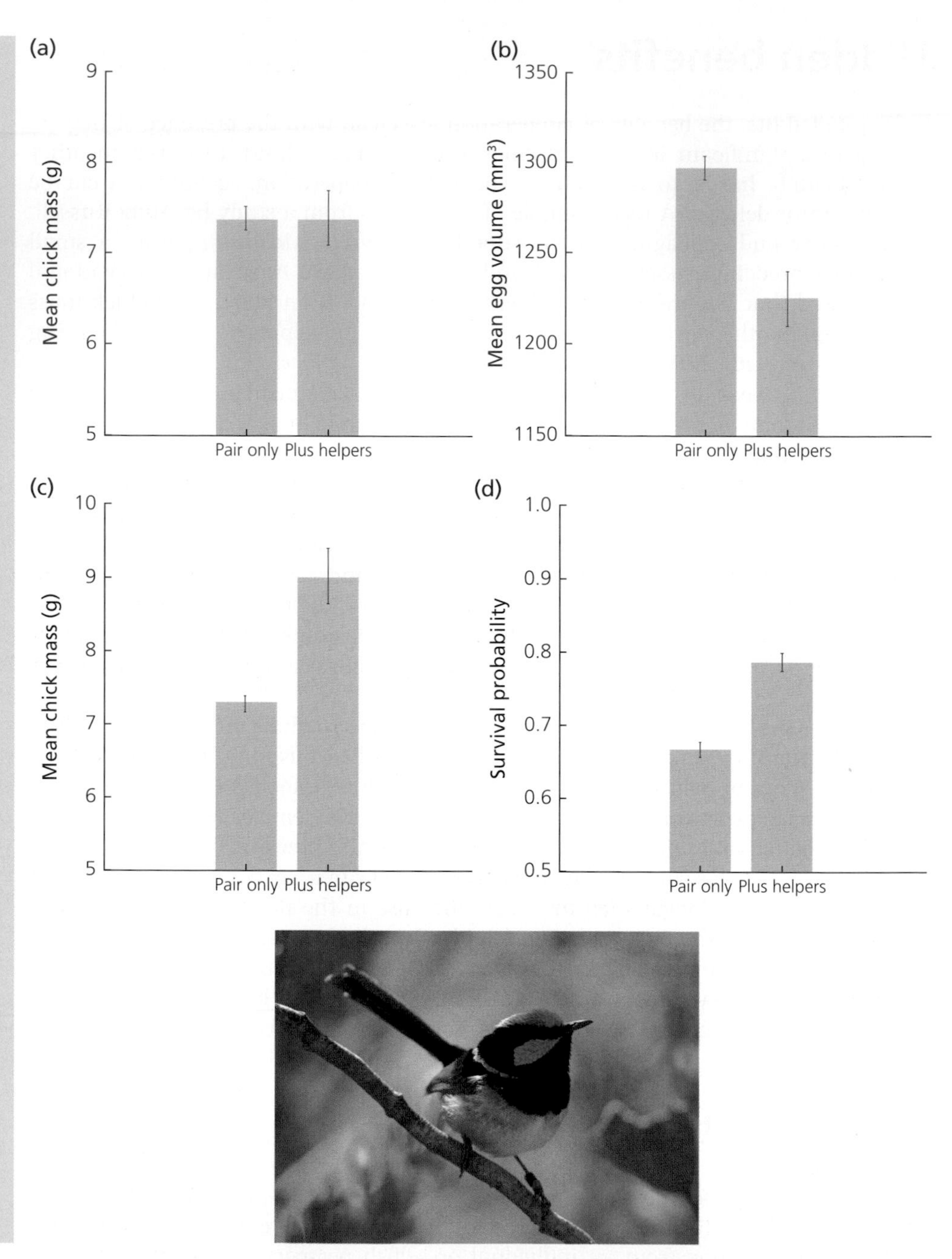

Fig. 12.4 Concealed helper effects in the superb fairy-wren (*Malurus cyaneus*). (a) The presence of helpers had no influence on the size of chicks reared, but (b) led to the dominant female laying smaller eggs. (c) In a cross-fostering experiment, where groups with same size eggs could be compared, the presence of helpers led to an increase in the size of chicks. (d) The probability of a breeding female surviving to breed in the next year was greater when she had received help. From Russell *et al*. (2007). Reprinted with permission from AAAS. Photograph of a male. Photo © Geoffrey Dabb.

equates to four units of energy, which is divided equally, giving each player two units. In this case, the hunter would get a pay-off of $2-1=1$ (the minus one is the cost of hunting) and the free rider would get a pay-off of $2-0=2$ (no cost of hunting). If both players take part, the synergistic benefit of cooperative hunting makes the hunt a huge success, such that the food returned equates to ten units of energy, which is divided

		Player B	
		Hunt (cooperate)	**Don't hunt (free ride)**
Player A	**Hunt (cooperate)**	$C=4$ Share from a hugely successful hunt	$S=1$ Share of a mildly successful hunt
	Don't hunt (free ride)	$F=2$ Free riding on a mildly successful hunt	$N=0$ No hunt

Table 12.3 A cooperative hunting game.; the pay-off to player A is shown with illustrative numerical values

equally, giving each player five units, and so an overall pay-off of $5-1=4$ after the cost of hunting is subtracted. With these pay-offs, taking part in the hunt (cooperation) is always the best option irrespective of what the other player does. Consequently, hunt is the ESS.

In a cooperative hunting game, hunt (cooperate) is the ESS ...

Hunt is the ESS, despite the fact that free riding on a hunt ($F=2$) is better than being the hunter free ridden on ($S=1$). To show why this is so, consider the extreme cases of when either hunting or not hunting are incredibly rare. If hunting is rare, then hunters will be with non hunters, and so their average pay-off will be one. Most non hunters will also be with non hunters and so their average pay-off will be just over zero (just over, because a small fraction will be with hunters, increasing their pay-off). So, when hunting is rare, it does better than not hunt and will invade the population. In contrast, if not hunt is rare, the non hunters will tend to be with hunters and so their average pay-off will be two, whereas hunters will tend to be with other hunters and so their pay-off will be just under four (just under, because a small fraction will be with non hunters). So, when not hunt is rare, it does worse than hunt, and cannot invade the population. This illustrates the general point that what matters for natural selection is how an individual (or strategy) does relative to the whole population, not just how it does relative to the partner (or subset of the population) with which it interacts (Grafen, 2007; see also table 15.1).

... even though it provides a benefit to others

Cooperative nest founding in ants

A remarkable example of by-product benefit, which comes with a gory ending, is provided by several ant species where unrelated queens join together to found nests, cooperating to excavate and build nests (Bernasconi & Strassmann, 1999). This behaviour occurs in a number of territorial species, where workers from mature colonies destroy new nests, and workers from newly founded nests steal brood from other newly founded nests. Cooperative nest founding provides clear benefits, with colonies founded by more queens growing at a quicker rate, which leads to them being better able to both raid their nests and defend their own nest from such raids. However, this cooperative behaviour is unstable, because the advantage of having multiple queens ends when adult workers emerge. Queens do not forage – instead they seal themselves in the nest and produce their first brood of workers from their body reserves (fat, protein and gylycogen obtained by digesting wing muscles). When the workers emerge, this ends

Unrelated queen ants join together and cooperatively form nests ...

... but later fight to the death

the period when brood production is dependent upon the body reserves of the queen. At this point, each queen no longer benefits from the presence of the others, and she can gain an enormous advantage from monopolizing reproduction. This leads to females who have previously shown no aggression to each other, fighting to the death.

Group augmentation in meerkats

By-product benefits can also be important in many cooperatively breeding vertebrates. Tim Clutton-Brock and colleagues carried out a long-term study on meerkats, a small (<1 kg) mongoose that is found in arid regions of southern Africa. Meerkats live in groups of up to 20 adults, accompanied by their dependent young. Each group is comprised of a dominant pair, subordinates of both sexes that were born in that group and, in some cases, subordinate males who have immigrated into the group. Successful breeding by meerkats is completely reliant on help from the subordinate members of the group, who help feed and guard the young at the burrow, while the rest of the group spends the day foraging elsewhere. This contrasts with species such as long-tailed tits, where helpers are a bonus that only occurs at some nests and not a requirement. The helping behaviours of meerkats are extremely costly – over the 12-hour period following the start of a full day babysitting, babysitters lost 1% of their body weight, while foraging members increased their body weight by 5.9% (Clutton-Brock *et al.*, 1998). Over the total time taken for a breeding attempt, the top babysitters lost up to 11% of their body weight.

Helping by meerkats is costly in the short term ...

Through increasing group size, the combined effects of which are sometimes called 'group augmentation', this helping behaviour can provide a future benefit to the babysitters in at least two ways. Firstly, a larger group size can be beneficial to all the members of the group, because large groups do better. For example, larger groups can watch out for predators more efficiently, spend a larger proportion of time foraging and are more likely to win territorial conflicts with other groups (Fig. 12.5a). This leads to mortality rates being lower in larger groups (Fig. 12.5b). Secondly, subordinate females, and immigrant subordinate males may inherit the dominant, breeding position; by helping they ensure that they will have helpers present in the future. One way of testing this idea is by examining whether there are sex differences in helping rates because future fitness benefits of helping would be greatest for the sex that is most likely to remain and breed in the natal group. Consistent with this, females are the most helping sex in meerkats, and the helping rate of males drops when they become adults and are about to leave the group (Fig. 12.5c).

... but can provide benefits in the future

Males and females help at different rates

The data from meerkats also illustrate how it can be difficult to separate the direct and indirect (kin selected) benefits of cooperation. The above discussion makes it clear that helping can provide direct fitness benefits to helpers. However, most members of a meerkat group are related to both the pups that they help raise, and the rest of the group (Fig. B11.3a). This means they gain indirect benefits through: (a) helping raise relatives and (b) subordinate relatives also gaining the benefits of increased group size. An implication of this is that the relative importance of various factors that favour helping may vary with both sex and age. For example, natal males may help due primarily to the indirect benefits of helping relatives, and the direct survival benefits of being in a larger group, whereas immigrant males may help due to the survival benefits of being in a larger group and to produce subordinates who would help them if they obtain dominance in that group.

Both direct and indirect fitness benefits can contribute to the evolution of helping behaviours

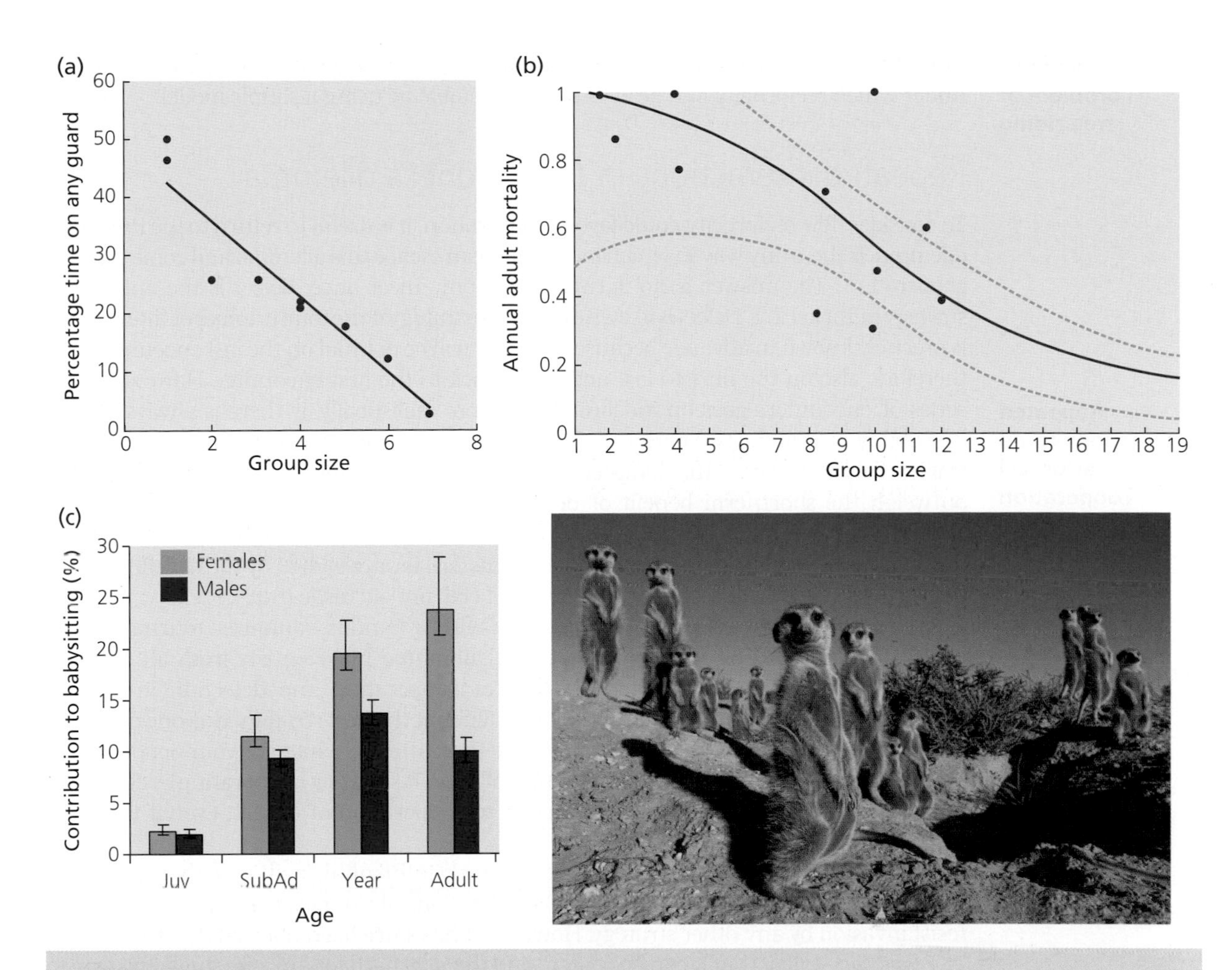

Fig. 12.5 Group augmentation in meerkats. In larger groups, (a) individuals spent less time on guard, looking out for predators and (b) the annual adult mortality rate was lower. From Clutton-Brock *et al.* (1999a, b). Reprinted with permission from AAAS. (c) Females help more than males, and the help provided by males drops when they are adults who are about to leave the group. From Clutton-Brock *et al.* (2002). Reprinted with permission from AAAS. Photo © Andrew Young.

Reciprocity

Paying back a favour in the future ...

In an extremely influential paper, Robert Trivers (1971) argued that if individuals preferentially help those that have helped them (reciprocal helping), then any short-term cost of helping another individual can be more than repaid by the help that is returned in exchange at a later point. For example, A helps B today and then B helps A tomorrow, as encapsulated by the well-known phrase 'you scratch my back and I'll scratch yours'. The problem for the evolution of reciprocity is the possibility of free riding. Because of the time delay between one individual gaining and the other doing so, B may accept help

... and the problem of free riding

from A today but refuse to repay the favour tomorrow. We will explore the conditions under which reciprocity can be evolutionarily stable by using a simple model.

Repeated interactions in the Prisoner's dilemma

To consider how reciprocity could favour cooperation, it is useful to return to the Prisoner's dilemma. Is there any way in which individuals can escape this dilemma and come to stable cooperation? The answer is no if two players only meet once; defect is the only stable strategy in Table 12.1. Defect is likewise the stable strategy if the total number of interactions is precisely known in advance because defection will be optimal on the last encounter and, therefore, also on the next-to-last and so on back to the first encounter. However, if the series of encounters goes on indefinitely or, more realistically, if there is always a finite probability, w, that the two players will meet again, then some form of cooperation may be stable. This is because the long-term benefits of cooperation between players could outweigh the short-term benefit of defecting. Although such behaviour is sometimes referred to as 'reciprocal altruism', this is misleading, because although helping is costly in the short term, it will only be favoured when this cost is outweighed by the benefit of being helped in the future, and so it is mutually beneficial, not altruistic (Box 11.1).

Repeated interactions could make reciprocal cooperation stable

Axelrod (1984) investigated this problem with a famous computer tournament in which he competed 62 different strategies, submitted by scientists from all over the world. These strategies involved mixtures of cooperation and defection in various sequences, and Axelrod's simulations suggested that the best strategy was one called 'tit for tat' (TFT): cooperate on the first move and thereafter do whatever your opponent did on the previous move. TFT is a combination of nice (it starts by cooperating), retaliatory (punishes defection) and forgiving (respond to cooperation of others, even if they had defected previously).

Tit for tat was the winner of Axelrod's computer simulations ...

Axelrod's computer experiment has been hugely influential, leading to the widespread assumption that reciprocity via TFT is an evolutionarily stable strategy (ESS) that can resist invasion by any other strategy. However, it has since been realized that the case for TFT was overstated, that it can be beaten, and that it is hard for any one single strategy to dominate. For example, other strategies which can perform well include 'suspicious TFT', which defects in the first interaction and then plays TFT, or 'tat for tit', which starts by not cooperating, and only switches to cooperation in response to the cooperation of others (Boyd & Lorberbaum, 1987; Binmore, 1994, 1998). Consequently, while reciprocity is theoretically possible, we shouldn't necessarily expect simple TFT rules in nature.

... but it can be beaten

Reciprocity in animals

Humans

Reciprocity seems common in the human world. One example comes from the Antwerp diamond market, where a relatively small group of experts trade diamonds. An individual will give another a bag of diamonds to take home to study. These bags are often very valuable indeed, but no receipts are exchanged and no contracts are signed. If anyone cheats, he is expelled from the community. So, here we have an example of reciprocal cooperation being sustained in a repeated game by use of a strategy that is termed GRIM, where an individual starts by cooperating, but then never cooperates again after a single defection. There is also an extensive literature using 'economic

GRIM reciprocity in humans

games' to test whether reciprocity can lead to cooperation between humans. In such experiments, individuals play games such as the prisoner's dilemma, and the more points they gain the greater cash reward they are given at the end of the game. It has been shown numerous times in such games that repeated interactions, with the potential for reciprocity lead to higher levels of cooperation (Carmerer, 2003).

Humans cooperate at higher levels in repeated interactions ...

If reciprocity is important in humans, then we would predict that individuals should be more likely to cooperate if they are being watched, because this would increase the chance of reciprocal helping. Evidence for this comes from an elegant field study by Melissa Bateson and colleagues (2006) in a university coffee room, where there was an honesty box to pay for tea and coffee. Above the honesty box, there was a notice that explained the system of payment for drinks. In alternate weeks, Bateson added a small (150 × 35 mm) image of either a pair of eyes or flowers. They found that people paid approximately three times as much for their drinks when eyes were displayed, rather than flowers, suggesting an automatic and unconscious tendency to cooperate in response to cues of being observed (Fig. 12.6). However, caution should be taken when interpreting this study as evidence for reciprocity, as the results could be explained equally by a number of other enforcement mechanisms, such as cooperating to avoid punishment. For example, economic games have shown that if humans are given the option of punishing free riders, that they do so, and that this leads to higher levels of cooperation (Fehr & Gachter, 2002). More generally, Trivers (1971) has suggested that it is selection for enforcement mechanisms such as reciprocity which provide an evolutionary basis for the human 'sense of fairness'.

... and in response to cues of being watched

Non-humans

Reciprocity has been suggested to be important in numerous cases, from cooperation between cleaner fish and their clients, the warning cries given by many birds in response to predators, the sharing of blood meals in vampire bats, fish inspecting predators and food sharing in chimpanzees. However, in all or at least most examples given, cooperation can usually be explained by a more simple mechanism, such as by-product benefit (Hammerstein, 2003; Clutton-Brock, 2009c). Consequently, whilst it used to be assumed that reciprocity was of widespread importance, it is now thought to be rare or even absent in animals. The point here isn't that reciprocity is not possible, just that despite lots of empirical attention there is a lack of conclusive examples, and so even in the best case, it is rare. This is illustrated by discussing two specific cases.

Vampire bats

Wilkinson (1984) studied a population of individually marked vampire bats, *Desmodus rotundus*, in Costa Rica. They roost during the day in stable groups of 8–12 that include mothers, their young, relatives and some non-relatives. Adults forage at night for animal blood but foraging is risky, and around a quarter of all bats return to the roost without having fed. These unfed individuals beg for food from those in their group that have obtained blood, and commonly receive some. Wilkinson discovered that regurgitation occurred only between close relatives or between unrelated individuals who were frequent roost-mates, and suggested that these latter cases involved reciprocity. However, in order to demonstrate that this really represent reciprocity, the following conditions are needed.

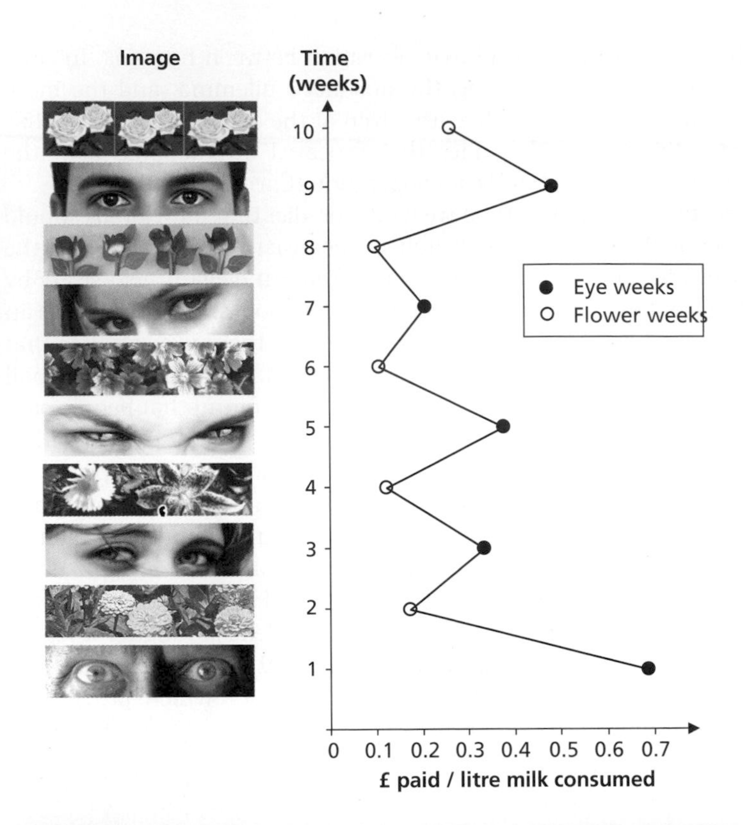

Fig. 12.6 The amount of money (pounds sterling) paid per litre of milk consumed as a function of week and the image placed next to a notice requesting payment. From Bateson *et al.* (2006). With permission of the Royal Society. Photograph shows a crime prevention poster produced by the West Midlands Police, UK, following this research. With permission of the West Midlands Police.

Vampire bats share food with both relatives and non-relatives

Repeated interactions occur …

(1) *Sufficient repeated pairwise interactions, so that there is the possibility for individuals to take turns helping each other.* In the field it was found that some unrelated individuals were constant companions in the roosts, sometimes for several years.

(2) *The benefit of receiving aid must outweigh the cost of donating it.* Fig. 12.7 shows that the bats lost weight with increasing time since their last meal. The decelerating curve means that a donation of a small amount of blood by a well-fed individual results in little cost in terms of time moved along the bottom axis towards the threshold of

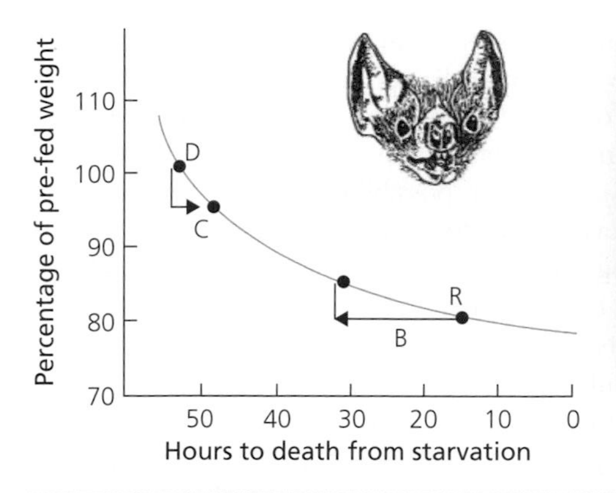

Fig. 12.7 In vampire bats, weight loss after feeding follows a negative exponential decline, with death from starvation occurring at 75 per cent of pre-fed weight at dusk. Therefore a donation of 5 per cent of pre-fed weight when at weight D should cause a donor to lose C hours but will provide B hours to a recipient at weight R. From Wilkinson (1984). Reprinted with permission from the Nature Publishing Group. Photo © Deitmer Nill/naturepl.com

death. However, this same amount can bring an enormous benefit to a starving individual, moving its position considerably to the left along the time axis. The blood meal, therefore, has little cost to the donor and great benefit to the recipient; indeed, it may save the recipient's life and enable it to survive until it has the chance to forage again for itself the next night.

... and the benefits far outweigh the cost ...

(3) *Individuals adjust their help to others dependent on the extent to which they have helped them previously – that is they must be able to recognize cooperators and cheats, feeding those that had fed them and refusing to feed previous recipients who fail to reciprocate.* Wilkinson carried out some clever experiments in the laboratory where he formed a group of bats, some individuals (all unrelated) coming from one roost and others (also unrelated) from another roost. In a series of trials, one bat, chosen at random, was removed and kept hungry while all the others had access to blood. The hungry bat was then reintroduced. It was found that 12 of the 13 regurgitations occurred between individuals from the same roost in the field, in other words individuals which were familiar with each other. Furthermore, the starved bats which received blood later reciprocated the donation significantly more often than expected had the exchanges occurred randomly. However, these experiments just show that individuals are more likely to feed bats that they usually associate with, which can have alternative explanations (as explained in the next section). Most importantly, it was not shown that individuals preferentially fed those who had previously fed them, or refused to feed those who had not fed them.

... but there is no evidence that cooperation is preferentially directed at more helpful individuals ...

(4) *Cooperation cannot be explained more simply via some mechanism that does not rely on reciprocation.* While Wilkinson's results are consistent with reciprocity, it is also possible to come up with a number of simpler explanations. One possibility

is that cooperation is favoured with group members because they will usually be close relatives (i.e. indirect benefits by limited dispersal; Foster, 2004). This possibility is emphasized by the fact that only five of the 98 observations of sharing between individuals of known genetic relatedness involved individuals less related than grandparent to grandchild ($r < 0.25$). Another possibility is that begging may represent a form of harassment that prevents well fed individuals settling into torpor for the day's rest, in which case feeding could be favoured for the direct benefit of reducing harassment. A final possibility is that there can be a direct benefit to keeping other individuals in the same group alive. For example, keeping other group members alive would increase the number of individuals that could be begged from in the future, providing a direct benefit analogous to group augmentation. Whilst this involves helping individuals because help could be obtained from them in the future, it does not rely upon preferentially directing help towards more helpful individuals.

... and simpler explanations are possible based on kin selection and by-product benefits.

Food and mate sharing in primates

Craig Packer (1977) studied a possible example of reciprocity in the olive baboon, *Papio anubis*. When a female comes into oestrus a male forms a consort relationship with her, following her around wherever she goes, awaiting the opportunity to mate. Sometimes pairs of males use identifiable signals to form coalitions to attack competitors that are consorting receptive females. This frequently led to the solicited male engaging the consort male in a fight and, while they are busy doing battle, the male who enlisted help goes off with the female! This was argued to be reciprocity, on the grounds that on a later occasion the roles are reversed, with the male who gave help being assisted by the one who received help previously. However, there is little evidence that individuals are more likely to 'allow' the other male to mate the female, if they have helped them in the past (condition 3 above). Furthermore, more recent observations suggests that after females have been 'liberated' from their previous consorts, both partners run for the female and do their best to monopolize her (Bercovitch, 1988), suggesting that cooperation can be explained more simply by the immediate direct benefits (condition 4) – forming a pair leads to an increased probability of mating with the female.

Male olive baboons pair up to compete for females

Another commonly cited example of reciprocity is the sharing of meat between male chimpanzees. Male chimpanzees hunt in groups for small and medium sized mammals, such as monkeys and small pigs. After a successful hunt, individuals will surround whichever animal has the prey item, begging for food. It is often argued that this represents a form of reciprocity, where individuals preferentially exchange meat with allies. However, more detailed studies have shown little evidence that individuals preferentially give meat to those that have fed them previously (condition 3). Instead, the probability that meat is shared is mainly dependent upon the amount of harassment that is received from beggars (Gilby, 2006), suggesting that cooperation is more simply explained as a mechanism to reduce harassment (condition 4).

Chimpanzees share meat

Enforcement

Our discussion of reciprocity considered how cooperation could be favoured between individuals who preferentially help each other ('help those that help you'). Whilst this may be relatively unimportant, there are many other ways to enforce cooperation, which have been referred to by terms such as 'punishment' 'policing', and 'sanctions' (Frank, 2003). The general point here is that is if there is a mechanism that rewards cooperators and/or punishes free riders (cheats), then this can alter the benefit/cost ratio of helping, and hence favour cooperation.

There are many way in which cooperation can be enforced

This can be illustrated by adding some form of enforcement to the Prisoner's dilemma. For example, imagine that an individual who defected could be punished by the other player. If the cost of this punishment outweighed the benefit of exploitation, then this punishment mechanism would remove any benefit of free riding, and hence make cooperation the best strategy, irrespective of what the other player does.

We discussed reciprocity first, and in such detail, even though it is a relatively unimportant enforcement mechanism, because it provides a good way of introducing the basic issues, and has attracted the most attention. Now, we will use specific examples to illustrate how some other enforcement mechanisms can work.

Infanticide and eviction in meerkats

One way to enforce cooperation is to harass potential helpers and, hence, reduce their ability to breed independently. Andy Young and colleagues investigated an example of this in meerkats. It had been previously observed that about one month before a dominant female gives birth she becomes aggressive towards some of the subordinates in the group and drives them from the group until her litter is born. Young *et al.* (2006) found that this aggression was not random – it was directed towards the subordinate females in the group who were pregnant, or most likely to become pregnant (older and less related to the dominant). This has a large negative impact on the likelihood that the subordinates will reproduce and so produce offspring that will compete for resources with the dominant's young. Whilst they are evicted, females spend their time alone or with other evictees, leading to elevated levels of stress hormones, a higher probability that a litter is aborted, a reduced rate of conception and a decrease in body weight (Figs. 12.8a–12.8c). In addition, this removes the subordinate females when they may be a danger to the offspring of the dominant female. In cases where pregnant subordinates are not evicted, they have been observed to attack and kill, and then usually eat, pups that were recently born to the dominant female (Young & Clutton-Brock, 2006). This effect is substantial, with the probability that the dominant's young survive being reduced by approximately 50% if one or more subordinate females are pregnant when the litter is born (Fig. 12.8d).

Dominant meerkats evict subordinates to prevent them breeding

Punishment in birds and fish

Another way in which cooperation could be enforced is by punishing individuals that do not cooperate. This increases the relative benefit of cooperation, because it leads to a cost of not cooperating. In many cooperative breeding birds, young individuals temporarily delay dispersal and independent reproduction, and instead remain in their natal nest to help rear other offspring. Raoul Mulder and Naomi Langmore (1993) tested whether punishment could play a role in explaining helping behaviour amongst superb fairy-wrens. To test this, they removed helpers and held them in captivity for

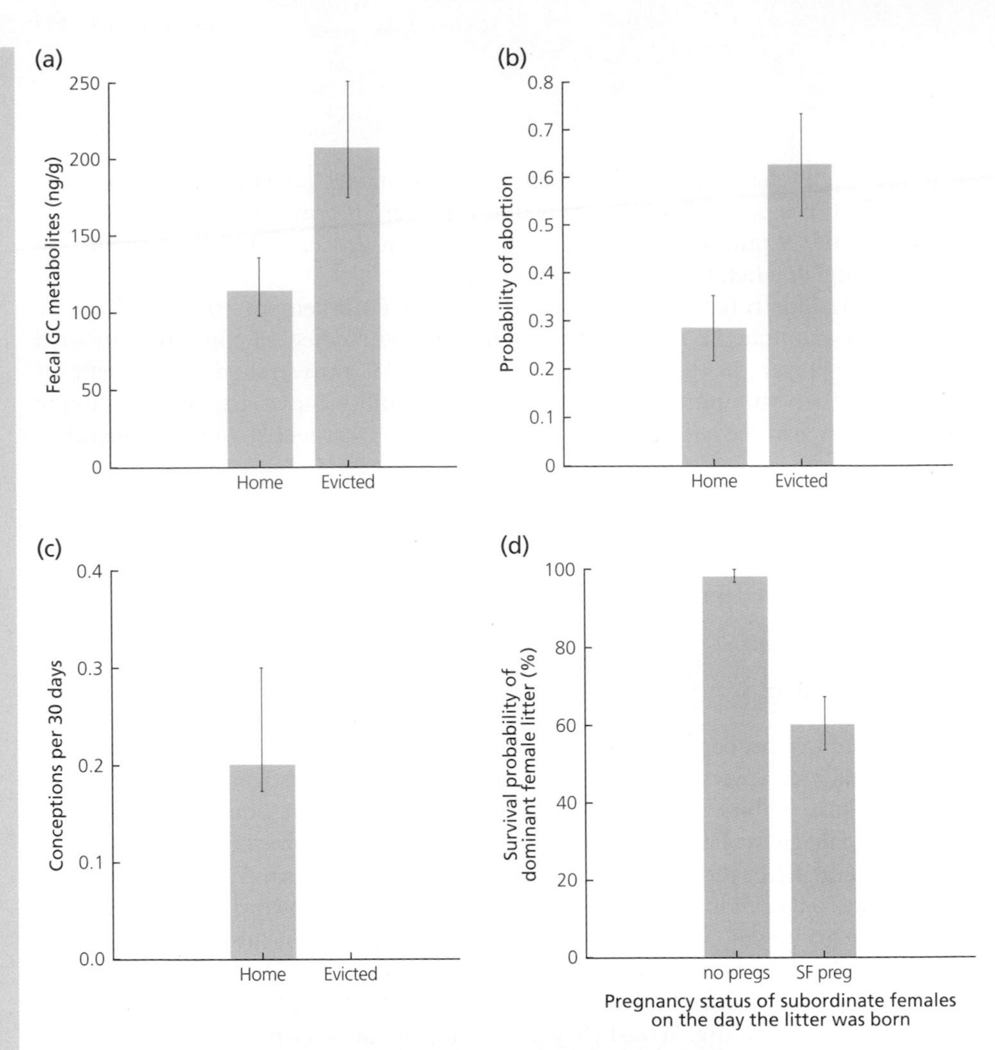

Fig. 12.8 Eviction in meerkats. Evicted females (a) show higher levels of glucocorticoid adrenhal hormone metabolites in their fecal samples, (b) have higher abortion rates and (c) have lower conception rates. From Young *et al.* (2006). (d) The litters of dominant females (SF) had lower survival rates when one or more of the subordinate females in the group was pregnant when the litter was born. From Young and Clutton-Brock (2006). With permission of the Royal Society. Photograph shows a dominant pinning down a subordinate. Photo © Andrew Young.

24 hours, before returning them to their natal group. When removal was carried out in the breeding season, hence preventing helping behaviour, the returning helpers were subjected to extreme harassment by the dominant male in the form of prolonged chases and pecking (9/14 cases). In contrast, when removal was carried out in the non-breeding season, when helping was not being performed, this never resulted in aggression towards the returned individual (0/12 cases). However, whilst these results are suggestive of a role for punishment, it has not also been shown that individuals adjust their behaviour in response to being punished by increasing their levels of helping.

Individuals who are removed and prevented from helping are punished upon their return

An example where the consequences of punishment have been elucidated is provided by the work of Redouan Bshary and colleagues on the cleaner fish *Labroides dimidiatus*. This species lives on coral reefs where it removes and eats ectoparasites from its 'clients', which refrain from consuming this potential prey while it performs the service. Although parasite removal and food acquisition are clearly beneficial to the client and cleaner, respectively, there is a conflict because the cleaners would prefer to eat the tissue or mucus of their hosts, which is costly to the host. Field observations suggested that when the cleaner does this, by taking a bite of its client, the client fish respond by aggressively chasing the cleaner fish and/or fleeing away (Bshary & Grutter, 2002). This punishment also led to a change in behaviour of the cleaner, with it making them less likely to take a bite of their client in future interactions. Bshary & Grutter (2005) then tested this role of punishment experimentally, by using Plexiglas plates to simulate the feeding opportunities offered by clients. They allowed cleaners to feed on mashed prawn or fish flakes from Plexiglas plates. The cleaners showed a strong preference for prawn over fish flakes, so Bshary & Grutter tested how the fish changed their feeding behaviour if they were punished for feeding on prawns, by removing the plate (to simulate fleeing) or by chasing the cleaner with the plate. They found that in response to both removal of or chasing by the plate, the cleaner fish adjusted their feeding behaviour and were more likely to feed on the food type that did not lead to this behaviour – fish flakes (Fig. 12.9).

Client fish punish cleaners who feed on them, rather than their parasites, by chasing them or fleeing away

Cleaner fish are more likely to feed on parasites, and less likely to feed on their clients, after they have been punished

Soybeans sanction non-cooperative bacteria

Another way to enforce cooperation is to terminate interactions with relatively non-cooperative individuals, so that cooperation is favoured to avoid such terminations. Toby Kiers and colleagues tested for the possibility of such 'sanctions' in the interaction between soybeans and the rhizobia bacteria that colonise their roots. Legume plants, such as beans and peas, contain nodules within their roots which house rhizobia bacteria that fix atmospheric nitrogen and then provide it to the plants which use it for growth and synthesis. Nitrogen fixation is energetically costly to the bacteria, reducing the resources that could be allocated to their own growth and reproduction, begging the question of why they carry out this cooperative behaviour. Kiers *et al.* (2003) forced the rhizobia to cheat (not cooperate) by replacing air (approximately 80% nitrogen, 20% oxygen) with a gas mixture (approximately 80% argon, 20% oxygen) that contains only traces of nitrogen. In this case, the lack of atmospheric nitrogen meant that the rhizobia could not cooperatively fix nitrogen. Kiers *et al.* repeated this experiment, forcing cheating at the level of the whole plant, one-half of the root system and at the

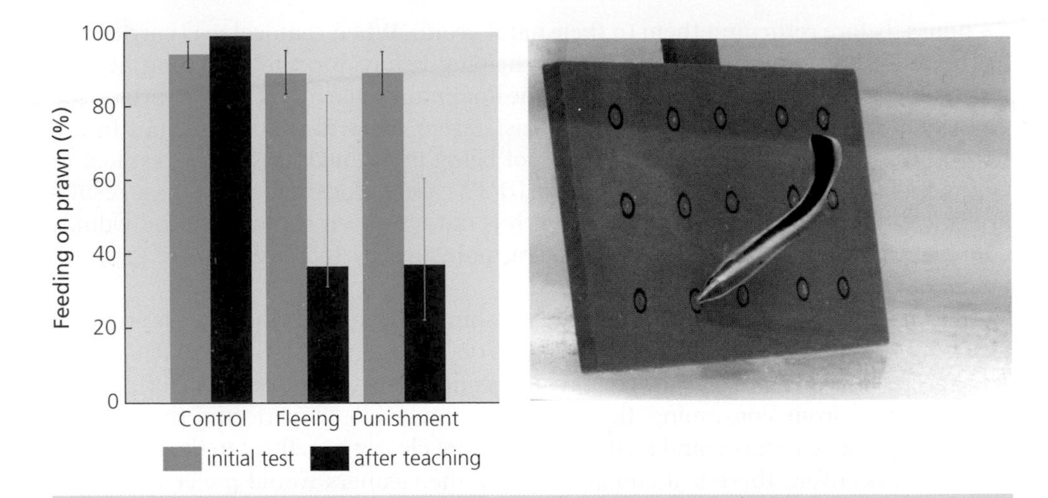

Fig. 12.9 Punishment and the cleaner fish *Labroides dimidiatus*. The figure shows the percentage of prawn items eaten from the Plexiglas plate by individuals during the initial preference test (light blue columns) and after the experimental treatment (dark blue columns). When the removal of prawns led to removal of the plate (to stimulate fleeing) or chasing with the plate (to simulate punishment), the cleaner fish were more likely to feed on the other food type, fish flakes. From Bshary and Grutter (2005). With permission of the Royal Society. The photograph shows an individual feeding on an experimental Plexiglas plate. Photo © Redouan Bshary.

If rhizobia in a root nodule do not supply nitrogen to their host plant, then the host cuts off the supply of resources to that nodule

level of individual nodules. They found that in all cases, when nitrogen fixation was prevented, it led to a large and significant decrease in growth of the bacteria (Fig. 12.10). Physiological monitoring showed that this was due to the plant reducing the supply of oxygen to nodules where nitrogen was not being supplied. Each nodule tends to be colonized by a single clonal bacterial lineage, and so the rhizobia are favoured to cooperate for a mixture of direct and indirect benefits, to avoid resources being cut off to both themselves and their clone-mates in the same nodule.

A case study – the Seychelles Warbler

In the above examples, we have often emphasized when cooperation leads to either a direct or kin selected benefit. However, we have also pointed out that multiple factors may play a role in a single species, sometimes in unexpected and even initially hidden ways. In this section, we illustrate this with a discussion of research on the Seychelles warbler (*Acrocephalus sechellensis*) by Jan Komdeur and colleagues, where the accumulating results of a long-term study have changed drastically the perception of why cooperation occurs. This species is like the long-tailed tit in that helpers are not required for breeding successfully, so only occur at some nests. When they do occur, the helpers aid in territory defence, predator mobbing, nest-building, incubation and feeding young.

Early work by Komdeur (1992) suggested that helping behaviour is driven by kin selection and a lack of vacant breeding territories. Helpers are offspring that were

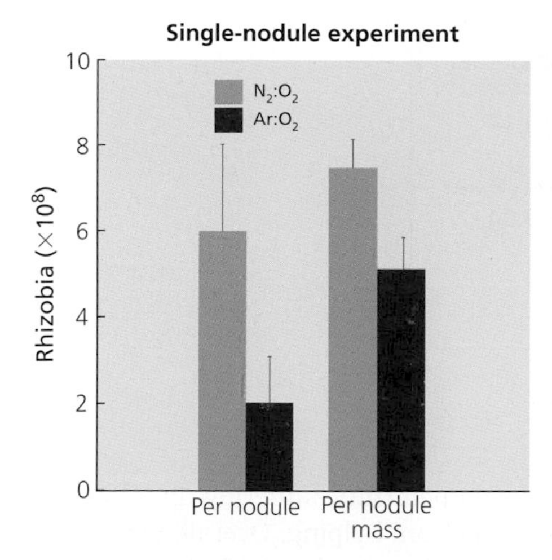

Fig. 12.10 Rhizobia which were able to fix nitrogen showed greater growth than rhizobia who were prevented from fixing. From Kiers *et al.* (2003). Reprinted with permission from the Nature Publishing Group. Photograph shows a split root experiment, where one half of the root system was supplied with air and the other was supplied with a gas mixture where nitrogen had been replaced with argon. Photo © Ford Denison.

previously produced on the territory where they help, so it was assumed that they would often be helping raise full siblings to whom they would be relatively highly related ($r=0.5$). A role of breeding territory availability was suggested by observations following a conservation programme in the 1970s that led to a spectacular rise in bird numbers on Cousin Island, from 26 to approximately 300 birds. During this time, as the population increased, a number of individuals became helpers at the nest, rather than attempt independent breeding. One way of thinking about this is that habitat saturation prevents individuals from breeding, so drastically reduces the cost (C) of helping – helping relatives can then be favoured to make the best of a bad situation (as with long-tailed tits). Komdeur (1992) confirmed the role of habitat saturation experimentally by removing 58 birds from Cousin to the previously unoccupied Aride and Cousine islands. As predicted by the habitat saturation hypothesis, all the vacancies created on Cousin Island were filled immediately, some within hours, by helpers from other territories, and all the birds moved to unsaturated new islands formed independent breeding pairs, with none acting as helpers.

Helping can be favoured due to a lack of opportunity to breed independently

When territories become available helpers move to fill them and breed independently

However, the idea that cooperative breeding in the Sychelles warbler was driven primarily by kin selection was overturned by a genetic analysis of who breeds and within-group relatedness, using microsatellite markers (Richardson *et al.*, 2002). These analyses revealed three big surprises. Firstly, 40% of the offspring were sired by males from other groups, so the chicks being raised are not usually full siblings of the helpers. On average, the relatedness between helpers and chicks was $r=013$, making the indirect benefits of helping much less than previously assumed. Secondly, female helpers often laid eggs in the nest, producing an average of 0.46 offspring per subordinate per year. Consequently, female subordinates gain substantial direct benefits from being helpers at

Almost half the offspring are sired by males from other groups

Table 12.4 The direct and indirect benefits of being a helper in the Seychelles warbler, as measured in offspring equivalents (Richardson *et al.*, 2002)

	Female helper	Male helper
Direct fitness (own offspring)	0.46	0.14
Indirect fitness (through raising primary breeder's offspring)	0.07	0.04

Both male and female helpers breed on the territory where they are helping

the nest. Thirdly, male helpers can sometimes gain paternity, fathering an average of 0.14 offspring per year, so also gain direct benefits from helping. Overall, when these different effects are added up, the direct benefits of being a helper far outweigh the indirect benefits (Table 12.4).

These results could give the impression that direct fitness alone drives the evolution of helping in the Seychelles warbler. However, more recent work has shown that this is not the whole story, because even though the indirect component is small, selection has still favoured individuals to maximize their indirect fitness, with kin discrimination. Richardson *et al.* (2003) examined whether subordinate Seychelles warblers preferentially directed their help at closer relatives by following the provisioning behaviour of 32 subordinate helpers on 23 different territories. Overall, they found that females subordinates fed closer relatives at a greater rate, but that male subordinates did not (Fig. 12.11a). It is not known why kin discrimination does not occur with male helpers, although helping by males is rare in this species. The mechanism of kin discrimination in females appears to be that helpers only help feed young when they are being raised by the same female that raised them (Fig. 12.11b), which should be a relatively reliable cue for kin discrimination, indicating that they are helping raise a sibling. In contrast, female subordinates do not adjust their level of helping in response to the presence of the male that raised them (Fig. 12.11b). The presence of the same male is likely to be a much less reliable cue of relatedness, because females often mate with males from other territories, making individuals less likely to share the same father. However, these results are observational and so it is hard to disentangle different cues of relatedness. Ideally, cross-fostering experiments should be carried out, that would allow the extent and mechanism of kin discrimination to be examined in greater detail, as described above with long-tailed tits.

Female subordinates preferentially help closer relatives ...

... by directing help towards young produced by their mother

Manipulation

Before concluding, a final complication is that some behaviours appear to be cooperation but, in fact, turn out to be manipulation by the recipient (Dawkins, 1982). This is most obvious, for example, when a parent bird feeds the offspring of a brood parasite, such as a cuckoo. The host gains nothing from this; it has simply been tricked by the cuckoo into feeding the wrong species (Chapter 4). Less well known, but equally striking are lycaenid

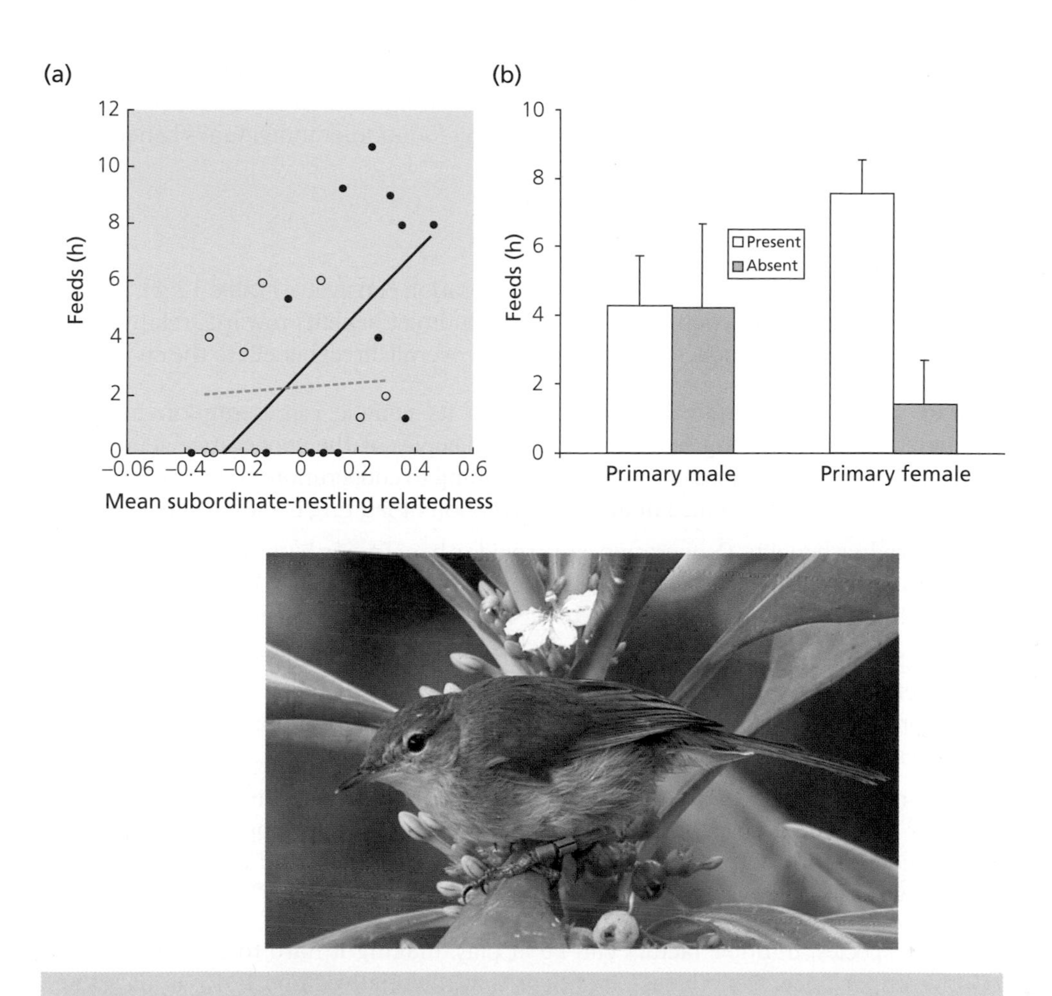

Fig. 12.11 Cooperation and kin discrimination in the Seychelles warbler. (a) The provisioning rate of female subordinates (filled circles and solid line) showed a positive relationship with relatedness to the nestlings being fed, whereas the provisioning rate of male subordinates (empty circles and dashed line) did not. (b) The provisioning rate of female subordinates was significantly higher when the dominant female at the nest was the dominant female at the time of their birth, but showed no relationship with the identity of the dominant male. From Richardson *et al.* (2003). Photo © Martijn Hammers.

butterfly larvae, which ant workers carry into their nests and then feed, thanks to the butterflies ability to mimic the chemical scents and sounds of ants (Barbero *et al.*, 2009).

Individuals may be tricked into helping others

Manipulation also occurs within a species. For example, some female birds lay eggs in the nests of conspecifics, thus avoiding the costs of incubation and parental care (Yom-Tov, 1980). Female starlings, *Sturnus vulgaris*, who 'dump' eggs in other female's nests first remove a host egg before laying their own. They then deposit the host egg on the ground nearby. It was at first thought that these eggs which appeared on the ground were laid by females who were unable to get back to their nests in time to lay normally! Then, when eggs were marked in nests as they were laid, it was found that the eggs which appeared on the ground were often marked ones, in other words eggs which had been removed from nests (Feare, 1984). In this case, just as with interspecific brood parasitism, the host is being tricked into behaving for another individual's benefit.

Summary

We have distinguished four ways in which cooperation can evolve (Table 12.2). The first (kin selection) relies on cooperation providing indirect benefits towards relatives, and the other three rely on cooperation providing an overall direct benefit to the cooperator.

(1) Kin selection. An individual can increase its genetic representation in future generations by helping relatives, who share copies of the cooperative gene. There are numerous examples of kin selection leading to cooperation, such as cooperative breeding in the long-tailed tit and the formation of fruiting bodies in slime moulds.

(2) By-product benefits. Cooperation can provide a benefit, as a by-product, or automatic consequence, of an otherwise 'self-interested' act. Examples include cooperative nest founding in ants and helping to increase group size in meerkats.

(3) Reciprocity. Reciprocal cooperation can be favoured, if individuals preferentially direct help towards those that have previously helped them. Although this is a theoretically appealing idea, and likely to play a role in humans, it is thought to be generally unimportant in other animals.

(4) Enforcement. If cooperation is rewarded and/or free riding punished, then this can alter the benefit/cost ratio of helping, and hence favour cooperation. Examples include eviction in meerkats, punishment of cleaner fish and the sanctioning of rhizobia by soybeans.

In many species, multiple factors can be at play, making it hard to disentangle their relative importance. Nonetheless, it is clear that the relative importance of direct and indirect benefits varies hugely across species.

Further reading

Lehman and Keller (2006) provide a comprehensive review of the evolutionary models that have been developed to explain cooperation, followed by 15 commentaries discussing their overview. Sachs *et al.* (2004) and West *et al.* (2007b) discuss how different explanations can be important in different organisms. There are a variety of reviews focusing on cooperation in specific taxa, including birds (Koenig & Dickinson, 2004; Hatchwell, 2009), mammals (Clutton-Brock, 2009b), primates (Silk, 2009), fish (Taborsky 1994) and microbes (West *et al.*, 2006). Classic early works on birds are Brown (1987) and Emlen and Wregge (1988, 1989). Rubenstein & Lovette (2007, 2009) show by comparative analyses that cooperative breeding in African starlings is associated with temporally variable environments, and that increased female- female competition for reproductive opportunities in these cooperative breeders has selected for more ornamented females and hence less sexual dimorphism. Arnold & Owens (1998, 1999) discuss the ecological and life history correlates of cooperative breeding in birds. The importance of queuing for dominance, in both insects and vertebrates, is reviewed by Field and Cant (2009). The social insects are discussed in more detail in Chapter 14.

Kin discrimination occurs in some cooperative breeding vertebrates, such as long-tailed tits and the Seychelles warbler, but not in others, such as meerkats and kookaburras. Cornwallis *et al.* (2009) show that this variation can be explained by variation in the relative benefit of kin discrimination. Specifically, kin discrimination is more likely when helping provides a greater benefit, and less likely when relatedness is higher within groups (i.e. when indiscriminate helping within group will be directed at relatives, so there is less to gain from discrimination).

Another potential example of by-product benefits is provided by the wasp *Polistes dominulus*, where genetic analyses of natural nests have showed that 15–35% of subordinates were unrelated to the dominant female. In this species, the helping behaviour of these unrelated subordinates appears to be explained by the direct fitness benefit obtained from nest inheritance (Queller *et al.*, 2000; Leadbeater *et al.*, 2011).

Clutton-Brock (2009c) discusses the lack of empirical support for the importance of reciprocity. Boyd and Richerson (1988) show that there are theoretical problems for reciprocity when interactions are between more than two individuals. Stevens and Hauser (2004) discuss why psychological mechanisms may prevent reciprocity in animals. Raihani *et al.* (2010) provide an example of punishment across the sexes in cleaner fish. Jander and Herre (2010) provide another example of enforcement in a between species mutualism – fig trees sanction fig wasps that provide poor pollination. Frank (2003) provides an overview of enforcement.

TOPICS FOR DISCUSSION

1. Discuss the relative merits of long-term field studies and field experiments for evaluating the costs and benefits of cooperation.
2. It is clear that behaviours such as punishment or sanctions can favour cooperation. However, if they are costly to perform, how can natural selection favour such enforcement mechanisms?
3. Read Krams *et al.* (2008), Russell and Wright (2009) and Wheatcroft and Krams (2009) and discuss whether the mobbing behaviour of pied flycatchers is reciprocal cooperation. What experiments would you carry out to resolve this controversy?
4. Why are cooperation and sexual selection so rarely studied in the same species (Boomsma, 2007)?
5. Discuss where the five explanations for cooperation given in Nowak (2006) fit onto Fig. 12.2 or into Table 12.2.
6. 'A whole generation of scholars swallowed the line that the Prisoners' dilemma embodies the essence of the problem of human cooperation' (Binmore, 2007, p. 18). Discuss.
7. Do any non-human organisms have a 'sense of fairness'?
8. Discuss the potential medical applications of bacterial behavioural ecology (Andre & Godelle, 2005; Brown *et al.*, 2009).
9. Discuss whether studies of human behaviour when playing economic games can be applied to help tackle the problem of global warming (Milinski, 2006).
10. MacLean and Gudelj (2006) show how the free rider problem arises with respect to the metabolism of sugar by yeast. How is the problem solved?
11. Can the soybean-rhizobia interaction be thought of as reciprocity?
12. Does free-loading occur in the fruiting body behaviour of microbes (Strassmann *et al.* 2000; Velicer *et al.* 2000; Buttery *et al.* 2009)?

Photo © Alex Wild

CHAPTER 13

Altruism and Conflict in the Social Insects

The social insects

The problem

Two problems: the evolution of sterility and the evolution of specialized castes

Cooperation and helping in vertebrates pales into insignificance beside what happens in the social insects. In these insects apparent self-sacrifice reaches the point where large numbers of individuals are completely sterile; they never reproduce themselves but instead spend their whole adult lives devoted to rearing the young of others. This is ultimate altruism! As Darwin himself and many other biologists since his time realized, this presents a real paradox, for if natural selection favours traits that increase the genetic contribution to future generations, how can it lead to the development of totally sterile individuals that never reproduce? What is more, these sterile individuals are often specialized for various tasks associated with helping (Figs. 13.1 and 13.2). This raises a further problem: if workers do not reproduce, how can their specialized traits evolve? In the last two chapters we have seen how altruistic cooperation can be favoured by kin selection, when cooperation occurs between relatives. However, one of the strengths of kin selection theory is that it not only predicts cooperation, but also when and why conflicts can arise in social groups. In this chapter we will consider to what extent kin selection can be used to understand how sterile castes and helping have evolved in the social insects, as well as conflicts that can occur within colonies, over the sex ratio and who breeds.

The definition of 'social insect'

What exactly is meant by the term 'social insect'? To be more precise, this chapter is largely about the *eusocial insects*, which Wilson (1971) originally characterized by three features: (i) they have cooperative care of the young involving more individuals than

An Introduction to Behavioural Ecology, Fourth Edition. Nicholas B. Davies, John R. Krebs and Stuart A. West.

Fig. 13.1 Eusocial species exhibit considerable variation within species between the different castes. (a) Castes of the carpenter ant, *Camponotus discolor*: male (left), queen (right) and worker (bottom). Photo © Alex Wild. (b) The three female castes of the leafcutter ant *Acromyrmex echinatior* on their fungus garden – a small worker (garden maintenance and brood nursing), large worker (foraging and defence) and winged virgin queen (who will disperse and mate during a nuptial flight, before shedding her wings and founding a new colony. Photo © David Nash. (c) A queen of the termite *Macrotermes bellicosus*, in her royal chamber, with the king and numerous members of two worker castes (major and minor). Photo © Judith Korb. (d) *Pheidologeton affinis* marauder ants have one of the most pronounced size differences among workers. This photograph shows a supermajor worker and several minor workers. Photo © Alex Wild. (e) In the social aphid *Colophina arma*, females develop into either soldiers (left) or reproductives (right). Photo © Harunobu Shibao. (f) In the eusocial Australia gall thrips, *Kladothrips morrisi*, a female (right) founds the gall, and some of her offspring develop into soldiers (left) that defend the gall from invaders. Photo © Laurence Mound.

Fig. 13.2 The existence of different castes in the social insects has allowed evolution to produce an amazing range of morphs that are specialized to particular tasks. Spectacular soldier morphs include: (a) *Zootemopsis nevadensis,* a dampwood termite whose large jaws are used to fight against competing colonies of the same species; (b) Nasute termite soldiers (Termitidae) can squirt a noxious sticky substance out of their snouts; (c) army ant (*Eciton* burchelli) soldiers have hooked mandibles to protect against vertebrate predators (humans have been known to use these jaws as 'stitches' to seal wounds); note the smaller worker – this species has a relatively complex caste system with at least four types of workers. Other specializations include: (d) soldiers of the turtle ant *Cephalotes varians*, which use the bizarre disc on their head as a living door to block the nest entrance; (e) the mandibles of the leaf cutter ant *Atta texana* are used to cut through leaves with a repeated scissoring motion, where the leading mandible is anchored into the leaf and pulls the trailing mandible to make the cut; (f) replete workers of the honeypot ant, *Myrmecocystus mexicanus,* become engorged with food and hang from the ceilings of chambers deep underground, acting as 'living storage vessels'. All photos © Alex Wild.

just the mother; (ii) they have sterile castes, and (iii) they have overlap of generations so that mother, adult offspring and young offspring are all alive at the same time. More recently, Crespi and Yanega (1995) have argued that this definition is too vague from an evolutionary perspective and can include species which are better characterized as cooperative breeders. They suggest that the key factor which should define eusociality is the presence of specialized castes, where different groups of individuals become irreversibly behaviourally distinct at some point prior to reproductive maturity. Individuals of one caste have higher rates of reproduction (breeders) and they are helped by at least one or more other caste members (helpers). This definition makes no use of overlapping generations, the presence or absence of which need not be linked to the level of altruism in or complexity of a society.

Eusocial insects have castes

Crespi and Yanega (1995) and Boomsma (2007, 2009) also stressed the special importance of whether eusociality is obligate. The key factor here is whether castes are permanently fixed or can be switched, and whether full reproduction is still possible for at least some of the helpers. If individuals of a caste retain, throughout their life, the ability to perform the full range of behaviours open to all the castes (including reproduction), then this is termed totipotency, literally *total potential*. In obligately eusocial societies, totipotency has been lost, so castes are permanently fixed. This is special from an evolutionary perspective because it leads to a complete mutual dependence, with the breeding caste dependent on the help of the helping caste (or castes), and the helping caste being dependent on the presence of some breeders to help. This contrasts with cooperative breeding or facultatively eusocial species, where at least some of the workers are also able to breed if the opportunity arises.

Obligate eusociality involves permanent castes

Up until the mid 1970s it was thought that eusociality only occurred in the social Hymenopetra (ants, bees and wasps) and the termites. Since then a number of other eusocial species have been found, including other insects (aphids, gall-forming thrips, *Austrolplatypus* ambrosia beetles; Aoki, 1977; Crespi, 1992; Kent & Simpson, 1992), sponge-swelling shrimps (Duffy, 1996) and even two mammals, the naked and Damaraland mole-rats (Jarvis, 1981). The number of times that eusociality is thought to have evolved depends upon how exactly it is defined (and the quality of phylogenetic trees!). For example, in the social Hymenoptera, while it is sometimes argued that eusociality has evolved up to 11 times, only three to five of these represent obligate eusociality. These are once in the ants (Brady *et al.*, 2006), once or twice in the bees (Cameron & Mardulyn, 2001; Danforth *et al.*, 2006), once in the vespine wasps (Hines *et al.*, 2007) and possibly once in the polistine wasps (Boomsma, 2009). Amongst the other cases, obligate eusociality is limited to the termites (one to three origins; Inward *et al.*, 2007; Boomsma, 2009) and possibly the aphids, thrips or Ambrosa beetles (Boomsma, 2009). However, in these last three cases, eusociality has not led to the successful radiations that have occured in the eusocial hymenoptera and termites.

Eusociality has been observed in several taxa ...

... but obligate eusociality has only been confirmed in the eusocial hymenoptera and the termites

The importance of social insects

The social insects are not only important because of their central role in the attempts of evolutionary theorists to understand the origin of altruism, but they are also extremely impressive in terms of their natural history. In a persuasive piece of salesmanship E.O. Wilson (1975) advertises that there are more than 12 000 species of social insects

Diversity of social insects

in the world, which is approximately equivalent to all the known species of birds and mammals. The staggering natural history of social insects can be illustrated by the following small sample of facts. In terms of size, a colony of African driver ants (*Dorylus wilverthi*) may contain up to 22 million individuals weighing a total of 20 kg. In terms of communication, the honeybee dance language, in which successful foragers tell other worker bees about the direction and distance to a food source, provide a rare example of a communication system in a wild animal where an abstract code (the speed and orientation of the dance) is used to transmit information about remote objects (see Fig. 14.14). In terms of feeding ecology, the diets of social insects include seeds, animal prey, fungus grown in special gardens on collections of leaves or caterpillar faeces and the excreta ('honeydew') of tended herds of aphids. The impact of this feeding can be great – for example, in some areas of Latin American tropical forest, leaf-cutter ants are the main herbivores, eating over 5% of the leaves produced each year (Leigh, 1999).

Specialized worker castes

Social insect colonies are often populated by individuals specialized to perform different tasks (so-called castes). Sometimes castes have bizarre morphological modifications to help them carry out their jobs (Fig. 13.2). For example, the head of soldier termites of the species *Nasutitermes exitiosus* is modified into a 'water pistol' used for squirting defensive sticky droplets at enemies, while the head of soldiers in the ant species *Cephalotes varians* has a disk on top which fits neatly into the nest entrance to keep out intruders.

In the following sections, we will first describe the life history of one example of a eusocial insect to provide a background, then discuss the various genetic and ecological factors which have been put forward to account for the evolutionary origins of sterile castes, before going on to also discuss how conflicts are expressed and resolved within colonies.

The life cycle and natural history of a social insect

In *Lasius niger* a single queen founds the colony

Lasius niger is a species of ant commonly found in Europe in woodlands, farmland and gardens (Fig. 13.3). It builds a nest, the chambers of which are excavated in the ground underneath flat stones or in open soil. The nest is started by a single fertile queen. She is fertilized during a 'nuptial flight' in July or August in which very large numbers of winged reproductive females and males swarm around in the air and copulate (only the sexual forms can fly, and then only at this stage in their life cycle). The queen stores the sperm obtained on this nuptial flight to use throughout her life, which may be many years. After the nuptial flight, the queen loses her wings and spends her first winter sealed inside the nest chamber, which she has built by excavating a hole in the ground. During the following spring, eggs which she has laid develop into larvae and mature into adult workers (the typical ants one sees scurrying around near ants' nests) before the autumn.

In the Hymenoptera workers are female

Up to the time of maturity of the first clutch of 5–20 workers, the queen survives on her own fat reserves and breaks down her flight muscles (which are not needed anymore) to provide the proteins to produce trophic eggs that the developing larvae eat. When the workers mature they start to care for their younger siblings and collect food for them and their mother. The workers are female, but they are sterile. They never develop wings,

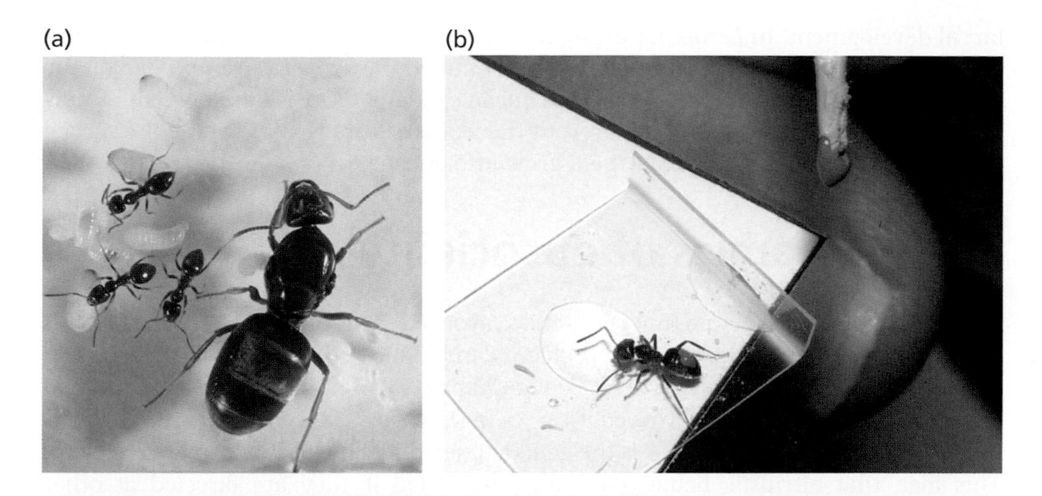

Fig. 13.3 (a) An incipient colony of *Lasius niger* with the larvae, pupae and first workers that a founding queen produces after about three months, solely from her own body reserves and a bit of drinking water. Photo © David Nash. (b) As with many insects, dots of paint can be used to distinguish individual ants in colonies being observed. Photo © Francis Ratnieks.

their ovaries do not mature and they never take part in a nuptial flight. In successive years the colony and its nest continue to grow, until after some years it contains several thousands of workers and still a single queen who lays the eggs. At this stage the colony starts to produce a new cohort of reproductives, winged females and males that eventually leave the colony on their nuptial flights. The old colony may continue reproducing in this way for another decade, but as soon as the old queen dies and stops laying eggs to replenish the worker force, the colony dwindles in size and dies off.

Typical worker life cycles

This picture of the life cycle is typical of many temperate zone ants, although of course the details vary greatly from species to species. The details of worker behaviour are also very variable, but the following generalizations are typical of many species. Workers usually spend the first few weeks of their lives inside the colony handling dead prey which have been brought back by foragers, feeding the larvae and the queen with regurgitated food, cleaning the nest and guarding the entrance. Later in life (this change takes place at an age of about 40 days in *Formica polyctena*, a species which has been studied in detail) workers begin to do jobs outside the colony, mainly foraging and defence against enemies. The total length of life of ant workers is not very well known but probably ranges from a few weeks to a few years. In wasps and bees, workers usually live for about 3–10 weeks. In addition to changes in worker behaviour with age, in some species of ants there are two castes of worker (both sterile females): soldiers and normal workers. Soldiers are usually larger and have large heads with jaws or glands for producing defensive secretions. As their name implies they are specialists in colony defence.

The females belonging to different castes (queen, worker, soldier) do not usually differ genetically; the determination of caste depends on environmental conditions during

Caste differentiation is usually non-genetic

larval development. In *Lasius*, for example, whether a larva develops into a queen or a worker appears to depend on factors such as nutrition, temperature and age of the queen who laid the egg. In honeybees the queen can suppress the development of new queens by chemical signals which prevent the workers from feeding larvae the special diet ('Royal Jelly') needed to make them grow into queens.

The economics of eusociality

Kin selection theory tells us how eusociality could have evolved ...

... but not what factors led to a high enough *r* and *B*/*C* ratio

Before discussing the factors that could have favoured the evolution of eusociality, it is useful to remind ourselves how the problem can be phrased in terms of kin selection theory, with Hamilton's rule, as described in Chapter 11. Hamilton's rule states that an altruistic behaviour can be favoured if $rB - C > 0$, where C is the cost to the actor, B is the benefit to the recipient and r is the genetic relatedness of the actor to the recipient. This idea that altruistic behaviours can be selected if they are directed at other individuals with the altruistic genes is the only plausible current explanation for altruism. Consequently, if we want to explain eusociality, we must determine what factors would have led to both a sufficiently high relatedness (r) between interacting individuals, and a sufficiently high benefit/cost ratio (B/C) of helping to raise the offspring of other individuals rather than breeding independently. In the next three sections we will consider factors which could have influenced r, B and C.

The pathway to eusociality

The two possible pathways along which the evolution of sterile castes could have proceeded are considered in this section. The hypotheses described refer to evolutionary history and, therefore, cannot be tested by direct experiments. In fact, both pathways involve intermediates that are observed in present-day 'primitively eusocial' hymenoptera, so the two hypotheses attempt to generalize from present patterns to evolutionary history.

The subsocial route: staying at home to help your mother

The parasocial route: sharing a nest with sisters

The first possible route to eusociality is via offspring remaining at their natal nest after reaching maturity, and then helping their mother rather than breeding independently. This is termed the subsocial route and intermediates can be observed in some 'primitively eusocial' species, such as amongst the *Polistes* paper wasps and the Stenogastrinae hover wasps. The second possible route is via multiple females coming together to found a nest, termed the parasocial route. In the most primitively social case each female would lay her own eggs and rear her own young, but this could then evolve to a situation where one female gained dominance and the others became her workers. Shared nest founding is also observed in several primitively eusocial species, such as allodapine bees and other paper wasps. In termites the two routes would also involve helping fathers or brothers respectively.

How can we test between these two alternate hypotheses? To do this it is necessary to use the comparative approach and look at the distribution of different types of social groups, maping them onto phylogenies. When this is done, all the evidence points towards eusociality having evolved via the subsocial route. In the Hymenoptera, there is no documented example where parasocial breeding is ancestral to the evolution of obligate sterile castes (Bourke & Franks, 1995; Boomsma, 2007; Hughes *et al.*, 2008). Furthermore, examining the cases where both types of nest occur in the same species, the helpers in subsocial groups tend to be more altruistic than the helpers in parasocial groups, in terms of giving up greater amounts of their own reproductions (Reeve & Keller, 1995). In the termites, both workers and soldiers are specialized juveniles, as expected from the subsocial route, and there is no evidence that reproductives move between colonies in a way that would allow parasocial breeding. Furthermore, the parasocial route is also inconsistent with the existing support for a key role of monogamy in the evolution of eusociality, which we shall discuss in greater detail later.

Eusociality evolved by the subsocial route

The haplodiploidy hypothesis

The masters of eusociality are clearly the Hymenoptera. Despite the fact that they constitute only approximately 6% of all insect species, eusociality has evolved more times in the Hymenoptera than in any other taxa (Crozier, 2008). Bill Hamilton (1964, 1972) was the first to suggest that this might be because the Hymenoptera have a genetic predisposition to the evolution of sterile castes. The special feature is *haplodiploidy:* males develop from unfertilized eggs and are haploid, while females develop from normally fertilized eggs and are, therefore, diploid.

Haplodiploidy: a special feature of Hymenoptera ...

A haploid male forms gametes without meiosis, so that every one of his sperm is genetically identical. This means that each of his daughters receives an identical set of genes to make up half her total diploid genome. With a diploid father, a female would stand a 50% chance of sharing any particular one of his genes with her sisters, but with a haploid father she is certain to share all of them (assuming the mother mated only once). The other half of a female hymenopteran's genes come from her diploid mother, so she has a 50% chance of sharing one of her mother's genes with a sister. If we now think about the total degree of relatedness between sisters we come to a remarkable conclusion. Half their genome is always identical, and the other half has a 50% chance of being shared, so the total relatedness is $0.5 + (0.5 \times 0.5) = 0.75$. In other words, because of haplodiploidy, full sisters are more closely related to one another than are parents and offspring in a normal diploid species. Hymenopteran queens are diploid and are, therefore, related to their sons and daughters by the usual 0.5 (Box 13.1, Table 13.1). A sterile female worker can, therefore, make a greater genetic profit by rearing a reproductive sister than she could if she suddenly became fertile and produced a daughter!

... results in unusual patterns of relatedness ...

The potential consequences of this for the evolution of altruism can be illustrated with Hamilton's rule. If we measure the costs and benefits in terms of offspring lost and gained, this leads to the following form of Hamilton's rule:

$$\frac{B}{C} > \frac{r_{\text{donor to own offspring}}}{r_{\text{donor to recipient' soffspring}}}$$

If we then compare the relative advantage to a worker of either producing a daughter ($r_{daughter} = 0.5$) or helping raise a sister ($r_{sister} = 0.75$) then Hamilton's rule would be satisfied as long as $B/C > 2/3$. Put into words, this means that a worker would make a genetic profit if, by helping her mother to reproduce, she could raise slightly more than

BOX 13.1 CALCULATING COEFFICIENTS OF RELATEDNESS, *r*, IN HAPLODIPLOID SPECIES

General point

Males develop from unfertilized eggs and so are haploid; all of a male's sperm are genetically identical so the probability of sharing a copy of a gene via the father is one. Females develop from fertilized eggs and so are diploid; the probability of sharing a copy of a gene via the mother is 0.5, because of meiosis.

Method

Draw out a pedigree, linking the two individuals through their recent common ancestors. To determine the coefficient of relatedness between individual A and individual B, draw arrows along the pathways, pointing from A to B. Indicate on each link in the pathway the probability that a copy of a gene will be shared.

Examples

(a) Sister–sister

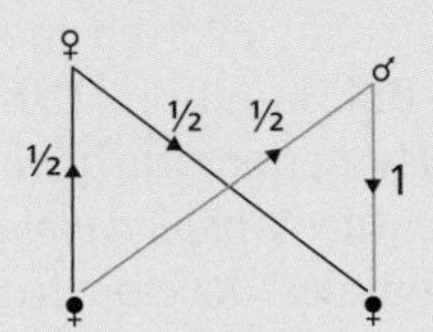

Half of a female's genes come from the father; the probability that a copy of one of these is shared with the sister is one. The other half come from the mother; the probability that a copy of one of these is shared is 0.5.

Via mother = (0.5×0.5) + Via father = (0.5×1); $r = 0.75$

(b) Sister–brother

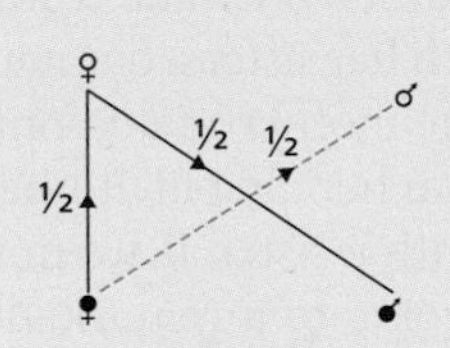

A female is linked to her brother only via her mother, as her brother develops from an unfertilized egg. Half of her genes come from her mother; the probability that a copy of one of these is shared is 0.5. The other half come from her father; the probability that a copy of one of these is shared is zero.

Via mother (0.5×0.5) + Via father (0.5×0); $r = 0.25$

(c) Brother–brother

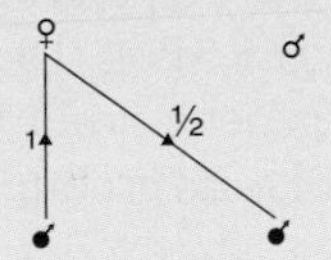

All of a male's genes come from his mother. There is a 0.5 chance of sharing a copy of a particular gene with his brother.

Via mother (1×0.5); $r = 0.5$

(d) Brother–sister

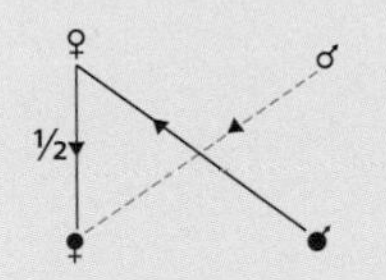

All of a male's genes come from his mother. There is a 0.5 chance of sharing a copy of a particular gene with his sister.

Via mother (1×0.5); $r = 0.5$

Note that haplodiploidy can lead to relatedness values being asymmetric. For example, a brother is more related to his sister ($r = 0.5$) than she is to him ($r = 0.25$) (compare b with d).

	Mother	Father	Sister	Brother	Son	Daughter	Niece or nephew (via sister)
Female	0.5	0.5	0.75	0.25	0.5	0.5	0.375
Male	1	0	0.5	0.5	0	1	0.25

Table 13.1 Degrees of relatedness between close relatives in a haplodiploid species (assuming females mate only once).

two sisters for every three offspring that she would have raised on her own. The magic thing here is that the B/C ratio required is <1, so a sterile worker could potentially evolve even if helping is less efficient than raising its own offspring! In contrast, with diploids, the relatedness to full siblings and offspring are both given by $r = 0.5$, so $B/C > 1$ is required. In other words it pays a diploid to help if it can replace one lost offspring with just over one sibling.

... and could predispose this group to evolve eusociality

But sadly, things aren't so simple. Robert Trivers and Hope Hare (1976) pointed out that Hamilton's haplodiploidy hypothesis wouldn't necessarily hold, because haplodiploidy also leads to a female being less related to a brother. Specifically, a female is related to her brother by only 0.25, since the 50% of her genes that come from her father have no chance of being shared with a brother, and the other half of her genes have a 50% chance of being shared: $0.5 \times 0.5 = 0.25$ (Box 13.1). The key point is that we need to compare the relative value of sons and daughters with that of brothers and sisters, not just sisters versus daughters. Assuming that queens produce equal numbers of male and female reproductives (drones and queens), as expected from Fisher's theory of equal investment (Chapter 10) this would then mean that the average relatedness of a worker to their brothers and sisters would be only 0.5 (the mean of 0.75 and 0.25), exactly the same as they would have to their own progeny if they had decided to leave home and have their own offspring. This reduced relatedness to brothers exactly cancels the benefit of increased relatedness to sisters, suggesting that haplodiploidy doesn't help (i.e. $B/C > 1$ would be required, as with diploids).

The increased relatedness to sisters is cancelled by a decreased relatedness to brothers

Trivers and Hare suggested, though, that the haplodiploidy hypothesis could be saved if the workers rear more queens (sisters) than drones (brothers). Workers are favoured to rear more queens than drones, because they are more related to sisters than brothers (Table 13.1). Specifically, as we shall describe in a later section, the evolutionarily stable strategy (ESS) sex ratio of reproductives, from a workers perspective, is to produce three queens for every one drone. With such a female bias, the average relatedness to the offspring of the queen is $(3/4 \times 3/4) + (1/4 \times 1/4) = 5/8$. This represents the relatedness to sisters multiplied by the proportion of brood that are sisters, plus the relatedness to brothers multiplied by the proportion of brood that are brothers. Plugging this into Hamilton's rule, the critical value of B/C is $1/2/5/8 = 4/5$. In other words, instead of having to rear just over one sibling for every potential offspring lost, the worker has to raise just over four siblings for every five offspring sacrificed. Consequently, if the sex ratio is female biased it would appear that haplodiploidy makes it easier for helping to evolve (i.e. the critical $B/C < 1$).

If sex ratios are female biased, then workers will be more related to the offspring of their mother, than their own offspring...

This line of reasoning is, however, still too simple. To see why we have to introduce the concept of the *value* of males and females, the proportion they will contribute to the gene pool of future generations. Recall that when the sex ratio is 3:1 in favour of females, the expected reproductive success of a male, or in other words his value as a machine for making grandchildren, is three times that of a female. If you include this factor, the pay-off for raising siblings and offspring must be calculated as (number raised × value × relatedness). Taking the example where the sex ratio of the population as a whole is 3:1 (this applies, therefore, to the average ratio of offspring and of siblings), we get the following answers.

For helping to raise a sibling:

$$(3/4\times1\times3/4)+(1/4\times3\times1/4)=12/16$$

For raising an offspring:

$$(3/4\times1\times1/2)+(1/4\times3\times1/2)=12/16$$

(In each line the first bracket is for females, the second for males. So on the top line the first bracket means 'three-quarters of siblings are females, with a value of one and a relatedness of three-quarters'.) Therefore, with a population sex ratio set at three reproductive females for every male, the critical value for helping to pay is $B/C>(12/16)/(12/16)=1$, the same as for a diploid species! In other words, *when the sex ratio in the population as a whole* is female biased to reflect the helper optimum haplodiploidy gives no advantage to helping: the extra relatedness to females is counter-balanced by the higher value of males (Trivers & Hare, 1976; Craig, 1979).

... but this is exactly balanced by a reduced value of females

So can haplodiploidy ever tip the balance in favour of helping, or is it totally irrelevant? Trivers and Hare (1976) realized the above complications and suggested that a female bias might still favour helping if it occured in a way that does not lower the value of females quite as much. This could happen if the sex ratio within the nest is female biased while the overall population sex ratio is not. For example, if the bias in the nest was 3:1 and the population sex ratio was 1:1, the value of males and females would be equal and we would be back to the simple calculation of B/C based on relatedness which gave $B/C > 4/5$. This could occur when worker control of the sex ratio spreads through population or through other mechanisms that lead to 'split sex ratios', with a relative excess of females produced in some nests and a relative excess of males produced in others (Trivers & Hare, 1976; Seger, 1983; Grafen, 1986).

Split sex ratios favour helping at nests which produce relatively female biased sex ratios...

However, there are two potential problems with even these mechanisms. Firstly, the increased value of helping in the female-biased colonies will be negated by a reduced value of helping in the male-biased colonies (Gardner *et al.*, 2012). The point here is that if a mutation leads to increased quality or amount of helping, this will be favoured when in relatively female-biased broods, but we must also consider that it will be selected against in relatively male-biased broods (Gardner *et al.*, 2012). Secondly, our above calculations have assumed that any potential worker would also produce an equally female-biased same-sex ratio if they bred independently. As the population sex ratio is female biased, males are worth more than females, so independently breeding females would be selected to produce only sons. The increased reproductive value of males means that sons would be of more value than a female-biased mixture of siblings (J. Alpedrinha *et al.*, unpublished). In the extreme, if the sex ratio is biased 3:1 in favour of females, then the pay-off from raising sons is $(1\times3\times1/2)=3/2$, and so the critical value for

... but disfavours helping at nests which produce relatively male biased sex ratios

helping to pay is $B/C > \frac{3/2}{12/16} = 2$. This is twice the value for a diploid species, showing that as worker control of the sex ratio spreads through a population, the resultant female bias in the population can select against helping in haplodiploids! This is because it makes sons more valuable than a female-biased mixture of sisters and brothers. The consequences of haplodiploidy for the evolution of eusociality are unresolved, and this remains a contentious area.

Haplodiploidy can even predispose groups against the evolution of eusociality

Let us summarize the complicated ups and ups downs of the haplodiploid hypothesis. Hamilton's (1964, 1972) haplodiploid hypothesis was that haplodiploidy would lead to the relatedness to siblings being greater than that to offspring, and so would lead to a genetic predisposition that favoured the evolution of eusociality. However, in the 1970s it was realized that things were not so simple, and that an increased relatedness to sisters would be exactly cancelled by a decreased relatedness to brothers. Biased sex ratios cannot rescue the haplodiploidy hypothesis, unless there are also 'split sex ratios', but even they are unlikely to have been hugely important, and may even have hindered the evolution of eusociality. So, overall, the haplodiploidy hypothesis may have been a bit of a red herring. Instead, we must turn to aspects of mating system and ecology to explain why eusociality has arisen multiply in the Hymenoptera. It is also important to not confuse the haplodiploidy hypothesis with kin selection (e.g. Wilson & Hölldobler, 2005). The kin selection explanation for eusociality does not rely on the haplodiploidy hypothesis, which was just a suggestion for how to make the evolution of eusociality especially easy in haplodiploids.

A red herring?

The monogamy hypothesis

Numerous authors have suggested that monogamy could have been important in the evolution of eusociality. The reason for this is that if queens mate multiply, then this reduces the relatedness between their offspring (Box 13.2; Fig. 13.4), hence lowering the *r* term in Hamilton's rule, making it harder to satisfy. However, it was Koos Boomsma (2007, 2009) who realized monogamy couldn't just help, but was actually crucial! In particular, Boomsma argued a role for strict lifetime monogamy, in which females only mate with one male in their entire life.

Monogamy leads to a potential worker being equally related ($r = 0.5$) to her own offspring and to the offspring of her mother (siblings). In this case, any small efficiency benefit for rearing siblings over their own offspring ($B/C > 1$) will favour cooperation that could eventually evolve towards eusociality if the benefit persists uninterrupted over many generations (Fig. 13.5). This holds for both haplodiploids and diploids. In contrast, even a low probability of multiple mating means that potential workers would be more related to their own offspring. In this case, costly helping would require a significant efficiency advantage to rearing siblings over own offspring ($B/C >> 1$; Fig. 13.5). Until group living is established, allowing the evolution of specialized cooperative behaviour and division of labour, the ratio B/C cannot be expected to greatly exceed one. For example, feeding a sibling is unlikely to be hugely more beneficial than feeding an offspring by the same amount. Consequently, in the absence of strict monogamy the

Kin selection theory predicts that lifetime monogamy can greatly aid the evolution of eusociality

BOX 13.2 COEFFICIENTS OF RELATEDNESS, r, IN HAPLODIPLOID SPECIES WHEN FEMALES MATE MULTIPLY

When haplodiploid queens mate multiple males, this reduces the relatedness of her daughters to their sisters, but not to their brothers. The methods are as in Box 13.1, with the only difference being that females are assumed to have mated a large number of males, such that offspring are half-siblings, who only share genes through their mother.

Examples

(a) Sister–sister

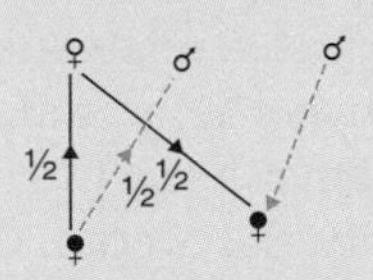

Half of a female's genes comes from the father; the probability that a copy of one of these is shared with the sister is zero because the sister has another father. The other half comes from the mother; the probability that a copy of one of these is shared is 0.5.

Via mother $= (0.5 \times 0.5)$ + Via father $= (0.5 \times 0)$; $r = 0.25$

(b) Sister–brother

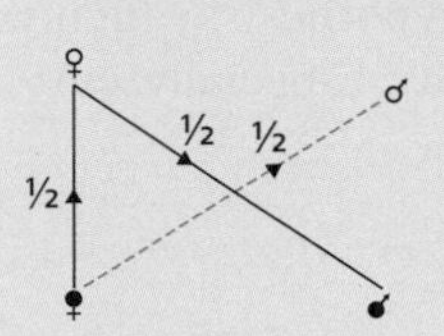

A female is linked to her brother only via her mother, as her brother develops from an unfertilized egg. Half of her genes come from her mother; the probability that a copy of one of these is shared is 0.5. The other half come from her father; the probability that a copy of one of these is shared is zero.

Via mother (0.5×0.5) + Via father (0.5×0); $r = 0.25$

The relatedness of brothers to their sisters and brother is unchanged from that in Box 13.1 (part b) because males obtain all of their genes from their mothers.

(c) Sister–nephew (or niece)

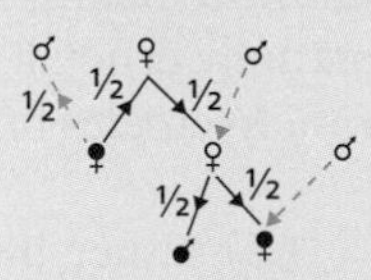

A female shares 0.25 of her genes with her sister, who passes on one half of her genes to her son.

Via reproduction of sister $= (0.25 \times 0.5)$; $r = 0.125$

population cannot even get started on the road to eusociality, although looser forms such as cooperative breeding can be maintained.

A possible role of monogamy is supported by the observation of lifetime monogamy in many eusocial species (Boomsma, 2007). In most termites physical lifetime monogamy is the default option, with a queen committing to a single male when founding a nest, after which they remain together. Many ants, bees and wasps have a functionally equivalent form of lifetime monogamy, where the queen mates a single male, who then dies before colony foundation. In these species, the queen continues to use the sperm of this single male for her entire lifetime, in some cases for periods of over 30 years. However, it is also true that multiple mating occurs in some social insects, such as honeybees, where females will mate with 10–20 males on their nuptial flight, and, ideally, we would thus like a more formal test of Boomsma's monogamy hypothesis.

Many social insects are monogamous ...

...but not all

Bill Hughes and colleagues did exactly this by collecting data on female mating frequency for 267 hymeopteran species. Amongst these species, most were monogamous, but about a quarter showed some levels of multiple mating (polyandry). Hughes *et al.* (2008) then mapped these data onto a

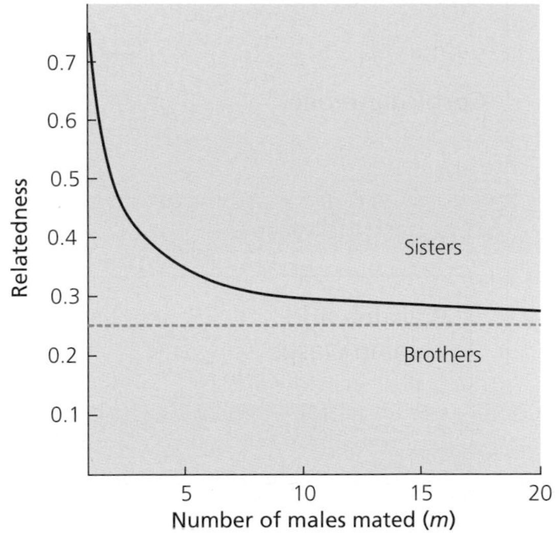

Fig. 13.4 The relatedness of a worker to her sisters and brothers, plotted against the number of times that her mother (the queen) has mated. The relatedness of a worker to her sisters declines from 0.75 to 0.25, whereas the relatedness to a brother is always 0.25 (Boxes 13.1 and 13.2).

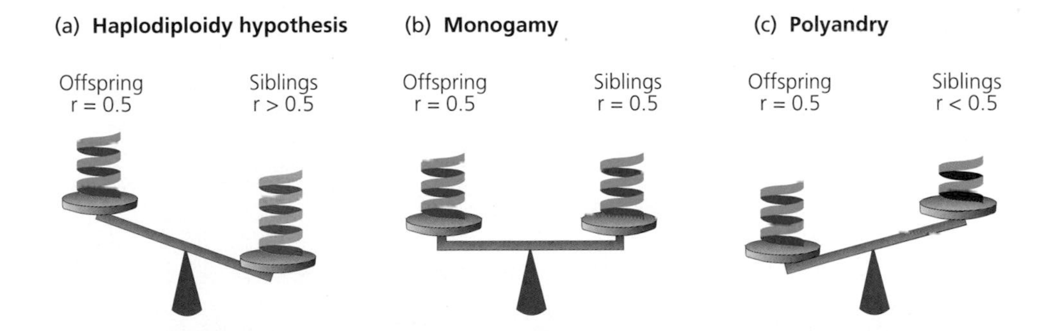

Fig. 13.5 Monogamy paves the way to eusociality. (a) The haplodiploidy hypothesis relies on individuals being more related to siblings than offspring, making siblings worth more than offspring. As originally envisioned, this appears to have mostly been a red herring. (b) The monogamy hypothesis emphasizes that if an individual is equally related to its siblings and its offspring, even a very slight but consistent efficiency benefit for raising siblings translates into a continuous selective advantage for helping. (c) Without strict monogamy, individuals are more related to their offspring than they are to their siblings, so that a large efficiency benefit is required in order for rearing siblings to be favoured. From West and Gardner (2010).

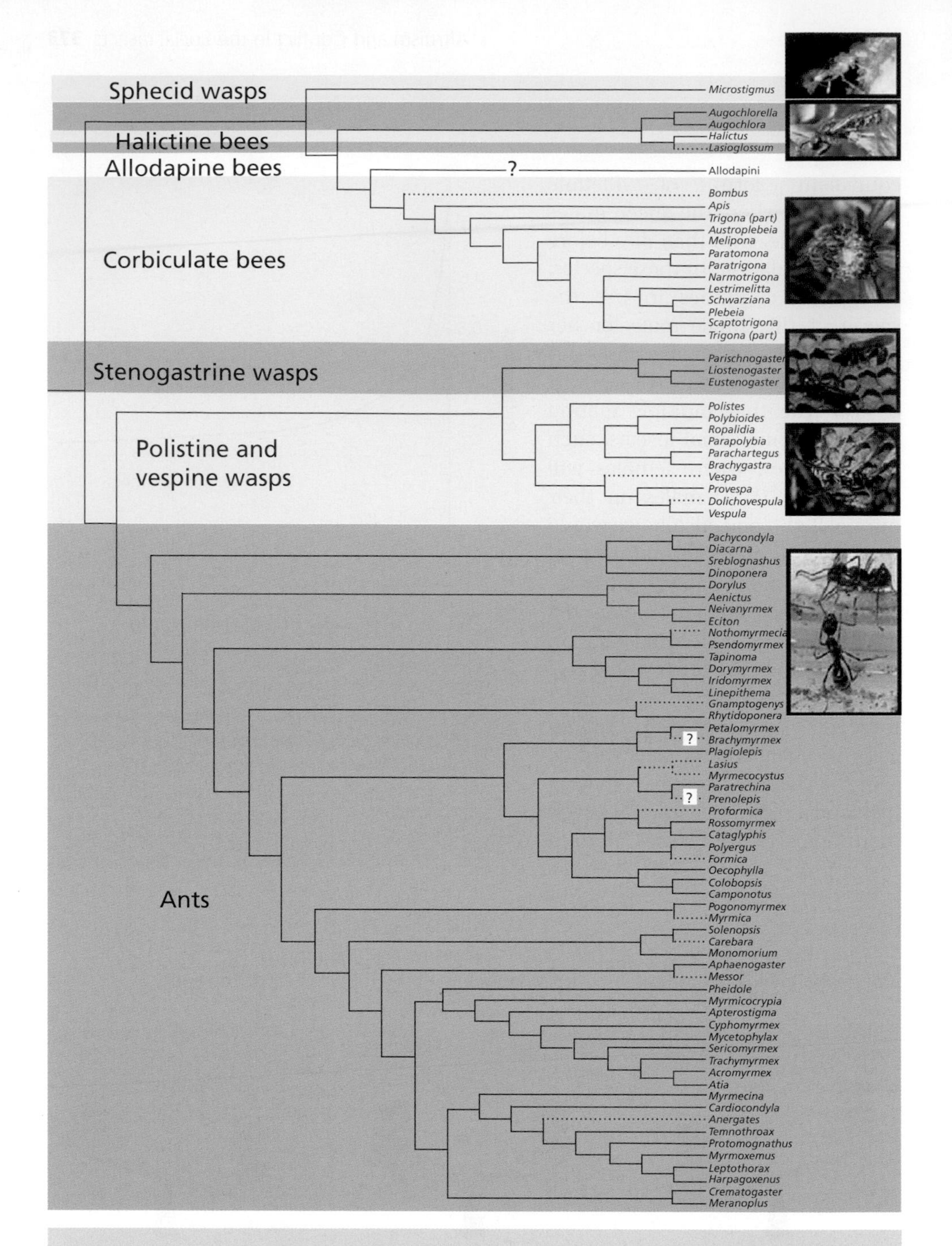

Fig. 13.6 Monogamy and the evolution of eusociality in the Hymenoptera. Shown is a phylogeny of the eusocial Hymenoptera for which female mating frequency data are available. Each independent origin of eusociality is indicated by alternately coloured clades. Clades exhibiting high (i.e. obligate) polyandry (>2 mates) have solid red branches, those exhibiting facultative low polyandry (many mate singly but some mate with two or three males) have dotted red branches and entirely monandrous genera have solid black branches. Mating frequency data are not available for the allodapine bees. From Hughes *et al*. (2008). Photos from top to bottom: *Microstigmus comes* by R. Matthews; *Lasioglossum malachurum* by C. Polidori; *Apis mellifera* by F.L.W. Ratnieks; *Liostenogaster flavolineata* by J. Fields; *Polistes dominulus and Diacamma sp*. by W.O.H. Hughes.

phylogeny and found two important results. Firstly, monogamy appeared to be the ancestral state in all the independent transitions to eusociality that they examined (Fig. 13.6). This suggests that monogamy originated first, giving a high relatedness, and then when ecological conditions led to a consistently favourable B/C, eusociality evolved. The same pattern of monogamy being ancestral to the evolution of eusociality has also been found in other eusocial taxa such as the termites and sponge-dwelling shrimps (Boomsma, 2007; Hughes *et al.*, 2008; Duffy & Macdonald, 2010).

Monogamy is the ancestral state when eusociality evolved

The second result found by Hughes *et al.* (2008) was that all the instances of multiple mating were in derived lineages, where castes had already evolved. In these cases, the reduction in relatedness due to multiple mating has not led to the loss of eusociality, because workers had already lost the ability to mate and realize full reproductive potential. Furthermore, these species had also already evolved division of labour and specialized helping behaviours, potentially giving them a substantial B/C that would allow Hamilton's rule to be satisfied, even when relatedness (r) becomes lower.

Multiple mating evolved after the transition to eusociality

The ecological benefits of cooperation

The monogamy hypothesis shows how a constantly high relatedness ($r = 0.5$) could occur between potential helpers and the offspring they would help raise (full siblings). However, for cooperation and eusociality to arise it is also required that the ecological conditions lead to a high enough benefit/cost ratio to make cooperation worthwhile. Specifically, that it is more efficient for a female to raise her siblings than her own offspring ($B/C > 1$). In this section we will discuss three ecological factors which have been suggested to be of key importance: life insurance, fortress defence and food distribution.

The benefits of life insurance

Dave Queller (1989, 1994) pointed out that in species where there is a period of extended parental care, such as ants, bees and wasps, helping could be favoured as a kind of 'life insurance'. If a solitary female dies during the period when she is raising a brood, the offspring dependent upon her care will die as well. If, however, the same female was part of a group helping to rear the brood, her death would not condemn the offspring to die, because others would continue to carry out brood care. Thus, as part of a group, a female has at least some 'assured fitness returns'. This advantage could, in theory, make it better to help others breed than to breed as a solitary female. In an attempt to quantify the importance of this 'insurance effect', Raghavendra Gadagkar (1991) studied the social wasp *Ropalidia marginata*. On average, the developmental period from egg to adult in this species lasts 62 days. An individual reproductive female has a probability of only 0.12 of surviving for 62 days. Thus, her expected reproductive success from solitary nesting is not very high. In fact, Gadagkar estimated that a female *Ropalidia marginata* could increase her expected success by 3.6-fold as a result of nesting in a group! This is clearly much greater than the value of just over 1.0 required when there is monogamy.

Death of females before the end of brood care can favour cooperative breeding

Jeremy Field and colleagues tested the importance of insurance effects experimentally. With observational data such as that collected by Gadagkar, it cannot be ruled out that some other confounding factor might vary between nests with and without helpers. For example, lower quality females might be less able to attract helpers, leading to a lower brood survival rate, independent of the effects of helpers. To avoid

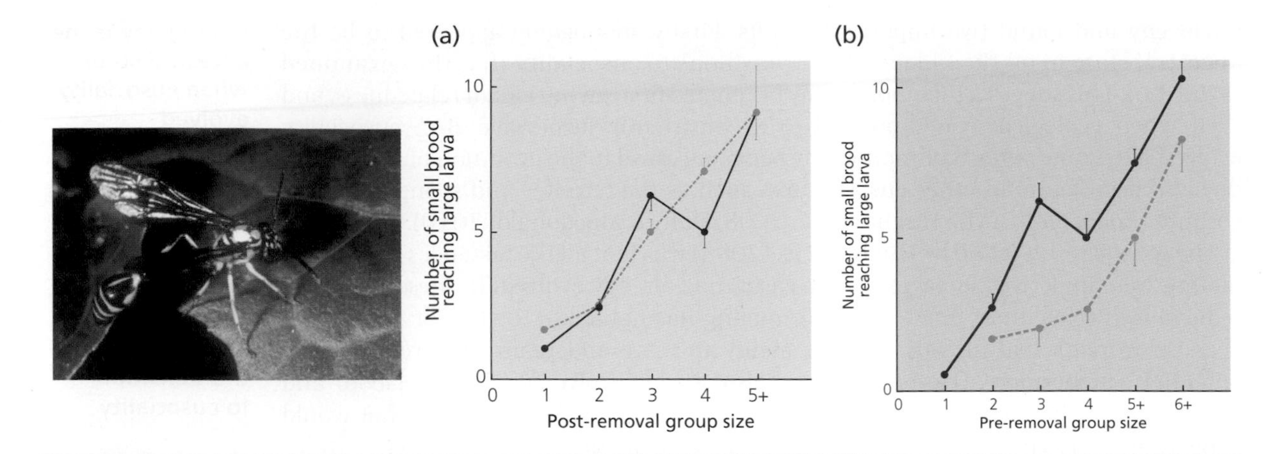

Fig. 13.7 Life insurance in the tropical hover wasp, *L. flavolineata*. The relationship between the number of small brood developing successfully to become large larvae plotted against (a) the post removal group size and (b) the pre-removal group size. Data are shown for control (dark blue) and removal (light blue) nests. From Field *et al.* (2000). Reprinted with permission from the Nature Publishing Group. Photograph of a female marked with dots of paint. Photo © Maurizio Casiraghi.

such potential problems, Field *et al.* (2000) experimentally removed helpers from nests of the tropical hover wasp *Liostenogaster flavolineata*. They found that the removal of helpers reduced the number of small larvae that were reared to a large size during the observation period, but that groups with more helpers post-removal still reared more larvae (Fig. 13.7). Overall their data suggest that a female could increase the number of offspring successfully raised by 2.4-fold as a result of helping rather than breeding independently, which is again substantially greater than 1.0.

Experimental removal of hover wasp helpers shows that life assurance benefits could be enough to favour the evolution of eusociality

The benefits of fortress defence

Having a defendable source of food can favour cooperation

Eusociality could also have been favoured by the potential benefit of staying at the natal nest and helping defend a valuable resource. This advantage of 'fortress defence' is likely to be important in species which live in protected, expandable sites, where food is obtained, such as the wood galleries of termites, the plant galls of social aphids and thrips, the galleries of ambrosia beetles and the sponges of sponge-dwelling shrimps. Remaining to help at the natal 'fortress' avoids the risk of death associated with migration and, because food is available locally, little feeding care is required, which allows the first worker specialists to generally be soldiers, specialized for defence (Figs. 13.1 & 13.2).

Emmett Duffy and colleagues tested the advantages of fortress defencc across sponge-dwelling shrimps in the genus *Synalpheus*. Each species within this genus appears to be specialized to live within, and consume, one or a few species of sponge. Within the sponge, individuals live in social groups, the nature of which ranges across species from heterosexual pairs, to groups with multiple breeders, to eusocial colonies that contain a single queen and more than 300 sterile workers. Because few predators can enter the narrow canals of the sponges, the greatest competition for resources appears to come from individuals of the same or closely related species. This competition for territories

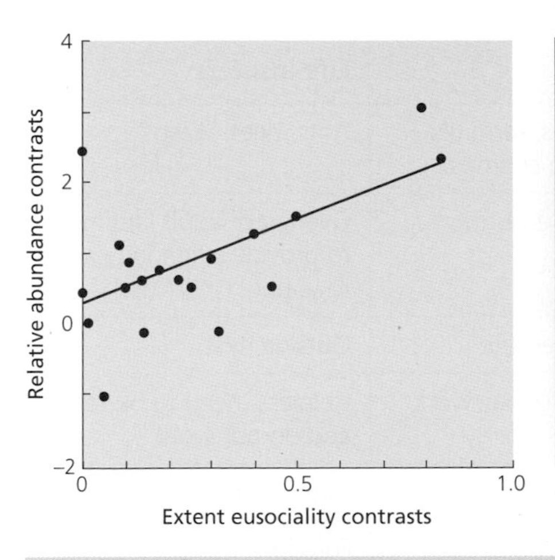

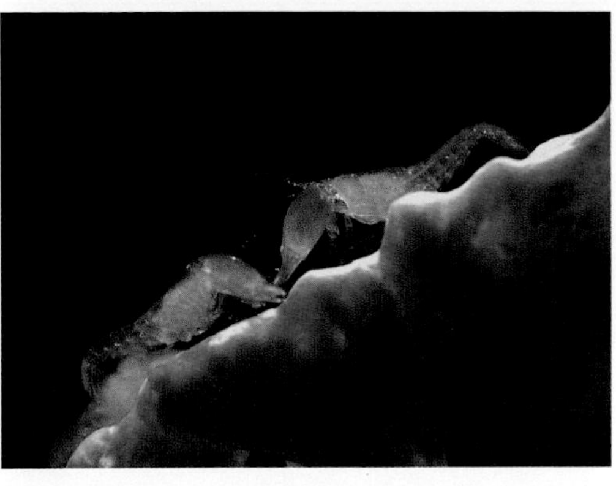

Fig. 13.8 Fortress defence in sponge-dwelling shrimps. Plotted are the phylogentically independent comparisons in relative abundance against the independent comparisons in the extent of sociality, using data from 20 species. From Duffy and Macdonald (2010). With permission of the Royal Society. Photograph shows non-breeding workers in the shrimp *Synalpheus regalis*. Photo © Emmett Duffy.

appears to be intense, and all species are territorial and equipped with a fighting claws that is used in communication and combat. When intruding individuals attempt to enter an already occupied sponge, the local shrimps suddenly begin cracking their claws in unison, producing a distinctive crackling noise that lasts for tens of seconds (Tòth & Duffy, 2004). This 'coordinated snapping' appears to be used to warn intruders away and helps explain why these species are sometimes called 'snapping shrimp'. If fortress defence plays an important role in favouring cooperation amongst sponge-dwelling shrimps, then we can predict that eusociality would enhances the ability to acquire, defend and retain limited host resources relative to less social species. Duffy and Macdonald (2010) tested this prediction, examining how shrimp abundance correlated with sociality among species in Belize. Consistent with a role of fortress defence, they found that eusocial species are more abundant, occupy more sponges and have broader host ranges than non-social sister species (Fig. 13.8). Ideally, this hypothesis would also be tested further with experiments that manipulated host availability or shrimp density.

Competition for sponges is intense in sponge-dwelling shrimps ...

... and eusocial species appear to be better at acquiring and keeping sponges

Experimental evidence for a role of fortress defence comes from William Foster's work on the gall-forming aphid, *Pemphigus spyrothecae*. In the eusocial aphids, there is a soldier caste that defends the colony against insect predators, such as ladybird and hoverfly larvae (Fig. 13.1e; Stern & Foster, 1996). Foster (1990) manipulated the composition of aphids in galls so that they contained either ten soldiers or ten non-soldiers, and then introduced a single predator. The results were striking – in colonies with soldiers, the predators were usually killed, with the loss of a few soldiers, whereas in colonies with non-soldiers the predators survived and all the aphids were usually killed and eaten. It has sometimes been suggested that the gall-forming aphids might have been predisposed to eusociality because they reproduce clonally, so relatedness to siblings is given by $r = 1$. However, the key point here is that a potential worker would be equally related to its siblings or its own offspring, so again what is required is $B/C > 1$, as with monogamy.

Soldier aphids kill insect predators, preventing the entire colony from being consumed

Table 13.2 The two types of social insects, as divided by whether the evolution of eusociality was driven by life insurance or fortress defence (adapted from Queller & Strassmann, 1998).

Characteristic	Fortress Defenders	Life Insurers
Taxa	Thrips, aphids, beetles, termites (and sponge-dwelling shrimps)	Ants, bees, wasps
Main advantage of social living	Valuable, defensible resource	Overlap of adult lifetimes to provide extended care to young
Food	Inside nest or protected site	Outside nest
Juveniles	Active, feed selves and may work	Helpless, need to be fed and do not work
Non-social ancestors	Not necessarily parental	Highly parental
First specialized caste to evolve	Soldiers	Foragers
Colony size	Usually small	Often large
Ecological success	Usually limited	Excessive

Dave Queller and Joan Strassmann (1998) have argued that the social insects can be divided into two types, depending upon whether the primary ecological benefit of sociality was life insurance or fortress defence. These types and their defining characteristics are given in Table 13.2.

Food distribution

Several researchers have argued that a major factor driving the evolution of cooperation and eusociality in mole-rats is the distribution of food (Jarvis *et al.*, 1994). Mole-rats are small African rodents that spend their lives underground, digging tunnels in search of subterranean storage organs of plants (roots and tubers), which are their source of food. The mole-rats include solitary species, group living species (with maximum group sizes of approximately 15) and two eusocial species, the naked mole-rat and the Damaraland mole-rat (with maximum group sizes of approximately 300 and 40 respectively; Fig. 13.9). In these eusocial species, only one pair in the group breeds. The other females have undeveloped ovaries and the males, while they may have active sperm, are apparently non-breeders.

A possible role of food distribution was suggested by the distribution of species across different habitats. The solitary species inhabit mesic areas with a well-balanced supply of water that leads to a more even distribution of food, and damper soil, that is easier to dig. In contrast, the eusocial species live in arid areas where rainfall is low and unpredictable. In the latter kind of habitat, food resources are more patchy, making foraging more risky, and the harder soil means that the energetic cost of digging is higher. Furthermore, when food is found, it is usually enough to feed a colony for a long time. This has led to the suggestion that a patchy and hard to find food source, that is plentiful when found,

(a) 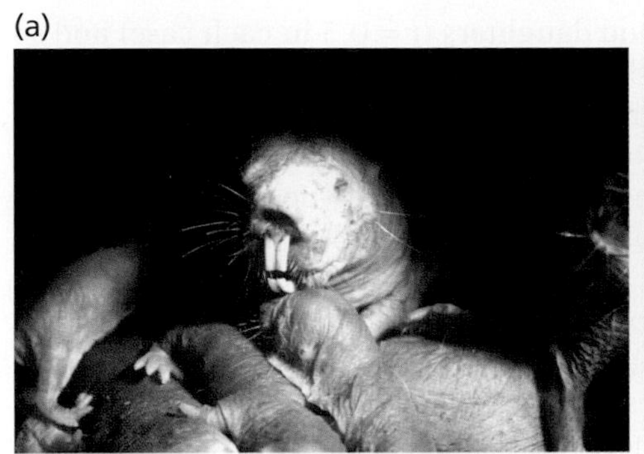

(b) 

Fig. 13.9 (a) Naked mole-rat queen with young and (b) a Damaraland mole-rat. Photo (a) © Neil Bromhall; (b) © Andrew Young.

could favour group living and cooperation, to more efficiently forage for food. Chris Faulkes and colleagues (1997) tested this hypothesis with a comparative study across 12 species of mole-rat. Across species, they found that, as predicted, larger groups are associated with a lower density of food and a greater variation in rainfall.

Sociality in mole-rats is correlated with conditions where it is hard to obtain food

Conflict within insect societies

In the previous sections of this chapter, we have examined how kin selection theory provides an explanation for the evolution of the sterile worker caste in the social insects. However, it also predicts when conflict could occur. The key point here is that when relatedness is greater than zero there is the potential for cooperation, but as long as it is also less than 1.0 there is also the potential for conflict, because individuals can be selected to obtain a disproportionate share of the reproductive success. In the following sections of this chapter two areas are discussed where conflict arises within colonies of hymenopteran social insects: the sex ratio and who should produce male eggs. Our general question will be: 'Given that there are queen and worker castes, how do queens and workers maximize their genetic contribution to the next generation?'. Put another way, when do conflicts arise, how are they resolved and who wins? In both of the areas examined, we will find that kin selection theory makes clear predictions about when conflict will arise, that the precise form of the conflicts depend upon the asymmetrics that result from haplodiploid genetics, and that there is a remarkable level of empirical support for these predictions.

Kin selection theory predicts both cooperation and conflict

Conflict over the sex ratio in the social hymenoptera

Queen–worker conflict

In Chapter 10 we discussed how individuals could be selected to adjust the sex of their offspring in response to local conditions. Let us now consider selection on sex ratio within the hymenopteran social insects; firstly, from the point of view of the queen.

The queen is equally related to her sons and daughters ($r = 0.5$ in each case) and so Fisher's theory of equal investment says she should produce equal numbers of male and female reproductive offspring (Chapter 10). To be more precise, the queen's ESS is to *invest equally* in the two sexes. So, for example, if the production of queens requires twice as many resources as the production as drones, then the Queen should produce twice as many drones as queens. It is important to emphasize that we are referring to equal investment in *reproductive* offspring, not sterile workers. Recall that in Chapter 10 the argument was that a 50:50 sex ratio was stable because the expected *reproductive success* of a male and a female is the same. Hence, the discussion of sex ratios is only pertinent to reproductives.

Queens prefer a 1:1 investment ratio, investing equally in male and female reproductives

Now for the twist: because workers are more related to their sisters ($r = 0.75$) than to their brothers ($r = 0.25$), they would rather rear a higher proportion of sisters. But how much bias in favour of reproductive sisters should they show? Once again we search for the ESS sex ratio, this time from the workers' point of view. If the workers rear too many sisters then the sex ratio in the population will become so female biased that a drone will have very much greater reproductive success than a queen. It turns out that the stable sex ratio for the workers is a 3:1 investment in favour of reproductive females (Trivers & Hare, 1976). When female reproductives are exactly three times as common as males, drones have three times the expected success of queens because on average each drone has three times the chance of finding a mate. From the workers' point of view this would exactly compensate for the fact that brothers are only one third as closely related as are sisters: a worker expects to get three nieces or nephews from her brothers for every one she gets from her sisters. Nieces and nephews on her sister's side are three times as closely related to her, so the total gain per unit investment via brothers and sisters is the same.

Workers prefer a female-biased sex ratio, investing three times as much in females as males

To summarize, the queen prefers an equal investment in male and female reproductive offspring, but the workers prefer a ratio biased 3:1 in favour of females. There is a direct conflict of interest over the sex ratio between workers and the queen. Who wins? And how?

Queens and workers disagree about the optimal ratio of investment in reproductives

Tests of worker–queen conflict

Robert Trivers and Hope Hare (1976) attempted to test whether the queen or workers win by analysing the ratio of dry-weight investment (more accurate than simply looking at numbers) in male and female offspring in 21 species of ants. The ant species were chosen because they were ones in which the conditions for the hypothesis were likely to hold (one queen, monogamous). Despite a considerable amount of scatter in their data, Trivers and Hare found that, on average, the ratio of investment was much closer to 3:1 than to 1:1 (Fig. 13.10). They concluded that the workers win the conflict and successfully manipulate the sex ratio towards their own optimum and away from that of the queen. To put it bluntly, the workers are successfully farming the queen as a producer of sisters and brothers: a far cry from the idea of workers as subordinate females making the best of a bad job! Trivers and Hare suggest that the workers win simply because they have both practical and numerical power; they are usually many and do all the offspring provisioning, and are thus in a position to selectively prioritize new queens rather than males.

Measuring the ratio of investment in ant nests shows a female bias ...

... workers appear to win the battle over the investment ratio

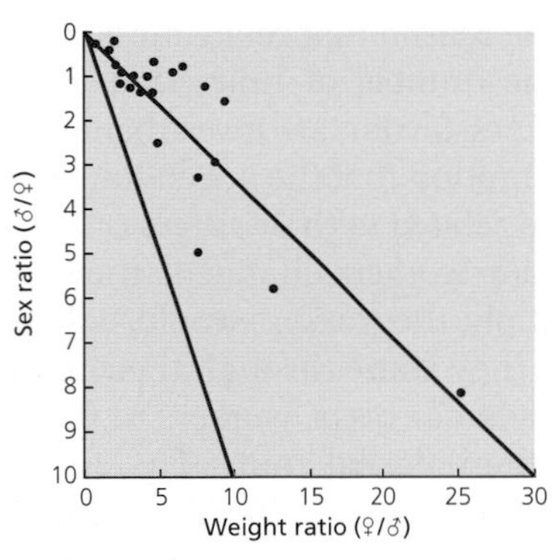

Fig. 13.10 Ratio of investment (measured by dry weight) in 21 species of ants. The x-axis is the ratio of female:male weight and the *y*-axis is the ratio of numbers of males:females in the colony. The lower line is the prediction if the investment ratio is 1:1 and the upper line is 3:1 in favour of females. The data are closer to the 3:1 line, as predicted if workers control the sex ratio. However, some analyses have suggested that things are not so simple, because: (a) dry weight overestimates investment in females, so the average investment is actually closer to 2:1, and (b) queens mate with multiple males in some species, which leads to workers favouring an investment less biased towards females. (To understand how the lines are drawn take the example of a ♀:♂ weight ratio of 6:1. Equal investment would mean six ♂ per ♀, and a 3:1 investment ratio in favour of ♀ would mean a ratio of 2 ♂ per ♀.) From Trivers and Hare (1976). Reprinted with permission from AAAS. Photograph shows a mating pair of the rover ant, *Brachymyrmex patagonicus*, in which females are considerably larger than males. Photo © Alex Wild.

Alternative explanations are possible

Although Trivers and Hare's results were highly impressive, several workers pointed out a number of potential problems with how the data were analysed and interpreted. One of these was that female-biased sex ratios could also be caused potentially by other reasons, such as local mate competition or local resource competition (Alexander & Sherman, 1977), as discussed in Chapter 10. Although these alternative explanations were later refuted (Nonacs, 1986; Boomsma, 1991), it also became clear that across species data were unlikely to provide definitive evidence for worker control of sex allocation. Much stronger evidence has since been provided by within species studies on sex ratio variation between colonies.

Split sex ratios

Some colonies specialize in producing males, while others specialize in producing females

When examining the sex ratios produced by individual colonies, rather than at the species level, it is often observed that some colonies tend to produce all or predominantly males (drones), while other colonies tend to produce all or predominantly females (new winged queens; often referred to as gynes). Koos Boomsma and Alan Grafen (1990, 1991; Boomsma, 1991) showed that such 'split sex ratios' could be predicted when workers are in control of the sex ratio, and if the relatedness structure varies across colonies.

One reason that relatedness structure could vary across colonies is due to variation in the number of times that the queen has mated. Our above argument, that the workers favour an investment ratio biased 3:1 towards sisters, was based on the assumption that the queen had mated only once, leading to workers being three times more related to their sisters ($r = 0.75$) than their brothers ($r = 0.25$). How would the ESS for workers change if their queens had mated multiply? If a queen has mated multiply, the workers would still be related to their brothers by $r = 0.25$, via the genes that they share through their mother (Box 13.1). However, multiple mating reduces the relatedness of workers to their sisters, because they are less likely to share genes via the paternal route (Fig. 13.4). In the extreme, if two sisters were fathered by different males (half-siblings), then they would only be related by $r = 0.25$, which represents the sum of the relatedness through their mother ($r = 0.25$) and their fathers ($r = 0$) (Box 13.2).

Workers are relatively less related to sisters when their queen has mated multiply

Consider now the consequence of when, in some colonies, the queen has mated singly but in other colonies the queen has mated multiply. A key assumption is that workers know how many times their queen has mated –evidence for this is provided later. In the colonies where the queen has mated singly, we would expect the workers to favour a sex ratio biased 3:1 towards females. In contrast, in the colonies where the queen has mated multiply, the extent to which the workers are more closely related to sisters has been reduced, and so a less biased sex ratio is favoured. However, things get more complicated because we have to consider the consequences of what is happening at the population level. The female bias favoured in the singly mated colonies increases the relative value of males, because each male will on average mate with more females. This selects on the colonies with multiply mated queens to invest a higher proportion of resources in males. At the same time, this lower investment in females by multiply mated colonies decreases the value of producing males, and hence selects on the colonies with singly mated queens to invest a higher proportion of resources in females. These effects feedback on each other, leading to the situation where the ESS is for colonies with singly mated queens to produce only or predominantly females, and colonies with multiply mated queens to produce only or predominantly males.

Workers favour the production of females in nests where the queen has mated singly, and males in nests where the queen has mated multiply

To summarize this fairly complicated argument, the key point is that variation across colonies in number of times that the queen has mated will lead to variation in the extent to which workers are more related to sisters. Workers could capitalize on this variation in the 'relatedness asymmetry' by conditionally producing the sex which maximizes their genetic contribution to the next generation. In colonies where queens mated singly, the workers are relatively more related to sisters, so they would do best to produce only or predominantly females. In colonies where queens mated mutiply, the workers are relatively less related to their sisters, so they would do best to produce only or predominantly males.

Some woods ants adjust their brood sex ratio in response to the number of times their queen has mated

The most amazing thing is that workers of some ants appear to do exactly this! Lotta Sundström (1994) studied the wood ant *Formica truncorum* on a number of small islands just off the southwest coast of Finland. She used allozyme markers to determine how many times each queen had mated and correlated this to the sex ratio of reproductives produced. As predicted by Boomsma and Grafen, she found that colonies with singly mated queens produced predominantly females, and colonies with multiply mated queens produced predominantly males (Fig. 13.11). Overall, Sundström was able to explain an impressive 66% of the variation in sex ratios across colonies.

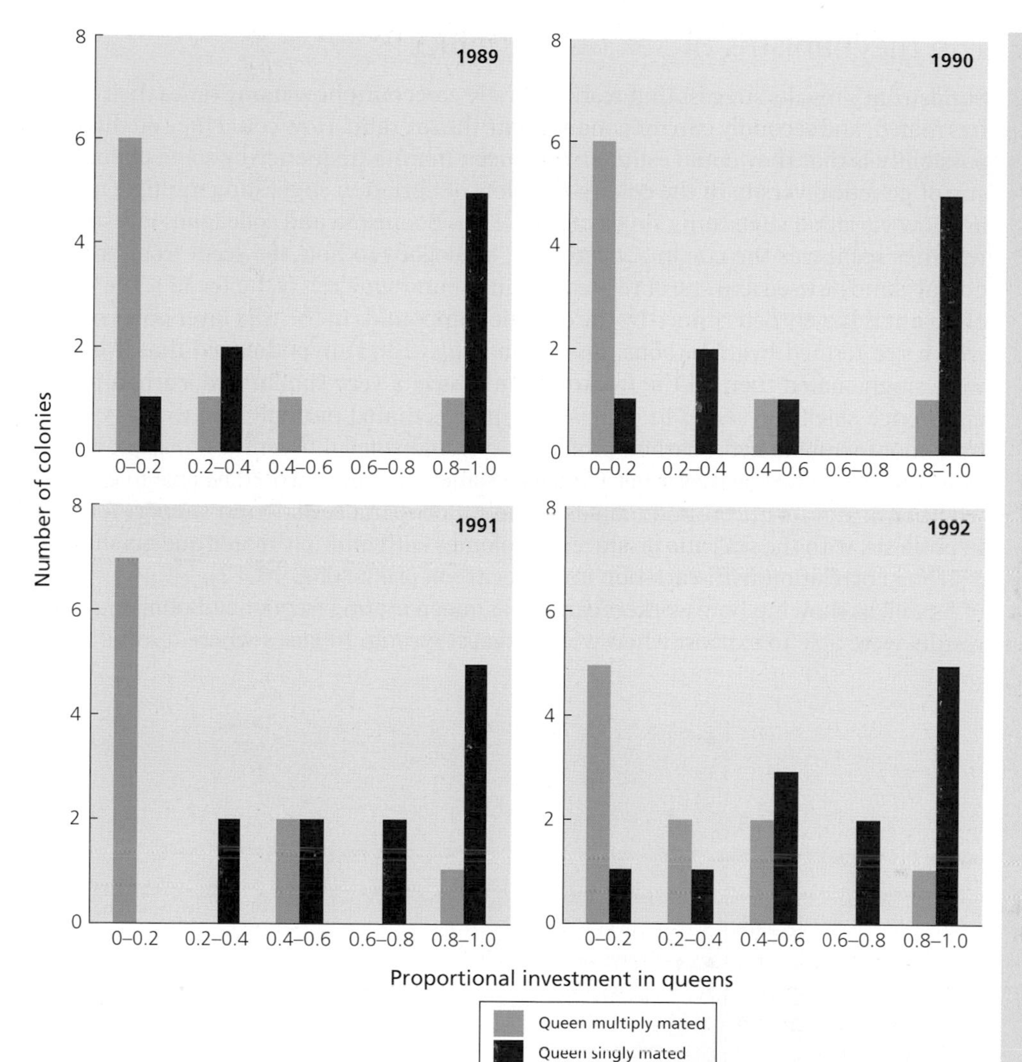

Fig. 13.11 Split sex ratios in the wood ant *F. truncorum*. The distribution of sex allocation (proportional investment in female reproductives) is shown for colonies where the queen had either singly (dark blue) or multiply (light blue) mated. Across data from four years, colonies with singly mated queens tended to produce females and colonies with multiply mated queens tended to produce males. From Sundstrom (1994). Reprinted with permission from the Nature Publishing Group. Photograph of a queen prior to her mating flight. Photo © Lotta Sundström.

The mechanisms of sex ratio conflict

Workers can count how many times their queen has mated ...

Sundström's results suggest that workers firstly can count how many times their queen has mated, and secondly can then manipulate the sex ratio. How could they do this? One possibility is that they could estimate the queen mating frequency by some phenotypic cue of genetic diversity in the colony, with lots of variation suggesting multiple mating and low variation suggesting single mating. Koos Boomsma and colleagues investigated whether scent was the cue in *F. truncorum*, analogous to how the scent from oral and dorsal glands is used as a cue of relatedness in ground squirrels (Chapter 11). The 'smell' of an ant is largely determined by the organic compounds in the wax layer on its cuticle, which are termed hydrocarbons. Boomsma *et al.* (2003) hypothesized that if queens were singly mated then all the workers would have a very similar hydocarbon profile, and hence smell the same. In contrast, if queens mated multiply, the workers would have more variable hydrocarbon profiles and would smell differently. Consequently, by examining variation across other workers in smell, a worker would be able to estimate whether or not its queen had mated multiply. Boomsma *et al.* found support for their hypothesis, with the sex ratio produced in colonies with multiply mated queens showing a strong correlation with variation in hydrocarbon profile (Fig. 13.12).

... by the variance in the smell of other workers

As well as showing how workers can assess queen mating frequency, Boomsma *et al.*'s results were able to explain when workers got it wrong. In cases where queens mated

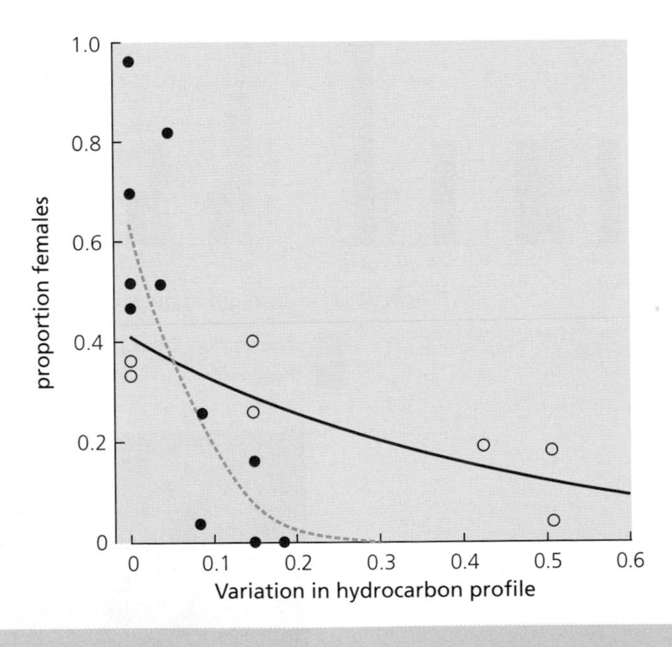

Fig. 13.12 Hydrocarbons and sex ratios in the wood ant *F. truncorum*. Across colonies with multiple mated queens, the proportion of females in the reproductive brood was negatively correlated with the variance in the cuticular hydrocarbon profile of the workers. Open circles and solid line are colonies sampled in 1994. Solid circles and dashed line are colonies sampled in 2000. Curves are logistic regression lines. From Boomsma *et al.* (2003).

with males that had similar hydrocarbon profiles, the sex ratio was unbiased or female biased. Put simply, if a queen mates with two males who smell the same by chance, then her offspring will vary little in smell, leading to the workers thinking their queen only mated once, and hence produce the wrong sex ratio. This provides a very elegant demonstration of the general problem that information processing can constrain how perfectly animals would be expected to behave.

Workers make mistakes when queens mate males who smell the same

Sundström and Boomsma's results also raise the problem of how could the workers manipulate the sex ratio to their advantage? The queen controls the initial sex ratio of the brood produced, so the workers must be doing something else. To determine what was going on, Sundström and colleagues (1996) turned to another wood ant, *Formica exsecta*, which also has split sex ratios in response to queen mating frequency. They compared the sex ratio of eggs with that of pupae and found that whilst these ratios were not different in colonies with multiply mated females, the proportion of males decreased significantly between the egg and the pupal stages in colonies with singly mated queens. This suggests that the workers are selectively neglecting or destroying males. Eliminating males is just one way to manipulate sex ratio; others have since been found. For example, in *Leptothorax acervorum* workers bias the sex ratio of reproductives by adjusting the proportion of females that develop as queens or workers (Hammond *et al.*, 2002). This non-destructive option may not be open to *F. excsecta* workers, because workers and reproductives are reared at different times in that species.

Workers manipulate the sex ratio by destroying males …

… or adjusting the proportion of females that develop as queens or workers

To summarize so far, workers are selected to manipulate the sex ratio of reproductives in their colonies in response to queen mating frequency. In at least some well studied species workers appear to do this, leading to split sex ratios, by either destroying males or adjusting the proportion of females that develop as queens or workers. This provides strong support for Trivers and Hare's original hypothesis that the workers win the conflict, albeit via a more complicated route than they imagined. So, is that the end of the story? No, because it seems that workers do not win all the time – sometimes the queens win!

Luc Passera and colleagues studied the red fire ant *Solenopsis invicta*. This species is so named because ants will swarm up a person and only start inflicting their painful sting in response to a pheromone released by the first ant to attack. This leads to the ants stinging in concert, causing a level of pain analogous to being burnt, which can cause death in small animals by overloading their immune system. Passera *et al.* (2001) experimentally switched queens between male- and female-producing colonies, and found that the sex ratios produced in a colony post-switching was predicted by the colony from which the queen came. Colonies produced predominantly females if the queen came from a female-producing colony, and vice versa, irrespective of whether the host colony (workers) was previously producing predominantly females or males (Fig. 13.13). This suggests that, in this species, the queens have a significant amount of control over the sex ratio! The queens appear to do this by varying the proportion of eggs that are male. In some colonies, they produce almost no males, forcing workers to rear only females, whereas in other colonies they produce only a small proportion of females, forcing the workers to rear them as workers and the males as reproductives.

But sometimes queens win

Overall, these results suggest that sex ratio conflict is a constant tug of war in the hymenopteran social insects. The workers appear to often gain the upper hand, but not always. This is a very active area of research and a major outstanding problem is explaining the variation in who wins, and how different players try to gain the upper hand.

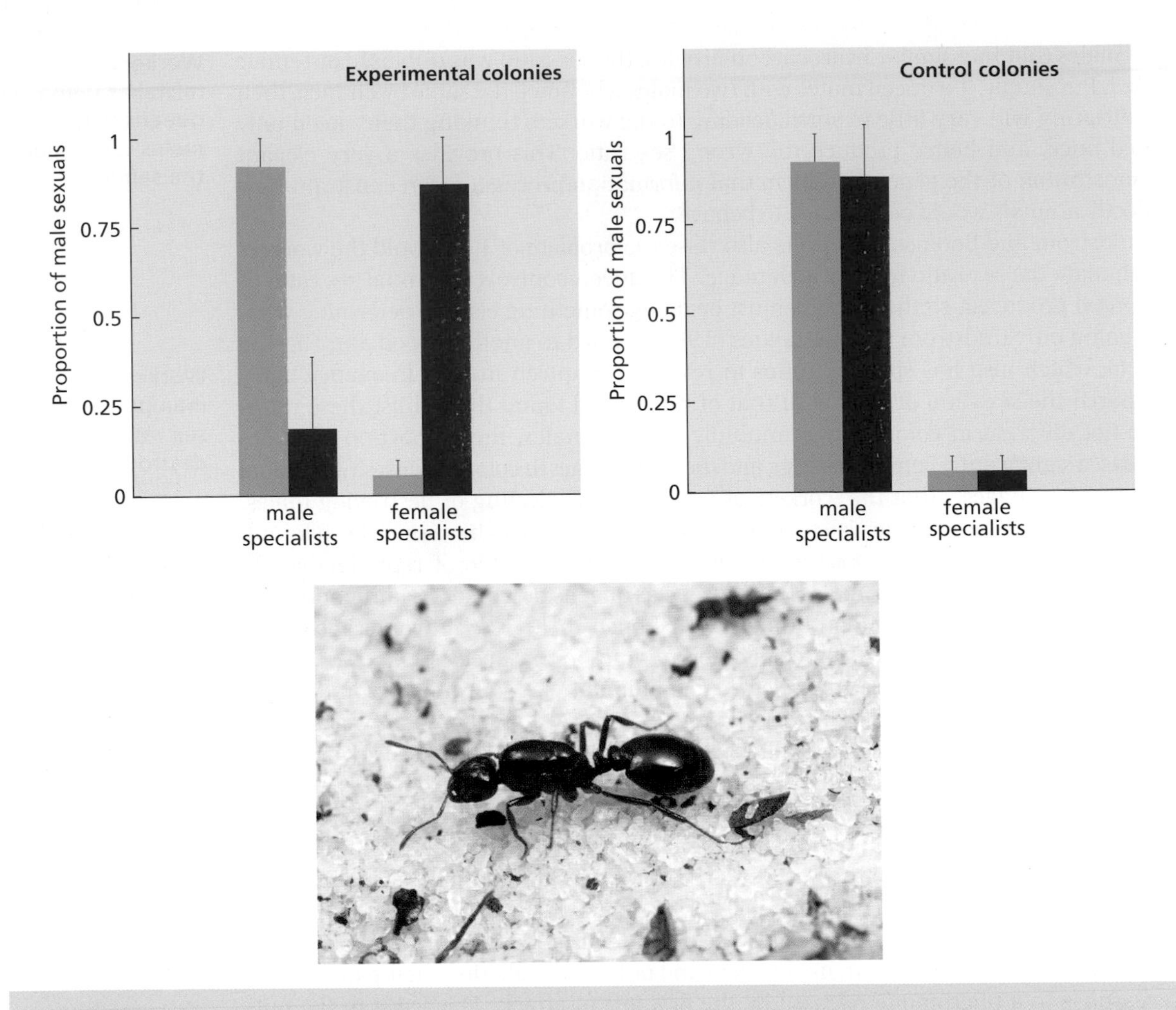

Fig. 13.13 Queen control in the fire ant *Solenopsis invicta*. The proportion of male sexuals raised in colonies both before (light blue bars) and five to six weeks after (dark blue bars) queen exchange experiments. In experimental colonies, queens were exchanged between colonies that had been producing primarily male reproductives and colonies that had been producing primarily female reproductives. In control colonies, queens were swapped between colonies producing primarily the same sex. Overall, following the swapping of queens, the colony sex allocation strategy followed that of the queen, not the workers. From Passera *et al*. (2001). Reprinted with permission from AAAS. Photograph shows a queen looking for a suitable place to start a new colony after her mating flight. Photo © Alex Wild.

Worker policing in the social hymenoptera

In the previous section, we focused on conflict between queens and their workers over the sex ratio. However, disagreements can also occur amongst workers and between workers and their queen over who should produce males. As with sex ratios, the key to understanding this conflict also comes from examining the relatedness structures that result from haplodiploid genetics.

Worker policing in the honeybee

Francis Ratnieks and Kirk Visscher studied conflict over who should produce males in the honeybee, *Apis mellifera*. The possibility for this conflict arises in many species of ants, bees and wasps because whilst workers never mate they are able to lay unfertilized eggs which, being haploid, develop into males (drones). On the basis of genetic relatedness, it is possible to calculate, all other things being equal, which colony members would benefit from worker egg laying. Let us look at the benefits for the queen, the worker that lays an egg and the other workers.

In the honeybee, queens mate multiply with 10–20 males during their nuptial flight. If we assume that offspring are half-siblings, who only share genes through their mother, the relatedness coefficients in Box 13.2 show that:

(1) The queen would prefer her sons to her grandsons (the sons of the workers/her daughters) ($r = 0.5 > r = 0.25$).

(2) The laying worker would prefer her sons to her brothers (the sons of the queen) ($r = 0.5 > r = 0.25$).

(3) Other workers would prefer their brothers (the sons of the queen) to their nephews (the sons of other workers) ($r = 0.25 > r = 0.125$) [remember that multiple mating reduces the relatedness between sisters from 0.75 to 0.375].

Workers would rather rear their own sons, than those of the queen

When a queen mates multiply, workers would rather rear the queen's sons, than those of their sisters

These analyses show that, queens are expected to try and suppress worker reproduction (result 1). However, the more interesting result is that while workers can be favoured to produce sons (result 2), other workers will be selected to suppress this reproduction (result 3) (Woyciechowski & Lomnicki, 1987; Ratnieks, 1988).

Ratnieks and Visscher (1989) showed that such 'worker policing' occurs in the honeybee. They experimentally introduced male eggs that had been laid by either the queen or a worker into colonies and found that while the worker-laid eggs were quickly removed and eaten, the majority of the queen-laid eggs were not (Fig.13.14). The possibility that the queen was responsible for this removal was excluded because Ratnieks and Visscher placed a wire mesh around the combs with their introduced eggs – the workers could pass through but not the larger queens. It is not known how workers determine whether an egg has been laid by the queen or another worker, although it is probably a chemical cue on the queen laid eggs (Martin *et al.*, 2005). Worker policing provides an explanation for worker laid eggs being very rare (<0.1% of males) in species such as the honeybee – although workers can and do lay eggs, they are destroyed by other workers. More generally, worker policing in the honeybee was predicted before it was observed, so this provides a nice example of how theory can lead to the discovery of new, amazing natural history.

In honeybees, workers remove or 'police' the eggs laid by other workers

A further prediction that follows from kin selection theory is that worker policing is not predicted when a queen only mates once. In this case, using the relatedness coefficients given in Table 13.1, whilst results (1) and (2) from above still hold, result (3) does not. The reason for this is that the workers will now be full sisters, related by $r = 0.75$, meaning that a worker would prefer the sons of the other workers (nephews) to the sons of the queen (brothers) ($r = 0.375 > r = 0.25$). Consequently, if the queen only mates once workers are not expected to police the sons produced by other workers.

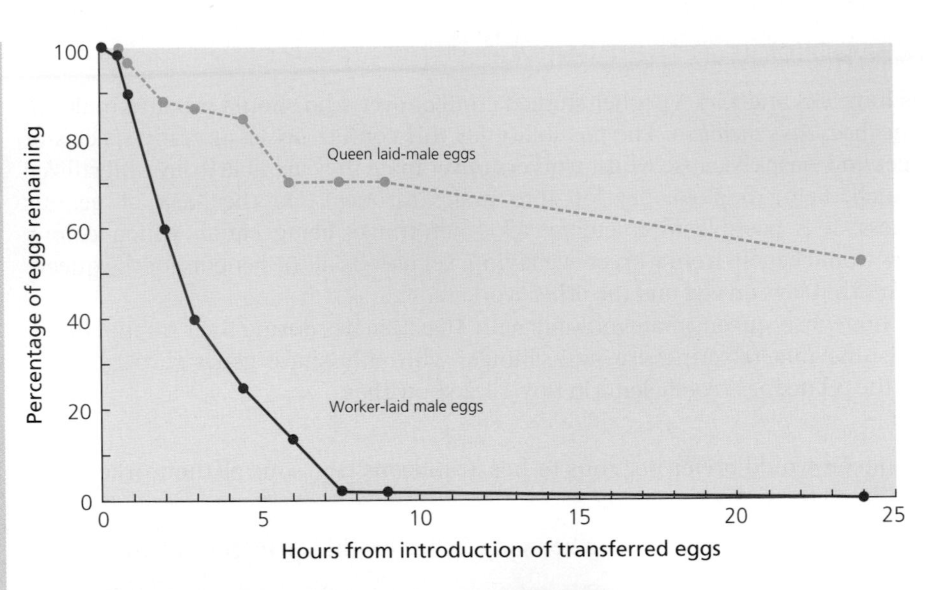

Fig. 13.14 Worker policing in the honeybee. The time course of worker- and queen-laid eggs remaining after experimental introduction into a colony. All worker-laid eggs were rapidly removed. Ratnieks and Visscher (1989). Reprinted with permission from the Nature Publishing Group. Photograph shows a worker inspecting and then removing (policing) an egg laid by another worker. Photo © Francis Ratnieks.

Worker policing is both predicted and observed to be less common when queens mate once

Tom Wenseleers and Francis Ratnieks (2006a) tested this hypothesis in a comparative study across 48 species of ants, bees and wasps by examining whether worker policing was more common in species where queens had greater mating frequencies. As predicted by kin selection theory, they found that in species where queens mated more frequently, and hence workers were less related to the sons produced by other workers, that worker policing was more common (Fig. 13.15).

Worker policing and enforced altruism

Should worker policing influence the egg laying behaviour of workers? When policing is common, this means that any egg laid by a worker is likely to be destroyed. Consequently, this reduces the relative benefit of producing sons, compared with investing time and resources into instead helping rear the sons of the queen. Another way of thinking about this is that policing removes the possibility for direct reproduction and hence

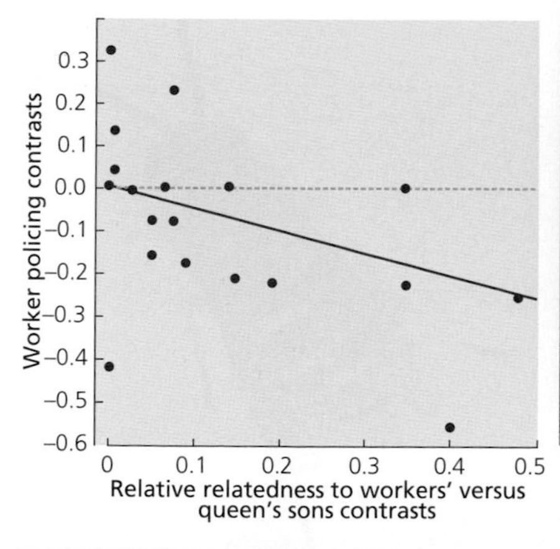

Fig. 13.15 Variation in level of worker policing across 48 species. Plotted are the phylogentically independent contrasts in the level of worker policing behaviour against the constrasts in relatedness to workers versus queens sons. Higher levels of worker policing are observed in species where workers are less related to the sons of other workers (nephews). From Wenseleers and Ratnieks (2006). With permission of the University of Chicago Press. Photograph shows a worker of the leafcutter ant *Acromyrmex echinatior*, in which worker policing occurs. Photo © Alex Wild.

reduces the cost (C) of cooperating – direct reproduction would be failure, so you might as well help. Wenseleers and Ratnieks (2006b) tested this hypothesis across ten bee and wasp species where the effectiveness of policing had been estimated by measuring the proportion of worker-laid eggs that were removed. As predicted, they found that the proportion of workers who produced eggs was negatively correlated with the effectiveness of policing (Fig. 13.16). In other words, policing can be an effective way to enforce altruism amongst relatives because it decreases the relative cost of sterility!

Policing can help select for altruism ...

These results do not suggest an alternative explanation for altruism to kin selection because the reproductive altruism of the workers still relies on the fact that they are helping raise relatives. Instead, they show how enforcement behaviours such as policing can help favour cooperation or altruism by influencing the associated benefits (B) or costs (C). In addition, the work on policing in the social hymenoptera illustrates the general point that different factors can be involved in the evolution and maintenance of behaviours. Eusociality initially evolved in species where queens mated singly, where worker policing is not expected. Then, in lineages where multiple mating arose, this led to worker policing, which could then play a role in maintaining eusociality.

... because it increases the relative kin selected benefit of helping relatives

Superorganisms

Colonies of social insects are sometimes referred to as 'superorganisms' (Wilson & Hölldobler, 2009). This is because the different members of the colony appear to be behaving largely for the good of the colony as a whole, just as different cells in the body of individual organisms act for the good of that individual organism. There are some

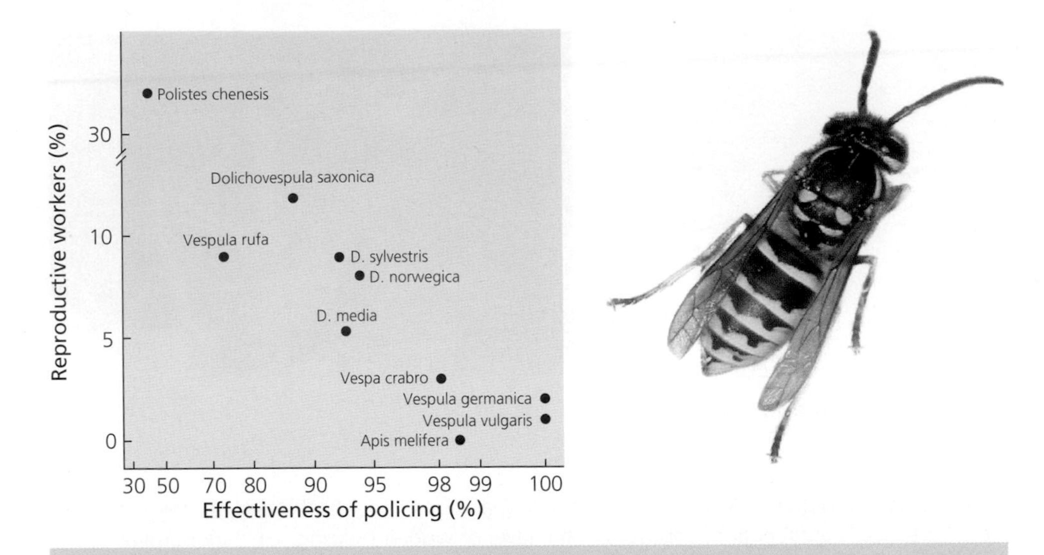

Fig. 13.16 Policing enforces altruism. Lower levels of worker reproduction are observed in wasp and bee species where worker policing is more effective. From Wenseleers and Ratnieks (2006). With permission of the University of Chicago Press. Photograph shows the common wasp *Vespula vulgaris*, which has highly effective worker policing (queens mate multiply) and low levels of worker reproduction. Photo © Tom Wenseleers.

Can we think of social insect colonies as a superorganism?

conflicts within colonies but the importance of these can be reduced by policing. Furthermore, many traits, such as foraging, waste removal and so on, can be completely understood from the perspective of the good of the colony (i.e. as if workers were trying to maximize colony productivity). However, this concept of a superorganism sounds dangerously like Wynne-Edwards' idea of group selection, that we explained was wrong in Chapter 2. How can we reconcile this apparent contradiction?

Andy Gardner and Alan Grafen (2009) addressed this problem theoretically, by asking when natural selection would lead to individuals behaving in a way that maximizes the fitness of the colony or group. They found that this could occur, but only in the restrictive conditions where either relatedness within colonies was sufficiently high (as would be the case with monogamous mating) or policing was so effective that it completely removed any benefit of more selfish behaviour such as worker reproduction within colonies (as would be the case with high levels of policing). So, given the high relatedness and/or effective policing that can occur in many of the eusocial insects (especially those that are obligately eusocial), it does seem valid to think of colonies of species such as the honey bee as a superorganism! The key point here is that behaving for the good of the colony can be selected for under restrictive conditions, but it is not a general evolutionary principle – the general principle is that individuals should maximize their inclusive fitness.

Comparison of vertebrates with insects

With the possible exception of naked mole-rats, there are no known examples of sterile castes in vertebrates, but in other respects there are some close parallels between the

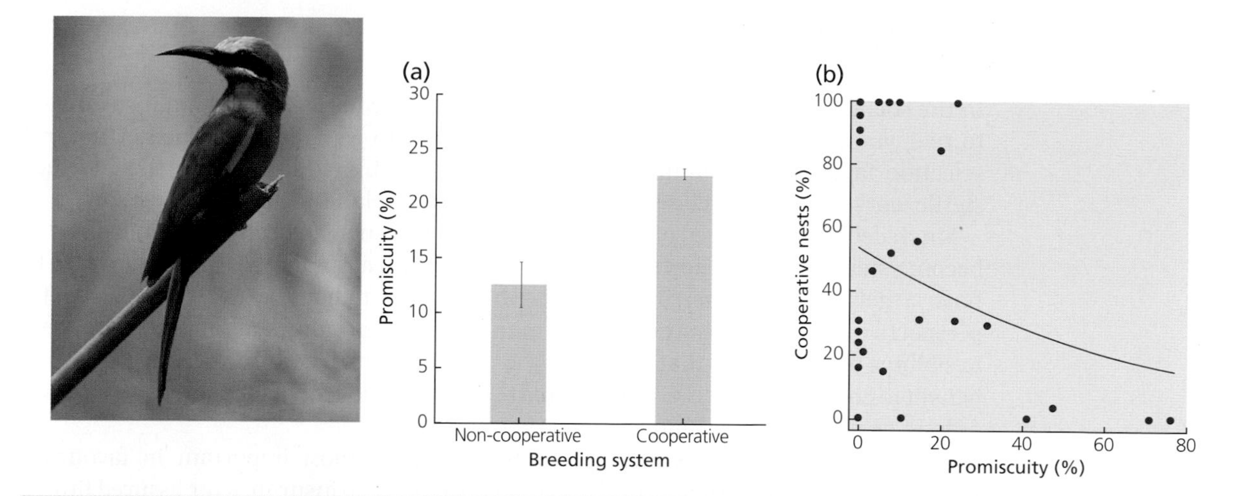

Fig. 13.17 Promiscuity and cooperation in birds. (a) The rate of promiscuity was significantly higher in non-cooperative than in cooperative species. (b) In cooperative species, helpers were present at a lower percentage of nests in species with higher rates of promiscuity. Cornwallis *et al.* (2010). Reprinted with permission from the Nature Publishing Group. Photograph shows the white-fronted bee-eater, a species with low levels of promiscuity (6%) where approximately 50% of nests have helpers. Photo © Erik Svensson.

Insects versus vertebrates: similarities ...

conclusions of this chapter and those of Chapter 12. For example: in both vertebrates and insects, whilst subsocial and parasocial cooperation occur, subsocial appears to be the most important, as observed in long-tailed tits and meerkats; helpers in species such as Seychelles warblers and workers in social insects play a similar role in defence of the nest against predators or parasites, and helping to provision young.

... and differences

Nonetheless, in spite of these similarities there must also be important differences which have led insects but not vertebrates to evolve sterile castes.

Firstly, strict lifetime monogamy seems unlikely to occur for entire populations of vertebrates, reducing the relatedness between helpers and the offspring they could help raise. Consequently, a significant and long term consistent efficiency benefit to cooperation would be required to start down the route to eusociality (i.e. $B/C >> 1$). In species such as long-tailed tits cooperation is only favoured when attempts to breed independently fail, so there is little to lose by becoming a helper. The importance of female promiscuity for retarding the evolution of cooperation in vertebrates has been demonstrated with a comparative study across 267 bird species. Charlie Cornwallis, Ashleigh Griffin and colleagues (Cornwallis *et al.*, 2010) showed that: (1) cooperative breeding is less likely to occur in species with high levels of promiscuity and (2) in cooperative breeding species helping is less common in more promiscuous species (Fig. 13.17).

Promiscuity leads to less cooperation in vertebrates

Secondly, both ecological factors which have been argued to favour the evolution of eusociality in insects are likely to be much less important in vertebrates. The insurance benefits of cooperating to rear a brood through to independence are far lower in vertebrates, where mortality rates are relatively low during the period it takes to rear a brood. The benefits of fortress defence are also much less important in vertebrates, because species such as meerkats and Seychelles warblers do not live in protected sites, where food is obtained. Instead, they must range over wide areas to find food.

Summary

In the social insects there are sterile workers that never have offspring but instead help to rear younger siblings. This appears at first sight to go against the idea of natural selection favouring maximum efficiency at passing on genes. However, the fact that sterile workers help rear close relatives provides the explanation for this altruism.

Much debate has focused on whether haplodiploid genetics predisposes individuals to become sterile workers. However, despite much enthusiasm for this idea, it appears to have largely been a red herring. Instead, what appears to have been key from a genetic perspective is the need for strict lifetime monogamy. Monogamy leads to the relatedness to siblings being equal to that to offspring ($r = 0.5$), in which case only a small efficiency benefit to cooperation ($B/C > 1$) is required to start a species down the route to eusociality as long as it persists over many generations.

The ecological factors which appear to have been most important in favouring eusociality are life insurance and fortress defence. The life insurance (or assured fitness returns) benefits of allowing helpers to complete parental care after the death of the mother are likely to have been important in ants, bees, and wasps. The fortress defence benefits of remaining to help use or defend a food source, when opportunities for successful migration are low, are likely to have been important in aphids, beetles, lower termites, thrips and shrimps.

In the Hymenoptera, haplodiploid genetics leads to conflicts over the sex ratio of reproductives (queens and drones) and who should produce males. Queens and workers disagree over the sex ratio. Whilst there is much data showing how workers win this conflict, they do not win all the time, and there are cases where the queens win. Workers can disagree over who should produce males, leading to the scenario that some workers lay male eggs, but these eggs are destroyed by other workers when most of them are not full sisters. This policing can help enforce altruistic sterility within colonies.

Kin selection is central to explaining both cooperation and conflict in the social insects. It is perhaps ironic that some of the clearest quantitative support for kin selection theory comes from its ability to explain and predict conflicts over the sex ratio and who produces males. The reason is that, in these situations, clear predictions can be made on the basis of relatedness (r) alone, and that we don't have to worry so much about the details of the harder to measure costs (C) and benefits (B).

A comparison of cooperation in insects and vertebrates shows both similarities (e.g. subsocial route, nest defence, provisioning of young, and promiscuity matters) and differences (e.g. strict lifetime monogamy, life insurance and fortress defence are all less important in vertebrates).

Further reading

Strassmann and Queller 2007 review the factors influencing cooperation and conflict in social insects, relating them to the extent to which a colony behaves as an organism. There are a number of reviews focusing on cooperation and conflict in specific social taxa, including the ants (Bourke & Franks, 1995), bees (Schwarz *et al.*, 2007), wasps (Gadagkar, 2009), termites (Thorne, 1997; Korb, 2010), aphids (Stern & Foster, 1996), thrips (Chapman *et al.*, 2008), shrimps (Duffy, 2003) and mole-rats (Bennett & Faulkes, 2000). Currie *et al.* (2010) show how the comparative method can be used to examine the evolution of complexity of human societies.

Hölldobler and Wilson (1990) give a grand tour of ant biology, whilst Helantera *et al.* (2009) discuss the ecology and evolution of unicolonial ants, where populations can consist of a single 'supercolony'. Gordon (1996) discusses variation in the tasks performed by workers, at both the individual and colony level. Powell and Franks (2007) show how individual ants sacrifice themselves for the good of others, even in their foraging behaviour. Mueller *et al.* (2005) review the behavioural ecology of agriculture in the fungus growing social insects.

Trivers and Hare (1976) is a tour de force blend of theory and data, that founded one of the most quantitatively successful areas of behavioural ecology – sex ratio conflict in the social insects. Boomsma's (2007, 2009) papers provide an excellent exposition of the importance of monogamy, and played a key role in simplifying our understanding of how eusociality evolved. Ratnieks *et al.* (2006) review how a number of conflicts are resolved in the social insects, including the sex ratio, rearing of queens, production of males, caste and breeding conflicts between queens. Sex ratio conflict in the social hymenoptera is reviewed in chapter nine of West (2009).

TOPICS FOR DISCUSSION

1. Discuss whether polyembryonic wasps and flatworms are eusocial (Grbic *et al.*, 1992; Crespi & Yanega, 1995; Hechinger *et al.*, 2011)? Could you test your ideas?
2. Why are all workers female in the hymenoptera?
3. Discuss the implications of there being a genetic component to caste determination in a eusocial species (Hughes *et al.*, 2003).
4. Given what we know about mating systems in natural populations, has the monogamy hypothesis made it too easy to explain eusociality?
5. Would worker production of sons help or hinder the way to eusociality (Charnov, 1978)?
6. Extended parental care can favour the evolution of eusociality through 'insurance benefits'. But why should extended parental care have evolved in the first place (Field & Brace, 2004)?
7. Why would the replacement of lost queens by one of her daughters lead to split sex ratios (Boomsma, 1991; Mueller, 1991)?
8. Split sex ratios lead to some colonies where only male reproductives (drones) are produced. This means that the males that mated with the queen gain no reproductive success (haplodiploidy means they only pass genes to daughters). Can the males do anything about this (Boomsma, 1996; Sundström & Boomsma, 2000)?
9. Worker policing occurs in some hymenopteran species where the workers are equally related to their nephews and their sons. Why would this have evolved? How would you test your ideas?
10. Compare the factors that have favoured the evolution of eusociality and multicellularity (Queller, 2000; Grosberg and Strathmann, 2007; Herron and Michod, 2008; Boomsma, 2009).
11. How would you test empirically whether haplodiploidy has favoured the evolution of eusociality?
12. Nowak *et al.* (2010) suggest that kin selection theory delivers only 'hypothetical explanations' and 'does not deliver any additional biological insight' into the social insects. Discuss.

Photos © Elizabeth Tibbetts

CHAPTER 14
Communication and Signals

Most of the interactions between individuals described in this book involve communication. Males attract females with showy ornaments or repel rivals with loud roars, offspring beg from their parents, bacteria release chemicals to coordinate cooperative behaviours at the population level and poisonous caterpillars warn their predators away with bright colours. All of these cases involve the use of special *signals* or *displays* that appear to have been designed for use in communication. This chapter is about how natural selection shapes signals.

Communication involves a sender and a receiver

The most obvious characteristic of a signal is that it is a feature of one individual (sender) that in some way modifies the behaviour of another (receiver). The receiver's response may be immediate and obvious (a male firefly rapidly flies towards a flashing female of the correct species); it may be subtle and difficult to detect (a male antelope slightly alters its direction of walking to avoid crossing a territorial boundary when it detects the scent marks of a resident); it may be delayed in time (the ovaries of a female budgerigar gradually develop as the result of the stimulus of male song); or it may not occur all the time (a territorial male blackbird sings for several hours during which time only one or two intruders hear the song and retreat).

Signals will only be evolutionarily stable if, on average, both sender and receiver benefit from the signalling system

It used to be assumed that signals evolved for the efficient transfer of information between sender and receiver. Dawkins and Krebs (1978; Krebs & Dawkins, 1984) pointed out that this idea will usually be incorrect, because when there is a conflict of interest between sender and receiver, each will be selected to pursue their own selfish interests. This led to them developing a more antagonistic idea that communication is an arms race between senders trying to manipulate receivers, and receivers mind-reading senders. However, whilst there will often be a conflict of interest between sender and receiver, it is also true that if natural selection cannot lead to a situation where, on average, both parties benefit from the exchange, then the signalling system would not be stable (Maynard Smith & Harper, 2003). If senders manipulated receivers too well, then receivers would be selected to ignore senders. So, signals must carry information

An Introduction to Behavioural Ecology, Fourth Edition. Nicholas B. Davies, John R. Krebs and Stuart A. West.

of interest to the receiver, about the state or future intentions of the signaller, and this information must be correct often enough for the receiver to be selected to respond to it.

The problem from an evolutionary perspective is what keeps a signal honest, such that the receiver benefits by responding to it? An honest signal would be one which conveys correct or useful information. For an alarm call, this would be information about a predator. For a displaying male, this would be information about their genetic or phenotypic quality as a mate. The problem is that senders could do better if they could lie or exaggerate their signals, so we must ask what prevents this and keeps signals honest? What stops a male peacock from producing such a large and impressive tail that all females choose to mate with it, or why don't all insects signal that they are poisonous to their vertebrate predators? If individuals did lie and exaggerate, then their signals would not carry useful information, there would be no advantage in paying attention to such signals and so the communication system would break down.

A signal is honest if, on average, it conveys correct or useful information

Consequently, the central problem of communication is what maintains the honesty of signals? Before we can address this question, we must first distinguish between the different types of interaction and communication.

What maintains the honesty of signals?

The types of communication

It is useful to distinguish between two types of interaction: cues and signals (Maynard Smith & Harper, 2003). A cue is when the receiver uses some feature of the sender to guide their behaviour, but this feature has not evolved for that purpose. An example of a cue is when a mosquito that is searching for a mammal to bite will fly up wind if it detects carbon dioxide. In this case, the carbon dioxide acts a cue to the mosquito (indicating the presence of a source of blood), but mammals do not produce carbon dioxide to signal their presence to mosquitoes (they would rather not be bitten!).

A cue is a feature of the world that can be used as a guide to future action

Signals are acts or structures produced by the sender that alter the behaviour of the receiver; they have evolved because of that effect and are effective because the receiver's response has evolved (Maynard Smith & Harper, 2003). This definition may seem convoluted but it has two important consequences. Firstly, because the receiver has evolved to respond to the signal, the response must, on average, benefit the receiver. Otherwise the receiver would be selected to not respond to the signal. The term 'on average' is key here because some individuals can produce deceptive signals, which are successful because they exploit the honest signalling of others; we shall return to this later in the chapter.

A signal is an act or structure that alters the behaviour of another organism, which evolved because of that effect and which is effective because the receiver's response has also evolved.

On average, the response to a signal provides a benefit to the receiver

The second consequence of this definition is that, because the signal has evolved owing to its effect on others, this distinguishes a signal from a cue. Consider contests between funnel-web spiders, *Agelenopsis aperta*, over web sites (Fig. 14.1). Susan Riechert (1978, 1984) found that when spiders differed in weight, the smaller spider retreated rather than fighting over the site. She confirmed this experimentally by gluing a flattened piece of lead shot to the back of some spiders and showing that this made larger spiders retreat from them. If the spiders were able to directly assess weight, independent of the actions of the other spider, then weight would just be a cue of size and fighting ability. This is because weight did not evolve to signal fighting ability – larger spiders just have an advantage in fights. However, spiders actually signal their size by vibrating the web. This act of vibrating the web would be

Signals must be able to evolve independently of the feature they carry information about

Fig. 14.1 The North American funnel-web spider, *Agelenopsis aperta*. Photo © Visuals Unlimited.

a signal if it has evolved because it influences the behaviour of another spider by providing information about weight. The key point here is that the signal must be able to evolve independently of the feature of the signaller about which it conveys information – spiders can vary if and how they vibrate a web irrespective of their weight and, hence, fighting ability. In contrast, fighting ability is tightly linked to weight.

In the rest of this chapter we will focus on signals, discussing how they can evolve and the form that they take. The major problem that we must consider is what keeps signals reliable or honest, such that receivers will be selected to respond to them? Put another way, what stops senders exaggerating their signals to deceive the receivers to their own advantage?

The problem of signal reliability

Consider an extreme signal such as the tail of the male Indian peafowl *Pavo cristatus*. On page 199 we described how females choose to mate with the males that have the most eyespots in their tail, and that the number of eyespots is a reliable signal of the genetic quality of males. However, this raises the problem of why don't lower quality males just put more eyespots in their train, to deceive females into thinking they are higher quality, and therefore obtain more mates? In this section we shall discuss three hypothetical answers to this problem, following the conceptual framework laid out by John Maynard Smith and David Harper (2003). Our aim is not to argue which one is correct, but rather to use the peafowl as a means to describing the three possible ways that the honesty of signal can be maintained.

An index is a signal that cannot be faked

One possibility is that the size of a tail is constrained by the size and, therefore, quality of a male, such that poor quality males just aren't big enough to be able to carry extra eyespots. In this case, the number of eyespots is a reliable signal because it cannot be faked – only high quality males are able to carry extra eyespots. Such a signal is termed an *index*, which is formally defined as a signal whose intensity is causally related to the quality being signalled and which cannot be faked.

A handicap is a signal that is costly to fake

Another possibility is that poor quality males can make the extra eyespots but that it would be very costly for them to do so. Perhaps if they invest resources into building extra eyespots, then they would have reduced resources to put into other actions, such as immune function, which would lead to them being plagued by parasites. In this case, the number of eyespots is a reliable signal because whilst it could be faked, the cost of doing so makes it relatively inefficient for low quality males (i.e. cost of producing a larger signal would outweigh the benefit of increased mating success). In contrast, high quality males have sufficient resources that the benefit of making extra eyespots (increased mating success) outweighs the cost. Such a signal is termed a *handicap*, which is formally defined as a signal whose reliability is ensured because it is costly to produce or has costly consequences.

The final possibility is that males and females have a shared interest in males honestly signalling their quality to females. Perhaps males and females form monogamous pairs for life (peafowl do not, but this a hypothetical example), and pairs that are matched in quality are able to raise more offspring. Possibly because they are genetically matched their offspring are better at avoiding parasites. In this case, the number of eyespots is a reliable signal of quality because while it could be faked, the sender and receiver both do better if the sender signals honestly to the receiver. The honesty is therefore ensured because the sender and receiver have a *common (or coincident) interest* in honest signalling.

Signals can also be reliable if the sender and receiver have a common interest

To summarize, there are three possible reasons to ensure signal honesty: index, handicap and common interest. With an index, dishonesty is not possible. With a handicap or when there is a common interest, dishonesty is possible but unprofitable. The unprofitability arises either through the cost of signalling (handicaps) or through other means (common interest). Put simply, indices can't be faked, handicaps are too costly to be faked and there is nothing to be gained from faking a signal when there is common interest.

There are three ways to explain reliable signals – indices, handicaps and common interest...

Whilst it is very useful to distinguish between these different ways in which signals can be reliable, it is unfortunately not always easy to disentangle them empirically. For example, the Hamilton–Zuk hypothesis for sexual displays described on page 196 could work by either an index or a handicap mechanism, with both predicting that sexually attracted traits would be more exaggerated in species where parasite pressure is greater.

... – but they can be hard to distinguish between

In the following sections, examples of these different kinds of signals, and the ways in which their reliability is maintained, are discussed.

Indices

As discussed in Chapters 5 and 7, individuals frequently have to compete with others for scarce resources such as food, territories or mates. When this happens, individuals often use sequences of various displays, in which they assess each other's fighting ability, after which the individual with the lower fighting ability backs down, rather than engage in a serious fight (Box 14.1). We shall find that the cues used to signal and assess fighting ability are often reliable because they cannot be faked, and are therefore indices.

Assessment is a key feature of many contests ...

Consider how red deer stags, *Cervus elaphus*, compete for females in the autumn rut. A male's reproductive success depends on fighting ability; the strongest stags are able to command the largest harems and enjoy the most copulations. Although fighting brings great potential benefits, it also entails serious costs. Almost all males suffer some slight injuries and between 20 and 30% of stags will become permanently injured sometime during their lives, through broken legs or being blinded by an antler point, for example. Fighting costs are minimized in competing stags by assessment of each other's fighting potential, which avoids heavily unmatched contests (Clutton-Brock & Albon, 1979).

... which helps prevent escalation to costly fights

In the first stage of the display, the harem holder and challenger roar at each other (Fig. 14.2a). Roaring is a good signal of fighting ability because it provides information about the size of a male and to roar well a stag has to be in good physical condition. If the defender can out-roar the intruder, the intruder retreats. If the challenger out-roars the defender, or matches him, then the contest moves to the second stage, where both stags engage in a parallel walk (Fig. 14.2b). This presumably enables them to assess each other more closely. Many fights end at this stage, but if the contestants are still equally matched then a serious fight ensues where they interlock antlers and push against each other (Fig. 14.2c). Body weight and skilful footwork are important determinants of victory, but there is a

Red dear assess fighting ability with roaring and parallel walks ...

... only escalating to physical conflict when equally matched

BOX 14.1 SEQUENTIAL ASSESSMENT

Magnus Enquist and Olof Leimar (1983, 1987, 1990) suggest that information accumulates during a contest in a way similar to statistical sampling. The outcome of a single bout contains a random error and to get a more accurate estimate contestants must 'increase the sample size' by repeating the behaviour. On this view we would expect the least costly displays to be used first of all, followed by the more costly but more accurate means of assessment. An individual will give up when it assesses its fighting ability to be less than that of its opponent. With more closely matched opponents, fights should last longer and escalate further simply because it will take longer to assess which is the strongest, just as a statistician needs to sample more to detect a small difference than a large one.

This idea makes good sense of contests involving sequences of displays, such as those of the South American cichlid fish *Nannacara anomala*. In nature, males compete for females who are about to spawn and contests can be readily staged in laboratory tanks. The order of displays in a contest is very consistent, with constant rates of behaviour within each phase (Fig. B14.1.1). (a) Initially

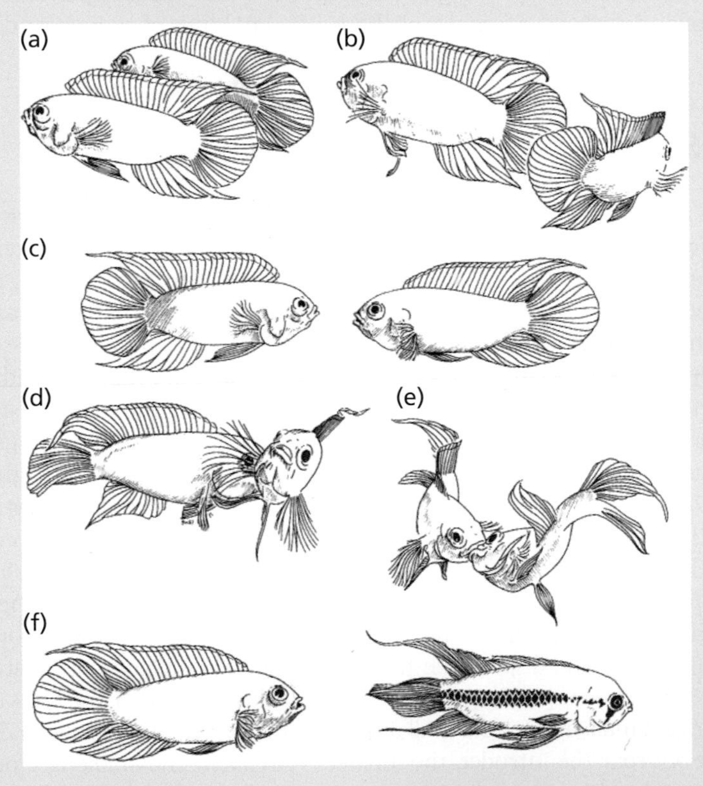

Fig. B14.1.1 Fighting sequence of male cichlid fish, *Nannacara anomala*. (a) Lateral orientation. (b) Tail beating. (c) Frontal orientation. (d) Biting. (e) Mouth wrestling. (f) The loser (right) gives up. Jakobsson *et al*. (1979). Drawing by Bibbi Mayrhofer.

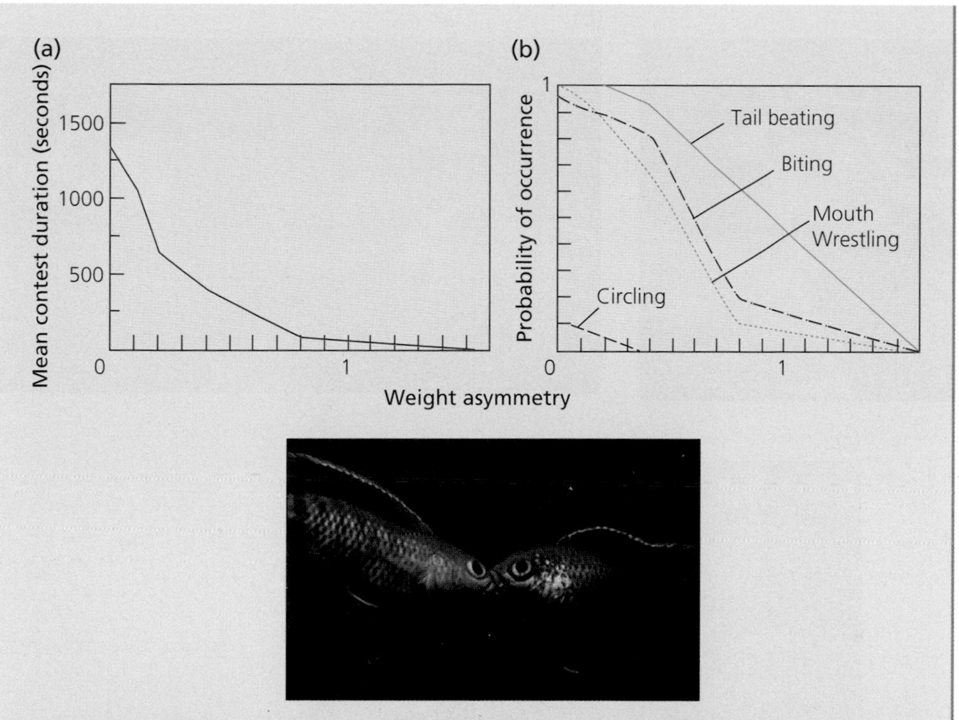

Fig. B14.1.2 Fighting in the cichlid fish, *Nannacara anomala*, is more prolonged (a) and escalates to more dangerous stages (b) the more closely the contestants are matched in size. Weight asymmetry is measured as the logarithm of the weight of the heavier fish divided by the weight of the lighter fish (0 = weight ratio of 1, i.e. equal weights). From Enquist and Leimar (1990). With permission from Elsevier.

the males orientate laterally with erect fins. (b) Then they engage in tail beating, pushing a stream of water towards the flank of an opponent. This involves alternating roles in which each fish takes turns between performing a tail beat and manoeuvring into a lateral position to receive the stream from its opponent's beat. (c) Then biting increases in frequency and the contestants begin to orientate frontally. (d) This is followed by mouth wrestling, where the males get a firm grip of each other's jaws and have a pushing and pulling contest, allowing more accurate assessment of relative strength. (e) Finally, circling occurs where both fish swim fast in a tight circle and try to bite the back of their opponent. At the end of the contest, losing fish signal their defeat by folding their fins, changing colour and retreating. Larger males were more likely to win, and with very large differences in size the smaller fish gave up more or less immediately at stage (a). As the difference in size decreased, however, fights became longer and escalated through the sequence with increasing risks of injury at successive stages (Fig. B14.1.2). The contests described in the main text between red deer stags provide another example of sequential assessment.

Fig. 14.2 Stages of a fight between two red deer stags. The harem holder roars at the challenger (a). Then the pair engage in a parallel walk (b). Finally they interlock antlers and push against each other (c). Photos by Tim Clutton-Brock.

chance that even the winner may get injured. The important point is that these escalated fights are rare and most contests are settled at an earlier stage by displays.

Cues of fighting ability are often reliable because they are indices

This raises the question of why individuals do not exaggerate their fighting ability in displays such as roars and parallel walks to make even stronger individuals back down? In this section, we will consider two examples in detail, toads and red deer, where the answer is that fighting ability cannot be exaggerated, so the displays are honest signals because they are indices.

Fighting assessment and deep croaks in toads

The croak of the common toad, *Bufo bufo,* may provide an index of size and fighting ability. Each spring, toads mass together in ponds to breed, with all spawning taking place in one to two weeks. Females only visit the pond for a small number of days, so there are always substantially more males than females, leading to intense competition for mates. In the field, males would sometimes attempt to dislodge mating males from the back of a female, by pushing in between the copulating pair. These takeover attempts were more likely to be successful when the interrupting male was larger than the paired male, and most takeover attempts involved a larger male trying to dislodge a smaller male.

Male toads displace smaller males

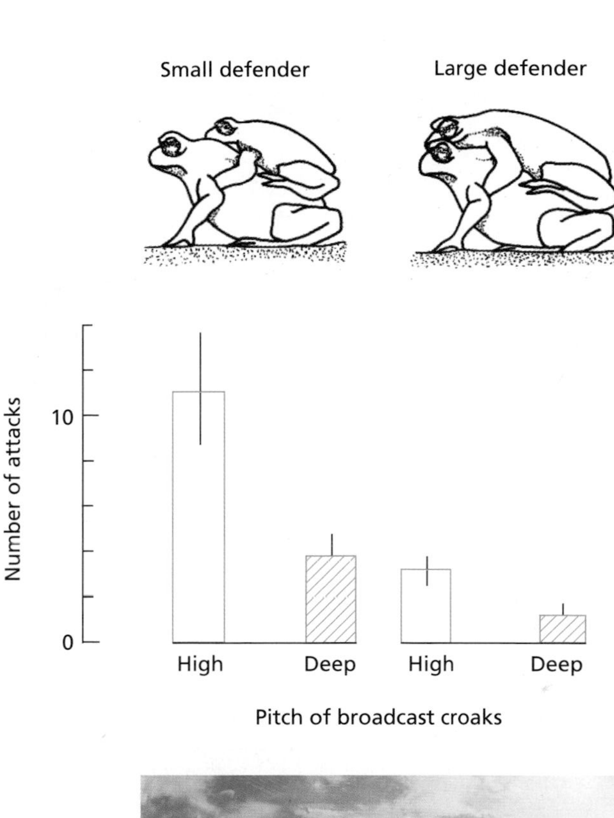

Fig. 14.3 Medium size males are more likely to attack silenced males when the high-pitched croak of a smaller male was played. However, croaks cannot be the only assessment cue because for either croak pitch there are fewer attacks at large defenders. The strength of a defender's kick may also be important. From Davies and Halliday (1978). Photo © Jurgen Freund/ naturepl.com

How does the attacking male assess the size and fighting ability of his rival? Whenever a male attacked a pair, the mating male always called. In this and many other species of frogs and toads, the pitch of a male's croak is closely related to his body size – the larger the male, the larger the vocal cords and so the deeper the croak. Consequently, croak could be a reliable index of size, and hence fighting ability. Davies and Halliday (1978) tested this idea with an experiment in which they allowed medium-sized males to attack either small or large paired males, which were silenced by means of a rubber band placed behind their arms and passing through their mouths like a horse's bit. During the attack, they used a loudspeaker next to the mating pair, to broadcasted tape-recorded croaks of either a large or a small male. For both sizes of defender they found that there were fewer attacks when the deep croaks of a large male were played than when the high-pitched croaks of a small male were broadcast (Fig. 14.3). Therefore, croak pitch is used to assess a rival's body size, which is a good predictor of fighting ability because large males are more difficult to displace.

The pitch of a croak is a reliable index of body size

Males are more likely to try and displace males when they are played a high-pitched croak

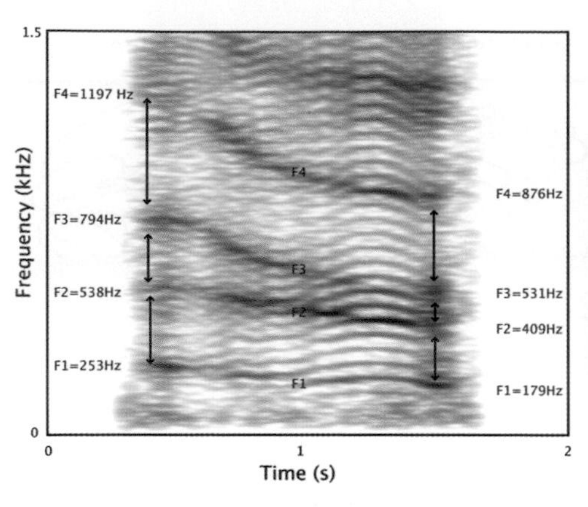

Fig. 14.4 Spectrogram of a red deer roar representing the distribution of the energy (in grey levels) across time (x axis) and frequency (y axis). The first four formants are visible as dark bands of energy (labelled F1 to F4), which decrease throughout the vocalization. The spacing between the formants is shown by the arrowed lines. The overall spacing between the formants (or 'formant dispersion') is estimated by linear regression and changes from 339 Hz at the beginning of the roar (corresponding to a 51.7 cm vocal tract) to 243 Hz at the end of the roar (corresponding to a 72 cm vocal tract). Red deer stags have a descended and mobile larynx that enables them to lengthen their vocal tract during roaring, causing the observed drop in formant frequencies. The minimum formant frequencies attained at the end of the roar reflect the fully extended vocal tract and, therefore, communicate information about body size. From Reby and McComb (2003). With permission from Elsevier. Photo © David Reby.

Red Deer

We now return to the example of red deer roars, which have been studied in depth by David Reby, Karen McComb and colleagues. As discussed above, roaring is used in the assessment process by which stags choose whether to escalate conflicts to physical conflict. Do stags use some acoustic property of the roar, as with pitch in toads, to assess the size of competitors and, therefore, their likely fighting ability?

Things are slightly more complex in mammals such as red deer, because the size of larynx can be varied independently of body size and so, in contrast to toads, the pitch of sound produced by the vocal cords is not a reliable index of body size. However, acoustic theory predicts that the resonant frequencies of the vocal tract (termed formants) and the average distance between them (termed formant dispersion) should be negatively correlated with vocal tract length (Fig. 14.4; Fitch & Reby, 2001). Consequently, if vocal tract length were constrained by body size, then formant frequencies and dispersion would provide a reliable index of body size. Reby and McComb (2003) tested this by comparing the roars of different stags from an intensively studied population on the island of Rum (Inner Hebrides, UK). They found that, as predicted, larger stags produced lower formant frequencies (Fig. 14.5a). Furthermore, the long-term data on the population showed that stags

Formant frequencies are a reliable index of body size in red deer

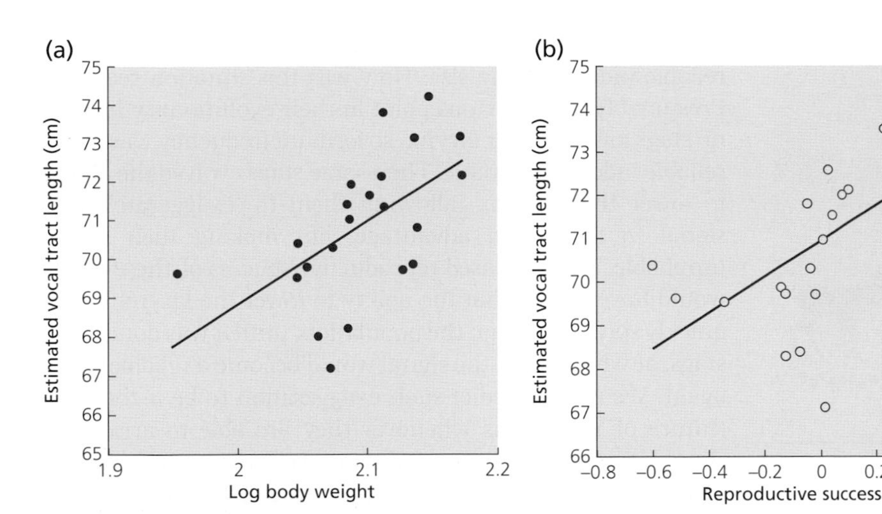

Fig. 14.5 Formant frequency, body size and reproductive success in male red deer. The length of the vocal tract can be estimated by the formant frequencies, as longer tracts lead to lower frequencies. The estimated vocal tract length is greater in (a) larger deer and (b) deer with a higher lifetime reproductive success. From Reby and McComb (2003). With permission from Elsevier.

with lower formant frequencies also had greater reproductive success, in terms of number of offspring fathered (Fig. 14.5b).

While, these results show that formant frequency is a reliable index of body size, they do not show that other deer use this information as a signal. Reby *et al.* (2005) tested this by playing synthesized vocalizations to harem holding stag holders to simulate the intrusion of a new challenger stag during the rutting season. They used recordings made from four mature stags in the 1980s whose roars had never been heard by the males on which the experiments were run, and re-synthesized them to produce roars with formant frequencies characteristic of a small, medium or large male. When these were played at close range to harem-holding stags, the stags paid more attention to, and roared more frequently, in response to roars suggesting larger opponents. Furthermore, when harem-holding stags are challenged with roars of larger opponents, they produce roars with lower formant spacing, suggesting a more fully extended vocal tract. So not only do stags roar more in response to the challenge of a larger male, but they make themselves sound bigger. More recently, Charlton *et al.* (2007) have shown that females also use formant frequencies as a signal of body size. When they were played the synthesized roars of small and large stags, females preferentially moved towards the speakers that were playing the roar of a large male.

Formant frequency is used as a signal of body size by stags ...

... and female deer

A puzzling feature of the roars produced by stags is that they are of much lower frequency than those produced by females, or by other similar sized animals. Fitch and Reby (2001) showed that this is because stags have a low resting larynx position relative to other mammals, and that they retract their larynx further when roaring. This retraction elongates the vocal tract and lowers the formant frequencies of their roar

Stags have dropped their larynx and elongated their vocal tract.

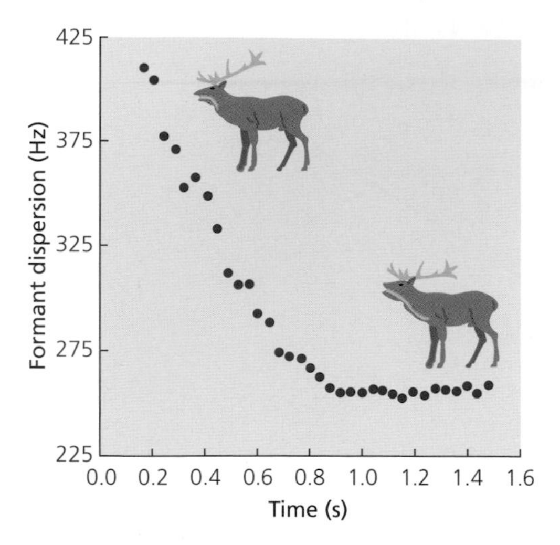

Fig. 14.6 As red deer stags roar, they retract their larynx, and hence increase the length of their vocal tract. This results in the production of lower formant frequencies. From Fitch and Reby (2001). With permission of the Royal Society.

(Fig. 14.6). Since all stags do this, formant dispersion is a reliable index of body size. How was this situation reached? Presumably, at a previous point in their evolutionary history, no stags lowered their larynx, so formant frequency was also a reliable index of body size. Then some stags evolved the ability to lower their larynx, allowing them to 'exaggerate' their signal to their own advantage, but making their signal unreliable. The increased reproductive success of these males would have meant that the ability to lower the larynx would quickly spread through the population, until it was done by all stags, at which point the signal would become a reliable index again. We might predict such exaggeration to be a common feature of indices, as whenever they are able to arise they should spread rapidly to fixation. An even more extreme example of an extended larynx is provided by the strange looking male hammerhead bat, *Hypsignathus monstrosus*, where over half its body is its larynx (Fig. 14.7; Bradbury, 1977). Other possible examples of traits that could have evolved to exaggerate size include when mammals such as dogs raise the fur along their neck and back as part of an aggressive display, or when threatened puffer fish expand by filling their elastic stomachs with water.

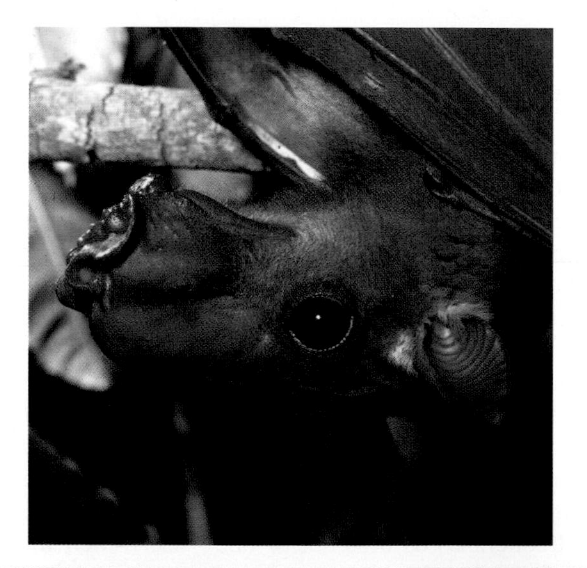

Fig. 14.7 A male hammerhead bat, *Hypsignathus monstrosus*. Males compete for mates in leks that contain 30–150 males, by signalling to females with loud honking calls. These calls appear to have a large influence on mating success, with just 6% of the males obtaining 79% of the copulations. Males have a bizarre morphology that was presumably selected because it exaggerates their calls, with an enormous bony larynx that fills their chest cavity and a head with enlarged cheek pouches, inflated nasal cavities and a funnel shaped mouth. Photo © CNRS Photothèque/Devez, Alain R.

Fig. 14.8 Extravagant signals. (a) The Royal flycatcher, *Onychorhynchus coronatus*. Photo © Joseph Tobias. (b) Bird of paradise, *Paradisaea rubra*. Photo © Tim Laman/nature/pl.com (c) The bower of a Vogelkop Gardener Bowerbird, *Amblyornis inornata*, consists of a cone-shaped hut, in front of which is an area that is kept clear of debris and decorated with items such as flowers, fruit, beetle wings and leaves. Photo © Richard Kirby/naturepl.com. (d) Male superb Lyrebirds, *Menura novaehollandiae*, have an amazing mimicking ability, producing a song that is a rich mixture of their own song and other sounds that they have heard, such as the songs of other birds, and human noises, ranging from camera shutters to chainsaws to car alarms. Photo © J. Hauke/Blickwinkel/Specialist Stock.

Handicaps

Animal signals can be costly and extravagant

Animal signals are often obvious to us because they are extravagant and showy. The volume and diversity of the dawn chorus, the flamboyant feathers and dances of birds of paradise, or the colours and sweet scents of flowers (Fig. 14.8). How could such extravagant and presumably costly displays be favoured by natural selection? Amotz

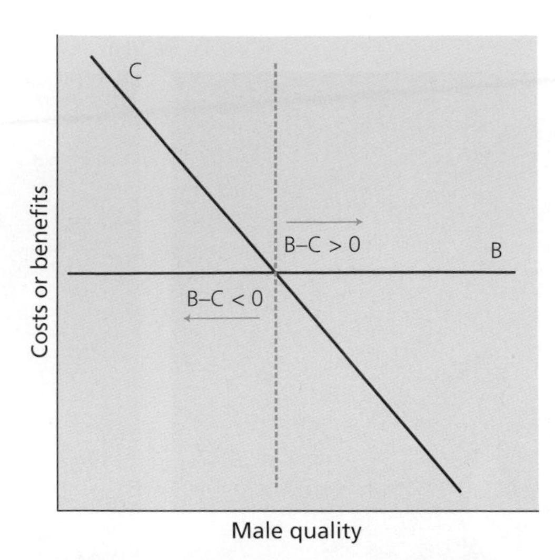

Fig. 14.9 The handicap principle and sexual selection. The benefit (*B*) of producing a costly signal, such as the increased mating success from having an ornament, is assumed to be roughly equal for all males. The cost (*C*) of producing the costly ornament is assumed to be lower for higher quality males, because they are in better condition and have additional resources to invest in ornament production. In this case, the benefit of producing the ornament only outweighs the cost ($B - C > 0$) for high quality males, so only high quality males are selected to produce the ornament, making the ornament a reliable signal of male quality. More generally, the handicap principle requires that the cost to benefit ratio is lower for individuals giving stronger signals.

Zahavi (1975) proposed that such displays would be favoured precisely because they are costly, and that it is the cost that makes them reliable. He called this the handicap principle.

Zahavi's idea was extremely controversial throughout the 1970s and 1980s. Indeed, one important theoretical paper was simply titled 'The handicap mechanism of sexual selection does not work' (Kirkpatrick, 1986). Much discussion was in the context of sexual selection, and whether males produced costly ornaments to signal quality to females. It was argued that if a female mated with a high quality male who produced costly ornaments, then their offspring would gain the benefit of good genes from their father, but that this would be negated by the cost of then producing ornaments themselves. Consequently, there would be no fitness advantage to either producing the ornament or basing mate choice upon it (Maynard Smith, 1976b).

Costly signals are reliable if the relative cost of signalling is greater for lower quality individuals

Alan Grafen (1990a, 1990b) solved this debate with a decisive pair of theoretical papers. He showed that the handicap principle could work but that it would only do so when the fitness cost of producing the costly ornament was greater for low quality males. The idea here is that both high and low quality males could produce costly ornaments, but that the ornament would be so costly for low quality males that the cost would outweigh the benefit. In contrast, with high quality males the benefit would outweigh the cost, so only high quality males would be selected to produce the ornament (Fig. 14.9).

Stalk-eyed flies

The handicap principle predicts strong condition dependence

In Chapter 7, we described how in the stalk-eyed fly, *Teleopsis dalmanni*, females choose to mate with males with relatively larger eye spans. How can we distinguish between male eye span being an index or a handicap? A clear prediction of the handicap model

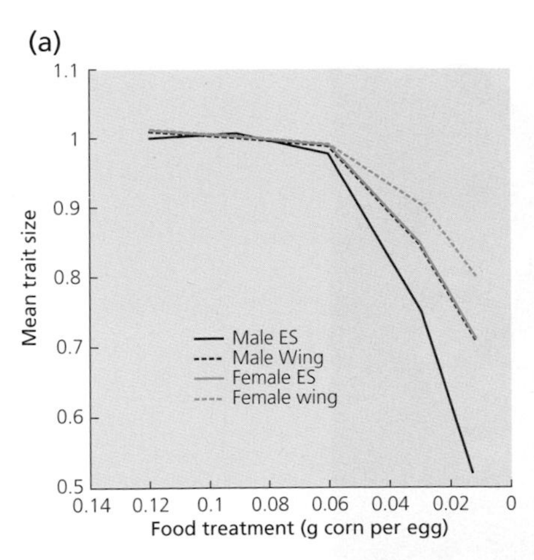

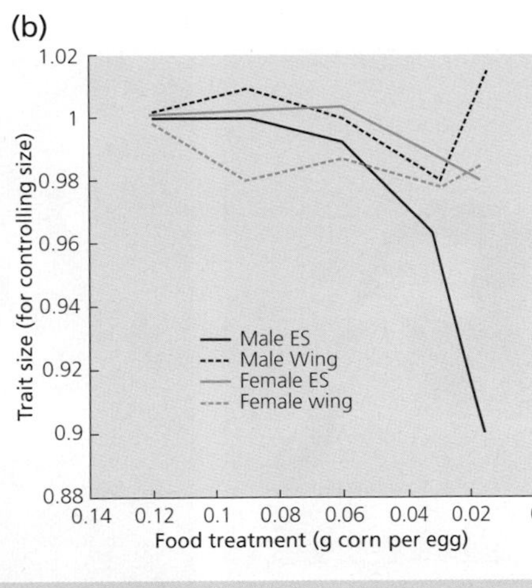

Fig. 14.10 Male eye span and wing size in response to food treatment, in the stalk eyed fly, *T. dalmanni*. Male eye span (ES) showed a much steeper decline with reduced food than male wing size, female eye span or female wing size. This result held when analysing absolute eye span and wing size (a), as well as when variation in body size was controlled for with multiple regressions (b). From Cotton *et al.* (2004). With permission of the Royal Society. Photograph of a flying male. Photo © Samuel Cotton.

is that high-quality individuals should be favoured to produce the signal (e.g. ornament), whereas low-quality individuals should not. This 'strong condition dependence' contrasts with indices, which are expected to vary more gradually with quality, alongside other measures of size or quality, as with the croaks of toads and roars of deer (the fact that this has to occur is what defines an index).

Cotton *et al.* (2004a) tested this with a laboratory population of *T. dalmanni*, manipulating condition by varying the amount of pureed corn that they fed to larvae. They found that absolute male eye span decreased sharply in the lower condition treatments, and that this decline was much greater than that with female eye span or wing traits (Fig. 14.10a). This greater decline in male eye span also persisted when they statistically controlled for variation in body size across the treatments, demonstrating that poorer condition males had relatively smaller eye spans (Fig. 14.10b). Overall, these results suggest that, as predicted by the handicap model, high-quality males allocate more resources to the sexual signal of eye span.

The eye stalks of male stalk-eyed flies show strong condition dependence, with larger males having relatively larger eye spans

Badges of status

Signals of quality or status can sometimes have low productions costs. In the Harris sparrow (*Zonotrichia querula*), individuals with larger black bibs on their chest are the dominants and they always displace pale birds from food supplies (Rohwer & Rohwer, 1978). Similar visual signals occur in a number of other bird species (Fig. 14.11). Given that such 'badges of status' are likely to have low production costs, why don't subordinate sparrows grow bigger bibs and enjoy a rise in status? In other words, what maintains the honesty of these signals? Dawkins and Krebs (1978; Krebs & Dawkins, 1984) suggested that individuals who falsely signalled their ability would pay social costs,

Relatively cost-free signals of status ...

Fig. 14.11 Possible badges of status. (a) The black throat patch or bib of male house sparrows (*Passer domesticus*). Photo © Tony Barakat. (b) The white forehead patch of the collared flycatcher, *Ficedula albicollis*. Photo © Thor Veen.

... could be maintained by the social cost of dishonest signals

through increased aggression or punishment. Consequently, what would maintain honesty is not the cost of producing the signal but rather the cost that would have to be paid if a dishonest signal was produced.

Elizabeth Tibbetts and colleagues studied the facial patterns of female paper wasps, *Polistes dominulus*. This is a social species, where females often cooperate to found colonies. Dominance determines the amount that each female reproduces, as well as settling conflict in many other scenarios, such as the order of queen succession, sharing of food and the probability of becoming a future queen. There is remarkable variation across individuals in the number, size and shape of black spots on the otherwise yellow region just above the mouth, the clypeus (Fig. 14.12). Tibbetts and Dale (2004) showed that this variation was highly correlated to dominance – more dominant individuals had more broken colouration, with more black spots. To an extent this was because larger individuals had more spots, but the relationship held even when size was controlled for, with the dominance rank between two wasps of the same size predicted by facial colouring. Furthermore, wasps show clear behavioural differences in response to facing markings. Tibbetts and Lindsay (2008) took pairs of same sized wasps with one black spot, and used yellow paint to make one of the wasps have zero spots and the other two. They then killed these wasps, placed each of them as a 'guard' of two sugar cubes in a corner of a triangular arena, and introduced a third wasp to see where she would go. In 39 out of 48 cases, the introduced wasp chose the food source with the guard who had less spots on her face.

Black facial spots are a signal of status in hover wasps

These results beg the question of why don't wasps just exaggerate their size and dominance by producing more spots? The production of spots is not likely to be costly, because the black pigment required accounts for <1% of the total black pigment on a wasp, and what matters anyway is how the black pigment is broken up into spots, not the total amount. Tibbetts and Dale (2004) tested whether the apparent honesty of facial markings could be explained by social costs that are imposed when individuals do not signal honestly. They took pairs of similar sized wasps that had not previously interacted, painted the face of one of them, to manipulate the number of spots, and then placed them together. The wasps then battled for dominance, with the eventual winner easily identified by 'mount' displays, where the loser lowers its antennae and allows

Fig. 14.12 Portraits of nine *P. dominulus* foundresses collected in Ithaca, New York, representing some of the diversity in facial patterns. The central wasp has no black clypeus pigmentation, the remaining wasps have one to three spots. In 158 randomly collected foundresses, 19.6% of foundresses had an entirely yellow clypeus, 65.8% had a single black spot, 12.7% had two spots and 1.9% had three spots. On average, 13% of a wasp's clypeus was pigmented black; this value ranged broadly from 0 to 39%. Photos © Elizabeth Tibbetts.

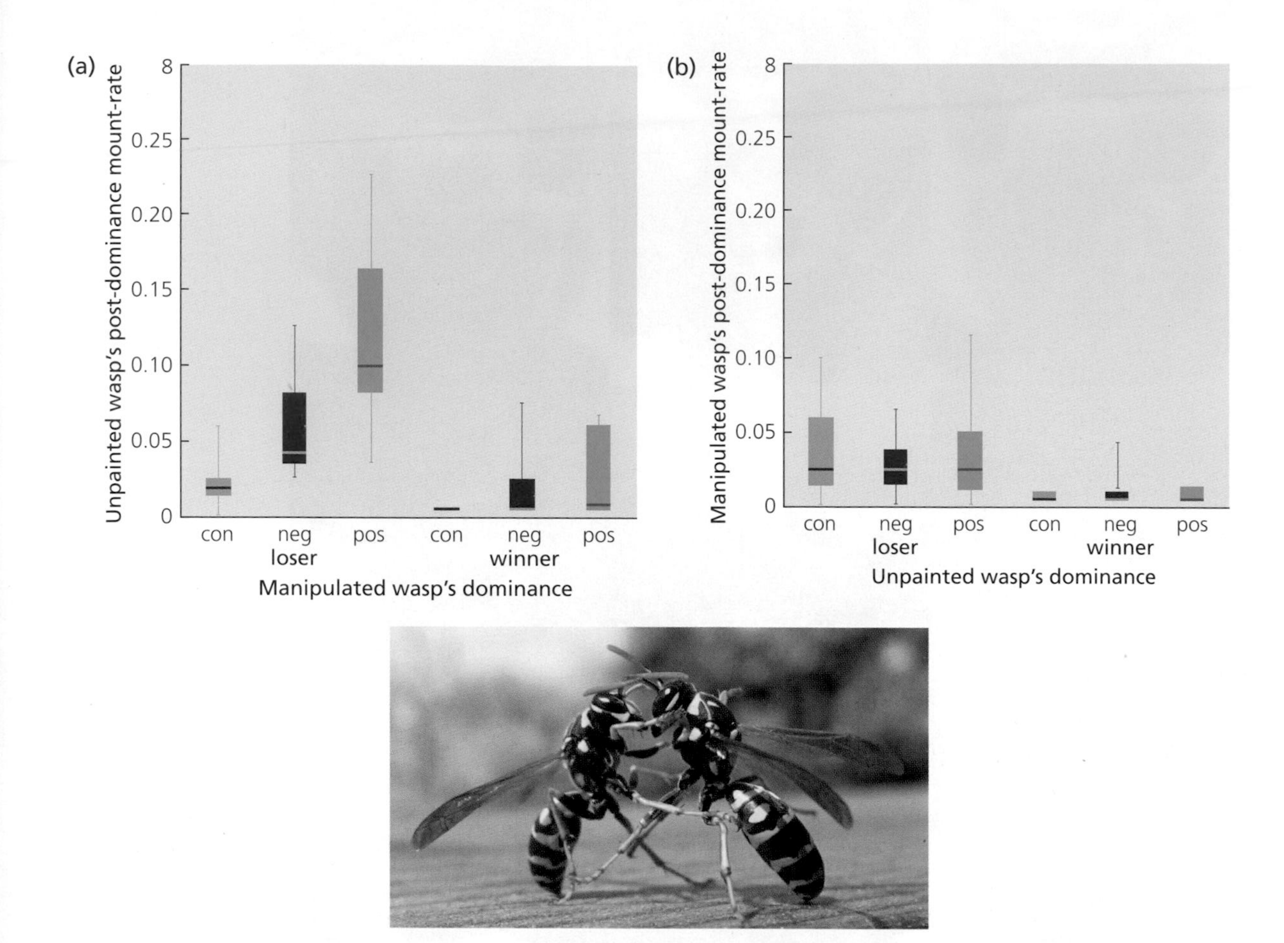

Fig. 14.13 The social cost of a dishonest badge of status in paper wasps. The mounting rate per minute (aggression) of the (a) manipulated (painted) and (b) unmanipulated wasps after dominance was sorted. The different columns correspond to: pairs in which the manipulated wasp was painted to have more (pos), less (neg) or the same (con) number of black spots when the manipulated wasp was the winner or loser of the contest for dominance. The manipulation caused no change in the behaviour of the manipulated wasp (b), but wasps given more spots that lost the contest for dominance experienced higher levels of aggression from the other wasp (a). From Tibbetts and Dale (2004). Reprinted with permission from the Nature Publishing Group. Photograph shows two females battling for dominance. Photo © Elizabeth Tibbetts.

Wasps that signal above their status are punished with aggression

the winner to climb on its head. Whilst the painting manipulation had no influence on who won dominance, it had a significant influence on behaviour after the battle for dominance. Losers who had been manipulated to have more spots received approximately six times more aggression than controls who had not been manipulated (Fig. 14.13a). In contrast, when manipulated individuals won dominance, the painting manipulation did not influence the level of aggression that they showed to the loser (Fig. 14.13b). Thus, while manipulation of face markings did not alter how wasps behaved, it did influence how they were treated. Overall, these results suggest that it is the social cost which maintains the honesty of facial signals in hover wasps.

The cost (or not) of handicaps and other signals

Honest handicap signals need not be costly ...

... and costly signals are not necessarily handicaps

In the previous section we showed how the honesty of a relatively cost-free signal could be maintained by a social cost (handicap) imposed on dishonest signallers. The key point here is that the handicap principle requires that the marginal cost of producing a larger signal would be greater for lower quality individuals, not that the signal is costly for all to produce. Consequently, honest handicap signals are not necessarily costly to produce. However, it is also true that when a signal is costly, it will not necessarily be a handicap. To illustrate this it is useful to distinguish between the efficacy and strategic costs of signals (Guilford & Dawkins, 1991). Efficacy costs are those required to ensure the information is reliably perceived, such as the energy required for a toad to croak or a human to speak. Strategic costs are those required to maintain the honesty of a handicap signal. Consequently, all forms of signal (index, handicap and common interest) will have some efficacy costs, the importance of which will vary, whereas only handicaps will entail strategic costs.

Efficacy costs ...

... and strategic costs

Other possible examples of handicap signalling are discussed in Chapter 7, in the context of sexual selection, and Chapter 8, in the context of offspring begging for food from their parents. This is still a very active area of research, where it has often proved much harder to decisively distinguish between indices and handicaps in empirical studies than in theoretical models.

Common interest

Honest signalling and common interest

The final possible explanation for honest signalling is that the sender and receiver have a common interest, such that there is no benefit to be had from deceiving the receiver. The easiest way in which this can occur is if the sender and receiver are genetically related, such that the sender gains a kin selected benefit by signalling honestly to the receiver.

A breathtaking example of this phenomenon is provided by the waggle dance of the honeybee (Fig. 14.14). When a foraging honeybee returns to its colony, it performs a dance to inform the other workers in the colony where food (pollen) can be found. The duration and the orientation of the dance provide information on both the distance and direction of the food source. That honeybees are selected to signal this information honestly is not surprising, because the worker gains a kin selected benefit telling other workers where to find food: those workers will then feed to relatives of the dancer. Whilst the waggle dance provides an example of a signal that can be explained by common interest, it does not allow the influence of relatedness to be tested directly, as honeybees are related and dance in all colonies. However, it has been possible to examine the consequences of variation in relatedness with a form of signalling between bacteria that is termed quorum sensing.

Genetic relatedness (kin selection) is one way to get common interest

Quorum sensing in bacteria

Quorum sensing is a form of between cell signalling in bacteria, used to regulate the production of molecules which are then released into the environment to aid growth. In many bacterial species, cells produce small diffusible signalling molecules that they

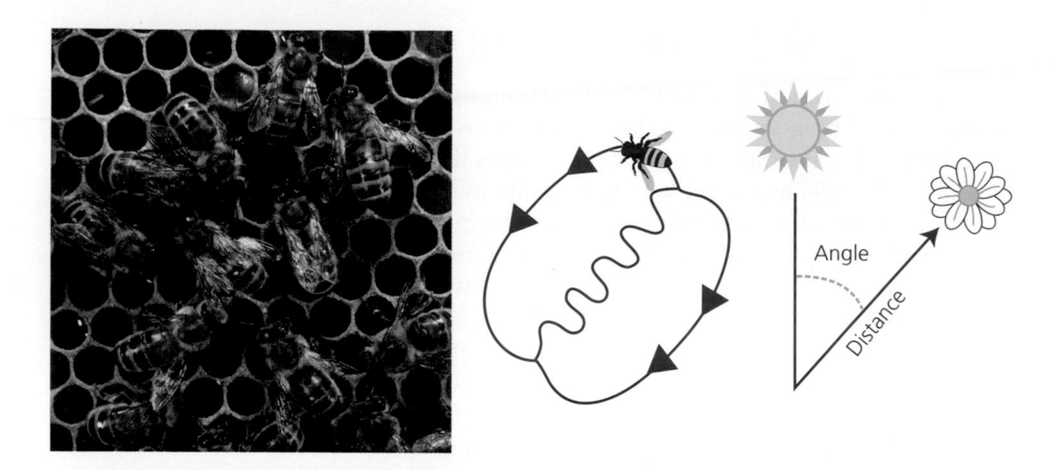

Fig. 14.14 The honey bee waggle dance provides information about both the distance and direction of nectar sources. After returning to the hive, successful foragers dance in a figure of eight pattern on the vertical comb. Distance to the nectar source is encoded by length of the straight line part of the dance, during which the workers waggle from side to side. The direction of the straight line part of the dance relative to the vertical plane of the honeycomb shows the direction of the nectar source relative to the sun. Photo © Kim Taylor/naturepl.com

release into the environment (Williams *et al.*, 2007). These molecules are then taken up by other bacterial cells where they have two consequences. Firstly, the uptake of signalling molecules stimulates the production and release of a number of other molecules, termed exoproducts. These exoproducts have a variety of uses that facilitate bacterial growth and success, including: enzymes that break down proteins, polymers to provide structure for growth, surfactants to facilitate movement and toxins to break down host tissues. Secondly, the uptake of signalling molecules stimulates the production of the signal molecule itself. This leads to positive feedback at high cell densities (when the quorum is reached), which results in a marked increase in production of both signal and exoproducts.

Quorum sensing allows bacteria to switch on certain behaviours at high population densities

It has been argued that the process of quorum sensing represents signalling between bacterial cells to coordinate the production of exoproducts. The idea here is that the production of exoproducts is only worthwhile at high cell densities, where they are likely to provide a significant benefit to the local cells. In contrast, at low densities, exoproducts would disperse into the environment before they can benefit cells, making their production relatively inefficient (Fig. 14.15). In the microbiology literature it has long been assumed that quorum sensing would be favoured to do this, because it would provide a benefit at the population level (e.g. Shapiro, 1998; Henke & Bassler, 2004).

However, as explained in Chapter 1, natural selection does not work in this way. Specifically, quorum sensing cells could be exploited by cells that avoided the cost of producing the signal themselves, and just responded to the signals of others, or by cells

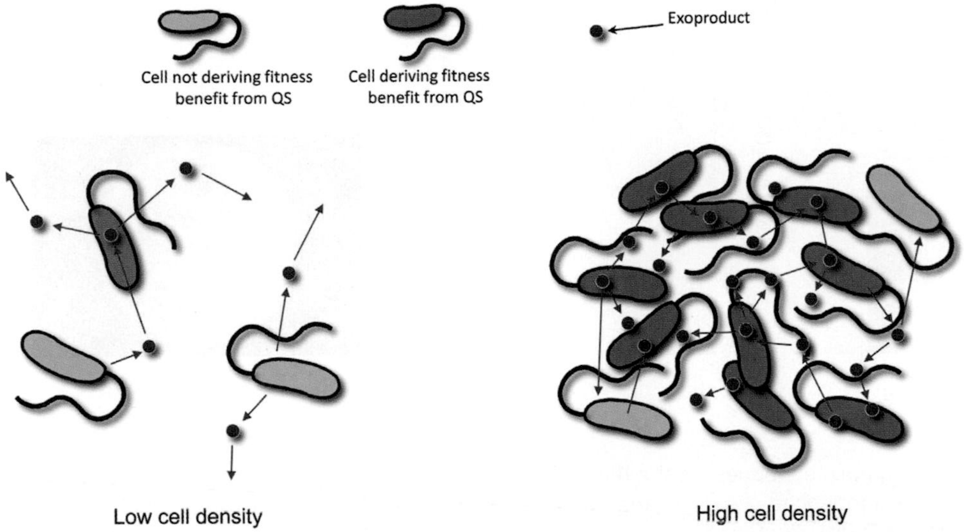

Fig. 14.15 The hypothesized function of quorum sensing. At low cell densities, a large proportion of the extracellular factors (common goods or exoenzymes) disperse before they can be used, so their production provides little fitness benefit. At high cell densities, a much greater proportion of the extracellular common goods (or the products they produce) can be used. Consequently, the production of extracellular common goods is more efficient and beneficial at higher population densities.

that avoided the cost of responding to signals, or even by cells that signalled at a greater rate to make other cells produce more exoproducts. What prevents cells from doing this, and hence keeps them signalling honestly? Sam Brown and Rufus Johnstone (2001) developed a theoretical model, which showed that quorum sensing could be evolutionarily stable but that this required cells having a common interest due to being related. When cells are growing alongside close relatives, kin selection provides a shared interest in signalling to coordinate their behaviours in order to help each other.

Quorum sensing requires that we can explain both honest signalling and the cooperative production of exoproducts ...

... which can be done by kin selection if cells are related

Steve Diggle and colleagues tested this experimentally with the bacterium *Pseudomonas aeruginosa*. They exploited the fact that by working with bacteria you can do things which just are not possible with animals such as peacocks or deer, such as make genetic mutants and follow selection in multigeneration experiments. Diggle *et al.* (2007) knocked out genes in the signalling pathway to create two genetic mutants, one that did not produce signal (signal negative) and one that did not respond to signal (signal blind). They first used these mutants to test whether signalling occurred between cells. This was necessary because it is plausible that all or most of the signal and benefits from exoproducts just flow back to the cell that produced them, in which case quorum sensing would not be between-cell signalling. To resolve this question, they grew normal quorum sensing (wild type) cells and mutants, either alone (monoculture) or in mixtures. When grown in monoculture, the wild type grew to larger densities than either of the mutants because it allowed the production of

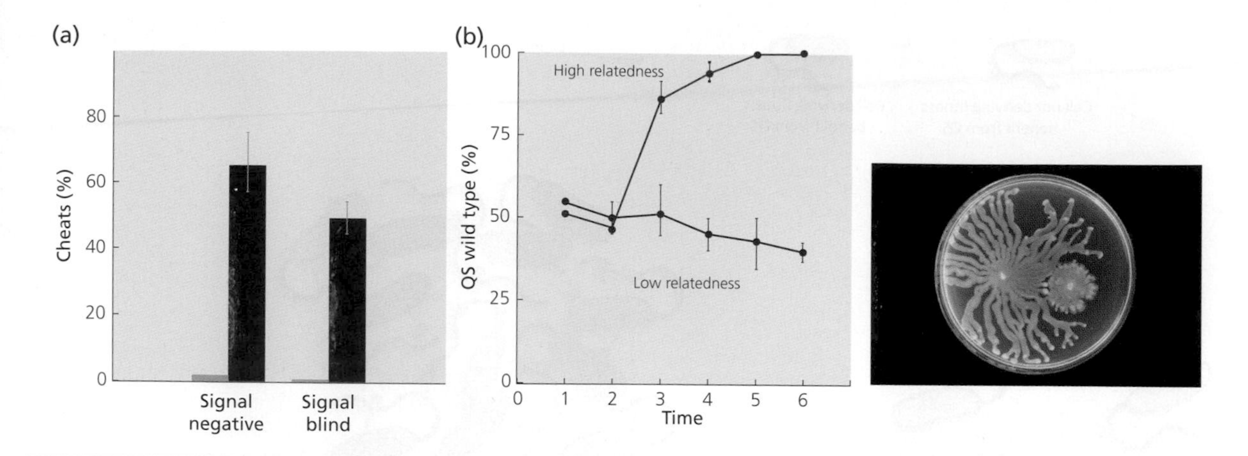

Fig. 14.16 Quorum sensing and relatedness in the bacterium *Pseudomonas aeruginosa*. (a) Quorum sensing signal negative and signal blind cheats invade populations of wild type cooperators over 48 h of growth. The mutants were distinguished from the wild type by labelling with a green fluorescent protein. Light blue and dark blue bars represent the starting and final percentage of cheats in the population, respectively (± s.e.m). (b) The proportion (± s.e.m.) of quorum sensing individuals (wild type) is plotted against time (rounds of growth). Blue points represent relatively low relatedness and red points represent relatively high relatedness. The experiment was started with an equal mixture of the wild type and signal blind mutants that didn't respond to signal. High relatedness selects for quorum sensing, whereas low relatedness allows mutants that do not respond to signal to be maintained in the population. From Diggle *et al.* (2007). Reprinted with permission from the Nature Publishing Group. Photograph shows a normal quorum sensing colony (left) and a colony consisting of (signal blind) mutants that do not quorum sense (right). Photo © Steve Diggle.

Individuals that don't signal, or don't respond to signal, can invade populations of quorum sensing cells

exoproducts to be switched on at high population densities. In contrast, when mutants and the wild type were grown in a mixture, both types of mutant were able to increase in frequency, showing that they can exploit the signalling or response to signal of the wild type, without paying the cost (Fig. 14.16a). This showed that quorum sensing can be exploited by individuals who do not signal honestly and, hence, that the problem of honest signalling occurs.

Quorum sensing is favoured when interacting cells are more highly related

Diggle *et al.* then tested whether the stability of quorum sensing can be explained by a common interest that results from a high relatedness. To do this they initiated populations with a mixture of wild type cells that quorum sensed, and signal blind mutants that did not respond to signal. They then maintained the populations under conditions that would lead to either a high or low average relatedness. They did this by splitting the population into subpopulations and then initiating each subpopulation, in each round of growth, with either a single bacterial clone (relatively high relatedness) or with a 20 μl sample that contained about 1.2×10^9 cells (lower relatedness). This experiment was, therefore, very similar in design to the experiment described in Chapter 11, which tested how relatedness influenced the cooperative production of iron scavenging siderophore molecules. As predicted by Brown and Johnstone's model, the signal blind mutant was rapidly lost from the

Fig. 14.17 The fruit fly *Drosophila subobscura*. Photo © Stephen Dalton/ naturepl.com.

population when relatedness was relatively high, but increased in frequency when relatedness was lower (Fig. 14.16b).

Courtship and receptivity in fruit flies

A common interest between signallers does not have to rely upon genetic relatedness. John Maynard Smith (1956) provided a clear and simple example of this in the fruit fly *Drosophila subobscura* (Fig. 14.17). Males of this species are incredibly enthusiastic about mating, and will even attempt to mate with lumps of wax about the size of a fly if they are moved in appropriate manner. When placed with an actual female, a male will perform a courtship dance in front of her for up to an hour, occasionally tapping her with his front legs, even if the female rejects him. However, if a male is placed with a mated female, she will extrude her ovipositor and bend her abdomen towards him, which causes him to cease his courtship immediately. The reason for this appears to be that, in this species, forced copulation is impossible and females never mate multiply. The female wants to signal to the male that she is already mated, because she does not want to be harassed by a male. The male wants to know when the female is mated, so that he doesn't waste time trying to court here. Consequently, both partners have a shared interest in the female signalling when she has already been mated.

If females only mate once, then females and males have a common interest in females honestly signalling when they have mated

Food calls

Another potential example of signalling with a common interest between non-relatives is provided by the vocalizations made by individuals who have found a source of food.

These 'food calls' have been observed in a wide range of birds and primates, and a number of possible explanations have been given, which include attracting others to increase safety from predation, attracting potential mates and claiming ownership of a resource (Searcy & Nowicki, 2005).

Chirruping attracts other house sparrows

Mark Elgar (1986a, 1986b) observed that house sparrows often gave 'chirrup' calls when they found food and that when others joined them they then flew down to feed together. Playback of these calls confirmed that they attracted others. Furthermore, sparrows adjust their level of calling in response to the likely benefit of attracting others – individuals chirruped less when food was harder to share (a single lump of bread rather than crumbs) and when predation risk was lower (nearer to cover; see also the section on feeding and danger in Chapter 4).

Human language

The evolution of human language was a major evolutionary transition that represents a crucial difference between humans and other animals (Maynard Smith & Szathmary, 1995). Humans have a vocabulary of thousands of words, which can be arranged into sentences capable of carrying an indefinitely large number of meanings. This has led to the possibility for rapid cultural evolution, with traits being passed from one generation to the next through teaching or social learning, rather than through genes. Most research into human language has come from the field of linguistics, which has focused on more mechanistic and proximate issues. However, we can also examine language from a behavioural ecology perspective, asking questions such as what factors would have favoured the evolution of language, and what maintains the honesty of language?

Honesty in human language could be maintained by the social cost of lying and/or common interest

Considering what maintains the honesty of human language, it is possible to imagine that both social costs of lying (punishment or reputation; a form of handicap analogous to badges of status) and common interest could have played a role, possibly in different situations. For example, one individual might tell another which restaurant they are going to because they want to meet there; whereas another individual might give information about whether they have spare food because their reputation would suffer if it was later found out that they had lied. Recent research has started to open up possibilities for examining such issues, by examining the evolution of language in controlled laboratory settings. Bruno Galantucci (2005) tested whether novel communication systems could rapidly evolve in a laboratory setting. People were put into pairs and made to play a game on the computer where the financially rewarded aim was to get two 'agents' into the same room. The players were not allowed to interact directly and instead could only communicate by drawing on a small digitizing pad, which then sent the drawing to the other players. Galantucci found that whilst the common interest of the financial reward led to most pairs solving the game by developing a communication system, some did not, even after 160 minutes. In the pairs where a communication system did arise, it took different forms. In some cases communication was icon based, with individuals using different symbols to label the rooms, whereas in others communication was map based, with individuals using the position or orientation of a symbol to indicate the room. This area of research is still in its infancy. It remains to be seen how such experiments can be scaled up to more complex language and more natural settings, or whether different selective factors can lead to different forms of communication.

Dishonest signals

A signal is dishonest when the sender does something that manipulates the behaviour of the receiver to the benefit of the sender and the detriment of the receiver

In the examples so far, we have focused on asking when and why signals would be honest. However, nature is also full of examples of dishonest or coercive signalling. A classic example of this is the lure that an anglerfish uses to attract other fish, which it can then prey on (Fig. 14.18). The anglerfish (sender) clearly benefits from the behavioural response to the lure by attracting prey, whereas the fish that respond to the prey (receiver) pays the cost of being eaten. In Chapter 4 we discussed another example of dishonesty, termed Batesian mimicry, where palatable species mimic the appearance of unpalatable species. For example, potential predators avoid hoverflies because they look like wasps.

Why don't receivers just ignore dishonest signals? Presumably the answer is that the response is, on average, a beneficial thing to do. So, for example, fish need to eat, and worm-like things are usually worms, not the lures of anglerfish, so it pays to try to head towards them. Similarly, wasp-like things are usually wasps not hoverflies, so it pays to avoid them. This suggests that dishonest signalling will only be evolutionarily stable when it occurs at a relatively low frequency.

Leena Lindström and colleagues (1997) tested how the benefit of dishonest signalling varies with its frequency, by experimentally mimicking an unpalatable prey and a palatable Batesian mimic. They allowed great tits to feed on mealworms, some of which were made unpalatable by dipping in a chloroquinine solution. They attached small blue cake decorations to these dipped mealworms, as a signal of unpalatability, but also attached these decorations to a variable proportion of the undipped mealworms, to make a Batesian mimic. When the great tits were allowed to feed on the mealworms, it was found that they were more likely to eat the decorated mealworms when a higher proportion of them were mimics. Hence, the fitness of those deceptive mimics decreased as they became more common.

Fig. 14.18 Photograph of an anglerfish. Photo © David Shale/naturepl.com.

This example raises a number of hard questions that we will examine with a couple of more specific examples. Firstly, when we observe deception, are we looking at a stable equilibrium of honest and dishonest signalling, or are we looking at signalling system that is breaking down and on its way out? Secondly, if we are looking a stable equilibrium, what keeps deception at a low enough frequency to prevent the system breaking down?

Fork-tailed drongos make deceptive alarm calls

A particularly impressive example of deception comes from Tom Flower's (2011) work on alarm calls (Box 14.2) by fork-tailed drongos in the South African Kalahari Desert. Fork-tailed drongos normally forage alone, catching insects on the wing, or lizards and crickets on the ground. However, they sometimes follow other species, such as cooperative groups of pied babblers or meerkats, catching food flushed by these other species, or stealing food that has already been caught. This stealing amounts to almost a quarter of the drongo's food intake and was achieved in two ways, by either directly attacking or by taking food the forager had abandoned after the drongos made a call from a nearby perch. Three sources of evidence suggested that these calls from the perch were dishonest alarm calls, made to make the forager flee from their food.

Fork-tailed drongos steal food from pied babblers and meerkats

BOX 14.2 ALARM CALLS

Animals often give alarms when they detect a predator (Zuberbühler, 2009; Magrath *et al.*, 2010). These alarms may warn mates, offspring or other kin (Chapter 11); they may attract others who join in mobbing the predator; or they may signal to the predator 'I've seen you' and so dissuade an attack which depends on surprise for success.

Different alarms are often given for different predators. For example, vervet monkeys have different alarm calls for leopards, eagles and snakes. Playback of the calls elicits different responses to each alarm, suited to the avoidance of the different modes of attack of each predator: running into a tree after leopard alarms; looking up and seeking shelter in a dense bush after eagle alarms; and looking down at the surrounding ground after snake alarms (Seyfarth *et al.*, 1980). Alarms may be varied, too, in response to the level of urgency, becoming noisier or more rapid as the predator approaches (Manser, 2001).

Amimals also often eavesdrop on the alarms of other species. For example, vervet monkeys respond appropriately to both ground and aerial predator alarms of superb starlings (*Spreo superbus*) (Seyfarth & Cheney, 1990). The ability to eavesdrop on other species' alarms may sometimes arise because of their acoustic similarity to conspecific alarms. However, learning is often likely to be involved because species may respond to heterospecific alarms with very different acoustic structure compared to their own alarms. Furthermore, they do so only when familiar with heterospecific alarms, not when isolated from them geographically (Magrath *et al.*, 2007, 2009).

Firstly, observations on foraging Drongos showed that almost all of the suspected alarm calls were made either in response to the presence of a predator (raptors, owls, foxes and mongooses) or when an individual being followed was handling a food item. The calls were not made when directly attacking for food or in non-alarm contexts. Secondly, the calls made by drongos involved a mixture of drongo-specific calls and what appeared to be the mimicked alarm calls of other species, especially glossy starlings. A structural analysis of the drongo-specific calls showed that the calls made when a predator was approaching (true alarms) did not differ from the calls made when there was no predator (false alarms). Furthermore, the mimicked glossy starling alarm calls made by drongos when there was no predator (false alarms) did not differ in structure from the calls made by glossy starlings in response to predators (true alarms). Taken together, these first two results suggest that drongos sometimes deceptively signal that a predator is approaching. Thirdly, when recorded drongo calls were played to meerkats, the foragers were more likely to scan for predators and abandon their food in response to alarm calls than other types of call that drongos make, but showed no difference in their response to true or false alarms (Fig. 14.19). This suggests that meerkats are deceived by the false alarm calls made by drongos.

Fork-tailed drongos make false alarm calls when there is no predator ...

... which make meerkats abandon food and flee for cover

Overall, these results show that drongos dishonestly signal the approach of a predator, and that meerkats are deceived by these calls, making them abandon their food to the drongos. Drongos obtain approximately 10% of their total food intake through the use of these alarm calls. This impressively high proportion is comparable to a human who could obtain one out of every 10 meals free of charge, by walking into a restaurant and emptying it with a dishonest shout of 'fire'. It is hard to imagine that anyone could get very far with such a strategy, raising the analogous question of why should meerkats continue to respond to drongo calls? One factor is that not all alarm calls are false – approximately half were true alarms, in which the drongos also fled for cover and an approaching predator could usually be observed. It remains an open question whether drongos are producing an evolutionarily stable strategy (ESS) proportion of false calls, above which the reduced response to their alarm calls would outweigh any benefit of making more false calls. Another factor is that the cost of ignoring a potentially true alarm call is relatively large, as being eaten is far worse than losing a food item. However, it would be wrong to think that meerkats are powerless in this game of deceit, as observations have shown that they are less likely to flee when drongos make a false alarm call in the absence of a predator. This suggests that whether or not a forager flees depends partly on whether a drongo makes an alarm call as well as upon whether the forager or one of their group members can see a predator.

Dishonest Weapon Displays in Mantis Shrimp

Our above examples of dishonest signalling involve deception between species, where one species is deceiving another. Another possibility is that dishonest signalling can occur within species, such that some individuals signal honestly, but others exploit this and signal dishonestly or 'bluff'. Mantis shrimps are fearsome

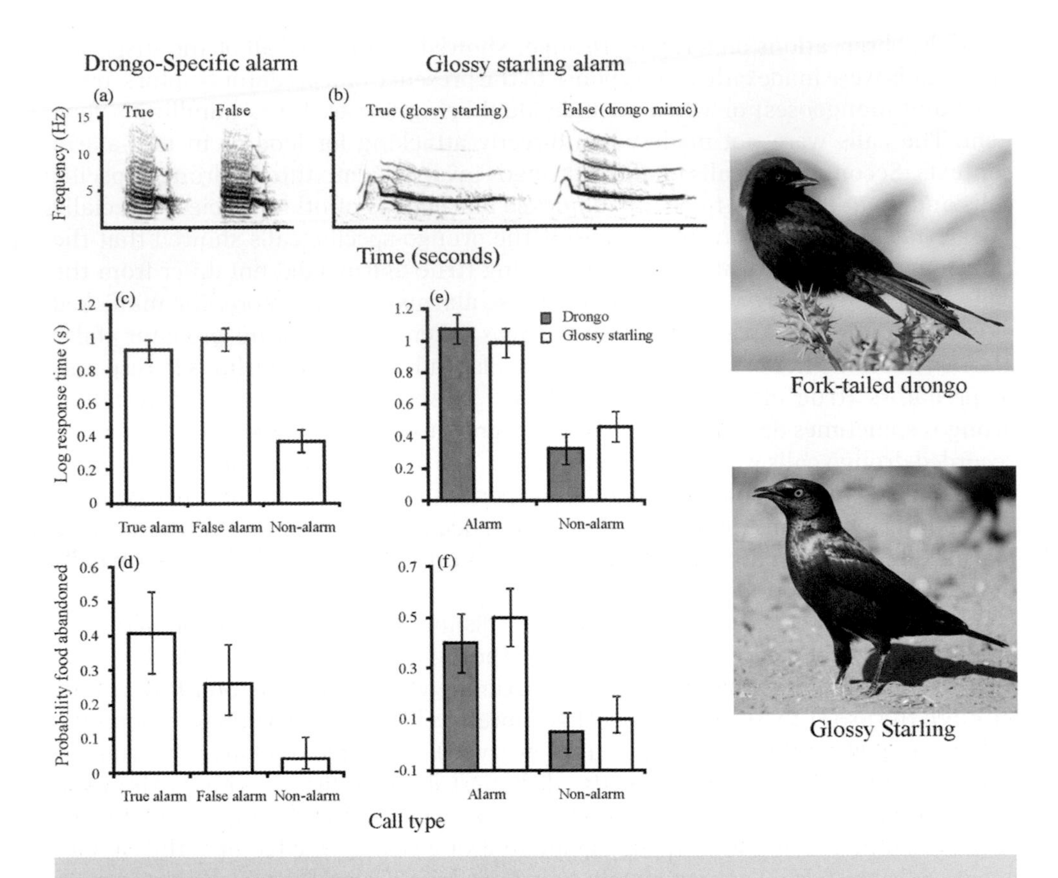

Fig. 14.19 Deceptive alarm calls by drongos. Sonograms from recordings of: (a) drongo-specific alarm calls made in a true (predator present) and false context (no predator present and stealing food); (b) glossy starling alarm calls made by a glossy starling in a true context and mimicked by a drongo in a false context. Meerkats responded for longer (c) and were more likely to abandon food (d) in response to playback of drongo-specific alarm calls than to non-alarm calls, but did not differ in their response to the true or false drongo-specific alarm calls. Meerkats responded for longer (e) and were more likely to abandon food (f) in response to playback of glossy starling alarm calls than to non-alarm calls, but did not differ in their response to the false (drongo-mimicked) and true glossy starling alarm calls. Means ±1 s.e. From Flower (2011). With permission of the Royal Society.

animals (Fig. 14.20). They use their powerful forelimbs, covered with a hard exoskeleton, to smash opponents to pieces (large species are able to smash their way out of an aquarium!). Mantis shrimps do not always fight by hitting each other: they also use a threat display in which the powerful forelimbs are spread out, which can lead to the individual with smaller limbs backing down. Work on one mantis shrimp, *Apheus*

Fig. 14.20 Photograph of a mantis shrimp *Gonodactylus bredini*. Photo © Jurgen Freund/naturepl.com

heterochaelis, has shown that larger individuals are more likely to win fights; they also have larger forelimbs, so the size of forelimbs appears to be a reliable index of fighting ability (Hughes, 1996).

Size of forelimbs appears to be a reliable index of fighting ability in mantis shrimps ...

However, in another species, *Gonodactylus bredini*, this threat signal is sometimes used in a dishonest way. Every two months, a mantis shrimp moults its hard exoskeleton and has to spend three days as a soft, impotent, warrior before the new skeleton hardens up. During the moult, individuals still use the threat to deter intruders, even though it is no longer an honest signalling of fighting ability (Steger & Caldwell, 1983). Furthermore, newly moulted individuals are more likely to display when facing smaller opponents, and hence when the opponent is more likely to back down (Adams & Caldwell, 1990). Bluffing against larger individuals can be more costly, because if the opponent does not back down and instead attacks, there is a high likelihood that the soft shrimp will be killed. This dishonest signal is relatively rare because only a small proportion of any individuals are moulting at any one time (<15%), ensuring that forelimb size is, on average, a relatively reliable index of fighting ability.

... but moulting mantis shrimps bluff

Summary

Signals are acts or structures that alter the behaviour of others, which have evolved because of that effect, and which are effective because the receiver's response has also evolved. This distinguishes signals from cues, which are a feature of the world that can be used as a guide to future action. So, for example, body size is a cue, but when a spider vibrates a web to give information about its body size, this is a signal.

The major problem of signalling is to explain what keeps signals reliable or honest. We have distinguished three ways in which this can occur:

(1) Indices are signals that cannot be faked. Examples include the croaks of toads, and the roars of deer.

(2) Handicaps are signals that could be faked, but where doing so is not economically beneficial. One possibility is that signals are costly to produce and the cost is greater for poor quality individuals. Examples include many of the extravagant sexual ornaments discussed in Chapter 7, such as the eye span in stalk-eyed flies, and how offspring beg for food from their parents, as discussed in Chapter 8. Another possibility is that dishonest signals incur social costs such as punishment. Examples include badges of status such as the black bib of sparrows or the facial markings of hover wasps.

(3) The sender and receiver can have a common interest in the interpretation of the signal. Examples include the waggle dance of the honeybee, quorum sensing in bacteria, receptivity in fruit flies and the chirrup calls of sparrows.

Language is a crucial difference between humans and other animals. Honesty in human language could be explained by a common interest and/or the social cost of dishonesty.

Dishonest signalling will only be evolutionarily stable if, on average, it pays the receiver to respond to the signal. Examples include the lures of anglerfish, Batesian mimicry, the false alarm calls of fork-tailed drongos and soft weapon displays in mantis shrimp.

Further reading

The evolution of communication and signals has been reviewed in three excellent books (Maynard Smith & Harper, 2003; Searcy & Nowicki, 2005; Bradbury & Vehrencamp, 2011). The differences between these books illustrate how the overarching conceptual framework, in terms of how the different explanations for signalling are carved up, is still a matter of debate.

Nakagawa *et al.* (2007) show how meta-analysis can be used to assess whether bib size acts as a badge of status in the house sparrow, *Passer domesticus*. Cotton *et al.* (2004b) review evidence that condition dependent expression (as required by handicaps) occurs and find the data lacking. Scott-Phillips & Kirby (2010) review laboratory experiments on the evolution of human language.

Keller and Surette (2006) provide an excellent discussion of signalling in bacteria from a behavioural ecology perspective. Kohler *et al.* (2009) and Rumbaugh *et al.* (2009) demonstrate the importance of quorum sensing for the damage that parasites do their hosts (virulence) and discuss how signalling theory can illuminate clinical patterns.

Jackson *et al.* (2004) provide a fascinating example of how ants supply information about trail direction through the use of geometry. Tobias and Seddon (2009) show how environmental condition can move a signal from common interest to conflict, with duetting and jamming in a pair-living antbird (*Hypocnemis peruviana*). Radford *et al.* (2011) show that the interaction between drongos and babblers may sometimes be mutually beneficial, despite kleptoparasitism, because the babblers get a net benefit from drongo vigilance.

TOPICS FOR DISCUSSION

1. Suggest examples of indices that have been exaggerated. How would you test your ideas?
2. Discuss whether the Hamilton and Zuk (1982) hypothesis, described in Chapter 7, is likely to revolve around indices or handicaps and how you would test this empirically.
3. Does the production of pheromones in yeast provide an example of a handicap signal of mate quality (Pagel, 1993; Smith & Greig, 2010)?
4. Lachmann *et al.* (2001) contrast situations in which honesty is maintained by the cost of producing signals, in which signals are costly, with those in which honesty is maintained by the social cost of dishonest signals, in which signals would be relatively cost free. How would you distinguish between these types experimentally? Do they both represent handicap signals?
5. Are elongated tails of birds likely to be costly signals (Norberg, 1994; Thomas & Rowe, 1997; Rowe *et al.*, 2001; Neuman *et al.*, 2007)?
6. Discuss whether quorum sensing would be expected to occur between species.
7. Backwell *et al.* (2000) suggest that 44% of fiddler crab individuals dishonestly signal claw strength. How can this high frequency of deception be maintained?
8. Discuss the use of the concept of information in categorizing and defining communication (Scott-Phillips, 2008, 2010; Rendall *et al.*, 2009; Font & Carazo, 2010; Scarantino, 2010; Seyfarth *et al.*, 2010).
9. Compare quorum sensing in bacteria (this chapter) with that in ants (Franks *et al.*, 2008).
10. Could animals be selected to deceive themselves (Trivers, 2000 , 2011)? How would you test this?

Photo © Arpat Ozgul

CHAPTER 15

Conclusion

The story of behaviour and adaptation that we have told in the last fourteen chapters is inevitably too simple. All the 'ifs' and 'buts' of an impeccably cautious and impregnable account would have made the book twice as long and half as easy to understand. However, we do not want to leave the impression that the ideas we have discussed are completely accepted by all evolutionary biologists. Far from it, even our basic assumptions are still sometimes challenged.

How plausible are our main premises?

Selfish genes or maximizing individuals?

Genes versus individuals

Our discussions of natural selection have been phrased in two alternate ways. One way is to emphasise selection on genes, with phrases such as: 'Imagine a gene for such and such behaviour; when would it tend to spread in a population?'. As we saw in Chapter 1, this approach does not imply that there are genes 'for' altruism, spite, long tails or whatever, but merely there are some genetic differences between individuals which are correlated with the behaviour or structure in question. The other way is to emphasize individuals maximising their fitness, with phrases such as 'How should individuals adjust such and such a behaviour, so as to maximize the number of offspring that they produce', or by developing evolutionarily stable strategy (ESS) models. As we also saw in Chapter 1, this approach does not assume that animals are consciously trying to maximize their fitness, just that natural selection will lead to organisms that appear to be doing this.

But how plausible are these views, and how justified is it to switch back and forth between the genetic and the individual level? Obviously the field biologist sees *individuals* dying, surviving and reproducing; but the evolutionary consequence is that the frequencies of *genes* in the population change. To clarify the link from gene to individual it is necessary to distinguish the process of natural selection from the design or purpose that this leads to in individuals. The process of natural selection is about gene dynamics over time. Specifically, genes which increase the fitness of the individuals carrying them by influencing the form of some heritable trait, such as a behaviour, will increase in frequency over time. A consequence of this is that heritable traits which increase fitness will accumulate over time, with natural selection leading to an increase in the mean fitness of individuals in a population (note that conflicts between individuals mean that this isn't the same as the overall population fitness). This will lead to organisms behaving or appearing as if they were designed to maximize their fitness.

Natural selection on genes will lead to individuals which behave as if they are trying to, or were designed to, maximize their fitness ...

The gene and individual viewpoints are, therefore, not competing but just different ways of looking at the same thing. If we were to ask 'what kind of organism would we expect natural selection acting on genes to produce?', then we would answer 'organisms that are designed to maximize their fitness', with the most general definition of fitness being inclusive fitness (Box 15.1). The beauty of this is that it takes evolutionary theory based on genes and translates it into a theory about how individuals behave, which is what fieldworkers can go out and observe, and what theoreticians can use ESS theory to model with ease. This also makes clear and justifies why intentional language is sometimes used, such as altruism, spite or 'trying to maximize fitness' – the dynamics of natural selection will lead to individuals behaving as if they have intention or purpose.

... and so the genetic and individual viewpoints are equivalent

However, although we can usually assume the equivalence of the genetic and the individual approach, there are rare occasions when they will disagree. If a gene can increase its own transmission, at a cost to the individual or other genes in the genome, then it may be selected to do so. In Chapter 10 we discussed how sex ratio distorters such as X drivers and male killers do this. In these cases, we have to take a gene's eye view to explain the behaviours. Nonetheless, such 'ultra-selfish' genes are extremely rare because the best way for genes to increase their transmission to the next generation is usually by increasing the reproductive success of the individual carrying them. Ultra-selfish genes tend to be associated with aspects of reproduction such as the sex ratio, which can be distorted to their own good. Contrast this with behaviours such as foraging or avoiding predators, where it is hard to imagine how a gene could increase its own transmission, other than to increase the success of the individual.

Genetic conflicts can occur ...

... but are relatively rare ...

There are two other reasons why genetic conflicts might have relatively rare consequences for behaviour. Firstly, even in cases where genes can distort things to their own favour, this can lead to the rest of the genome becoming united to suppress this distortion, in what Egbert Leigh (1971) eloquently called the 'parliament of the genes'. Secondly, the traits examined by behavioural ecologists tend to be relatively complex adaptations, built by multiple genes. The evolution of such traits has relied upon multiple genes all working in the same direction, in a way that would have been prevented by genetic conflicts. Consequently, if we are concerned with explaining adaptation then we are focusing on traits where genetic conflicts are likely to be relatively unimportant.

... and are unlikely to lead to complex adaptations

To sum up, the gene and individual approach are equivalent as long as we assume that genetic conflicts are relatively unimportant within individuals. This is expected on theoretical grounds, but also has been empirically validated by the success of the ESS

BOX 15.1 ADAPTATION AND DESIGN

The problem of adaptation is to explain the empirical fact that organisms appear designed. William Paley (1802) provided a particularly clear illustration of this problem in his book *Natural Theology*, discussing traits that ranged from the sting of a bee, marsupial pouch and stomach of a camel to the tongue of a woodpecker. Considering the eye, he marvelled at how the pupil and lens can be manipulated, so as to allow for variation in the level of light, or the distance to an object being viewed, and the superiority of the eye over the telescopes of the time. He then used what we would now call the comparative approach, to consider variation in

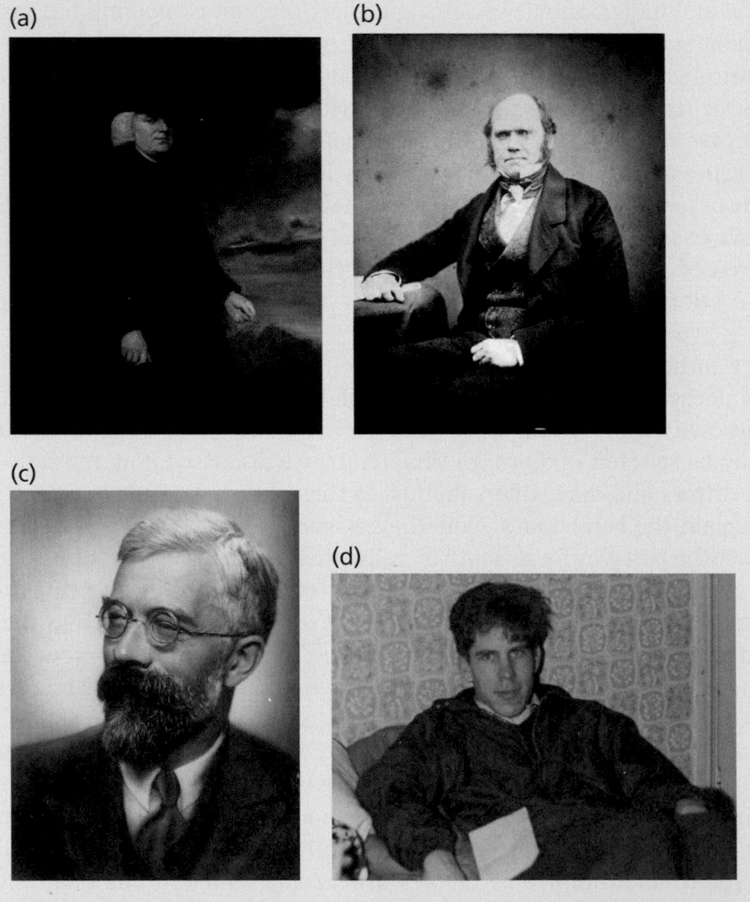

The development of adaptationism. (a) William Paley (1743–1805) formulated the problem of adaptation as the need to explain the apparent design of organisms. Photo © National Portrait Gallery. (b) Charles Darwin (1809–1882) solved the problem of adaptation with the theory of Natural Selection. (c) R.A. Fisher (1890–1962) showed how natural selection could be described by changes in gene frequencies. Photo © National Portrait Gallery. (d) Bill Hamilton (1936–2000) showed that the fitness which natural selection would lead to organisms maximizing is inclusive fitness. Image courtesy of the family of W.D. Hamilton.

eye anatomy across mammals, fish, birds and eels, suggesting that the variation is because each species has an eye that is more suitable for the environment in which it lives. Paley's explanation for this fit to the environment was that this provided evidence for the 'adapting hand' of an intelligent designer who knew 'the most secret laws of optics', and that this provided proof of a benevolent deity.

As explained in Chapter 1, Darwin (1859) showed that the appearance of design could arise though natural selection and does not require a divine designer. Importantly, Darwin's theory of natural selection explained both the process (or dynamics) by which adaptation occurs and how this leads to the appearance of design. The process is that heritable characters associated with greater reproductive success will be selected for and accumulate in natural populations. This leads to the apparent purpose of adaptation, that organisms will appear as if they were designed to maximize their reproductive success (fitness).

Since Darwin, there have been two major conceptual advances in the study of adaptation. Firstly, Fisher (1930) united Darwin's theory with Mendelian genetics by showing how natural selection could be described by changes in gene frequencies. He showed that genes associated with greater individual fitness will increase in frequency and that this would lead to an increase in mean fitness, such that individuals would appear as if they had been designed to maximize their fitness. Secondly, as discussed in Chapter 11, Hamilton (1964) showed that consequences for relatives have to be factored in, and that the fitness which organisms should appear designed to maximize is inclusive fitness, not just direct reproductive success. Since then, the field of behavioural ecology has provided one of the most fertile testing grounds for testing Darwin's theory.

approach. Put another way, the empirical success of the ESS approach has shown that it is often reasonable to ignore the possibility of genetic conflicts within individuals. Furthermore, even in cases where genetic conflicts are important, this is usually most easily discovered by deviations between ESS theory and empirical data.

Group selection

At the beginning of the book, we more or less dismissed group selection as a viable alternative to selection acting on individuals or selfish genes. We acknowledged that it could in principle work, but suggested that the conditions for group selection to be a powerful evolutionary force were not likely to be met in nature very often. However, this is not a universally accepted point of view. Every so often, papers or books claim that group selection is more important than we thought, can explain things that cannot be explained by selection at the individual level, and has been overly rejected (Sober & Wilson, 1998; Nowak, 2006; Traulsen & Nowak, 2006; Wilson & Wilson, 2007). How should we treat such claims?

One point to bear in mind from the start is that these claims are based upon models which are more subtle than the simple 'differential extinction of groups' model discussed in Chapter 1. The essential feature of these 'new' group selection models is that populations are divided into groups ('trait groups' or 'demes'), within which selection

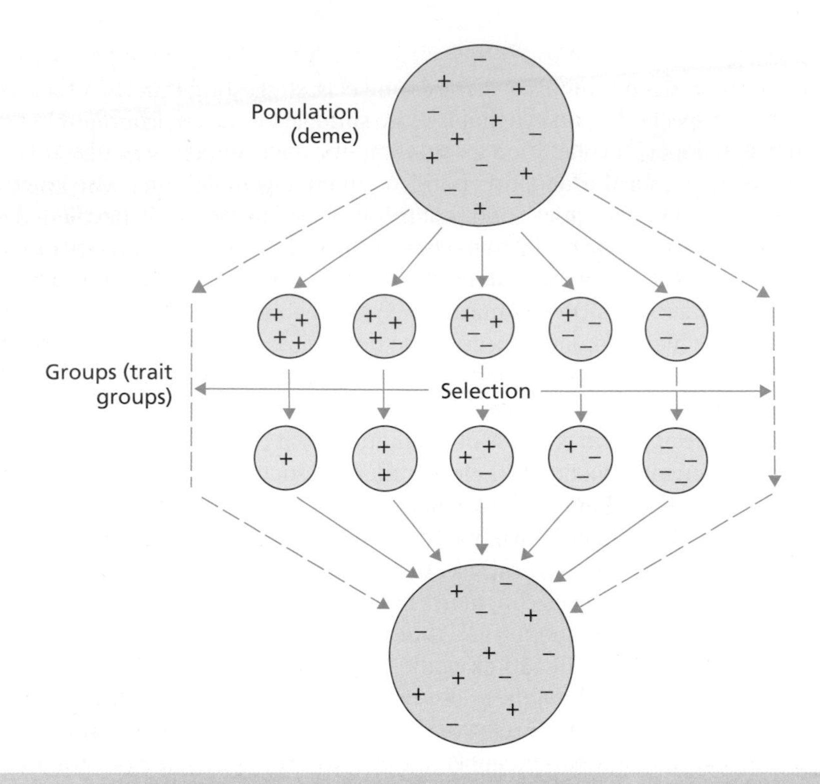

Fig. 15.1 New group selection. Individuals with the – allele are relatively cooperative, whereas those with the + allele are not. Within groups, individuals with the + allele do relatively better (increase in frequency). Between groups, groups with a higher frequency of the – allele do better and make a greater contribution to the next generation. Whether the – allele spreads depends upon the relative importance of these within and group components. From Harvey (1985).

for or against cooperative traits (or any other trait for that matter) occurs (Fig. 15.1). After selection has operated on the groups, the whole population mixes together before splitting up again into new groups for the next round of selection. A consequence of this population structuring is that cooperators are at a relative disadvantage *within* a group (because of their self-sacrifice) but groups with cooperators are more likely to contribute to the next generation than are trait groups with no cooperators. Cooperative traits spread when the between group selection is greater than the within selection component.

New group selection ...

The first question we should ask is whether such models give different predictions from inclusive fitness theory and can explain altruism in novel scenarios. No, they are just a different way of looking at the dynamics of natural selection, and give identical predictions (Hamilton, 1975; Grafen, 1984; Frank, 1986; Queller, 1992). The new group selection approach tells us that cooperation is favoured by: increasing group benefits; reducing individual cost; and increasing the proportion of genetic variance that is between-group as opposed to within-group. However, this is mathematically equivalent to the prediction from Hamilton's rule that altruism is favoured by high B, low C and high r.

... makes the same predictions as Hamilton's rule...

Given this mathematical equivalence, does this mean that the group selection approach is as equally useful as the inclusive fitness approach? No, because the group selection approach does not lead to an equivalent maximizing principle. Again, it is necessary to return to Darwin's distinction between the process (dynamics) and resultant product (design) of natural selection. Either inclusive fitness or group selection could be used to predict the dynamics of natural selection and each might be more useful for modelling certain situations. But, if we are interested in how organisms are designed to behave (adaptation), in terms of what they will appear to be maximizing, then the answer is that they will appear to be maximizing their inclusive fitness, and not group fitness. The empirical success of behavioural ecology is based on this idea of being able to treat organisms as if they are fitness maximizers, which flows from inclusive fitness theory. In contrast, individuals would only be selected to maximize group fitness in the extreme scenario where there is no conflict between individuals within groups, and so within group selection is negligible – and, of course, they are still selected to maximise their inclusive fitness in that scenario (Gardner & Grafen, 2009).

... but does not lead to a maximizing principle

In addition, the group selection approach is usually less easy to implement, and often spreads confusion (West *et al.*, 2007a).

The limitations of the new group selection approach ...

(1) From a theoretical perspective, individual level models are usually both easier to construct and extend to different biological scenarios. Consequently, while it is possible to develop an abstract group selection model for a trait like altruism or the sex ratio, it is much harder to develop models that can be applied and tested with specific species.

... less easy to develop models ...

(2) It is usually easier to link the individual approach with empirical data, because it emphasizes easily understood and measured parameters, such as B, C and r.

... less easy to link with empirical data ...

(3) Possibly because it lacks a direct link between selection and maximization, the group selection approach often leads to misconceptions. For example, incorrect suggestions that kin selection is a subset of group selection, or that group selection can apply in situations where inclusive fitness cannot.

... leads to misconceptions ...

(4) The group selection literature has also led to a strange use of terms such as altruism. The group selection literature redefines a behaviour as altruistic if it has a relative fitness disadvantage within the group, irrespective of the between group component. Because the latter can outweigh the former, such as when by-product benefits of helping flow to everyone within the group (Chapter 12), this means that a behaviour would be defined as altruistic even if it increased the direct reproductive success of the supposed 'altruist' (Table 15.1)!

... results in confusing use of terms ...

(5) The group selection approach often leads to the question: at what level does selection occur? This is usually not very useful or informative because selection can occur at multiple levels, with the relative importance of different levels (e.g. within or between group) depending upon the details of the system being modelled (and hence the details of a particular model and the parameter values of the model). In contrast, if we ask 'at what level does adaptation occur?', then we can give the very general answer, that it occurs at the level of the individual, in order to maximize inclusive fitness (Gardner & Grafen, 2009).

... and moves the focus from adaptation

To sum up, although the inclusive fitness (or individual) and new group selection approaches can both be used to give predictions, the inclusive fitness approach has proved much more useful. All of the advances in our understanding of behaviour that have

Table 15.1 The fitness of cooperative individuals who perform a cooperative trait (C) and defectors who do not (D). The calculation assumes that the cooperators (C) invest x resources in cooperation, the benefit to cost ratio is three, groups are composed of two individuals, and that benefits are shared amongst all group members. Cooperators would be defined as altruists or 'weak altruists' from a group selection perspective, because $x > 0$, irrespective of the benefit to cost ratio. From the selfish perspective of an individual, C always leads to a higher fitness, irrespective of whether it is in a group with a C ($2x > 3x/2$) or a D ($x/2 > 0$). This shows that a behaviour which would be classed as altruistic from a group selection perspective can be selected for because it increases an individual's direct fitness (West *et al.* 2007a).

Group	Two cooperators		One cooperator		No cooperators	
Type of individual	C	C	C	D	D	D
Baseline fitness	1	1	1	1	1	1
Individual cost of cooperating	x	x	x	0	0	0
Benefit of cooperation (shared within group)	$\frac{6x}{2} = 3x$	$3x$	$\frac{3x}{2}$	$\frac{3x}{2}$	0	0
Benefit – cost	$3x - x = 2x$	$2x$	$\frac{3x}{2} - x = \frac{x}{2}$	$\frac{3x}{2}$	0	0
Fitness	$1 + 2x$	$1 + 2x$	$1+\frac{x}{2}$	$1+\frac{3x}{2}$	1	1

been documented in the earlier chapters of this book have flowed from the individual/inclusive fitness approach, and not from group selection. This success of the inclusive fitness approach has arisen because it provides the maximization principle that is central to the behavioural ecology approach, is easier to develop and apply, and less likely to spread confusion.

The major evolutionary transitions

In the above discussion, we have emphasized that individuals are selected to maximize inclusive fitness and not group fitness. The only time we would expect individuals to behave as if they were trying to maximize group fitness, as originally envisaged by

Wynne-Edwards, would be when this also coincides with inclusive fitness. This requires extremely restrictive conditions, with extremely high relatedness within groups (e.g. clonality, $r = 1$) or some mechanism such as policing that completely aligns the interest of the individual and the group (Gardner & Grafen, 2009).

The alignment of inclusive and group fitness can lead to a major evolutionary transition

Although this alignment of inclusive and group fitness should happen extremely rarely, when it does it can have huge consequences because it can lead to a major evolutionary transition. A major evolutionary transition is when groups of individuals that were previously capable of independent reproduction before the transition can only replicate as a larger unit after it (Maynard Smith & Szathmary, 1995; Fig. 15.2). In Chapter 13 we discussed the transition to eusociality, where colonies of social insects have become 'superorganisms' – this represents one of only approximately eight major evolutionary transitions which have occurred (Table 15.2). The major transitions approach makes it clear that the behavioural ecology problem of cooperation has played a central role in evolutionary progress, because the problem of cooperation must be solved if a group of individuals is to come together to form a new, more complex, organism. Furthermore, similar to the examples discussed in Chapter 12, the major evolutionary transitions can be divided between those which have involved either relatives (e.g. the evolution of multicellularity and eusociality) or non-relatives (e.g. the evolution of chromosomes and the eukaryotic cell; Queller, 2000), and hence either indirect or direct benefits to cooperation.

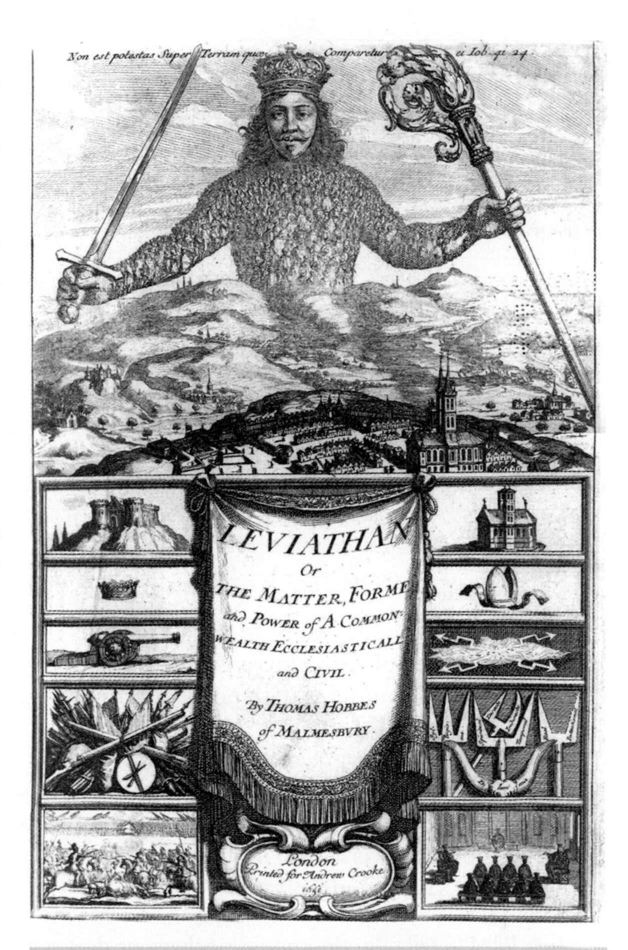

Fig. 15.2 Thomas Hobbes' *Leviathan* shows a giant crowned figure, which is composed from over 300 humans. The quote above is from the Book of Job and translates as 'There is no power on earth to be compared with him'. However, the major transition approach emphasizes that this is exactly what happens on earth, with groups of individuals coming together to form higher level individuals.

Table 15.2 The major evolutionary transitions, each of which has led to a new level of complexity (Maynard Smith & Szathmary, 1995)

Replicating molecules	⇒	Populations of molecules in compartments
Independent replicators	⇒	Chromosomes
RNA as gene and enzyme	⇒	DNA & protein (genetic code)
Prokaryotes	⇒	Eukaryotes (cell nucleus & organelles)
Asexual clones	⇒	Sexual populations
Protists	⇒	Multicelled animals, plants & fungi (with cell differentiation, i.e. organs)
Solitary individuals	⇒	Colonies (with non-reproductive worker castes)
Primate societies	⇒	Human societies (language)

Optimality models and ESSs

Criticisms of optimality models

In nearly every chapter we have used the ideas of optimality and ESSs. Here we will briefly recap some of the criticisms which have been levelled at optimality arguments and some of the limitations with putting them into practice.

1 *The idea that animals are optimal cannot be tested.*

As we saw in Chapter 3 this criticism is based on a mistaken notion. The aim of using an optimality model is not to test whether animals are optimal, but to test whether the particular optimality criterion and constraints used in the model give a good description of the animal's behaviour.

2 *It is hard to tell why the animal's behaviour does not fit the predictions exactly.*

Very often the simple models give an approximate but not exact description of the animal's behaviour. This could be because the model makes incorrect assumptions about constraints or goals, or because some component of cost has not been measured. There is no simple way of distinguishing between these possibilities.

One way around this problem is to use theory to make predictions for how traits should vary (across individuals or populations or species), rather than to predict a single value for a trait in a population. The former relies on a qualitative prediction from models that can be very simple and just need to capture a key aspect of the biology; whereas the latter relies on a quantitative prediction from a possibly very complex model that could depend upon an almost infinite number of biological details. Steve Frank (1998) has argued that this often makes the qualitative approach the best use of theory.

3 *Animals are not well enough adapted to optimize.*

Reasons for lack of adaptation ...

The main rationale for using optimality and ESS models is the assumption that natural selection produces well-adapted animals, the aim of the models being to find out how they are adapted. There are, however, at least four factors which can limit the extent to which animals are perfectly adapted.

(i) The physical or biological environment may fluctuate too rapidly for the animals to 'catch up' in their adaptations (Chapter 4). We provided an example of this in Chapter 1, when discussing how some populations of birds have not advanced their breeding times sufficiently to keep up with the rapid advancement of spring in recent decades.

... evolutionary lag ...

(ii) Adaptation could be constrained by the underlying genetics of the behaviour. One reason is that there may be insufficient genetic variation for new strategies to evolve. If the environment changes or if for some other reasons the optimal phenotype changes, animals can adapt to the new conditions only if there is genetic variation in the population. However, this is unlikely to be of general importance, because whenever genetic variation is looked for, it is usually found (Lynch & Walsh, 1998).

... genetic constraints ...

Another reason is that traits could be genetically linked, in which case selection on one trait could influence the other. An example of this is provided from work on Soay sheep living on the island of St Kilda, where coat colour is either dark brown or light tawny (Fig. 15.3). Variation in coat colour is controlled by a single locus called *TYRP1*, where the dark allele (G) is dominant: GG and GT give a dark coat, whereas TT gives a light coat. Over a 20-year period the proportion of sheep with dark coats and the G allele has been decreasing (Fig. 15.3). This cannot be due to selection for cryptic colouration (Chapter 4) or sexual selection (Chapter 9) because there are no predators on the island and mate choice does not depend upon coat colour.

Instead, Gratten *et al.* (2008) found that the decline in frequency of dark coats is due to selection on some other gene (or genes) for fitness that are physically close to *TRYP1* and which cause homozygous GG sheep to have a reduced fitness, relative to both GT and TT. The linked genes and why they affect fitness have yet to be identified. This study documents an unusual case, because the lack of predators and coat dependent mate choice means that coat colour *per se* is not under selection, and so it is free to be pulled along by linked genes. Presumably, if this was not the case, and there was a selective advantage to dark coats, then there would be selection to break the association between coat colour and the unknown fitness effect. Consequently, whilst the underlying genetics can constrain adaptation, this does not necessarily lead to a dead end, because the underlying genetics themselves are also subject to selection.

The way in which ESS models ignore the underlying genetics and assume all phenotypes are possible has been termed the 'phenotypic gambit' by Alan Grafen (1984). On one hand, the phenotypic gambit can be seen as a pragmatic approach, which makes it easier to develop models. However, it can also be argued that it is often the most useful approach. Working out an exact solution to a specific genetic model may not be very helpful, because there will be a 'cloud' of possible models, which differ in the genetic details (about which we are unlikely to ever know). The phenotypic gambit provides a robust approximation to a wide range of models, which can therefore tell us more than the exact solution of a more specific genetic model. The key point here is that the phenotypic gambit delivers approximate predictions that can be

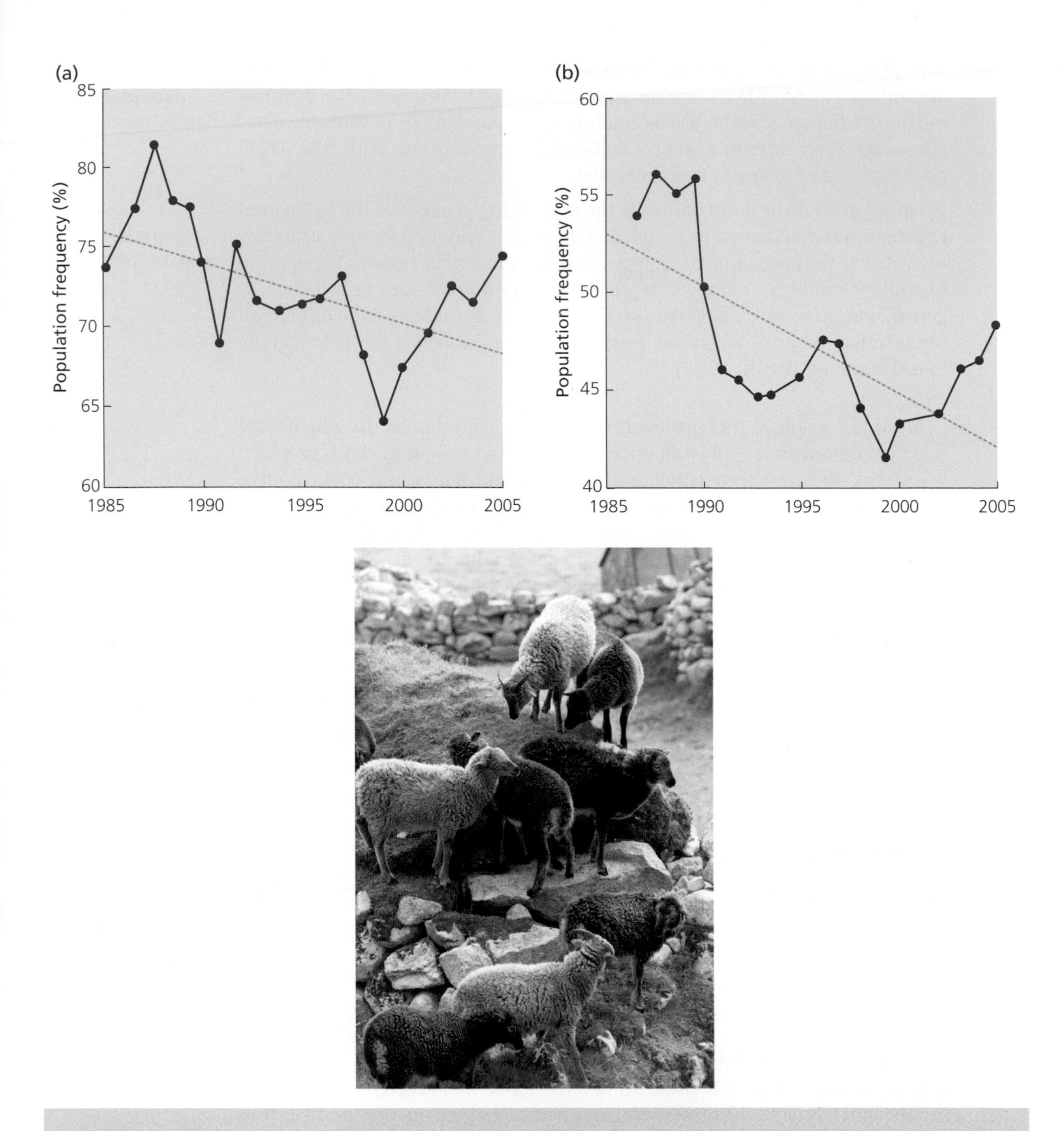

Fig. 15.3 Coat colour in Soay sheep. Estimated frequency of (a) dark sheep and (b) the *TRYP1* G allele (dominant for dark colour) in a population on St Kilda, from 1985 to 2005. From Gratten *et al*. (2008). Photograph shows individuals with both the dark and light colour morph. Photo © Arpat Ozgul.

applied across a range of organisms, facilitating empirical testing through both experiments and comparative studies.

(iii) There may be coevolutionary arms races (Chapter 4), as for example between predator and prey; if one side is ahead in the arms race, the other will appear to be poorly adapted to its environment, for example hosts that are killed or debilitated by pathogens and parasites (Rothstein, 1986).

... arms races ...

(iv) Individuals may be limited in the extent to which they can process relevant information. Most ESS models implicitly assume that individuals have complete information about the environment. For example, that the ducks and sticklebacks of Chapter 5 are able to assess the relative rate at which food can be acquired in different patches and adjust their foraging appropriately, or that the parasitoid wasps of Chapter 10 can estimate the number of other females laying eggs on a patch and adjust their offspring sex ratio correspondingly. If individuals cannot assess such variables perfectly, or make errors, then we would not expect them to behave 'perfectly'. For example, worker ants can only tell if their queen has mated multiply, and adjust the sex ratio accordingly, if the queen mated with males who smell differently (Fig. 13.12). Even if individuals could assess the relevant features of the environment perfectly, it may not be in their best interest to do so, if the resources that would have to be invested (e.g. time) are too costly. Incorporating such informational constraints into ESS models, and modelling information acquisition itself, remains a major task that would allow ESS models to be tested more quantitatively.

... and information constraints

A general point about all of the possible constraints on adaptation is that they do not invalidate the ESS approach, but rather suggest care in its application. Indeed, by emphasizing clear and testable predictions, the ESS approach provides a clear method for identifying constraints.

4 *Quantitative tests of theory are often not possible.*

The critical reader will have noticed that although we stressed the value of making quantitative predictions from optimality and ESS models, most of the tests of these predictions were qualitative. The animals were usually seen to do 'approximately the right thing': the dung flies in Chapter 3, for example, copulated for 36 minutes instead of the predicted 41 minutes. Some might ask whether it is worth developing quantitative arguments if the tests are only qualitative. There are three issues here.

Quantitative or qualitative tests of theory

(i) In some cases, tests are only qualitative because of limitations in our understanding or the techniques used to carry out the tests, and these can be overcome. Once the quantitative predictions can be tested accurately, discrepancies between observed and predicted results help to tell us what is wrong with the models. For example, in Chapter 1 we discussed how allowing for the cost of reproduction moved the predicted clutch size closer to the observed.

(ii) In some cases, quantitative predictions rely on so many biological details that we would be unlikely to ever make quantitative predictions, and it would not be very useful to try and do so. For example, in Chapter 2 we showed how variation in sexual dimorphism across primate species could be explained, without attempting to predict from an ESS model the sexual dimorphism in a particular species.

(iii) In some cases, we should not necessarily hold quantitative predictions as the Holy Grail, because the development and testing of qualitative models will be the more powerful approach (Frank, 1998). ESS models will always be oversimplifications of the real world. Furthermore, it may be possible to develop alternative theories that make the same quantitative prediction. In contrast, qualitative predictions about how traits should vary in response to variation in key parameters can usually be made much more unambiguously. For example, in Chapter 10, we discussed Hamilton's local mate competition (LMC) theory, which provides one of the greatest success stories in the field of behavioural ecology, having been supported in a huge range of taxa, from malaria parasites, to worms, to insects, to mites, to snakes. The majority of the empirical support for this theory has come from testing the qualitative prediction that females should lay more female-biased sex ratios when fewer females lay eggs on a patch (Fig. 10.4), and not from testing whether it quantitatively predicts average population sex ratios. This makes sense from a theoretical perspective – a slew of models have investigated the consequence of incorporating various life history details into Hamilton's model, with the general result being that, whilst they alter the quantitative predictions, the qualitative prediction is robust (West, 2009).

To give another example, Darwin's (1859) *Origin of Species* had little quantification. What made his argument so compelling, nevertheless, was that many independent lines of qualitative evidence (from population growth, geographical distributions, variation, embryology and so on) all pointed to the power of evolution by natural selection.

It is possible to carry on discussing the pros and cons of optimality models for a long time, but the strongest argument in their favour is that, over and over again, optimality arguments have helped us to understand adaptations. Although alternative approaches have been suggested (Gould & Lewontin, 1979), they just haven't proven very useful. We have illustrated the use of the optimality approach in the preceding chapters with behavioural examples – foraging, flock size, territory size and so on – but optimality arguments can equally well be used to understand adaptations at the physiological and biochemical level. For example, the familiar 'herring bone' arrangement of the swimming muscles of many fish is not merely an incidental design feature. This arrangement allows the muscles to contract at a rate which maximizes their power output (Alexander, 1975). At the biochemical level the energy for muscle contraction is generated by oxidation of carbohydrates or fats via the Krebs cycle. It would be chemically feasible to carry out the oxidation by a more direct route, but the advantage of the cycle is that it maximizes the net energy gain per molecule oxidized (Baldwin & Krebs, 1981).

Causal and functional explanations

Behavioural ecology is about functional explanations (the answers to 'why?' questions) of behaviour. As emphasized in Chapter 1, a great deal of misunderstanding can arise if functional and causal ('how?') explanations are confused.

Causal and functional explanations complement each other

Although it is important to be clear about the distinction between causal and functional explanations, it is equally valuable to realize that the two kinds of question are complementary and that asking 'why?' questions can often help to understand the answers to 'how?' questions, or vice versa. An example of how causal and functional explanations go hand in hand is illustrated in Fig. 15.4. Black-tailed prairie dogs (*Cynomys ludovicianus*) are colonial and live in underground tunnels which may be up to 15 m long. The tunnels are usually simple U-shaped passages with an opening to the surface at either end. It has been known for a long time that prairie dogs build little mounds of soil around the two entrances of the tunnel. These mounds were considered to function either as lookout points or to protect the tunnel against floods. However, a closer inspection revealed that the two ends of the burrow have different kinds of mounds. At one end there is a high steep sided 'crater' mound while at the other end there is a low rounded 'dome' (Fig. 15.4). If the mounds are simple lookouts or flood barriers, why should they be different shapes? The answer to this 'why?' question comes from an understanding of how air is exchanged in the tunnel (Vogel *et al.*, 1973). A prairie dog living in the long underground tunnel cannot survive without a regular supply of fresh air and it appears that the mounds around the two entrances are designed to ensure a continuous flow of air through the tunnel. The crater mound is higher and has steeper sides than the dome; as a consequence, air is sucked out of the crater end of the tunnel and into the dome end.

The forces causing the air flow are viscous sucking and the Bernoulli effect. Viscous sucking refers to the fact that when moving air passes a region of stationary air the still air is dragged along with the current. The effect is larger at the crater end because the crater is higher than the dome and so it is exposed to faster winds. The Bernoulli effect

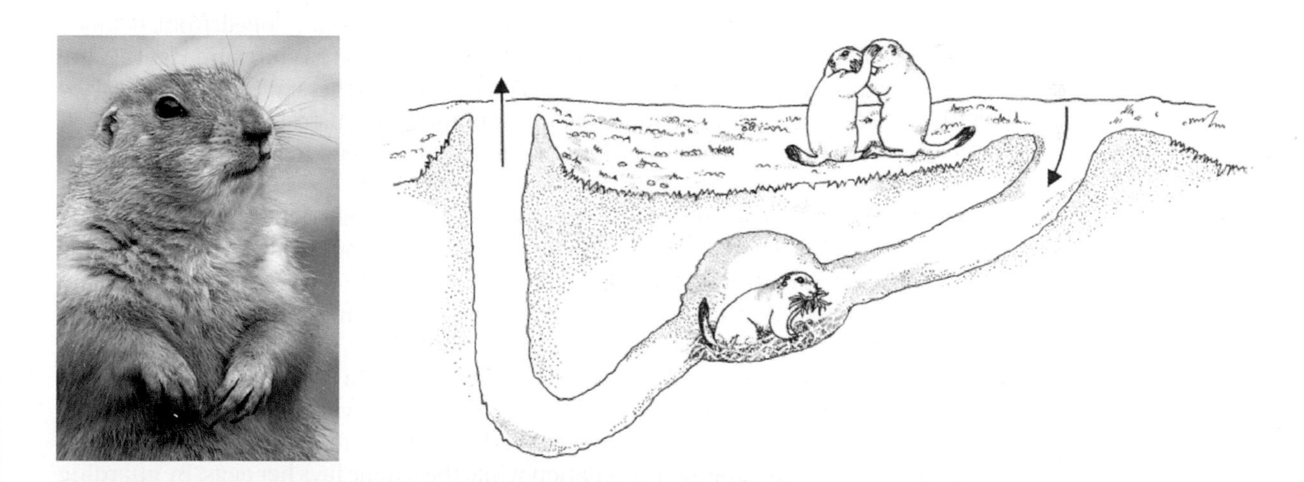

Fig. 15.4 A diagrammatic section of a prairie dog burrow. A typical burrow has two entrances, one with a low, round 'dome' at its entrance and the other with a taller, steeper-sided 'crater'. The different heights and shapes of the burrow entrances cause air to be sucked out of the crater end and, therefore, in through the dome. From Vogel *et al.* (1973). Photo © Elaine Miller Bond.

states that the pressure of a steadily moving fluid decreases when its velocity increases. The velocity of air above the crater is greater than above the mound and the crater has very still air inside because of its steep edges. The pressure drop between the inside and outside of the crater is therefore higher than in the case of the dome, so the Bernoulli effect causes air to be sucked out of the crater end of the tunnel. Vogel *et al.* (1973) demonstrated, by means of laboratory experiments with miniature model tunnels and by dropping smoke bombs down real tunnels in the field, that the mound system is so effective that it causes the air in the tunnel to change once every ten minutes, even in a very light breeze. The rate of air exchange is related to wind speed but is unaffected by wind direction since the mounds are symmetrical. This second feature is important because wind direction is unpredictable in the prairie dog's natural habitat.

Prairie dog burrows are designed to generate air flow

The Prairie dog burrow example illustrates that a functional question 'what are mounds for?' led to a detailed understanding of a mechanism question 'how do prairie dogs get enough fresh air?'. Examples of mechanistic questions improving our understanding of a functional question include: how the mechanism which worker ants use to count the number of times their queen has mated can constrain their ability to produce adaptive sex ratios (Fig. 13.12), or how parasitoid wasps determine the number of other females that have laid eggs on a patch (Fig. 10.5).

The optimality models we have encountered in this book bring together mechanisms (in the form of constraints) and functions (the currency) in building up an explanation of behaviour.

A final comment

From natural history to quantitative models

What we have described as behavioural ecology in this book is the present-day equivalent of natural history. It stands in a lineage which gradually developed from detailed descriptions of animal behaviour by naturalists such as Gilbert White and Henri Fabre to experimental studies of natural history by Tinbergen and others. We have emphasized the idea of making testable predictions about adaptation. To illustrate how this approach has developed from studies of natural history, let us construct a hypothetical lineage of studies of mating behaviour in dung flies.

A few hundred years ago naturalists would have been satisfied to discover that when two dung flies were seen together, one riding on the other's back, the top one was a male and the one underneath was a female and that the two were mating. A hundred years ago Darwin realized that males in general compete for females. A description of the natural history of dung fly mating at this stage would have included reference to the fact that males are larger than females and that this may be a result of sexual selection. Forty years ago, at the start of the 'behavioural ecology revolution', an evolutionary biologist stressed the idea that males ride on the backs of females not only to inject sperm but also for some time after copulation while the female lays her eggs. By guarding the female in this way the male guarantees that his sperm are not displaced by those of another male. A year or two later, an explanation was developed as to why it is that the male copulates for 40 minutes and not 10, 20 or 60. In the last twenty years or so, research has shown not only how the optimal copulation time relates to the sizes of the mating male and female but also suggests how these optima relate to the transfer of sperm from the male and its subsequent transport in the female. These behavioural studies are now inspiring studies at the molecular level of how sexual conflict leads to rapid evolution of seminal proteins.

In developing a theory to explain dung fly copula durations, it has become apparent that the same kind of analysis can be used for bumblebees sucking nectar out of flowers, parents investing in their offspring, bacteria secreting products that help them grow in the human lung and many other problems. This gradual reductionist progression from broad description to detailed quantitative analysis and simple generalizations is one of the major themes of development of the natural history lineage.

The field of behavioural ecology has flourished over the last 20 years, thanks to a combination of individual ingenuity and improved methods. This has allowed long standing questions to be resolved (e.g. can the handicap principle work?), led to the realization that some issues are far more important or complicated than was previously assumed (e.g. sexual conflict) and, in some cases, an overturning of conventional wisdom (e.g. the importance of haplodiploidy or reciprocity for the evolution of cooperation). However, in addition to this work on classical areas, behavioural ecology has also grown in novel and often unexpected directions. Examples include:

(1) **Parasitology.** The problem of how to exploit a limited resource (Chapter 5), and how this can be influenced by the relatedness between competitors (Chapter 11) applies to how parasites should exploit their hosts, and hence can help explain variation in the damage that parasites do to their hosts (Herre, 1993; Frank, 1996; Read *et al.* 2002; Boots & Mealor, 2007).

(2) **Medicine.** Conflict between parents can lead to selection for genomic imprinting (Chapter 8), which can explain complications during pregnancy (Moore & Haig, 1991) and disorders such as autism (Badcock & Crespi, 2006) or Prader–Willi syndrome (Úbeda, 2008).

(3) **Conservation.** Supplementary feeding can be detrimental to the growth of endangered populations if it increases female condition and makes them change their pattern of sex allocation (Fig. 15.5). Understanding mutualistic cooperation between species has been central to the conservation of a *Maculinea* butterfly (Thomas *et al.*, 2009).

(4) **Agriculture.** Plants use sanctions to enforce the cooperative supply of nitrogen from bacteria in their roots (Fig. 12.10). The use of artificial nitrogen fertilizers reduces the need for sanctions, and so the domestication of plants can lead to plants that are less good at getting nutrients via symbionts (Kiers *et al.*, 2007).

(5) **Microbiology.** The growth of pathogenic bacteria appears to rely on the production of cooperative public goods (Fig. 11.7). Consequently, the social dynamics of such traits can both explain clinical patterns (Kohler *et al.*, 2009) and be exploited in intervention strategies (Brown *et al.*, 2009).

(6) **Social and human sciences.** Although the application of ESS thinking applied to humans has had a turbulent history, there is currently a surge of interest in almost all areas of the social and human sciences, including economics, anthropology,

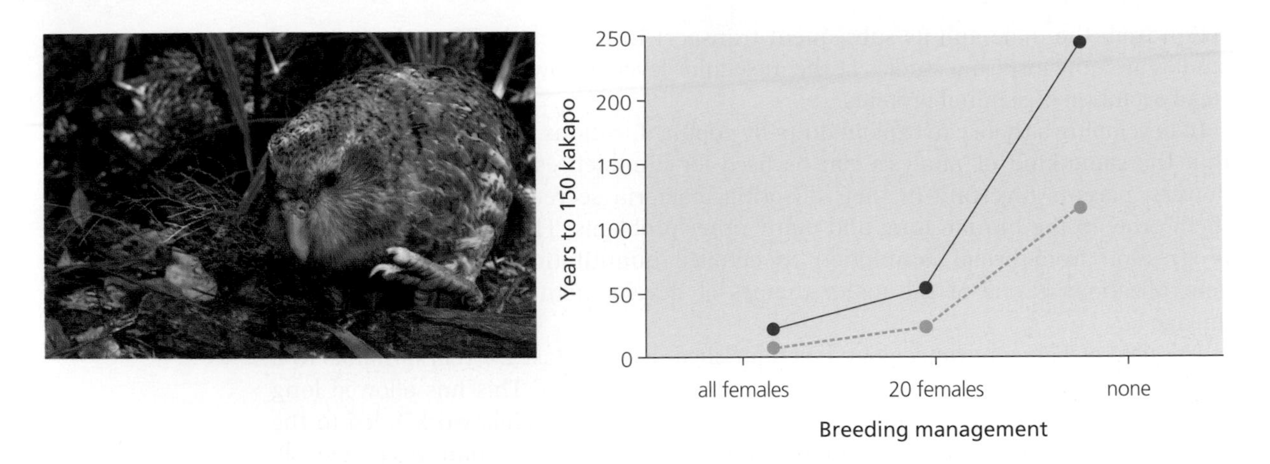

Fig. 15.5 Sex allocation and population recovery in the kakapo (*Strigops habroptilus*). The kakapo is a highly endangered flightless nocturnal parrot, endemic to New Zealand. Supplementary feeding of females led to the offspring sex ratio changing from 29% to 67% males. This can be explained by the Trivers and Willard (1973) hypothesis, because supplementary feeding would have led to females being in better condition, selecting them to produce sons, as with the red deer of Fig. 10.9. A new supplementary feeding programme was designed to raise the weight of females above the weight threshold (1.5 kg) above which they reproduce, but to keep them below the weight at which they produce an excess of males (2 kg). The figure shows the estimated number of years for the kakapo population to recover to 150 individuals, for different amounts of supplementary feeding, and from either *ad libitum* feeding (dark blue line) or the redesigned feeding programme (light blue dashed line). From Robertson *et al.* (2006). With permission of the Royal Society. Photo © Mark Carwardine / naturepl.com

psychology, linguistics, archeology and others (Gintis *et al.*, 2005; Henrich & Henrich, 2007; Dunbar & Barrett, 2009; Nettle, 2009).

(7) **The history of life on earth.** The view that major evolutionary transitions have been driven by an overcoming of the problem of cooperation (Chapter 11) offers a more profound vision of the history of life on earth than previous alternatives, such as the succession of different taxonomic groups (e.g. the Age of Fishes led to the Age of Reptiles, which led to the Age of Mammals; Bourke, 2011).

Who would have thought, just a few years ago, that we would be able to study sexual conflict at the molecular level; or that we would be using experiments with robotic fish to model human crowd control; or that microbes would become wonderful experimental models for studying the evolution of social behaviour; or that Hamilton's kin selection ideas could be applied to our understanding of human diseases? With the advent of these exciting ideas, and the use of new experimental and analytical techniques, it seems to us that these are the very best of times to be a behavioural ecologist.

Summary

This chapter is in three parts. In the first section, we assess the value and the limitations of the selfish gene and optimality views of evolution. We re-assess group selection as an alternative to individual selection and show how the behavioural ecology approach leads to a categorization of the major evolutionary transitions. The value of optimality arguments can be illustrated by studies of adaptations at the behavioural, physiological and biochemical levels.

In the second part of the chapter, we show how different kinds of questions (function and causation) should go hand in hand in studies of behaviour.

In the final section we discuss the growth of behavioural ecology, and its expansion to novel areas of research.

Further reading

The way in which we can move from gene dynamics to maximization and design at the individual level is reviewed by Dawkins (1982, Chapter 10) and Grafen (2007). Provine (1971) provides an excellent history of how Darwinism and Mendelism were seen as competing explanations for evolution, prior to the development of theoretical population genetics by Fisher, Wright and Haldane.

The role of group selection was recently debated by West *et al.* (2007a, 2008), Wilson (2008) and Wilson and Wilson (2007). Bourke (2011) argues that the 'major transitions view', framed by inclusive fitness theory and the behavioural ecology approach, provides the most profound and scientifically satisfying vision of life's evolutionary history.

Parker and Maynard Smith (1990) discuss the pros and cons of the ESS/optimality approach. Gould and Lewontin (1979) is a famous critique of this approach, which is entertainingly evaluated by Queller (1995). Grafen's (1984, 1991) chapters provide an excellent overview of the theoretical underpinnings of the behavioural ecology approach. Alcock (2003) provides a very readable account of the controversy surrounding the application of the behavioural ecology approach to humans. Wehner (1987) discusses how a functional (ultimate) explanation of a behaviours can help answer questions about the underlying causal (proximate) mechanisms.

There are a number of reviews on the implications of applying ESS thinking to different areas, including medicine (Williams & Nesse, 1991; Nesse & Williams, 1996), health and disease (Stearns & Koella, 2007), agriculture (Denison *et al.*, 2003) and conservation (Caro & Sherman, 2011).

Photo © Sophie Lanfear

References

Abele, L.G. & Gilchrist, S. (1977) Homosexual rape and sexual selection in acanthocephalan worms. *Science,* **197,** 81–83.

Abrahams, M. & Dill, L.M. (1989) A determination of the energetic equivalence of the risk of predation. *Ecology,* **70,** 999–1007.

Adams, E.S. & Caldwell, R.L. (1990) Deceptive communication in asymmetric fights of the stomatopod crustacean *Gonodactylus bredini. Animal Behaviour,* **39**, 706–716.

Agrawal, A.F. (2001) Sexual selection and the maintenance of sexual reproduction. *Nature,* **411**, 692–695.

Alatalo, R.V. & Mappes, J. (1996) Tracking the evolution of warning signals. *Nature,* **382**, 708–710.

Alatalo, R.V., Carlson, A., Lundberg, A. & Ulfstrand, S. (1981) The conflict between male polygamy and female monogamy: the case of the pied flycatcher, *Ficedula hypoleuca. American Naturalist,* **117,** 738–753.

Alatalo, R.V., Lundberg, A. & Rätti, O. (1990) Male polyterritoriality, and imperfect female choice in the pied flycatcher *Ficedula hypoleuca. Behavioral Ecology,* **1,** 171–177.

Alatalo, R.V., Höglund, J. & Lundberg, A. (1991) Lekking in the black grouse – a test of male viability. *Nature,* **352,** 155–156.

Alcock, J. (2003) *The Triumph of Sociobiology.* Oxford University Press, Oxford.

Alcock, J., Jones, C.E. & Buchmann, S.L. (1977) Male mating strategies in the bee *Centris pallida*, Fox (Anthophoridae: Hymenoptera). *American Naturalist,* **111,** 145–155.

Alexander, R. McN. (1975) *The Chordates.* Cambridge University Press, Cambridge.

Alexander, R.D. & Sherman, P.W. (1977) Local mate competition and parental investment in social insects. *Science* **196,** 494–500.

Allsop, D.J. & West, S.A. (2004) Sex-ratio evolution in sex changing animals. *Evolution,* **58**, 1019–1027.

Alonzo, S.H. & Sinervo, B. (2001) Mate choice games, context-dependent good genes, and genetic cycles in the side-blotched lizard, *Uta stansburiana. Behavioral Ecology and Sociobiology,* **49**, 176–186.

Altmann, S.A. (1974) Baboons, space, time and energy. *American Zoologist,* **14,** 221–240.

Altmann, S.A. Wagner, S.S. & Lenington, S. (1977) Two models for the evolution of polygyny. *Behavioural Ecology and Sociobiology*, **2**, 397–410.

Anderson, D.J. (1990) Evolution of obligate siblicide in boobies. I. A test of the insurance-egg hypothesis. *American Naturalist,* **135** 334–350.

An Introduction to Behavioural Ecology, Fourth Edition. Nicholas B. Davies, John R. Krebs and Stuart A. West.

Andersson, M. (1982) Female choice selects for extreme tail length in a widowbird. *Nature*, **299**, 818–820.

Andersson, M. (1994) *Sexual Selection*. Princeton University Press, Princeton, NJ.

Andersson, M. (2005) Evolution of classical polyandry: three steps to female emancipation. *Ethology*, **111**, 1–23.

Andersson, M. & Iwasa, Y. (1996) Sexual selection. *Trends in Ecology & Evolution*, **11**, 53–58.

Andersson, M. & Krebs, J.R. (1978) On the evolution of hoarding behaviour. *Animal Behavior*, **26**, 707–711.

Andersson, M. & Wicklund, C.G. (1978) Clumping versus spacing out: experiments on nest predation in fieldfares (*Turdus pilaris*). *Animal Behavior*, **26**, 1207–12.

Andersson, S., Pryke, S.R., Örnborg, J., Lawes, M.J. & Andersson, M. (2002) Multiple receivers, multiple ornaments, and a trade-off between agonistic and epigamic signalling in a widowbird. *American Naturalist*, **160**, 683–691.

André, J.-B. & Godelle, B. (2005) Multicellular organization in bacteria as a target for drug therapy? *Ecology Letters*, **8**, 800–810.

Aoki, S. (1977) *Colophina clematis* (Homoptera: Pemphigidae), and aphid species with soldiers. *Kontyu, Tokyo* **45**, 276–282.

Arak, A. (1983) Sexual selection by male–male competition in natterjack toad choruses. *Nature* **306**, 261–262.

Arak, A. (1988) Callers and satellites in the natterjack toad: evolutionarily stable decision rules. *Animal Behavior*, **36**, 416–432.

Arnold, K.E. and I.P.F. Owens. (1998) Cooperative breeding in birds: a comparative test of the life history hypothesis. *Proceedings of the Royal Society of London Series B*, **265**, 739–745.

Arnold, K.E. and I.P.F. Owens. (1999) Cooperative breeding in birds: the role of ecology. *Behavioral Ecology*, **10**, 465–471.

Arnqvist, G. & Rowe, L. (2002a) Antagonistic coevolution between the sexes in a group of insects. *Nature*, **415**, 787–789.

Arnqvist, G. & Rowe, L. (2002b) Correlated evolution of male and female morphologies in water striders. *Evolution*, **56**, 936–947.

Arnqvist, G. & Rowe, L. (2005) *Sexual Conflict*. Princeton University Press, Princeton, NJ.

Arnqvist, G. & Wooster, D. (1995) Meta-analysis: synthesizing research findings in ecology and evolution. *Trends in Ecology & Evolution*, **10**, 236–240.

Austad, S.N. (1983) A game theoretical interpretation of male combat in the bowl and doily spider, *Frontinella pyramitela*. *Animal Behavior*, **31**, 59–73.

Axelrod, R. (1984) *The Evolution of Cooperation*. Basic Books, New York.

Axelrod, R. & Hamilton, W.D. (1981) The evolution of cooperation. *Science*, **211**, 1390–1396.

Backwell, P.R.Y., Christy, J.H., Telford, S.R., Jennions, M.D. & Passmore, N.I. (2000) Dishonest signalling in a fiddler crab. *Proceedings of the Royal Society of London, Series B*, **267**, 719–724.

Badcock, C. & Crespi, B.J. (2006) Imbalanced genomic imprinting in brain development: an evolutionary basis for the etiology of autism *Journal of Evolutionary Biology* **19**, 1007–1032.

Badyaev, A.V., Hill, G.E., Beck, M.L., *et al.* (2001) Sex-biased hatching order and adaptive population divergence in a passerine bird. *Science*, **295**, 316–318.

Bailey, N.W. & Zuk, M. (2009) Same-sex sexual behaviour and evolution. *Trends in Ecology & Evolution*, **24**, 439–446.

Bain, R.S., Rashed, A., Cowper, V.J., Gilbert, F.S. & Sherratt, T.N. (2007) The key mimetic features of hoverflies through avian eyes. *Proceedings of the Royal Society of London, Series B*, **274**, 1949–1954.

Bakker, T.C.M. (1993) Positive genetic correlation between female preference and preferred male ornament in sticklebacks. *Nature*, **363**, 255–7.

Balda, R.P. & Kamil, A.C. (1992) Long-term spatial memory in Clark's Nutcracker *Nucifraga columbiana*. *Animal Behavior*, **44**, 761–769.

Balda, R.P. & Kamil, A.C. (2006) Linking life zones, life history traits, ecology, and spatial cognition in four allopatric southwestern seed caching corvids. In: *Animal Spatial Cognition: Comparative, Neural and Computational Approaches* (eds

M.F. Brown and R.G. Cook), pp. 761–769. Comparative Cognition Society (available on line).

Baldwin, J.E. & Krebs, H.A. (1981) The evolution of metabolic cycles. *Nature*, **291**, 381–382.

Balmford, A., Deutsch, J.C., Nefdt, R.J.C. & Clutton-Brock, T. (1993) Testing hotspot models of lek evolution: data from three species of ungulates. *Behavioral Ecology and Sociobiology*, **33**, 57–65.

Balshine-Earn, S. (1995) The costs of parental care in Galilee St. Peter's fish, *Sarotherodon galilaeus*. *Animal Behavior*, **50**, 1–7.

Balshine-Earn, S. & Earn, D.J.D. (1998) On the evolutionary pathway of parental care in mouth-brooding cichlid fish. *Proceedings of the Royal Society of London, Series B*, **265**, 2217–2222.

Barbero, F., Thomas, J.A., Bonelli, S., Balletto, E. & Schönrogge, K. (2009) Queen ants make distinctive sounds that are mimicked by a butterfly social parasite. *Science*, **323**, 782–785.

Barnard, C.J. & Sibly, R.M. (1981) Producers and scroungers: a general model and its application to captive flocks of house sparrows. *Animal Behavior*, **29**, 543–550.

Bart, J. & Tornes, A. (1989) Importance of monogamous male birds in determining reproductive success: evidence for house wrens and a review of male removal studies. *Behavioral Ecology and Sociobiology*, **24**, 109–116.

Basolo, A.L. (1990) Female preference predates the evolution of the sword in swordtail fish. *Science*, **250**, 808–810.

Basolo, A.L. (1994) The dynamics of Fisherian sex-ratio evolution: theoretical and experimental investigations. *American Naturalist*, **144**, 473–490.

Basolo, A.L. (1995) Phylogenetic evidence for the role of a pre-existing bias in sexual selection. *Proceedings of the Royal Society of London, Series B*, **259**, 307–311.

Bateman, A.J. (1948). Intra-sexual selection in *Drosophila*. *Heredity*, **2**, 349–368.

Bates, H.W. (1862) Contributions to an insect fauna of the Amazon valley. Lepidoptera: Heliconidae. *Transactions of the Linnean Society, London*. **23**, 495–566.

Bateson, M., Nettle, D. & Roberts, G. (2006) Cues of being watched enhance cooperation in a real-world setting. *Biology Letters*, **2**, 412–414.

Bateson, P. (1994) The dynamics of parent-offspring relationships in mammals. *Trends in Ecology & Evolution*, **9**, 399–403.

Baxter, S.W., Nadeau, N.J., Maroja, L.S., *et al.* (2010) Genomic hotspots for adaptation: the population genetics of Müllerian mimicry in the *Heliconius melpomene* clade. *PloS Genetics*, **6**, e1000794.

Bearhop, S., Fiedler, W., Furness, R.W., *et al.* (2005) Assortative mating as a mechanism for rapid evolution of a migratory divide. *Science*, **310**, 502–504.

Beck, C.W. (1998) Mode of fertilization and parental care in anurans. *Animal Behavior*, **55**, 439–449.

Bednekoff, P.A. (1997) Mutualism among safe, selfish sentinels: a dynamic game. *American Naturalist*, **150**, 373–392.

Bednekoff, P.A. & Krebs, J.R. (1995) Great tit fat reserves: effects of changing and unpredictable feeding day length. *Functional Ecology*, **9**, 457–462.

Beehler, B.M. & Foster, M.S. (1988) Hotshots, hot-spots and female preferences in the organization of mating systems. *American Naturalist*, **131**, 203–219.

Beekman, M., Fathke, R.L. & Seeley, T.D. (2006) How does an informed minority of scouts guide a honeybee swarm as it flies to its new home? *Animal Behavior*, **71**, 161–171.

Beissinger, S.R. & Snyder, N.F.R. (1987) Mate desertion in the snail kite. *Animal Behavior*, **35**, 477–487.

Bell, A.M. (2005) Behavioural differences between individuals and two populations of sticklebacks (*Gasterosteus aculeatus*). *Journal of Evolutionary Biology*, **18**, 464–473.

Bell, G. (1978) The evolution of anisogamy. *Journal of Theoretical Biology*, **73**, 247–270.

Bell, G. (1980) The costs of reproduction and their consequences. *American Naturalist*, **109**, 453–464.

Benabentos, R., Hirose, S., Sucgang, R., *et al.* (2009) Polymorphic members of the *lag* gene family mediate kin discrimination in *Dictyostelium*. *Current Biology*, **19**, 567–572.

Bennett, N.C. & Faulkes, C.G. (2000) *African Mole-Rats: Ecology and Eusociality*. Cambridge University Press, Cambridge.

Bennett, P.M. & Owens, I.P.F. (2002) *Evolutionary Ecology of Birds: Life Histories, Mating Systems and Extinction*. Oxford University Press, Oxford.

Bensch, S. & Hasselquist, D. (1992) Evidence for active female choice in a polygynous warbler. *Animal Behavior*, **44**, 301–311

Ben-Shahar, Y., Robichon, A., Sokolowski, M.B. & Robinson, G.E. (2002) Influence of gene action across different time scales on behaviour. *Science*, **296**, 741–744.

Benson, W.W. (1972) Natural selection for Müllerian mimicry in *Heliconius erato* in Costa Rica. *Science*, **176**, 936–939.

Bercovitch, F. (1988) Coalitions, cooperation and reproductive tactics among adult male baboons. *Animal Behavior*, **36**, 1198–209.

Berglund, A. & Rosenqvist, G. (2003) Sex role reversal in pipefish. *Advances in the Study of Behavior*, **32**, 131–167.

Berglund, A., Rosenqvist, G. & Robinson-Wolrath, S. (2006) Food or sex – males and females in a sex role reversed pipefish have different interests. *Behavioral Ecology and Sociobiology*, **60**, 281–287.

Bernasconi, G. & Strassmann, J.E. (1999) Cooperation among unrelated individuals: the ant foundress case. *Trends in Ecology & Evolution*, **14**, 477–482.

Berthold, P. & Querner, U. (1981) Genetic basis of migratory behaviour in European warblers. *Science*, **212**, 77–79.

Berthold, P., Mohr, G. & Querner, W. (1990) Steuerung und potentielle Evolutionsgeschwindigkeit des obligaten Teilzieherverhaltens: Ergebnisse eines Zweiwegselektions experiments mit der Mönchgrassmücke (*Sylvia atricapilla*). *Journal für Ornithologie*, **131**, 33–45.

Berthold, P., Helbig, A.J., Mohr, G. & Querner, U. (1992) Rapid microevolution of migratory behaviour in a wild bird species. *Nature*, **360**, 668–670.

Bertram, B.C.R. (1975) Social factors influencing reproduction in wild lions. *Journal of Zoology*, **177**, 463–482.

Bertram, B.C.R. (1980) Vigilance and group size in ostriches. *Animal Behavior*, **28**, 278–286.

Binmore, K. (1994) *Game Theory And The Social Contract Volume 1: Playing Fair*. MIT Press, Cambridge, MA.

Binmore, K. (1998) *Game Theory And The Social Contract Volume 2: Just Playing*. MIT Press, Cambridge, MA.

Binmore, K. (2007a) *Playing for Real: A Text on Game Theory*. Oxford University Press, Oxford.

Binmore, K. (2007b) *Game Theory: A Very Short Introduction*. Oxford University Press, Oxford.

Birkhead, T.R. (1977) The effect of habitat and density on breeding success in common guillemots, *Uria aalge*. *Journal of Animal Ecology*, **46**, 751–764.

Birkhead, T.R. (2000) *Promiscuity: An Evolutionary History of Sperm Competition and Sexual Conflict*. Faber and Faber, London.

Birkhead, T.R. & Møller, A.P. (1992) *Sperm Competition in Birds: Evolutionary Causes and Consequences*. Academic Press, London.

Birkhead, T.R. & Møller, A.P. (eds) (1998) *Sperm Competition and Sexual Selection*. Academic Press, London.

Birkhead, T.R., Pellatt, J.E. & Hunter, F.M. (1988) Extra-pair copulation and sperm competition in the zebra finch. *Nature*. **334**, 60–62.

Birkhead, T.R., Burke, T., Zann, R., Hunter, F.M. & Krupa, A.P. (1990) Extra-pair paternity and intraspecific brood parasitism in wild zebra finches, *Taeniopygia guttata*, revealed by DNA finger-printing. *Behavioral Ecology and Sociobiology*, **27**, 315–324.

Birkhead, T.R. & Pizzari, T. (2002) Postcopulatory sexual selection. *Nature Reviews, Genetics*, **3**, 262–273.

Biro, P.A. & Stamps, J.A. (2008) Are animal personality traits linked to life-history productivity? *Trends in Ecology & Evolution*, **23**, 361–368.

Black, J.M. (1988) Preflight signalling in swans – a mechanism for group cohesion and flock formation. *Ethology*, **79**, 143–157.

Blount, J.D., Speed, M.P., Ruxton, G.D. & Stephens, P.A. (2009) Warning displays may function as honest signals of toxicity. *Proceedings of the Royal Society of London, Series B*, 276, 871–877.

Bond, A.B. & Kamil, A.C. (1998) Apostatic selection by blue jays produces balanced polymorphism in virtual prey. *Nature*, **395**, 594–596.

Bond, A.B. & Kamil, A.C. (2002) Visual predators select for crypticity and polymorphism in virtual prey. *Nature*, **415**, 609–613.

Boomsma, J.J. (1989) Sex-investment ratios in ants: has female bias been systematically overestimated? *American Naturalist*, **133**, 517–532.

Boomsma, J.J. (1991) Adaptive colony sex ratios in primitively eusocial bees. *Trends in Ecology & Evolution*, **6**, 92–95.

Boomsma, J.J. (1996) Split sex ratios and queen-male conflict over sperm allocation. *Proceedings of the Royal Society of London, Series B*, **263**, 697–704.

Boomsma, J.J. (2007) Kin selection versus sexual selection: why the ends do not meet. *Current Biology* **17**, R673-R683.

Boomsma, J.J. (2009) Lifetime monogamy and the evolution of eusociality. *Philosophical Transactions of the Royal Society of London. Series B*, **364**, 3191–3208.

Boomsma, J.J. & Grafen, A. (1990) Intraspecific variation in ant sex ratios and the Trivers-Hare hypothesis. *Evolution* **44**, 1026–1034.

Boomsma, J.J. & Grafen, A. (1991) Colony-level sex ratio selection in the eusocial Hymenoptera. *Journal of Evolutionary Biology*, **4**, 383–407.

Boomsma, J.J., Nielsen, J., Sundström, L., *et al.* (2003) Informational constraints on optimal sex allocation in ants. *Proceedings of the National Academy of Sciences USA*, **100**, 8799–8804.

Boots, M. & Mealor, M. (2007) Local interactions select for lower pathogen infectivity. *Science*, **315**, 1284–1286.

Borgia, G. (1985) Bower quality, number of decorations and mating success of male satin bowerbirds (*Ptilinorynchus violaceus*): an experimental analysis. *Animal Behavior*, **33**, 266–271.

Both C. & Visser, M.E. (2001) Adjustment to climate change is constrained by arrival date in a long-distance migrant bird. *Nature*, **411**, 296–298.

Bourke, A.F.G. (2011) *Principles of Social Evolution*. Oxford University Press, Oxford.

Bourke, A.F.G. & Franks, N.R. (1995) *Social Evolution in Ants*. Princeton University Press, Princeton, NJ.

Boyd, R. & Lorberaum, J.P. (1987) No pure strategy is evolutionarily stable in the repeated Prisoner's Dilemma game. *Nature* **327**, 58–59.

Boyd, R. & Richerson, P.J. (1988) The evolution of reciprocity in sizable groups. *Journal of Theoretical Biology*, **132**, 337–356.

Bradbury, J.W. (1977) Lek mating behaviour in the hammer-headed bat. *Zeitschrift für Tierpsychologie*, **45**, 225–255.

Bradbury, J.W. & Gibson, R.M. (1983) Leks and mate choice. In: *Mate Choice* (ed. P. Bateson). pp. 109–138. Cambridge University Press. Cambridge.

Bradbury, J.W. & Vehrencamp, S.L. (2011) *Principles of Animal Communication*, 2nd Edn. Sinauer Associates, Inc., Sunderland, MA.

Bradbury, J.W., Gibson, R.M. & Tsai, I.M. (1986) Hotspots and the evolution of leks. *Animal Behavior*, **34**, 1694–1709.

Brady, S.G., Schultz, T.R., Fisher, B.L. & Ward, P.S. (2006) Evaluating alternative hypotheses for the early evolution and diversification of ants. *Proceedings of the National Academy of Sciences USA*, **103**, 18172–18177.

Brennan, P.L.R., Prum, R.O., McCracken, K.G., Sorenson, M.D., Wilson, R.E. & Birkhead, T.R. (2007) Coevolution of male and female genital morphology in waterfowl. *PLoS ONE*, **2** (5), e418.

Bretman, A., Newcombe, D. & Tregenza, T. (2009) Promiscuous females avoid inbreeding by controlling sperm storage. *Molecular Ecology*, **18**, 3340–3345.

Briskie, J.V., Naugler, C.T. & Leech, S.M. (1994) Begging intensity of nestling birds varies with sibling relatedness. *Proceedings of the Royal Society of London, Series B*, **258**, 73–78.

Brockmann, H.J. (2001) The evolution of alternative strategies and tactics. *Advances in the Study of Behavior*, **30**, 1–51.

Brockmann, H.J. (2002) An experimental approach to alternative mating tactics in male horseshoe crabs (*Limulus polyphemus*). *Behavioral Ecology*, **13**, 232–238.

Brockmann, H.J., Colson, T. & Potts, W. (1994) Sperm competition in horseshoe crabs *Limulus polyphemus*. *Behavioral Ecology and Sociobiology*, **35**, 153–160.

Brodin, A. (1994) The role of naturally stored food supplies in the winter diet of the boreal willow tit *Parus montanus*. *Ornis Svecica*, **4**, 31–40.

Brodin, A. (2010) The history of scatter-hoarding studies. *Philosophical Transactions of the Royal Society of London. Series B*, **365**, 869–881.

Brodin, A. & Ekman, J. (1994) Benefits of food hoarding. *Nature*, **372**, 510.

Bro-Jørgensen, J., Johnstone, R.A. & Evans, M.R. (2007) Uninformative exaggeration of male sexual ornaments in barn swallows. *Current Biology*, **17**, 850–855.

Brooke, M. de L. & Davies, N.B. (1988) Egg mimicry by cuckoos *Cuculus canorus* in relation to discrimination by hosts. *Nature*, **335**, 630–632.

Brooks, R. (2000) Negative genetic correlation between male sexual attractiveness and survival. *Nature*, **406**, 67–70.

Brooks, R. & Couldridge, V. (1999) Multiple sexual ornaments coevolve with multiple mating preferences. *American Naturalist*, **154**, 37–45.

Brown, C.R. (1988) Enhanced foraging efficiency through information centers: a benefit of coloniality in cliff swallows. *Ecology*, **69**, 602–613.

Brown, C.R. & Brown, M.B. (1986) Ectoparasitism as a cost of coloniality in cliff swallows (*Hirundo pyrrhonota*). *Ecology*, **67**, 1206–1218.

Brown, J.L. (1964). The evolution of diversity in avian territorial systems. *Wilson Bulletin*, **76**, 160–169.

Brown, J.L. (1969) The buffer effect and productivity in tit populations. *American Naturalist*, **103**, 347–354.

Brown, J.L. (1982) Optimal group size in territorial animals. *Journal of Theoretical Biology*, **95**, 793–810.

Brown, J.L. (1987) *Helping and Communal Breeding in Birds*. Princeton University Press, Princeton, NJ.

Brown, S.P. & Johnstone, R.A. (2001) Cooperation in the dark: signalling and collective action in quorum-sensing bacteria. *Proceedings of the Royal Society of London, Series B*, **268**, 961–965.

Brown, S.P., West, S.A., Diggle, S.P. & Griffin, A.S. (2009) Social evolution in micro-organisms and a Trojan horse approach to medical intervention strategies. *Philosophical Transactions of the Royal Society of London. Series B*, **364**, 3157–3168.

Brown, W.D. & Keller, L. (2000) Colony sex ratios vary with queen number but not relatedness asymmetry in the ant *Formica exsecta*. *Proceedings of the Royal Society of London, Series B-Biological Sciences*, **267**, 1751–1757.

Brown, W.D. & Keller, L. (2006) Resource supplements cause a change in colony sex-ratio specialization in the mount-building ant, *Formica exsecta*. *Behavioral Ecology & Sociobiology* **60**, 612–618.

Bshary, R. & Grutter, A.S. (2002) Asymmetric cheating opportunities and partner control in a cleaner fish mutualism. *Animal Behaviour* **63**, 547–555.

Bshary, R. & Grutter, A.S. (2005) Punishment and partner switching cause cooperative behaviour in a cleaning mutualism. *Biology Letters*, **1**, 396–399.

Buchanan, K.L. & Catchpole, C.K. (2000) Song as an indicator of male parental effort in the sedge warbler. *Proceedings of the Royal Society of London, Series B*, **267**, 321–326.

Buckling, A. & Rainey, P.B. (2002) Antagonistic coevolution between a bacterium and a bacteriophage. *Proceedings of the Royal Society of London, Series B*, **269**, 931–936.

Bull, J.J. (1980) Sex determination in reptiles. *The Quarterly Review of Biology*, **55**, 3–21.

Bull, J.J. (1983) *Evolution of sex determining mechanisms*. Benjamin/Cummings Publishing Co.

Bulnheim, H.P. (1967) On the influence of the photoperiod on the sex realization in *Gammarus duebeni*. *Helgolander Wissenschafliches Meeresuntersungen*, **16**, 69–83.

Bumann, D., Krause, J. & Rubenstein, D.I. (1997) Mortality risk of spatial positions in animal groups: the danger of being in the front. *Behaviour*, **134**, 1063–1076.

Burke, T., Davies, N.B., Bruford, M.W. & Hatchwell, B.J. (1989) Parental care and mating behaviour of polyandrous dunnocks *Prunella modularis* related to paternity by DNA fingerprinting. *Nature*, **338**, 249–251.

Burley, N. (1981) Sex ratio manipulation and selection for attractiveness. *Science*, **211**, 721–722.

Burley, N. (1986) Sexual selection for aesthetic traits in species with biparental care. *American Naturalist*, **127**, 415–445.

Burley, N. (1988) The differential allocation hypothesis: an experimental test. *American Naturalist*, **132**, 611–628.

Burt, A. & Trivers, R. (2006) *Genes in Conflict: The Biology of Selfish Genetic Elements*. Harvard University Press, Cambridge, MA.

Burton-Chellew, M., Koevoets, T., Grillenberger, B.K., *et al.* (2008) Facultative sex ratio adjustment in natural populations of wasps: cues of local mate competition and the precision of adaptation. *American Naturalist*, **172**, 393–404.

Buston, P. (2003) Size and growth modification in clownfish. *Nature*, **424**, 145–146.

Butchart, S.H.M., Seddon, N. & Ekstrom, J.M.M. (1999a) Polyandry and competition for territories in bronze-winged jacanas. *Journal of Animal Ecology*, **68**, 928–939.

Butchart, S.H.M., Seddon, N. & Ekstrom, J.M.M. (1999b) Yelling for sex: harem males compete for female access in bronze-winged jacanas. *Animal Behavior*, **57**, 637–646.

Buttery, N.J., Rozen, D.E., Wolf, J.B. & Thompson, C.R.L. (2009) Quantification of social behavior in *D. discoideum* reveals complex fixed and facultative strategies. *Current Biology*, **19**, 1373–1377.

Bygott, J.D., Bertram, B.C.R. & Hanby, J.P. (1979) Male lions in large coalitions gain reproductive advantage. *Nature*, **282**, 839–841.

Calvert, W.H., Hedrick, L.E. & Brower, L.P. (1979) Mortality of the monarch butterfly, *Danaus plexippus*: avian predation at five over-wintering sites in Mexico. *Science*, **204**, 847–851.

Cameron, E.Z. (2004) Facultative adjustment of mammalian sex ratios in support of the Trivers-Willard hypothesis: evidence for a mechanism. *Proceedings of the Royal Society of London, Series B*, **271**, 1723–1728.

Cameron, S.A. & Mardulyn, P. (2001) Multiple molecular data set suggest independent origins of highly eusocial behaviour in bees (Hymenoptera: Apinae). *Systematic Biology*, **50**, 194–214.

Cant, M.A. (2003) Patterns of helping effort in co-operatively breeding banded mongooses (*Mungos mungo*). *Journal of Zoology*, **259**, 115–121.

Cant, M.A. (2011) The role of threats in animal cooperation. *Proceedings of the Royal Society of London, Series B*, **278**, 170–178.

Cant, M.A. & Johnstone, R.A. (2009) How threats influence the evolutionary resolution of within-group conflict. *American Naturalist*, **173**, 759–771.

Cant, M.A., Hodge, S.J., Bell, M.B.V., Gilchrist, J.S. & Nichols, H.J. (2010) Reproductive control via eviction (but not the threat of eviction) in banded mongooses. *Proceedings of the Royal Society of London, Series B*, **277**, 2219–2226.

Caraco, T. & Wolf, L.L. (1975) Ecological determinants of group sizes of foraging lions. *American Naturalist*, **109**, 343–352.

Caraco, T., Blanekenhorn, W.U., Gregory, G.M. Newman, J.A., Recer, G.M. & Zwicker, S.M. (1990). Risk-sensitivity: ambient temperature affects foraging choice. *Animal Behavior*, **39**, 338–345.

Carayon, J. (1974) Insemination traumatique hétérosexuelle et homosexuelle chez *Xylocoris maculipennis* (Hem. Anthocoridae). *Comptes rendus des séances de l'Académie des sciences. Série D, Sciences naturelles*, **278**, 2803–2806.

Carere, C. & Eens M. (eds) (2005) Unravelling animal personalities: how and why individuals consistently differ. *Behaviour*, **142**, 1149–1431.

Carmerer. (2003) *Behavioral Game Theory, Experiments in Strategic Interaction*. Princeton University Press, Princeton, NJ.

Caro, T. (2005) *Antipredator Defences in Birds and Mammals*. University of Chicago Press, Chicago, IL.

Caro, T.M. & Hauser, M.D. (1992) Is there teaching in nonhuman animals? *Quarterly Review of Biology*, **67**, 151–174.

Caro, T. & Sherman. P.W. (2011) Endangered species and a threatened discipline: behavioural ecology. *Trends in Ecology & Evolution*, **26**, 111–118.

Cartar, R.V. & Dill, L.M. (1990) Why are bumblebees risk-sensitive foragers? *Behavioral Ecology and Sociobiology*, **26**, 121–127.

Cash, K. & Evans, R.M. (1986) Brood reduction in the American white pelican (*Pelecanus erythrorhynchos*). *Behavioral Ecology and Sociobiology*, **18**, 413–418.

Catchpole, C.K. (1980) Sexual selection and the evolution of complex songs among European warblers of the genus *Acrocephalus*. *Behaviour*. **74**, 149–166.

Catchpole, C.K., Dittami, J. & Leisler, B. (1984) Differential responses to male song repertoires in female songbirds implanted with oestradiol. *Nature*, **312**, 563–564.

Chaine, A.S. & Lyon, B.E. (2008) Adaptive plasticity in female mate choice dampens sexual selection on male ornaments in the lark bunting. *Science*, **319**, 459–462.

Chapman, T., Arnqvist, G., Bangham, J. & Rowe, L. (2003) Sexual Conflict. *Trends in Ecology & Evolution*, **18**, 41–47.

Chapman, T., Liddle, L.F., Kalb, J.M., Wolfner, M.F. & Partridge, L. (1995) Cost of mating in *Drosophila melanogaster* females is mediated by male accessory gland products. *Nature*, **373**, 241–244.

Chapman, T. W., Crespi, B. J. & Perry, S. P. (2008) The evolutionary ecology of eusociality in Australian gall thrips: a 'model clades' approach. In: *Ecology of Social Evolution* (eds J Korb & J Heinze), Princeton University Press, Princeton, NJ.

Charlat, S., Hornett, E.A., Dysen, E.A., *et al.* (2005) Prevalence and penetrance variation of male-killing *Wolbachia* across Indo-Pacific populations of the butterfly *Hypolimnas bolina*. *Molecular Ecology*, **14**, 3525–3530.

Charlat, S., Hornett, E.A., Fullard, J.H., *et al.* (2007) Extraordinary flux in sex ratio. *Science*, **317**, 214.

Charlton, B.D., Reby, D. & McComb, K. (2007) Female red deer prefer the roars of larger males. *Biology Letters*, **3**, 382–385.

Charmantier, A., McCleery, R.H., Cole, L.R., Perrins, C., Kruuk, L.E.B. & Sheldon, B.C. (2008) Adaptive phenotypic plasticity in response to climate change in a wild bird population. *Science*, **320**, 800–3.

Charnov, E.L. (1976a) Optimal foraging: the marginal value theorem. *Theoretical Population Biology*, **9**, 129–136.

Charnov, E.L. (1976b) Optimal foraging: attack strategy of a mantid. *American Naturalist*, **110**, 141–151.

Charnov, E.L. (1977) An elementary treatment of the genetical theory of kin selection. *Journal of Theoretical Biology* **66**, 541–550.

Charnov, E.L. (1978) Evolution of eusocial behavior: offspring choice or parental parasitism? *Journal of Theoretical Biology* **75**, 451–465.

Charnov, E.L. (1982) *The Theory of Sex Allocation*. Princeton University Press, Princeton, NJ.

Charnov, E.L. & Bull, J.J. (1977) When is sex environmentally determined? *Nature*, **266**, 828–830.

Charnov, E.L. & Bull, J.J. (1989) Non-fisherian sex ratios with sex change and environmental sex determination. *Nature*, **338**, 148–150.

Charnov, E.L. & Hannah, R.W. (2002) Shrimp adjust their sex ratio to fluctuating age distributions. *Evolutionary Ecology Research*, **4**, 239–246.

Charnov, E.L. & Krebs, J.R. (1974) On clutch size and fitness. *Ibis*, **116**, 217–219.

Chase, I.D. (1980) Cooperative and non-cooperative behaviour in animals. *American Naturalist*, **115**, 827–857.

Chippindale, A.K., Gibson, J.R. & Rice, W.R. (2001) Negative genetic correlation of adult fitness between sexes reveals ontogenetic conflict in *Drosophila*. *Proceedings of the National Academy of Sciences USA*, **98**, 1671–1675.

Cialdini, R.B. (2001) *Influence*. 4th edn. Allyn & Bacon, Boston, MA.

Clark, A.B. (1978) Sex ratio and local resource competition in a prosimian primate. *Science*, **201**, 163–165.

Clark, A.G., Begun, D.J. & Prout, T. (1999) Female x male interactions in *Drosophila* sperm competition. *Science*, **283**, 217–220.

Clayton, N.S. & Krebs, J.R. (1994) Hippocampal growth and attrition in birds affected by experience. *Proceedings of the National Academy of Sciences USA*, **91**, 7410–7414.

Clayton, N.S. & Dickinson, A. (1998) Episodic-like memory during cache recovery by scrub jays. *Nature*, **395**, 272–274.

Clifford, L.D. & Anderson, D.J. (2001) Experimental demonstration of the insurance value of extra eggs in an obligately siblicidal seabird. *Behavioral Ecology*, **12**, 340–347.

Clutton-Brock, T.H. (1983) Selection in relation to sex. In: *From Molecules to Men* (ed. D.S. Bendall). pp. 457–481. Cambridge University Press, Cambridge.

Clutton-Brock, T.H. (1989) Mammalian mating systems. *Proceedings of the Royal Society of London, Series B*, **236**, 339–372.

Clutton-Brock, T.H. (1991) *The Evolution of Parental Care*. Princeton University Press, Princeton, NJ.

Clutton-Brock, T.H. (2009a) Sexual selection in females. *Animal Behavior*, **88**, 3–11.

Clutton-Brock, T.H. (2009b) Structure and function in mammalian societies. *Philosophical Transactions of the Royal Society of London. Series B*, **364**, 3229–3242.

Clutton-Brock, T.H. (2009c) Cooperation between non-kin in animal societies *Nature* **462**, 51–57.

Clutton-Brock, T.H. & Albon, S.D. (1979) The roaring of red deer and the evolution of honest advertisement. *Behaviour*, **69**, 145–170.

Clutton-Brock, T.H. & Harvey, P.H. (1976) Evolutionary rules and primate societies. In: *Growing Points in Ethology* (eds P.P.G. Bateson & R.A. Hinde). pp. 195–237. Cambridge University Press, Cambridge.

Clutton-Brock, T.H. & Harvey, P.H. (1977) Primate ecology and social organisation. *Journal of Zoology*, **183**, 1–39.

Clutton-Brock, T.H. & Parker, G.A. (1992) Potential reproductive rates and the operation of sexual selection. *Quarterly Review of Biology*, **67**, 437–455.

Clutton-Brock, T.H., Guinness, F.E. & Albon, S.D. (1982) *Red Deer: The Behaviour and Ecology of Two Sexes*. Chicago University Press, Chicago, IL.

Clutton-Brock, T.H., Albon, S.D. & Guinness, F.E. (1984) Maternal dominance, breeding success and birth sex ratios in red deer. *Nature*, **308**, 358–360.

Clutton-Brock, T.H., Hiraiwa-Hasegawa, M. & Robertson, A. (1989) Mate choice on fallow deer leks. *Nature*, **340**, 463–465.

Clutton-Brock, T.H., Gaynor, D., Kansky, R., *et al.* (1998) Costs of cooperative behaviour in suricates (*Suricata suricatta*). *Proceedings of the Royal Society of London, Series B-Biological Sciences*, **265**, 185–190.

Clutton-Brock, T.H., O'Riain, M.J., Brotherton, *et al.* (1999a) Selfish sentinels in cooperative mammals. *Science*, **284**, 1640–1644.

Clutton-Brock, T.H., Gaynor, D., McIlrath, G.M., *et al.* (1999b) Predation, group size and mortality in a cooperative mongoose, *Suricata suricatta*. *Journal of Animal Ecology*, **68**, 672–683.

Clutton-Brock, T.H., Russell, A.F., Sharpe, L.L., Young, A.J., Balmforth, Z. & McIlrath, G.M. (2002) Evolution and development of sex differences in cooperative behavior in Meerkats. *Science*, **297**, 253–256.

Cockburn, A. (2006) Prevalence of different modes of parental care in birds. *Proceedings of the Royal Society of London, Series B*, **273**, 1375–1383.

Coker, C.R. McKinney, F., Hays, H., Briggs, S. & Cheng, K. (2002) Intromittent organ morphology and testis size in relation to mating system in waterfowl. *Auk*, **119**, 403–413.

Conradt, L. & Roper, T.J. (2005) Consensus decision making in animals. *Trends in Ecology & Evolution*, **20**, 449–456.

Conradt, L. & Roper, T.J. (2007) Democracy in animals: the evolution of shared group decisions. *Proceedings of the Royal Society of London, Series B*, **274**, 2317–2326.

Cook, P.A. & Wedell, N. (1999) Non-fertile sperm delay female remating. *Nature*, **397**, 486.

Coolen, I., van Bergen, Y., Day, R.L. & Laland, K.N. (2003) Species differences in adaptive use of public information in sticklebacks. *Proceedings of the Royal Society of London, Series B*, **270**, 2413–2419.

Cornwallis, C., West, S.A. & Griffin, A.S. (2009) Routes to cooperatively breeding vertebrates: kin discrimination and limited dispersal. *Journal of Evolutionary Biology*, **22**, 2245–2457.

Cornwallis, C., West, S.A., Davies, K.E. & Griffin, A.S. (2010) Promiscuity and the evolutionary transition to complex societies. *Nature*, **466**, 969–972.

Cott, H.B. (1940) *Adaptive Coloration in Animals*. Oxford University Press, Oxford.

Cotton, S., Fowler, K. & Pomiankowski, A. (2004a) Condition dependence of sexual ornament size and variation in the stalk-eyed fly *Cyrtodiopsis dalmanni* (Diptera: Diopsidae). *Evolution*, **58**, 1038–1046.

Cotton, S., Fowler, K. & Pomiankowski, A. (2004b) Do sexual ornaments demonstrate heightened condition-dependent expression as predicted by the handicap hypothesis? *Proceedings of the Royal Society of London, Series B*, **271**, 771–783.

Couzin, I.D. & Franks, N.R. (2003) Self-organised lane formation and optimized traffic flow in army ants. *Proceedings of the Royal Society of London, Series B*, **270**, 139–146.

Couzin, I.D. & Krause, J. (2003) Self-organisation and collective behaviour in vertebrates. *Advances in the Study of Behavior*, **32**, 1–75.

Couzin, I.D., Krause, J., James, R., Ruxton, G.D. & Franks, N.R. (2002) The collective behaviour of animal groups in three dimensional space. *Journal of Theoretical Biology*, **218**, 1–11.

Couzin, I.D., Krause, J., Franks, N.R. & Levin, S.A. (2005) Effective leadership and decision-making in animal groups on the move. *Nature*, **433**, 513–516.

Cowie, R.J. (1977) Optimal foraging in great tits. *Parus major. Nature*, **268**, 137–139.

Cox, C.R. & Le Boeuf, B.J. (1977) Female incitation of mate competition: a mechanism of mate selection. *American Naturalist*, **111**, 317–335.

Craig, R. (1979) Parental manipulation, kin selection, and the evolution of altruism. *Evolution*, **33**, 319–334.

Creel, S., Marusha Creel, N. & Monfort, S.L. (1998) Birth order, estrogen and sex ratio adaptation in African wild dogs (*Lycaon pictus*). *Animal Reproduction Science*, **53**, 315–320.

Crespi, B.J. (1992) Eusociality in Australian gall thrips. *Nature*, **359**, 724–726.

Crespi, B.J. & Yanega, D. (1995) The definition of eusociality. *Behavioral Ecology*, **6**, 109–115.

Cresswell, W. (1994) Flocking is an effective anti-predation strategy in redshanks, *Tringa totanus. Animal Behavior*, **47**, 433–442.

Crook, J.H. (1964) The evolution of social organisation and visual communication in the weaver birds (Ploceinae). *Behaviour* (Suppl.), **10**, 1–178.

Crozier, R. H. (2008) Advanced eusociality, kin selection and male haploidy. *Australian Journal of Entomology*, **47**, 2–8.

Cullen, E. (1957) Adaptations in the kittiwake to cliff nesting. *Ibis*, **99**, 275–302.

Cunningham, E.J.A. & Russell, A.F. (2000) Egg investment is influenced by male attractiveness in the mallard. *Nature*, **404**, 74–77.

Currie, T.E., Greenhill, S.J., Gray, R.D., Hasegawa, T. & Mace, R. (2010) Rise and fall of political complexity in island South-East Asia and the Pacific. *Nature*, **467**, 801–804.

Cuthill, I.C., Stevens, M., Sheppard, J., Maddocks, T., Parraga, C.A. & Troscianko, T.S. (2005) Disruptive coloration and background pattern matching. *Nature*, **434**, 72–74.

Dale, S., Amundsen, T., Lifjeld, J.T. & Slagsvold, T. (1990) Mate sampling behaviour of female pied flycatchers: evidence for active mate choice. *Behavioral Ecology and Sociobiology*, **27**, 87–91.

Dall, S.R.X. (2002) Can information sharing explain recruitment to food from communal roosts? *Behavioral Ecology*, **13**, 42–51.

Dally, J.M., Emery, N.J. & Clayton, N.S. (2006) Food-caching western scrub jays keep track of who was watching when. *Science*, **312**, 1662–1665.

Daly, J.W., Kaneko. T., Wilham, J. *et al.* (2002) Bioactive alkaloids of frog skin: combinatorial bioprospecting reveals that pumiliotoxins have an arthropod source. *Proceedings of the National Academy of Sciences USA*, **99**, 13996–14001.

Daly, M. (1979) Why don't male mammals lactate? *Journal of Theoretical Biology*, **78**, 325–345.

Danchin, E., Giraldeau, L-A., Valone, T.J. & Wagner, R.H. (2004) Public information: from nosy neighbours to cultural evolution. *Science*, **305**, 487–491.

Danforth, B.N., Sipes, S., Fang, J. & Brady, S.G. (2006) The history of early bee diversification based on five genes plus morphology. *Proceedings of the National Academy of Sciences USA*, **103**, 15118–15123.

Darst, C.R., Cummings, M.E. & Cannatella, D.C. (2006) A mechanism for diversity in warning signals: conspicuousness versus toxicity in poison frogs. *Proceedings of the National Academy of Sciences USA*, **103**, 5852–5857.

Darwin, C. (1859) *On the Origin of Species*. Murray, London.

Darwin, C. (1871) *The Descent of Man and Selection in Relation to Sex*. Murray, London.

Darwin, C. (1876) Sexual selection in relation to monkeys. *Nature*, **15**, 18–19.

Davies, N.B. (1989) Sexual conflict and the polygamy threshold. *Animal Behavior*, **38**, 226–234.

Davies, N.B. (1992) *Dunnock Behaviour and Social Evolution*. Oxford University Press, Oxford.

Davies, N.B. (2011) Cuckoo adaptations: trickery and tuning. *Journal of Zoology* 284, 1–14.,

Davies, N.B. & Brooke, M. de L. (1988) Cuckoos versus reed warblers: adaptations and counter-adaptations. *Animal Behavior*, **36**, 262–284.

Davies, N.B. & Brooke, M. de L. (1989a) An experimental study of co-evolution between the cuckoo *Cuculus canorus* and its hosts. I. Host egg discrimination. *Journal of Animal Ecology*, **58**, 207–224.

Davies, N.B. & Brooke, M. de L. (1989b) An experimental study of co-evolution between the cuckoo *Cuculus canorus* and its hosts. II. Host egg markings, chick discrimination and general discussion. *Journal of Animal Ecology*, **58**, 225–236.

Davies, N.B. & Halliday, T.R. (1978) Deep croaks and fighting assessment in toads *Bufo bufo*. *Nature*, **274**, 683–65.

Davies, N.B. & Houston, A.I. (1981) Owners and satellites: the economics of territory defence in the pied wagtail, *Motacilla alba*. *Journal of Animal Ecology*, **50**, 157–180.

Davies, N.B., Hatchwell, B.J., Robson, T. & Burke, T. (1992) Paternity and parental effort in dunnocks *Prunella modularis*: how good are male chick-feeding rules? *Animal Behavior*, **43**, 729–745.

Davies, N.B., Brooke, M. de L. & Kacelnik, A. (1996a) Recognition errors and probability of parasitism determine whether reed warblers should accept or reject mimetic cuckoo eggs. *Proceedings of the Royal Society of London, Series B*, **263**, 925–931.

Davies, N.B., Hartley, I.R., Hatchwell, B.J. & Langmore, N.E. (1996b) Female control of copulations to maximise male help: a comparison of polygynandrous alpine accentors, *Prunella collaris*, and dunnocks *P. modularis*. *Animal Behavior*, **51**, 27–47.

Davies, N.B., Kilner, R.M. & Noble, D.G. (1998) Nestling cuckoos *Cuculus canorus* exploit hosts with begging calls that mimic a brood. *Proceedings of the Royal Society of London, Series B*, **265**, 673–678.

Dawkins, M. 1971. Perceptual changes in chicks: another look at the 'search image' concept. *Animal Behavior*, **19**, 566–574.

Dawkins, R. (1976) *The Selfish Gene*. Oxford University Press, Oxford.

Dawkins, R. (1978) Replicator selection and the extended phenotype. *Zeitschrift für Tierpsychologie*, **47**, 61–76.

Dawkins, R. (1979) Twelve misunderstandings of kin selection. *Zeitschrift für Tierpsychologie*, **51**, 184–200.

Dawkins, R. (1980) Good strategy or evolutionarily stable strategy? In: *Sociobiology: Beyond Nature/Nurture* (eds G.W. Barlow & J. Silverberg). pp. 331–367. Westview Press, Boulder, CO.

Dawkins, R. (1982) *The Extended Phenotype*. W.H. Freeman, Oxford.

Dawkins, R. (1986) *The Blind Watchmaker*. Longman, London.

Dawkins, R. (1989) *The Selfish Gene*, 2nd edn. Oxford Paperbacks, Oxford.

Dawkins, R. & Carlisle, T.R. (1976) Parental investment, mate desertion and a fallacy. *Nature*, **262**, 131–133.

Dawkins, R. & Krebs, J.R. (1978) Animal signals: information or manipulation? In: *Behavioural Ecology: An Evolutionary Approach* (eds J.R. Krebs & N.B. Davies), 1st edn. pp. 282–309. Blackwell Scientific Publications, Oxford.

Dawkins, R. & Krebs, J.R. (1979) Arms races between and within species. *Proceedings of the Royal Society of London, Series B*, **205**, 489–511.

Dawson, A. (2008) Control of the annual cycle in birds: endocrine constraints and plasticity in response to ecological variability. *Philosophical Transactions of the Royal Society of London. Series B*, **363**, 1621–1633.

De Voogd, T.J., Krebs, J.R., Healy, S.D. & Purvis, A. (1993) Relations between song repertoire size and the volume of brain nuclei related to song: comparative evolutionary analyses amongst oscine birds. *Proceedings of the Royal Society of London, Series B*, **254**, 75–82.

Decaestecker, E., Gaba, S., Raeymaekers, J.A.M., *et al.* (2007) Host-parasite 'Red Queen' dynamics archived in pond sediment. *Nature*, **450**, 870–873.

DeLay, L.S., Faaborg, J., Naranjo, J., Paz, S.M., de Vries, Tj. & Parker, P.G. (1996) Paternal care in the cooperatively polyandrous Galapagos hawk. *Condor*, **98**, 300–311.

Denison, R.F., Kiers, E.T. & West, S.A. (2003) Darwinian agriculture: when can humans find solutions beyond the reach of natural selection. *Quarterly Review of Biology*, **78**, 145–168.

Dennett, D. (1983) Intentional sytems in cognitive ethology: The 'Panglossian paradigm defended'. *Behavioral and Brain Sciences*, **6**, 343–390.

Deschner, T., Heistermann, M., Hodges, K. & Boesch, C. (2004) Female sexual swelling size, timing of ovulation, and male behaviour in wild West African chimpanzees. *Hormones and Behavior*, **46**, 204–215.

Despland, E. & Simpson, S.J. (2005) Food choices of solitarious and gregarious locusts reflect cryptic and aposematic antipredator strategies. *Animal Behavior*, **69**, 471–479.

DeVos, A. & O'Riain, M.J. (2010) Sharks shape the geometry of a selfish seal herd: experimental evidence from seal decoys. *Biology Letters*, **6**, 48–50.

Dewsbury, D.A. (1982) Ejaculate cost and male choice. *American Naturalist*, **119**, 601–610.

Diggle, S.P., Griffin, A.S., Campbell, G.S. & West, S.A. (2007) Cooperation and conflict in quorum-sensing bacterial populations. *Nature*, **450**, 411–414.

Dingemanse, N.J. & Réale, D. (2005) Natural selection and animal personality. *Behaviour* **142**, 1159–1184.

Dingemanse, N.J., Both, C., Drent, P.J. & Tinbergen, J.M. (2004) Fitness consequences of avian personalities in a fluctuating environment. *Proceedings of the Royal Society of London, Series B*, **271**, 847–852.

Dingemanse, N.J., Both, C., Drent, P.J., van Oers, K. & van Noordwijk, A.J. (2002) Repeatability and heritability of exploratory behaviour in great tits from the wild. *Animal Behavior*, **64**, 929–937.

Dingemanse, N.J., Wright, J., Kazem, A.J.N., Thomas, D.K., Hickling, R. & Dawnay, N. (2007) Behavioural syndromes differ predictably between twelve populations of three-spined stickleback. *Journal of Animal Ecology*, **76** 1128–1138.

Domb, L.G. & Pagel, M. (2001) Sexual swellings advertise female quality in wild baboons. *Nature*, **410**, 204–206.

Drent, P.J., van Oers, K. & van Noordwijk, A.J. (2003) Realized heritability of personalities in the great tit (*Parus major*). *Proceedings of the Royal Society of London, Series B*, **270**, 45–51.

Drummond, H. & Chavelas, C.G. (1989) Food shortage influences sibling aggression in the blue-footed booby. *Animal Behavior*, **37**, 806–819.

Duffy, J.E. (1996) Eusociality in a coral-reef shrimp. *Nature*, **381**, 512–514.

Duffy, J.E. (2003) The ecology and evolution of eusociality in sponge-dwelling shrimp. In: *Genes, Behaviors and Evolution of Social Insects* (eds T KIkuchi, S Higachi, & N Azuma). Pp. 217–254. Hokkaido University Press, Japan.

Duffy, J.E. & Macdonald, K.S. (2010) Kin structure, ecology and the evolution of social organization in shrimp: a comparative analysis. *Proceedings of the Royal Society of London, Series B*, **277**, 575–584.

Dunbar, R.I.M. (1984) *Reproductive Decisions: An Economic Analysis of Gelada Baboon Social Strategies*. Princeton University Press, Princeton, NJ.

Dunbar, R.I.M. & Barrett, L. (2009) *Oxford Handbook of Evolutionary Psychology*. Oxford University Press, Oxford.

Duncan, P. & Vigne, N. (1979) The effect of group size in horses on the rate of attacks by blood-sucking flies. *Animal Behavior*, **27**, 623–625.

Dunford, C. (1977) Kin selection for ground squirrel alarm calls. *American Naturalist*, **111**, 782–5.

Dussourd, D.E., Harvis, C.A., Meinwald, J. & Eisner, T. (1991) Pheromonal advertisement of a nuptial gift by a male moth. *Proceedings of the National Academy of Sciences USA*, **88**, 9224–9227.

Dyer, J.R.G., Ioannou, C.C., Morrell, L.J., *et al.* (2008) Consensus decision making in human crowds. *Animal Behavior*, **75**, 461–470.

Eberhard, W.G. (1996) *Female Control: Sexual Selection by Cryptic Female Choice*. Princeton University Press, Princeton, NJ.

Eggert, A-K. & Müller, J.K. (1992) Joint breeding in female burying beetles. *Behavioral Ecology and Sociobiology*, **31**, 237–242.

Eggert, A-K. & Sakaluk, S.K. (1995) Female-coerced monogamy in burying beetles. *Behavioral Ecology and Sociobiology*, **37**, 147–153.

Eizaguirre, C., Yeates, S.E., Lenz, T.L., Kalbe, M. & Milinski, M. (2009) MHC-based mate choice combines good genes and maintenance of MHC polymorphism. *Molecular Ecology*, **18**, 3316–3329.

Elgar, M.A. (1986a) House sparrows establish foraging flocks by giving chirrup calls if the resources are divisible. *Animal Behavior*, **34**, 169–174.

Elgar, M.A. (1986b) The establishment of foraging flocks in house sparrows: risk of predation and daily temperature. *Behavioral Ecology & Sociobiology*, **19**, 433–438.

Elgar, M.A., McKay, H. & Woon, P. (1986) Scanning, pecking and alarm flights in house sparrows (*Passer domesticus*, L.). *Animal Behavior*, **34**, 1892–1894.

Elner, R.W. & Hughes, R.N. (1978) Energy maximization in the diet of the shore crab, *Carcinus maenas*. *Journal of Animal Ecology*, **47**, 103–116.

Emery, N.J. & Clayton, N.S. (2001) Effects of experience and social context on prospective caching strategies by scrub jays. *Nature*, **414**, 443–446.

Emlen, D.J. (1996) Artificial selection on horn length – body size allometry in the horned beetle *Onthophagus acuminatus* (Coleoptera: Scarabaeidae). *Evolution*, **50**, 1219–1230.

Emlen, D.J. (1997) Alternative reproductive tactics and male-dimorphism in the horned beetle *Onthophagus acuminatus* (Coleoptera: Scarabaeidae). *Behavioral Ecology and Sociobiology*, **41**, 335–341.

Emlen, D.J. & Nijhout, H.F. (1999) Hormonal control of male horn length dimorphism in the dung beetle *Onthophagus taurus* (Coleoptera: Scarabaeidae). *Journal of Insect Physiology*, **45**, 45–53.

Emlen, S.T. (1995) An evolutionary theory of the family. *Proceedings of the National Academy of Sciences USA*, **92**, 8092–8099.

Emlen, S.T. & Oring, L.W. (1977) Ecology, sexual selection and the evolution of mating systems. *Science*, **197**, 215–223.

Emlen, S.T. & Wrege, P.H. (1988) The role of kinship in helping decisions among white-fronted bee-eaters. *Behavioral Ecology & Sociobiology*, **23**, 305–315.

Emlen, S.T. & Wrege, P.H. (1989) A test of alternate hypotheses for helping behaviour in white-fronted bee-eaters of Kenya. *Behavioral Ecology & Sociobiology*, **25**, 303–320.

Emlen, S.T. & Wrege, P.H. (2004) Size dimorphism, intrasexual competition and sexual selection in wattled jacana *Jacana jacana*, a sex-role-reversed shorebird in Panama. *Auk*, **121**, 391–403.

Emlen, S.T., Wrege, P.H. & Webster, M.S. (1998) Cuckoldry as a cost of polyandry in the sex-role reversed wattled jacana *Jacana jacana*. *Proceedings of the Royal Society of London, Series B*, **265**, 2359–2364.

Emlen, S.T., Demong, N.J. & Emlen, D.J. (1989) Experimental induction of infanticide in female wattled jacanas. *Auk*, **106**, 1–7.

Endler, J.A. (1980) Natural selection on colour patterns in *Poecilia reticulata*. *Evolution*, **34**, 76–91.

Endler, J.A. (1983) Natural and sexual selection on color patterns in poeciliid fishes. *Environmental Biology of Fishes*, **9**, 173–190.

Endler, J.A. & Mappes, J. (2004) Predator mixes and the conspicuousness of aposematic signals. *American Naturalist*, **163**, 532–547.

Enquist, M. & Leimar, O. (1983) Evolution of fighting behaviour; decision rules and assessment of relative strength. *Journal of Theoretical Biology*, **102**, 387–410.

Enquist, M. & Leimar, O. (1987) Evolution of fighting behaviour; the effect of variation in resource value. *Journal of Theoretical Biology*, **127**, 187–205.

Enquist, M. & Leimar, O. (1990) The evolution of fatal fighting. *Animal Behavior*, **39**, 1–9.

Enquist, M., Leimar, O., Ljungberg, T., Mallner, Y. & Segerdahl, N. (1990) A test of the sequential assessment game: fighting in the cichlid fish *Nannacara anomala*. *Animal Behavior*, **40**, 1–14.

Ens, B.J., Choudhury, S. & Black, J.M. (1996) Mate fidelity and divorce in monogamous birds. In: *Partnerships in Birds. The Study of Monogamy* (ed. J.M. Black). pp. 344–401. Oxford University Press, Oxford.

Epstein, R., Kirshnit, C.E., Lanza, R.P. & Rubin, L.C. (1984) 'Insight' in the pigeon: antecedents and determinants of an intelligent behaviour. *Nature*, **308**, 61–62.

Erichsen, J.T., Krebs, J.R. & Houston, A.I. (1980) Optimal foraging and cryptic prey. *Journal of Animal Ecology*, **49**, 271–276.

Ezaki, Y. (1990) Female choice and the causes and adaptiveness of polygyny in great reed warblers. *Journal of Animal Ecology*, **59**, 103–119.

Faaborg, J., Parker, P.G., DeLay, L., *et al.* (1995) Confirmation of cooperative polyandry in the Galapagos hawk *Buteo galapagoensis*. *Behavioral Ecology and Sociobiology*, **36**, 83–90.

Faulkes, C.G., Bennett, N.C., Bruford, M.W., O'Brien, H.P., Aguilar, G.H. & Jarvis, J.U.M. (1997) Ecological constraints drive social evolution in the African mole-rats. *Proceedings of the Royal Society of London, Series B*, **264**, 1619–1627.

Feare, C. (1984) *The Starling*. Oxford University Press, Oxford.

Fedorka, K.M. & Mousseau, T.A. (2004) Female mating bias results in conflicting sex-specific offspring fitness. *Nature*, **429**, 65–67.

Fehr, E. & Gächter, S. (2002) Altruistic punishment in humans. *Nature*, **415**, 137–140.

Felsenstein, J. (1985) Phylogenies and the comparative method. *American Naturalist*, **125**, 1–15.

Felsenstein, J. (2008) Comparative methods with sampling error and within-species variation: contrasts revisited and revised. *American Naturalist*, **171**, 713–725.

Field, J. & Brace, S. (2004) Pre-social benefits of extended parental care. *Nature*, **428**, 650–652.

Field, J. & Cant, M.A. (2009) Social stability and helping in small animal societies. *Philosophical Transactions of the Royal Society of London. Series B*, **364**, 3181–3189.

Field, J., Shreeves, G., Sumner, S. & Casiraghi, M. (2000) Insurance-based advantage to helpers in a tropical hover wasp. *Nature*, **404**, 869–871.

Field, J., Cronin, A. & Bridge, C. (2006) Future fitness and helping in social queues. *Nature*, **441**, 214–217.

Fisher, R.A. 1930. *The Genetical Theory of Natural Selection*. Clarendon Press, Oxford.

Fisher, D.O., Double, M.C., Blomberg, S.P., Jennions, M.D. & Cockburn, A. (2006) Post-mating sexual selection increases lifetime fitness of polyandrous females in the wild. *Nature*, **444**, 89–92.

Fitch, W.T. & Reby, D. (2001) The descended larynx is not uniquely human. *Proceedings of the Royal Society of London, Series B* **268**, 1669–1675s.

FitzGibbon, C.D. (1989) A cost to individuals with reduced vigilance in groups of

Thomson's gazelles hunted by cheetahs. *Animal Behavior*, **37**, 508–510.

Fitzpatrick, J.L., Montgomerie, R., Desjardins, J.K., Stiver, K.A., Kolm, N. & Balshine, S. (2009) Female promiscuity promotes the evolution of faster sperm in cichlid fishes. *Proceedings of the National Academy of Sciences USA*, **106**, 1128–1132.

Fitzpatrick, M.J., Feder, E., Rowe, L. & Sokolowski, M.B. (2007) Maintaining a behaviour polymorphism by frequency-dependent selection on a single gene. *Nature*, **447**, 210–212.

Floody, O.R. & Arnold, A.P. (1975) Uganda Kob (*Adenota kob thomasi*): territoriality and the spatial distributions of sexual and agonistic behaviours at a territorial ground. *Zeitschrift für Tierpsychologie*, **37**, 192–212.

Flower, T. (2011) Fork-tailed drongos use deceptive mimicked alarm calls to steal food. *Proceedings of the Royal Society of London, Series B*, **278**, 1548–1555.

Foerster, K., Coulson, T., Sheldon, B.C., Pemberton, J.M., Clutton-Brock, T.H. & Kruuk, L.E.B. (2007) Sexually antagonistic genetic variation for fitness in red deer. *Nature*, **447**, 1107–1111.

Foitzik, S. & Herbers, J.M. (2001) Colony structure of a slavemaking ant: II Frequency of slave raids and impact on the host population. *Evolution*, **55**, 316–323.

Foitzik, S., De Heer, C.J., Hunjan, D.N. & Herbers, J.M. (2001) Coevolution in host-parasite systems: behavioural strategies of slavemaking ants and their hosts. *Proceedings of the Royal Society of London, Series B*,. **268**, 1139–1146.

Foitzik, S., Fischer, B. & Heinze, J. (2003) Arms-races between social parasites and their hosts: geographic patterns of manipulation and resistance. *Behavioral Ecology*, **14**, 80–88.

Font, E. & Carazo, P. (2010) Animals in translation: why there is meaning (but probably no message) in animal communication. *Animal Behaviour*, **80**, e1-e6.

Forbes, S. (2005) *A Natural History of Families*. Princeton University Press, Princeton, NJ.

Forsgren, E., Amundsen, T., Borg, A.A. & Bjelvenmark, J. (2004) Unusually dynamic sex roles in a fish. *Nature*, **429**, 551–554.

Foster, K.R. (2004) Diminishing returns in social evolution: the not-so-tragic commons. *Journal of Evolutionary Biology*, **17**, 1058–1072.

Foster, W.A. (1990) Experimental evidence for effective and altruistic colony defence against natural predators by soldiers of the gall-forming aphid *Pemphigus spyrothecae* (Hemiptera: Pemphigidae). *Behavioral Ecology and Sociobiology*, **27**, 421–430.

Foster, W.A. & Treherne, J.E. (1981) Evidence for the dilution effect in the selfish herd from fish predation on a marine insect. *Nature*, **295**, 466–467.

Frank, S.A. (1986) Hierarchical selection theory and sex ratios. I. General solutions for structured populations. *Theoretical Population Biology*, **29**, 312–342.

Frank, S.A. (1987) Individual and population sex allocation patterns. *Theoretical Population Biology*, **31**, 47–74.

Frank, S.A. (1990) Sex allocation theory for birds and mammals. *Annual Review of Ecology, Evolution and Systematics*, **21**, 13–55.

Frank, S.A. (1996) Models of parasite virulence. *Quarterly Review of Biology*, **71**, 37–78.

Frank, S.A. (1998) *Foundations of Social Evolution*. Princeton University Press, Princeton, NJ.

Frank, S.A. (2003) Repression of competition and the evolution of cooperation. *Evolution*, **57**, 693–705.

Franks, N.R. & Hölldobler, B. (1987) Sexual competition during colony reproduction in army ants. *Biological Journal of the Linnean Society*, **30**, 229–243.

Franks, N.R. & Richardson, T. (2006) Teaching in tandem-running ants. *Nature*, **439**, 153.

Franks, N.R., Dornhaus, A., Fitzsimmons, J.P. & Stevens, M. (2003) Speed versus accuracy in collective decision making. *Proceedings of the Royal Society of London, Series B*, **270**, 2457–2463.

Franks, N.R., Pratt, S.C., Mallon, E.B., Britton, N.F. & Sumpter, D.J.T. (2002) Information flow, opinion polling and collective intelligence in house-hunting social insects. *Philosophical Transactions of the*

Royal Society of London. Series B, **357**, 1567–1583.

Franks, N.R., Hardcastle, A., Collins, S., *et al.* (2008) Can ant colonies choose a far-and-away better nest over an in-the-way poor one? *Animal Behaviour* **76**, 323–334.

Freckleton, R.P. & Harvey, P.H. (2006) Non-Brownian trait evolution in adaptive radiations. *PLoS Biology*, **4**, 2104–2111.

Freckleton, R.P. (2009) The seven deadly sins of comparative analysis. *Journal of Evolutionary Biology*, **22**, 1367–1375.

Fretwell, S.D. (1972) *Populations in a Seasonal Environment*. Princeton University Press, Princeton, NJ.

Fricke, H. & Fricke, S. (1977) Monogamy and sex change by aggressive dominance in coral reef fish. *Nature*, **266**, 830–832.

Frith, C.B. & Frith, D.W. (2004) *Bowerbirds*. Oxford University Press, Oxford.

Gadagkar, R. (1991) Demographic predisposition to the evolution of eusociality: a hierarchy of models. *Proceedings of the National Academy of Sciences USA*, **88**, 10993–10997.

Gadagkar, R. (2009) Interrogating an insect society. *Proceedings of the National Academy of Sciences USA*, **106**, 10407–10414.

Galantucci, B. (2005) An experimental study into the emergence of human communication systems. *Cognitive Science*, **29**, 737–767.

Galef, B.G. & Wigmore, S.W. (1983) Transfer of information concerning distant foods: a laboratory investigation of the 'information centre hypothesis'. *Animal Behavior*, **31**, 748–758.

Gardner, A. (2009) Adaptation as organism design. *Biology Letters*, **5**, 861–864.

Gardner, A. & Grafen, A. (2009) Capturing the superorganism: a formal theory of group adaptation. *Journal of Evolutionary Biology*, **22**, 659–671.

Gardner, A. & West, S.A. (2004) Spite and the scale of competition. *Journal of Evolutionary Biology*, **17**, 1195–1203.

Gardner, A. & West, S.A. (2010) Greenbeards. *Evolution*, **64**, 25–38.

Gardner, A., West, S.A. & Buckling, A. (2004) Bacteriocins, spite and virulence. *Proceedings of the Royal Society of London, Series B*, **271**, 1529–2535.

Gardner, A., Hardy, I.C.W., Taylor, P.D. & West, S.A. (2007) Spiteful soldiers and sex ratio conflict in polyembryonic parasitoid wasps. *American Naturalist*, **169**, 519–533.

Gardner, A., Alpedrina, J. and West, S.A. (2012). Haplodiploidy and the evolution of eusociality: split sex ratios. *American Naturalist*, In press.

Gentle, L.K. & Gosler, A.G. (2001) Fat reserves and perceived predation risk in the great tit *Parus major*. *Proceedings of the Royal Society of London, Series B*, **268**, 487–491.

Getty, T. (2002) The discriminating babbler meets the optimal diet hawk. *Animal Behavior*, **63**, 397–402.

Ghalambor, C.K. & Martin, T.E. (2001) Fecundity-survival trade-offs and parental risk-taking in birds. *Science*, **292**, 494–497.

Ghiselin, M.T. (1969) The evolution of hermaphroditism among animals. *Quarterly Review of Biology*, **44**, 189–208.

Gibbs, H.L., Weatherhead, P.J., Boag, P.T., White, B.N., Tabak, L.M. & Hoysak, D.J. (1990) Realized reproductive success of polygynous red-winged blackbirds revealed by DNA markers. *Science*, **250**, 1394–1397.

Gibson, R.M. (1996) A re-evaluation of hotspot settlement in lekking sage grouse. *Animal Behavior*, **52**, 993–1005.

Gilbert, L.E. (1976) Postmating female odor in *Heliconius* butterflies: a male contributed anti-aphrodisiac? *Science*, **193**, 419–420.

Gilbert, O.M., Foster, K.R., Mehdiabadi, N.J., Strassmann, J.E. & Queller, D.C. (2007) High relatedness maintains multicellular cooperation in a social amoeba by controlling cheater mutants. *Proceedings of the National Academy of Sciences USA*, **104**, 8913–8917.

Gilby, I.C. (2006) Meat sharing among the Gombe chimpanzees: harassment and reciprocal exchange. *Animal Behaviour*, **71**, 953–963.

Gill, F.B. & Wolf, L.L. (1975) Economics of feeding territoriality in the golden-winged sunbird. *Ecology*, **56**, 333–345.

Gilliam, J.F. (1982) *Foraging under mortality risk in size-structured populations*. PhD thesis, Michigan State University, MI.

Gintis, H., Bowles, S., Boyd, R. & Fehr, E. (2005) *Moral Sentiments and Material Interests: The Foundations of Cooperation in Economic Life*. MIT Press, Cambridge, MA.

Giraldeau, L.-A. & Gillis, D. (1985) Optimal group size can be stable: a reply to Sibly. *Animal Behavior,* **33**, 666–667.

Giraldeau, L.-A. & Caraco, T. (2000) *Social Foraging Theory*. Princeton University Press, Princeton, NJ.

Giraldeau, L-A. & Dubois, F. (2008) Social foraging and the study of exploitative behaviour. *Advances in the Study of Behavior,* **38**, 59–104.

Giron, D. & Strand, M.R. (2004) Host resistance and the evolution of kin recognition in polyembryonic wasps. *Proceedings of the Royal Society of London, Series B (Suppl.)*, **271**, S395-S398.

Giron, D., Dunn, D.W., Hardy, I.C.W. & Strand, M.R. (2004) Aggression by polyembryonic wasp soldiers correlates with kinship but not resource competition. *Nature,* **430**, 676–679.

Gittleman, J.L. & Harvey, P.H. (1980) Why are distasteful prey not cryptic? *Nature,* **286**, 149–150.

Godfray, H.C.J. (1991) Signalling of need by offspring to their parents. *Nature,* **352**, 328–330.

Godfray, H.C.J. (1995) Evolutionary theory of parent-offspring conflict. *Nature,* **376**. 133–138.

Godfray, H.C.J., Partridge, L. & Harvey, P.H. (1991) Clutch Size. *Annual Review of Ecology and Systematics*, **22**, 409–429.

Gordon, D.M. (1996) The organization of work in social insect colonies. *Nature,* **380**, 121–124.

Gorrell, J.C., McAdam, A.G., Coltman, D.W., Humphries, M.M. & Boutin, S. (2010) Adopting kin enhances inclusive fitness in asocial red squirrels. *Nature Communications,* **1**, 22.

Gosler, A.G., Greenwood, J.J.D. & Perrins, C.M. (1995) Predation risk and the cost of being fat. *Nature,* **377**, 621–623.

Götmark, F., Winkler, D.W. & Andersson, M. (1986) Flock-feeding increases individual success in gulls. *Nature,* **319**, 589–591.

Gould, S.J. & Lewontin, R.C. (1979) The spandrels of San Marco and the Panglossian paradigm: a critique of the adaptationist programme. *Proceedings of the Royal Society of London, Series B,* **205**, 581–598.

Gowaty, P.A. (1981) An extension of the Orians-Verner-Willson model to account for mating systems besides polygyny. *American Naturalist,* **118**, 851–859.

Grafen, A. (1982) How not to measure inclusive fitness. *Nature,* **298**, 425–426.

Grafen, A. (1984) Natural selection, kin selection and group selection. In: *Behavioural Ecology: An Evolutionary Approach* (eds J R Krebs & N B Davies). pp. 62–84. Blackwell Scientific Publications, Oxford.

Grafen, A. (1985) A geometric view of relatedness. *Oxford Surveys in Evolutionary Biology*, **2**, 28–89.

Grafen, A. (1986) Split sex ratios and the evolutionary origins of eusociality. *Journal of Theoretical Biology*, **122**, 95–121.

Grafen, A. (1989) The phylogenetic regression. *Philosophical Transactions of the Royal Society of London. Series B,* **326**, 119–157.

Grafen, A. (1990a) Biological signals as handicaps. *Journal of Theoretical Biology*, **144**, 517–546.

Grafen, A. (1990b) Sexual selection unhandicapped by the Fisher process. *Journal of Theoretical Biology*, **144**, 473–516.

Grafen, A. (1991) Modelling in behavioural ecology. In: *Behavioural Ecology, an Evolutionary Approach* (eds J R Krebs & N B Davies). pp. 5–31. Blackwell Scientific Publications, Oxford.

Grafen, A. (2007) The formal Darwinism project: a mid-term report. *Journal of Evolutionary Biology*, **20**, 1243–1254.

Grant, P.R. & Grant, B.R. (2002) Unpredictable evolution in a 30-year study of Darwin's finches. *Science*, **296**, 707–711.

Gratten, J., Wilson A.J., McRae A.F., *et al.* (2008) A localized negative genetic correlation constrains microevolution of coat colour in wild sheep. *Science,* **319**, 318–320.

Gray, S.M., Dill, L.M. & McKinnon, J.S. (2007) Cuckoldry incites cannibalism: male fish turn to cannibalism when perceived certainty of paternity decreases. *American Naturalist*, **169**, 258–263.

Grbic, M., Ode, P.J. & Strand, M.R. (1992) Sibling rivalry and brood sex-ratios in polyembryonic wasps. *Nature*, **360**, 254–256.

Greenlaw, J.S. & Post, W. (1985) Evolution of monogamy in seaside sparrows *Ammodramus maritimus*: tests of hypotheses. *Animal Behavior*, **33**, 373–383.

Griffin, A.S., West, S.A. & Buckling, A. (2004) Cooperation and competition in pathogenic bacteria. *Nature*, **430**, 1024–1027.

Griffin, A.S., Sheldon, B.C. & West, S.A. (2005) Cooperative breeders adjust offspring sex ratios to produce helpful helpers. *American Naturalist*, **166**, 628–632.

Griffin, A.S., Pemberton, J.M., Brotherton, P.N.M., *et al.* (2003) A genetic analysis of breeding success in the cooperative meerkat (*Suricata suricatta*). *Behavioral Ecology* **14**, 472–480.

Griffith, S.C., Owens, I.P.F. & Thurman, K.A. (2002) Extra pair paternity in birds: a review of interspecific variation and adaptive function. *Molecular Ecology*, **11**, 2195–2212.

Griffith, S.C., Ornborg, J., Russell, A.F., Andersson, S. & Sheldon, B.C. (2003) Correlations between ultraviolet coloration, overwinter survival and offspring sex ratio in the blue tit. *Journal of Evolutionary Biology*, **16**, 1045–1054.

Grim, T., Kleven, O. & Mikulica, O. (2003) Nestling discrimination without recognition: a possible defence mechanism for hosts towards cuckoo parasitism? *Proceedings of the Royal Society of London, Series B*, **270**, S73-S75.

Grodzinski, U. & Lotem, A. (2007) The adaptive value of parental responsiveness to nestling begging. *Proceedings of the Royal Society of London, Series B*, **274**, 2449–2456.

Grosberg, R.K. & Strathmann, R.R. (2007) The evolution of multicellularity: a minor major transition. *Annual Review of Ecology and Systematics*, **38**, 621–654.

Gross, M.R. (1996) Alternative reproductive strategies and tactics: diversity within sexes. *Trends in Ecology & Evolution*, **11**, 92–98.

Gross, M.R. & Sargent, R.C. (1985) The evolution of male and female parental care in fishes. *American Zoologist*, **25**, 807–822.

Gross, M.R. & Shine, R. (1981) Parental care and mode of fertilization in ectothermic vertebrates. *Evolution*, **35**, 775–793.

Guilford, T. (1986) How do 'warning colours' work? Conspicuousness may reduce recognition errors in experienced predators. *Animal Behavior*, **34**, 286–288.

Guilford, T. & Dawkins, M.S. (1991) Receiver psychology and the evolution of animal signals. *Animal Behaviour*, **42**, 1–14.

Gustafsson, L. & Sutherland, W.J. (1988) The costs of reproduction in the collared flycatcher *Ficedula albicollis*. *Nature*, **335**, 813–815.

Gwynne, D.T. (1982) Mate selection by female katydids (Orthoptera Tettigoniidae. *Conocephalus nigropleurum*). *Animal Behavior*, **30**, 734–738.

Gwynne, D.T. & Simmons, L.W. (1990) Experimental reversal of courtship roles in an insect. *Nature*, **346**, 171–174.

Hadfield, J.D. & Nakagawa, S. (2010) General quantitative genetic methods for comparative biology: phylogenies, taxonomies and multi-trait models for continuous and categorical characters. *Journal of Evolutionary Biology*, **23**, 494–508.

Haig, D. (1997) The social gene. In: *Behavioural Ecology: An Evolutionary Approach* (eds J.R. Krebs & N.B. Davies), 4th edn. pp. 284–304. Blackwell Science, Oxford.

Haig, D. (2000) The kinship theory of genomic imprinting. *Annual Review of Ecology and Systematics*, **31**, 9–32.

Haig, D. (2004) Genomic imprinting and kinship: how good is the evidence? *Annual Review of Genetics*, **38**, 553–585.

Haig, D. & Graham, C. (1991) Genomic imprinting and the strange case of the insulin-like growth factor-II receptor. *Cell*, **64**, 1045–1046.

Hale, R.E. & St Mary, C.M. (2007) Nest tending increases reproductive success, sometimes; environmental effects of paternal care and mate choice in flagfish. *Animal Behavior*, **74**, 577–588.

Hamilton, I.M. (2000) Recruiters and joiners: using optimal skew theory to predict group size and the division of resources within groups of social animals. *American Naturalist*, **155**, 684–695.

Hamilton, W.D. (1963) The evolution of altruistic behaviour. *American Naturalist*, **97**, 354–356.

Hamilton, W.D. (1964) The genetical evolution of social behaviour, I & II. *Journal of Theoretical Biology*, **7**, 1–52.

Hamilton, W.D. (1967) Extraordinary sex ratios. *Science*, **156**, 477–488.

Hamilton, W.D. (1970) Selfish and spiteful behaviour in an evolutionary model. *Nature*, **228**, 1218–1220.

Hamilton, W.D. (1971) Geometry for the selfish herd. *Journal of Theoretical Biology*, **31**, 295–311.

Hamilton, W.D. (1972) Altruism and related phenomena, mainly in social insects. *Annual Review of Ecology and Systematics*, **3**, 193–232.

Hamilton, W.D. (1975) Innate social aptitudes of man: an approach from evolutionary genetics. In: *Biosocial Anthropology* (ed. R. Fox). pp. 133–155. John Wiley & Sons Ltd, Chichester.

Hamilton, W.D. (1979) Wingless and fighting males in fig wasps and other insects. In: *Sexual Selection and Reproductive Competition in Insects* (eds M.S. Blum & N.A. Blum). pp. 167–220. Academic Press, London.

Hamilton, W.D. (1996) *Narrow roads of gene land: I Evolution of social behaviour*. W.H. Freeman, Oxford.

Hamilton, W.D. & Zuk, M. (1982) Heritable true fitness and bright birds: a role for parasites? *Science*, **218**, 384–387.

Hammerstein, P. (2003) *Genetic and Cultural Evolution of Cooperation*. MIT Press, Cambridge, MA.

Hammond, R.L., Bruford, M.W. & Bourke, A.F.G. (2002) Ant workers selfishly bias sex ratios by manipulating female development. *Proceedings of the Royal Society of London, Series B*, **269**, 173–178.

Hanken, J. & Sherman, P.W. (1981) Multiple paternity in Belding's ground squirrel litters. *Science*, **212**, 351–353.

Hanlon, R. (2007) Cephalopod dynamic camouflage. *Current Biology*, **17**, R400-R404.

Hannonen, M. & Sundstrom, L. (2003) Worker nepotism among polygynous ants. *Nature*, **421**, 910.

Harcourt, A.H., Harvey, P.H., Larson, S.G. & Short, R.V. (1981) Testis weight, body weight and breeding system in primates. *Nature*, **293**, 55–57.

Harcourt, A.H., Purvis, A. & Liles, L. (1995) Sperm competition: mating system, not breeding season, affects testes size of primates. *Functional Ecology*, **9**, 468–476.

Harcourt, J.L., Ang, T.Z., Sweetman, G., Johnstone, R.A. & Manica, A. (2009) Social feedback and the emergence of leaders and followers. *Current Biology*, **19**, 248–252.

Hardy, I.C.W. (2002) *Sex ratios: concepts and research methods*. Cambridge University Press, Cambridge.

Harper, D.G.C. (1982) Competitive foraging in mallards: 'ideal free' ducks. *Animal Behavior*, **30**, 575–584.

Harrison, F., Barta, Z., Cuthill, I. & Szekely, T. (2009) How is sexual conflict over parental care resolved? A meta-analysis. *Journal of Evolutionary Biology*, **22**, 1800–1812.

Hart, N.S. (2001) The visual ecology of avian photoreceptors. *Progress in Retinal and Eye Research*, **20**, 675–703.

Harvey, P.H. (1985) Intrademic group selection and the sex ratio. In: *Behavioural Ecology: Ecological Consequences of Adaptive Behaviour* (eds R.M. Sibly & R.H. Smith). pp. 59–73. 25th Symposium of the British Ecological Society. Blackwell Scientific Publications, Oxford.

Harvey, P.H. & Pagel, M.D. (1991) *The Comparative Method in Evolutionary Biology*. Oxford University Press, Oxford.

Harvey, P.H. & Purvis, A. (1991) Comparative methods for explaining adaptations. *Nature*, **351**, 619–624.

Harvey, P.H., Kavanagh, M. & Clutton-Brock, T.H. (1978) Sexual dimorphism in primate teeth. *Journal of Zoology*, **186**, 475–486.

Harvey, P.H., Bull, J.J., Pemberton, M. & Paxton, R.J. (1982) The evolution of aposematic coloration in distasteful prey: a family model. *American Naturalist*, **119**, 710–719.

Harvey, P.H., Bull, J.J. & Paxton, R.J. (1983) Looks pretty nasty. *New Scientist*, **97**, 26–27.

Hasselquist, D., Bensch, S. & von Schantz, T. (1996) Correlation between male song repertoire, extra-pair paternity and offspring survival in the great reed warbler. *Nature*, **381**, 229–232.

Hatchwell, B.J. (2009) The evolution of cooperative breeding in birds: kinship, dispersal and life history. *Philosophical Transactions of the Royal Society of London. Series B*, **364**, 3217–3227.

Hatchwell, B.J. & Sharp, S.P. (2006) Kin selection, constraints, and the evolution of cooperative breeding in long tailed tits. *Advances in the Study of Animal Behavior*, **36**, 355–395.

Hatchwell, B.J., Russell, A.F., MacColl, A.D.C., Ross, D.J., Fowlie, M.K. & McGowan, A. (2004) Helpers increase long-term but not short-term productivity in cooperatively breeding long-tailed tits. *Behavioral Ecology*, **15**, 1–10.

Hauber, M.E., Moskat, C. & Ban, M. (2006) Experimental shift in hosts' acceptance threshold of inaccurate-mimic brood parasite eggs. *Biology Letters*, **2**, 177–180.

Heany, V. & Monaghan, P. (1995) A within-clutch trade-off between egg production and rearing in birds. *Proceedings of the Royal Society of London, Series B*, **261**, 361–365.

Hechinger, R.F., Wood, A.C. & Kuris, A.M. (2011) Social organization in a flatworm: trematode parasites form soldier and reproductive castes. *Proceedings of the Royal Society of London, Series B*, **278**, 656–665.

Heg, D., Bruinzeel, L.W. & Ens, B.J. (2003) Fitness consequences of divorce in the oystercatcher, *Haematopus ostralegus*. *Animal Behavior*, **66**, 175–184.

Heiling, A.M., Herberstein, M.E. & Chittka, L. (2003) Crab-spiders manipulate flower signals. *Nature*, **421**, 334.

Heinrich, B., Marzluff, J.M. & Marzluff, C.S. (1993) Common ravens are attracted by appeasement calls of food discoverers when attacked. *Auk*, **110**, 247–254.

Heinsohn, R. (2008) The ecological basis of unusual sex roles in reverse-dichromatic eclectus parrots. *Animal Behavior*, **76**, 97–103.

Heinsohn, R., Legge, S. & Endler, J.A. (2005) Extreme reversed sexual dichromatism in a bird without sex role reversal. *Science*, **309**, 617–619.

Helanterä, H., Strassmann, J.E., Carrillo, J. & Queller, D.C. (2009) Unicolonial ants: where do they come from, what are they and where are they going? *Trends in Ecology & Evolution*, **24**, 341–349.

Helfman, G.S. & Schultz, E.T. (1984) Social transmission of behavioural traditions in a coral reef fish. *Animal Behavior*, **32**, 379–384.

Heller, R. & Milinski, M. (1979) Optimal foraging of sticklebacks on swarming prey. *Animal Behavior*, **27**, 1127–1141.

Henke, J.M. & Bassler, B.L. (2004) Bacterial social engagements. *Trends in Cell Biology*, **14**, 648–656.

Henrich, N. & Henrich, J. (2007) *Why humans cooperate: a cultural and evolutionary explanation*. Oxford University Press, Oxford.

Herre, E.A. (1987) Optimality, plasticity and selective regime in fig wasp sex ratios. *Nature*, **329**, 627–629.

Herre, E.A. (1993) Population structure and the evolution of virulence in nematode parasites of fig wasps. *Science*, **259**, 1442–1445.

Herron, M.D. & Michod, R.E. (2008) Evolution of complexity in the volvocine algae: transitions in individuality through Darwin's eye. *Evolution*, **62**, 436–451.

Herron, M.D., Hackett, J.D., Aylward, F.O. & Michod, R.E. (2009) Triassic origin and early radiation of multicellular volvocine algae. *Proceedings of the National Academy of Sciences USA*, **106**, 3254–3258.

Hewison, A.J.M. & Gaillard, J.M. (1999) Successful sons or advantaged daughters? The Trivers-Willard model and sex-biased maternal investment in ungulates. *Trends in Ecology & Evolution*, **14**, 229–234.

Higham, J.P. Semple, S., MacLarnon, A., Heistermann, M. & Ross, C. (2009) Female reproductive signaling and male mating behaviour in the olive baboon. *Hormones and Behavior*, **55**, 60–67.

Hill, G.E. (1991) Plumage colouration is a sexually selected indicator of male quality. *Nature*, **350**, 337–339.

Hinde, C.A. (2006) Negotiation over offspring care? A positive response to partner-provisioning rate in great tits. *Behavioral Ecology*, **17**, 6–12.

Hinde, C.A., Buchanan, K.L. & Kilner, R.M. (2009) Prenatal environmental effects match offspring begging to parental provisioning. *Proceedings of the Royal Society of London, Series B*, **276**, 2787–2794.

Hinde, C.A., Johnstone, R.A. & Kilner, R.M. (2010) Parent-offspring conflict and coadaptation. *Science*, **327**, 1373–1376.

Hines, H.B., Hunt, J.H., O'Conner, T.K., Gillespie, J.J. & Cameron, S.A. (2007) Multigene phylogeny reveals eusociality evolved twice in vespid wasps. *Proceedings of the National Academy of Sciences USA*, **104**, 3295–3299.

Hitchcock, C.L. & Houston, A.I. (1994) The value of a hoard: not just energy. *Behavioral Ecology*, **5**, 202–205.

Hoare, D.J., Couzin, I.D., Godin, J-G.J. & Krause, J. (2004) Context-dependent group size choice in fish. *Animal Behavior*, **67**, 155–164.

Hodges, C.M. & Wolf, L.L. (1981) Optimal foraging bumblebees? Why is nectar left behind in flowers? *Behavioral Ecology and Sociobiology*, **9**, 41–44.

Hogan-Warburg, A.J. 1966. Social behaviour of the ruff *Philomachus pugnax* (L.). *Ardea*, **54**, 109–229.

Höglund, J. (1989) Size and plumage dimorphism in lek-breeding birds: a comparative analysis. *American Naturalist*, **134**, 72–87.

Höglund, J. & Alatalo, R.V. (1995) *Leks*. Princeton University Press, Princeton, NJ.

Högstedt, G. (1980) Evolution of clutch size in birds: adaptive variation in relation to territory quality. *Science*, **210**, 1148–1150.

Holland, B. & Rice, W.R. (1998) Perspective: Chase-away sexual selection: antagonistic seduction versus resistance. *Evolution*, **52**, 1–7.

Holland, B. & Rice, W.R. (1999) Experimental removal of sexual selection reverses intersexual antagonistic coevolution and removes a reproductive load. *Proceedings of the National Academy of Sciences USA*, **96**, 5083–5088.

Hölldobler, B. & Wilson, E.O. (1990) *The Ants*. Harvard University Press, Cambridge, MA.

Hölldobler, B. & Wilson, E.O. (1994) *Journey to the Ants: A story of scientific exploration*. Belknap Press, Harvard, MA.

Hollen, L.I., Bell, M.B.V. & Radford, A.N. (2008) Cooperative sentinel calling? Foragers gain increased biomass intake. *Current Biology*, **18**, 576–579.

Holman, L. & Snook, R.R. (2008) A sterile sperm caste protects brother fertile sperm from female-mediated death in *Drosophila pseudoobscura*. *Current Biology*, **18**, 292–296.

Holmes, W.G. & Mateo, J.M. (2006) Fostering clarity in kin recognition designs: reply to Todrank & Heth. *Animal Behaviour*, **72**, e5-e7.

Holmes, W.G. & Sherman, P.W. (1982) The ontogeny of kin recognition in two species of ground squirrels. *American Zoologist*, **22**, 491–517.

Holzer, B., Kümmerli, R., Keller, L. & Chapuisat, M. (2006) Sham nepotism as a result of intrinsic differences in brood viability in ants. *Proceedings of the Royal Society of London, Series B*, **273**, 2049–2052.

Hoogland, J.L. (1983) Nepotism and alarm calling in the black-tailed prairie dog, *Cynomys ludovicianus*. *Animal Behavior*, **31**, 472–479.

Hoogland, J.L. (1995) *The Black-Tailed Prairie Dog*. University of Chicago Press, Chicago, IL.

Hori, M. (1993) Frequency-dependent natural selection in the handedness of scale-eating cichlid fish. *Science*, **260**, 216–219.

Hornett, E.A., Charlat, S., Duplouy, A.M.R., *et al.* (2006) Evolution of male-killer suppression in a natural population. *PLOS Biology*, **4**, e283.

Hosken, D.J., Garner, T.W.J. & Ward, P.I. (2001) Sexual conflict selects for male and female reproductive characters. *Current Biology*, **11**, 489–493.

Houde, A.E. (1988) Genetic differentiation in female choice between two guppy populations. *Animal Behavior*, **36**, 511–516.

Houde, A.E. & Endler, J.A. (1990) Correlated evolution of female mating preferences and male colour patterns in the guppy *Poecilia reticulata*. *Science*, **248**, 1405–1408.

Houston, A.I. & Davies, N.B. (1985) The evolution of cooperation and life history in the dunnock *Prunella modularis*. In: *Behavioural Ecology: Ecological Consequences of Adaptive Behaviour* (eds R.M. Sibly & R.H. Smith). pp. 471–487. Blackwell Scientific Publications, Oxford.

Houston, A.I. & McNamara, J.M. (1982) A sequential approach to risk-taking. *Animal Behavior*, **30**, 1260–1261.

Houston, A.I. & McNamara, J.M. (1985) The choice of two prey types that minimises the probability of starvation. *Behavioral Ecology and Sociobiology*, **17**, 135–141.

Houston, A.I., Clark, C.W., McNamara, J.M. & Mangel, M. (1988) Dynamic models in behavioural and evolutionary ecology. *Nature*, **332**, 29–34.

Houston, A.I., Szekely, T. & McNamara, J.M. (2005) Conflicts between parents over care. *Trends in Ecology & Evolution*, **20**, 33–38.

Howard, R.D. (1978a) Factors influencing early embryo mortality in bullfrogs. *Ecology*, **59**, 789–798.

Howard, R.D. (1978b) The evolution of mating strategies in bullfrogs. *Rana catesbeiana*. *Evolution*, **32**, 850–871.

Hrdy, S.B. (1977) *The Langurs of Abu: Female and Male Strategies of Reproduction*. Harvard University Press, Cambridge, MA.

Hrdy, S.B. (1979) Infanticide among animals: a review, classification, and examination of the implications for the reproductive strategies of females. *Ethology and Sociobiology*, **1**, 13–40.

Hrdy, S.B. (1999) *Mother Nature: A History of Mothers, Infants and Natural Selection*. Pantheon, New York.

Huchard, E., Courtiol, A., Benavides, J.A., Knapp, L.A., Raymond, M. & Cowlishaw, G. (2009) Can fertility signals lead to quality signals? Insights from the evolution of primate sexual swellings. *Proceedings of the Royal Society of London, Series B*, **276**, 1889–1897.

Hughes, M. (1996) The function of concurrent signals: visual and chemical communication in snapping shrimp. *Animal Behaviour* **52**, 247–257.

Hughes, W.O.H., Sumner, S., Borm, S.V. & Boomsma, J.J. (2003) Worker caste polymorphism has a genetic basis in *Acromyrmex* leaf-cutting ants. *Proceedings of the National Academy of Sciences USA*, **100**, 9394–9397.

Hughes, W.O.H., Oldroyd, B.P., Beekman, M. & Ratnieks, F.L.W. (2008) Ancestral monogamy shows kin selection is the key to the evolution of eusociality. *Science*, **320**, 1213–1216.

Hunt, J. & Simmons, L.W. (2001) Status-dependent selection in the dimorphic beetle *Onthophagus taurus*. *Proceedings of the Royal Society of London, Series B*, **268**, 2409–2414.

Hunt, S., Cuthill, I.C., Bennett, A.T.D. & Griffiths, R. (1999) Preferences for ultraviolet partners in the blue tit. *Animal Behavior*, **58**, 809–815.

Hurly, T.A. (1992) Energetic reserves of marsh tits (*Parus palustris*): food and fat storage in response to variable food supply. *Behavioral Ecology*, **3**, 181–188.

Hurst, L.D. (1991) The incidences and evolution of cytoplasmic male killers. *Proceedings of the Royal Society of London, Series B*, **244**, 91–99.

Ims, R.A. (1987) Responses in spatial organization and behaviour to manipulations of the food resource in the vole *Clethrionomys rufocanus*. *Journal of Animal Ecology*, **56**, 585–596.

Ims, R.A. (1988) Spatial clumping of sexually receptive females induces space sharing among male voles. *Nature*, **335**, 541–543.

Inglis, R.F., Gardner, A., Cornelis, P. & Buckling, A. (2009) Spite and virulence in the bacterium *Pseudomonas aeruginosa*. *Proceedings of the National Academy of Sciences USA*, **106**, 5703–5707.

Inward, D., Beccaloni, G. & Eggleton, P. (2007) Death of an order: a comprehensive molecular phylogenetic study confirms that termites are eusocial cockroaches *Biology Letters* **3**, 331–335.

Iwasa, Y., Pomiankowski, A. & Nee, S. (1991) The evolution of costly mate preferences II. The 'handicap' hypothesis. *Evolution*, **45**, 1431–1442.

Jackson, D.E., Holcombe, M. & Ratnieks, F.L.W. (2004) Trail geometry gives polarity to ant foraging networks. *Nature*, **432**, 907–909.

Jaenike, J. (2001) Sex chromosome meiotic drive. *Annual Review of Ecology and Systematics*, **32**, 25–49.

Jakobsson, S., Radesäter, T. & Järvi, T. (1979) On the fighting behaviour of *Nannacara anomala* (Pisces, Cichlidae) males. *Zeitschrift für Tierpsychologie*, **49**, 210–220.

Jander, K.C. & Herre, E.A. (2010) Host sanctions and pollinator cheating in the fig tree-fig wasp mutualism. *Proceedings of the Royal Society of London, Series B*, **277**, 1481–1488.

Janzen, F.J. & Phillips, P.C. (2006) Exploring the evolution of environmental sex determination, especially in reptiles. *Journal of Evolutionary Biology*, **19**, 1775–1784.

Jarman, P.J. (1974) The social organization of antelope in relation to their ecology. *Behaviour*, **48**, 215–267.

Jarvis, J.U.M. 1981 Eusociality in a mammal: cooperative breeding in naked mole-rat colonies. *Science*, **212**, 571–573.

Jarvis, J.U.M., O'Riain, M.J., Bennett, N.C. & Sherman, P.W. (1994) Mammalian eusociality: a family affair. *Trends in Ecology & Evolution*, **9**, 47–51.

Jeffreys, A.J., Wilson, V. & Thein, S.L. (1985) Hyper-variable 'minisatellite' regions in human DNA. *Nature*, **314**, 67–73.

Johnstone, R.A. (2000) Models of reproductive skew – a review and synthesis. *Ethology*, **106**, 5–26.

Johnstone, R.A. (2004) Begging and sibling competition: how should offspring respond to their rivals? *American Naturalist*, **163**, 388–406.

Johnstone, R.A. & Hinde, C.A. (2006) Negotiation over offspring care – how should parents respond to each other's efforts? *Behavioral Ecology*, **17**, 818–827.

Jones, I.L. & Hunter, F.M. (1993) Mutual sexual selection in a monogamous seabird. *Nature*, **362**, 238–239.

Jones, J.C. & Reynolds, J.D. (1999) Costs of egg ventilation for male common gobies breeding in conditions of low dissolved oxygen. *Animal Behavior*, **57**, 181–188.

Jones, T.M. & Quinnell, R.J. (2002) Testing predictions for the evolution of lekking in the sandfly, *Lutzomyia longipalpis*. *Animal Behavior*, **63**, 605–612.

Jukema, J. & Piersma, T. (2006) Permanent female mimics in a lekking shorebird. *Biology Letters*, **2**, 161–164.

Kacelnik, A. (1984) Central place foraging in Starlings (*Sturnus vulgaris*). I. Patch residence time. *Journal of Animal Ecology*, **53**, 283–299.

Kacelnik, A. (2009) Tools for thought or thought for tools? *Proceedings of the National Academy of Sciences USA*, **106**, 10071–10072.

Kacelnik, A. & Bateson, M. (1997) Risk sensitivity: crossroads for theories of decision-making. *Trends in Cognitive Sciences*, **8**, 304–309.

Kacelnik, A., Krebs, J.R. & Bernstein, C. (1992) The ideal free distribution and predator–prey populations. *Trends in Ecology & Evolution*, **7**, 50–55.

Kazancioglu, E. & Alonzo, S.H. (2010) A comparative analysis of sex change in Labridae supports the size advantage hypothesis. *Evolution*, **64**, 2254–2264.

Keller, L. & Reeve, H.K. (1994) Partitioning of reproduction in animal societies. *Trends in Ecology & Evolution*, **9**, 98–102.

Keller, L. & Ross, K.G. (1998) Selfish genes: a green beard in the red fire ant. *Nature*, **394**, 573–575.

Keller, L. & Surette, M.G. (2006) Communication in bacteria: an ecological and evolutionary perspective. *Nature Reviews Microbiology*, **4**, 249–258.

Kelman, E.J., Tiptus, P. & Osorio, D. (2006) Juvenile plaice (*Pleuronectes platessa*) produce camouflage by flexibly combining two separate patterns. *Journal of Experimental Biology*, **209**, 3288–3292.

Kempenaers, B. (2007) Mate choice and genetic quality: a review of the heterozygosity theory. *Advances in the Study of Behavior*, **37**, 189–278.

Kempenaers, B., Verheyen, G.R. & Dhondt, A.A. (1997) Extrapair paternity in the blue tit (*Parus caeruleus*): female choice, male charactistics and offspring quality. *Behavioral Ecology*, **8**, 481–492.

Kendal, R.L., Coolen, I., van Bergen, Y. & Laland, K.N. (2005) Trade-offs in the adaptive use of social and asocial learning. *Advances in the Study of Behavior,* **35**, 333–379.

Kent, D.S. & Simpson, J.A. (1992) Eusociality in the Beetle *Austroplatypus incompertus* (Coleoptera: Curculionidae). *Naturwissenschaften*, **79**, 86–87.

Kenward, R.E. (1978) Hawks and doves: factors affecting success and selection in goshawk attacks on wood-pigeons. *Journal of Animal Ecology*, **47**, 449–460.

Kerr, B., Riley, M.A., Feldman, M.W. & Bohannan, B.J.M. (2002) Local dispersal promotes biodiversity in a real-life game of rock-paper-scissors. *Nature*, **418**, 171–174.

Kiers, E.T., Rousseau, R.A., West, S.A. & Denison, R.F. (2003) Host sanctions and the legume-rhizobium mutualism. *Nature*, **425**, 78–81.

Kiers, E.T., Hutton, M.G. & Denison, R.F. (2007) Human selection and the relaxation of legume defences against ineffective rhizobia. *Proceedings of the Royal Society of London, Series B*, **274**, 3119–3126.

Kilner, R.M. (1995) When do canary parents respond to nestling signals of need? *Proceedings of the Royal Society of London, Series B*, **260**, 343–348.

Kilner, R.M. (1997) Mouth colour is a reliable signal of need in begging canary nestlings. *Proceedings of the Royal Society of London, Series B*, **264**, 963–968.

Kilner, R.M. (1999) Family conflicts and the evolution of nestling mouth colour. *Behaviour,* **136**, 779–804.

Kilner, R.M. (2001) A growth cost of begging in captive canary chicks. *Proceedings of the National Academy of Sciences USA*, **98**, 11394–11398.

Kilner, R.M. & Hinde, C.A. (2008) Information warfare and parent-offspring conflict. *Advances in the Study of Behavior,* **38**, 283–336.

Kilner, R.M. & Langmore, N.E. (2011) Cuckoos versus hosts in insects and birds: adaptations, counter-adaptations and outcomes. *Biological Reviews*, **86**, 836–852.

Kilner, R.M., Noble, D.G. & Davies, N.B. (1999) Signals of need in parent-offspring communication and their exploitation by the common cuckoo. *Nature*, **397**, 667–672.

Kilner, R.M., Madden, J.R. & Hauber, M.E. (2004) Brood parasitic cowbird nestlings use host young to procure resources. *Science*, **305**, 877–879.

King, A.J., Douglas, C.M.S., Huchard, E., Isaac, N.J.B. & Cowlishaw, G. (2008) Dominance and affiliation mediate despotism in a social primate. *Current Biology*, **18**, 1833–1838.

King, A.J., Johnson, D.D.P. & Van Vugt, M. (2009) The origins and evolution of leadership. *Current Biology*, **19**, R911-R916.

Kirkpatrick, M. (1982) Sexual selection and the evolution of female choice. *Evolution* **36**, 1–12.

Kirkpatrick, M. (1986) The handicap mechanism of sexual selection does not work. *American Naturalist*, **127**, 222–240.

Kleiman, D.G. (1977) Monogamy in mammals. *Quarterly Review of Biology*, **52**, 39–69.

Klug, H. & Bonsall, M.B. (2007) When to care for, abandon, or eat your offspring: the evolution of parental care and filial cannibalism. *American Naturalist*, **170**, 886–901.

Klug, H., Heuschele, J., Jennions, M.D. & Kokko, H. (2010) The mismeasurement of sexual selection. *Journal of Evolutionary Biology*, **23**, 447–462.

Knowlton, N. (1974) A note on the evolution of gamete dimorphism. *Journal of Theoretical Biology*, **46**, 283–285.

Kodric-Brown, A. (1989) Dietary carotenoids and male mating success in the guppy: an environmental component to female choice. *Behavioral Ecology and Sociobiology*, **25**, 393–401.

Koenig, W.D. & Dickinson, J.L. (2004) *Evolution and Ecology of Cooperative Breeding in Birds*. Cambridge University Press.

Kohler, W. (1929) *The Mentality of Apes*. Vintage Books.

Kohler, T., Buckling, A. & van Delden, C. (2009) Cooperation and virulence of clinical *Pseudomonas aeruginosa* populations. *Proceedings of the National Academy of Sciences USA*, **106**, 6339–6344.

Kokko, H. (2003) Are reproductive skew models evolutionarily stable? *Proceedings of the Royal Society of London, Series B*, **270**, 265–270.

Kokko, H. & Jennions, M. (2003) It takes two to tango. *Trends in Ecology & Evolution*, **18**, 103–104.

Kokko, H. & Jennions, M.D. (2008) Parental investment, sexual selection and sex ratios. *Journal of Evolutionary Biology*, **21**, 919–948.

Kokko, H., Brooks, R., Jennions, M.D. & Morley, J. (2003) The evolution of mate choice and mating biases. *Proceedings of the Royal Society of London, Series B*, **270**, 653–664.

Kölliker, M., Brinkhof, M., Heeb, P., Fitze, P. & Richner, H. (2000) The quantitative genetic basis of offspring solicitation and parental response in a passerine bird with parental care. *Proceedings of the Royal Society of London, Series B*, **267**, 2127–2132.

Kölliker, M., Brodie III, E.D. & Moore, A.J. (2005) The coadaptation of parental supply and offspring demand. *American Naturalist*, **166**, 506–516.

Komdeur, J. (1992) Importance of habitat saturation and territory quality for evolution of cooperative breeding in the Seychelles warbler. *Nature*, **358**, 493–495.

Komdeur, J., Daan, S., Tinbergen, J. & Mateman, C. (1997) Extreme modification of sex ratio of the Seychelles Warbler's eggs. *Nature*, **385**, 522–525.

Komdeur, J., Magrath, M.J.L. & Krackow, S. (2002) Pre-ovulation control of hatchling sex ratio in the Seychelles warbler. *Proceedings of the Royal Society of London, Series B*, **269**, 1067–1072.

Kondrashov, A.S. (1988) Deleterious mutations as an evolutionary factor. III Mating preference and some general remarks. *Journal of Theoretical Biology*, **131** 487–496.

König, B., Riester, J. & Markl, H. (1988) Maternal care in house mice (*Mus musculus*): II. The energy cost of lactation as a function of litter size. *Journal of Zoology*, **216**, 195–210.

Korb, J. (2010) The ecology of social evolution in termites. In: *Ecology of Social Evolution* (eds J Korb & J Heinze). pp. 151–174. Springer-Verlag.

Kraaijeveld, K., Kraaijeveld-Smit, F.J.L. & Komdeur, J. (2007) The evolution of mutual ornamentation. *Animal Behavior*, **74**, 657–677.

Krakauer, A.H. (2005) Kin selection and cooperative courtship in wild turkeys. *Nature*, **434**, 69–72.

Kramer, D.L. (1985) Are colonies supraoptimal groups? *Animal Behavior*, **33**, 1031.

Krams, I., Krama, T., Igaune, K. & Mänd, R. (2008) Experimental evidence of reciprocal altruism in the pied flycatcher. *Behavioral Ecology & Sociobiology*, **62**, 599–605.

Krause, J. (1993a) The effect of 'Schreckstoff' on the shoaling behaviour of the minnow – a test of Hamilton's selfish herd theory. *Animal Behavior*, **45**, 1019–1024.

Krause, J. (1993b) The relationship between foraging and shoal position in a mixed shoal of roach (*Rutilus rutilus*) and chub (*Leuciscus leuciscus*): a field study. *Oecologia*, **93**, 356–359.

Krause, J. & Ruxton, G.D. (2002) *Living in Groups*. Oxford University Press, Oxford.

Krebs, E.A. & Putland, D.A. (2004) Chic chicks: the evolution of chick ornamentation in rails. *Behavioral Ecology*, **15**, 946–951.

Krebs, J.R. 1971. Territory and breeding density in the great tit. *Parus major* L. *Ecology* **52**, 2–22.

Krebs, J.R. (1990) Food storing birds: adaptive specialisation in brain and behaviour? *Philosophical Transactions of the Royal Society of London. Series B*, **329**, 153–160.

Krebs, J.R. & Dawkins, R. (1984) Animal signals: mind reading and manipulation. In: *Behavioural Ecology: An Evolutionary Approach* (eds J.R. Krebs & N.B. Davies), 2nd edn. pp. 380–402. Blackwell Scientific Publications, Oxford.

Krebs, J.R., MacRoberts, M.H. & Cullen, J.M. (1972) Flocking and feeding in the great tit *Parus major*: an experimental study. *Ibis*, **114**, 507–530.

Krebs, J.R., Erichsen, J.T., Webber, M.I. & Charnov, E.L. (1977) Optimal prey selection in the great tit. *Parus major. Animal Behavior*, **25**, 30–38.

Krebs, J.R., Kacelnik, A. & Taylor, P. (1978) Test of optimal sampling by foraging great tits. *Nature*, **275**, 27–31.

Krüger, O. (2011) Brood parasitism selects for no defence in a cuckoo host. *Proceedings of the Royal Society of London, Series B*, **278**, 2777–2783.

Kruuk, H. 1964. Predators and anti-predator behaviour of the black headed gull, *Larus ridibundus*. *Behaviour* (Suppl.), **11**, 1–129.

Kruuk, H. (1972) *The Spotted Hyena*. University of Chicago Press, Chicago, IL.

Kümmerli, R., Helms, K.R. & Keller, L. (2005) Experimental manipulation of queen number affects colony sex ratio investment in the highly polygynous ant *Formica exsecta*. *Proceedings of the Royal Society of London, Series B*, **272**, 1789–1794.

Lachmann, M., Szamado, S. & Bergstrom, C.T. (2001) Cost and conflict in animal signals and human language. *Proceedings of the National Academy of Sciences USA*, **98**, 13189–13194.

Lack, D. (1947) The significance of clutch-size. Parts 1 and 2. *Ibis*, **89**, 302–352.

Lack, D. (1966) *Population Studies of Birds*. Clarendon Press, Oxford.

Lack, D. (1968) *Ecological Adaptations for Breeding in Birds*. Methuen, London.

Lahti, D.C. (2005) Evolution of bird eggs in the absence of cuckoo parasitism. *Proceedings of the National Academy of Sciences USA*, **102**, 18057–18062.

Lahti, D.C. (2006) Persistence of egg recognition in the absence of cuckoo brood parasitism: pattern and mechanism. *Evolution*, **60**, 157–168.

Laland, K.N. (2008) Animal cultures. *Current Biology*, **18**, R366-R370.

Laland, K.N. & Janik, V.M. (2006) The animal cultures debate. *Trends in Ecology & Evolution*, **21**, 542–547.

Laland, K.N. & Williams, K. (1997) Shoaling generates social learning of foraging information in guppies. *Animal Behavior*, **53**, 1161–1169.

Laland, K.N. & Williams, K. (1998) Social transmission of maladaptive information in the guppy. *Behavioral Ecology*, **9**, 493–499.

Lande, R. (1981) Models of speciation by sexual selection of polygenic traits. *Proceedings of the National Academy of Sciences USA*, **78**, 3721–3725.

Landeau, L. & Terborgh, J. (1986) Oddity and the confusion effect in predation. *Animal Behavior*, **34**, 1372–1380.

Langmore, N.E., Davies, N.B., Hatchwell, B.J. & Hartley, I.R. (1996) Female song attracts males in the alpine accentor, *Prunella collaris*. *Proceedings of the Royal Society of London, Series B*, **263**, 141–146.

Langmore, N.E., Hunt, S. & Kilner, R.M. (2003) Escalation of a coevolutionary arms race through host rejection of brood parasitic young. *Nature*, **422**, 157–160.

Langmore, N.E., Cockburn, A., Russell, A.F. & Kilner, R.M. (2009) Flexible cuckoo chick-rejection rules in the superb fairy-wren. *Behavioral Ecology*, **20**, 978–984.

Langmore, N.E., Stevens, M., Maurer, G., *et al.* (2011) Visual mimicry of host nestlings by cuckoos. *Proceedings of the Royal Society of London, Series B*, **278**, 2455–2463.

Lank, D.B., Oring, L.W. & Maxson, S.J. (1985) Mate and nutrient limitation of egg laying in a polyandrous shorebird. *Ecology*, **66**, 1513–1524.

Lank, D.B., Smith, C.M., Hanotte, O., Burke, T. & Cooke, F. (1995) Genetic polymorphism for alternative mating behaviour in lekking male ruff *Philomachus pugnax*. *Nature*, **378**, 59–62.

Lank, D.B., Smith, C.M., Hanotte, O., Ohtonen, A., Bailey, S. & Burke, T. (2002) High frequency of polyandry in a lek mating system. *Behavioral Ecology*, **13**, 209–215.

Lazarus, J. & Inglis, I.R. (1986) Shared and unshared parental investment, parent-offspring conflict and brood size. *Animal Behavior*, **34**, 1791–1804.

Leadbeater, E., Carruthers, J.M., Green, J.P., Roser, N.S., & Field, J. (2011) Nest inheritance is the missing source of direct fitness in a primitively eusocial insect. *Nature*, **333**, 874–876.

Le Boeuf, B.J. (1972) Sexual behaviour in the northern elephant seal. *Mirounga angustirostris*. *Behaviour*, **41**, 1–26.

Le Boeuf, B.J. (1974) Male–male competition and reproductive success in elephant seals. *American Zoologist*, **14**, 163–176.

Le Boeuf, B.J. & Reiter, J. (1988) Lifetime reproductive success in Northern elephant seals. In: *Reproductive Success* (ed. T.H. Clutton-Brock). pp. 344–362. Chicago University Press, Chicago, IL.

Lehmann, L. & Keller, L. (2006) The evolution of cooperation and altruism. A general framework and classification of models. *Journal of Evolutionary Biology*, **19**, 1365–1378.

Leigh, E.G. (1971) *Adaptation and Diversity*. Freeman, Cooper and Company, Cambridge.

Leigh, E.G.J. (1999) *Tropical Forest Ecology: A View from Barro Colorado Island*. Oxford University Press, Oxford.

Leonard, M.L. & Picman, J. (1987) Female settlement in marsh wrens: is it affected by other females? *Behavioral Ecology and Sociobiology*, **21**, 135–140.

Leonard, M.L. & Horn, A.G. (2001) Dynamics of calling by tree swallow (*Tachycineta bicolor*) nestmates. *Behavioral Ecology and Sociobiology*, **50**, 430–435.

Lightbody, J.P. & Weatherhead, P.J. (1988) Female settling patterns and polygyny: tests of a neutral-mate-choice hypothesis. *American Naturalist*, **132**, 20–33.

Lill, A. (1974) Sexual behaviour of the lek-forming white-bearded manakin (*Manacus manacus trinitatis*). *Zeitschrift für Tierpsychologie*, **36**, 1–36.

Lima, S.L. (1984) Downy woodpecker foraging behavior: efficient sampling in simple stochastic environments. *Ecology*, **65**, 166–174.

Lima, S.L. (1986) Predation risk and unpredictable feeding conditions: determinants of body mass in birds. *Ecology*, **67**, 377–385.

Lima, S.L. (1994) On the personal benefits of anti-predator vigilance. *Animal Behavior*, **48**, 734–736.

Lima, S.L. (1998) Stress and decision making under the risk of predation. Recent developments from behavioural, reproductive and ecological perspectives. *Advances in the Study of Behavior*, **27**,215–290.

Lima, S.L., Valone, T.J. & Caraco, T. (1985) Foraging-efficiency–predation-risk tradeoff in the grey squirrel. *Animal Behavior*, **33**, 155–165.

Lindstedt, C., Lindström, L. & Mappes, J. (2008) Hairiness and warning colours as components of antipredator defence: additive or interactive benefits? *Animal Behavior*, **75**, 1703–1713.

Lindstedt, C., Lindström, L. & Mappes, J. (2009) Thermoregulation constrains effective warning signal expression. *Evolution*, **63**, 469–478.

Lindstedt, C., Talsma, J.H.R., Ihalainen, E., Lindström, L. & Mappes, J. (2010) Diet quality affects warning coloration indirectly: excretion costs in a generalist herbivore. *Evolution*, **64**, 68–78.

Lindström, L., Alatalo, R.V. & Mappes, J. (1997) Imperfect Batesian mimicry – the effects of the frequency and the distastefulness of the model. *Proceedings of the Royal Society of London, Series B*, **264**, 149–153.

Lock, J.E., Smiseth, P.T. & Moore, A.J. (2004) Selection, inheritance and the evolution of parent-offspring interactions. *American Naturalist*, **164**, 13–24.

Lotem, A. (1993) Learning to recognize nestlings is maladaptive for cuckoo *Cuculus canorus* hosts. *Nature*, **362**, 743–745.

Lotem, A., Nakamura, H. & Zahavi, A. (1995) Constraints on egg discrimination and cuckoo-host co-evolution. *Animal Behavior*, **49**, 1185–1209.

Lynch, M. & Walsh, B. (1998) *Genetics and Analysis of Quantitative Traits*. Sinauer Associates, Sunderland, MA.

Lyon, B.E., Montgomerie, R.D. & Hamilton, L.D. (1987) Male parental care and monogamy in snow buntings. *Behavioral Ecology and Sociobiology*, **20**, 377–382.

Lyon, B.E., Eadie, J.M. & Hamilton, L.D. (1994) Parental choice selects for ornamental plumage in American coot chicks. *Nature*, **371**, 240–243.

Lyon, B.E., Chaine, A.S. & Winkler, D.W. (2008) A matter of timing. *Science*, **321**, 1051–1052.

MacColl, A.D.C. & Hatchwell, B.J. (2004) Determinants of lifetime fitness in a cooperative breeder, the long tailed tit *Aegithalos caudatus*. *Journal of Animal Ecology*, **73**, 1137–1148.

Mackinnon, J. (1974) The ecology and behaviour of wild orangutans. *Pongo pygmaeus. Animal Behavior,* **22**, 3–74.

MacLean, R.C. & Gudelj, I. (2006) Resource competition and social conflict in experimental populations of yeast. *Nature,* **441**, 498–501.

Madden, J.R. (2002) Bower decorations attract females but provoke other male spotted bowerbirds: males resolve this trade-off. *Proceedings of the Royal Society of London, Series B,* **269**, 1347–1352.

Madden, J.R. (2003a) Bower decorations are good predictors of mating success in the spotted bowerbird. *Behavioral Ecology and Sociobiology,* **53**, 269–277.

Madden, J.R. (2003b) Male spotted bowerbirds preferentially choose, arrange and proffer objects that are good predictors of mating success. *Behavioral Ecology and Sociobiology,* **53**, 263–268.

Magrath, R.D. (1989) Hatch asynchrony and reproductive success in the blackbird. *Nature,* **339**, 536–538.

Magrath, R.D., Pitcher, B.J. & Gardner, J.L. (2007) A mutual understanding? Interspecific responses by birds to each other's aerial alarm calls. *Behavioral Ecology,* **18**, 944–951.

Magrath, R.D., Pitcher, B.J. & Gardner, J.L. (2009) Recognition of other species' aerial alarm calls: speaking the same language or learning another? *Proceedings of the Royal Society of London, Series B,* **276**, 769–774.

Magrath, R.D., Haff, T.M., Horn, A.G. & Leonard, M.L. (2010) Calling in the face of danger: predation risk and acoustic communication by parent birds and their offspring. *Advances in the Study of Behavior,* **41**, 187–253.

Maguire, E.A., Gadian, D.G., Johnsrude, I.S., *et al.* (2000) Navigation-related structural change in the hippocampi of taxi drivers. *Proceedings of the National Academy of Sciences USA,* **97**, 4398–4403.

Magurran, A.E. & Higham, A. (1988) Information transfer across fish shoals under predation threat. *Ethology,* **78**, 153–158.

Magurran, A.E. (1990) The adaptive significance of schooling as an antipredator defence in fish. *Annals of Zoology, Fennica,* **27**, 51–66.

Magurran, A.E. & Seghers, B.H. (1991) Variation in schooling and aggression amongst guppy (*Poecilia reticulata*) populations in Trinidad. *Behaviour,* **118**, 214–234.

Magurran, A.E., Seghers, B.H., Carvalho, G.R. & Shaw, P.W. (1992) Behavioural consequences of an artificial introduction of guppies (*Poecilia reticulata*) in N. Trinidad: evidence for the evolution of antipredator behaviour in the wild. *Proceedings of the Royal Society of London, Series B,* **248**, 117–122.

Major, P.F. (1978) Predator–prey interactions in two schooling fishes. *Caranx ignobilis* and *Stolephorus purpureus. Animal Behavior,* **26**, 760–777.

Mallet, J. & Barton, N.H. (1989) Strong natural selection in a warning color hybrid zone. *Evolution,* **43**, 421–431.

Mallet, J. & Gilbert, L.E. (1995) Why are there so many mimcry rings – correlations between habitat, behaviour and mimicry in *Heliconius* butterflies. *Biological Journal of the Linnean Society,* **55**, 159–180.

Mallet, J. & Joron, M. (1999) Evolution of diversity in warning color and mimicry: polymorphisms, shifting balance and speciation. *Annual Review of Ecology and Systematics,* **30**, 201–233.

Mangel, M. & Clark, C.W. (1988) *Dynamic Modelling in Behavioural Ecology*. Princeton University Press, Princeton, NJ.

Manica, A. (2002) Alternative strategies for a father with a small brood: mate, cannibalise or care. *Behavioral Ecology and Sociobiology,* **51**, 319–323.

Manica, A. (2004) Parental fish change their cannibalistic behaviour in response to the cost-to-benefit ratio of parental care. *Animal Behavior,* **67**, 1015–1021.

Manser, M.B. (1999) Response of foraging group members to sentinel calls in suricates *Suricata suricatta. Proceedings of the Royal Society of London, Series B,* **266**, 1013–1019.

Manser, M.B. (2001) The acoustic structure of suricates' alarm calls varies with predator type and the level of response urgency. *Proceedings of the Royal Society of London, Series B,* **268**, 2315–2324.

Mappes, J., Marples, N. & Endler, J.A. (2005) The complex business of survival by aposematism. *Trends in Ecology & Evolution,* **20**, 598–603.

Marples, N.M. & Kelly, D.J. (1999) Neophobia and dietry conservatism: two distinct processes? *Evolutionary Ecology,* **13**, 641–53.

Marsh, B., Schuck-Paim, C. & Kacelnik, A. (2004) Energetic state during learning affects foraging choices in starlings. *Behavioral Ecology,* **15**, 396–399.

Marshall, N.J. (2000) Communication and camouflage with the same 'bright' colours in reef fishes. *Philosophical Transactions of the Royal Society of London. Series B,* **355**, 1243–1248.

Martin, S.J., Chaline, N.G., Ratnieks, F.L.W. & Jones, G.R. (2005) Searching for the egg-marking signal in honeybees. *Journal of Negative Results,* **2**, 1–9.

Marzluff, J.M., Heinrich, B. & Marzluff, C.S. (1996) Raven roosts are mobile information centres. *Animal Behavior,* **51**, 89–103.

Mateo, J.M. (2002) Kin-recognition abilities and nepotism as a function of sociality. *Proceedings of the Royal Society of London, Series B,* **269**, 721–727.

Mateo, J.M. & Holmes, W.G. (2004) Cross-fostering as a means to study kin recognition. *Animal Behaviour* **68**, 1451–1459.

Maynard Smith, J. (1956) Fertility, mating behaviour and sexual selection in *Drosophila subobscura. Journal of Genetics,* **54**, 261–279.

Maynard Smith, J. (1964) Group selection and kin selection. *Nature,* **201**, 1145–1147.

Maynard Smith, J. (1976a) Group selection. *Quarterly Review of Biology,* **51**, 277–283.

Maynard Smith, J. (1976b) Sexual selection and the handicap principle. *Journal of Theoretical Biology,* **57**, 239–242.

Maynard Smith, J. (1977) Parental investment – a prospective analysis. *Animal Behavior,* **25**, 1–9.

Maynard Smith, J. (1982) *Evolution and the Theory of Games.* Cambridge University Press, Cambridge.

Maynard Smith, J. & Harper, D. (2003) *Animal Signals.* Oxford University Press, Oxford.

Maynard Smith, J. & Price, G.R. 1973. The logic of animal conflict. *Nature,* **246**, 15–18.

Maynard Smith, J. & Szathmary, E. (1995) *The Major Transitions in Evolution.* W.H. Freeman, Oxford.

McCabe, J. & Dunn, A.M. (1997) Adaptive significance of environmental sex determination in an amphipod. *Journal of Evolutionary Biology,* **10**, 515–527.

McClintock, W.J. & Uetz, G.W. (1996) Female choice and pre-existing bias: visual cues during courtship in two *Schizocosa* wolf spiders (Araneae: Lycosidae). *Animal Behavior,* **52**, 167–181.

McDonald, D.B. (2010) A spatial dance to the music of time in the leks of long-tailed manakins. *Advances in the Study of Behaviour,* **42**, 55–81.

McDonald, D.B. & Potts, W.K. (1994) Cooperative display and relatedness among males in a lek-mating bird. *Science,* **266**, 1030–1032.

McNamara, J.M. & Houston, A.I. (1990) The value of fat reserves and the tradeoff between starvation and predation. *Acta Biotheoretica,* **38**, 37–61.

McNamara, J.M. & Houston, A.I. (1992) Evolutionarily stable levels of vigilance as a function of group size. *Animal Behavior,* **43**, 641–658.

McNamara, J.M., Gasson, C. & Houston, A.I. (1999) Incorporating rules for responding into evolutionary games. *Nature,* **401**, 368–371.

Mehdiabadi, N.J., Jack, C.N., Farnham, T.T., *et al.* (2006) Kin preference in a social microbe. *Nature,* **442**, 881–882.

Merrill, R.M. & Jiggins, C.D. (2009) Müllerian mimicry: sharing the load reduces the legwork. *Current Biology,* **19**, R687-R689.

Mery, F., Belay, A.T., So, A.K.-C., Sokolowski, M.B. & Kawecki, T.J. (2007) Natural polymorphism affecting learning and memory in *Drosophila. Proceedings of the National Academy of Sciences USA,* **104**, 13051–13055.

Mesterton-Gibbons, M. & Dugatkin, L.A. (1999) On the evolution of delayed recruitment. *Behavioral Ecology*, **10**, 377–390.

Metcalf, R. A. 1980 Sex ratios, parent-offspring conflict, and local competition for mates in the social wasps *Polistes metricus* and *Polistes variatus*. *American Naturalist*, **116**, 642–654.

Michl, G., Török, J., Griffith, S.C. & Sheldon, B.C. (2002) Experimental analysis of sperm competition mechanisms in a wild bird population. *Proceedings of the National Academy of Sciences USA*, **99**, 5466–5470.

Miles, D.B., Sinervo, B. & Frankino, W.A. (2000) Reproductive burden, locomotor performance and the cost of reproduction in free ranging lizards. *Evolution*, **54**, 1386–1395.

Milinski, M. (1979) An evolutionarily stable feeding strategy in sticklebacks. *Zeitschrift für Tierpsychologie*, **51**, 36–40.

Milinski, M. (1984) A predator's cost of overcoming the confusion effect of swarming prey. *Animal Behavior*, **32**, 1157–1162.

Milinski, M. (2006a) The major histocompatibility complex, sexual selection, and mate choice. *Annual Review of Ecology, Evolution and Systematics*, **37**, 159–186.

Milinski, M. (2006b) Stabilizing the Earth's climate is not a losing game: supporting evidence from public goods experiments. *Proceedings of the National Academy of Sciences USA*, **103**, 3994–3998.

Milinski, M. & Bakker, T.C.M. (1990) Female sticklebacks use male coloration in mate choice and hence avoid parasitized males. *Nature*, **344**, 330–333.

Milinski, M. & Heller, R. (1978) Influence of a predator on the optimal foraging behaviour of sticklebacks (*Gasterosteus aculeatus*). *Nature*, **275**, 642–644.

Milinski, M. & Parker, G.A. (1991) Competition for resources. In: *Behavioural Ecology: an Evolutionary Approach* (eds J.R. Krebs & N.B. Davies), 3rd edn. pp. 137–168. Blackwell Scientific Publications, Oxford.

Mock, D.W. (2004) *More than Kin and Less than Kind: The evolution of family conflict*. Harvard University Press, Cambridge, MA.

Mock, D.W. & Parker, G.A. (1997) *The Evolution of Sibling Rivalry*. Oxford University Press, Oxford.

Moczek, A.P. & Emlen, D.J. (2000) Male horn dimorphism in the scarab beetle, *Onthophagus taurus*: do alternative reproductive tactics favour alternative phenotypes? *Animal Behavior*, **59**, 459–466.

Moksnes, A., Røskaft, E., Braa, A.T., Korsnes, L., Lampe, H.M. & Pedersen, H.Ch. (1991) Behavioural responses of potential hosts towards artificial cuckoo eggs and dummies. *Behaviour*, **116**, 64–89.

Møller, A.P. (1988) Female choice selects for male sexual tail ornaments in the monogamous swallow. *Nature*, **332**, 640–642.

Møller, A.P. (1989) Viability costs of male tail ornaments in a swallow. *Nature*, **339**, 132–135.

Molloy, P.P., Goodwin, N.B., Côté, I.M., Reynolds, J.D. & Gage, M.J.G. (2007) Sperm competition and sex change: a comparative analysis across fishes. *Evolution*, **61**, 640–652.

Monaghan, P. & Nager, R.G. (1997) Why don't birds lay more eggs? *Trends in Ecology & Evolution*, **12**, 270–274.

Moore, T. & Haig, D. (1991) Genomic imprinting in mammalian development: a parental tug-of-war. *Trends in Genetics*, **7**, 45–59.

Mosser, A. & Packer, C. (2009) Group territoriality and the benefits of sociality in the African lion, *Panthera leo*. *Animal Behavior*, **78**, 359–370.

Mottley, K. & Giraldeau, L-A. (2000) Experimental evidence that group foragers can converge on predicted producer-scrounger equilibria. *Animal Behavior*, **60**, 341–350.

Mueller, U.G. (1991) Haplodiploidy and the evolution of facultive sex ratios in a primitively eusocial bee. *Science*, **254**, 442–444.

Mueller, U.G., Gerardo, N.M. Aanen, D.K., Six, D.L. & Schultz, T.R. (2005) The evolution of agriculture in insects. *Annual Review of Ecology, Evolution and Systematics* **36**, 563–595.

Mulder, R.A. & Langmore, N.E. (1993) Dominant males punish helpers for temporary defection in superb fairy wrens. *Animal Behaviour,* **45**, 830–833.

Müller, F. (1878) Über die Vortheile der Mimicry bei Schmetterlingen. *Zoologischer Anzeiger,* **1**, 54–55.

Müller, J.K., Braunisch, V., Hwang, W. & Eggert, A-K. (2006) Alternative tactics and individual reproductive success in natural associations of the burying beetle, *Nicrophorus vespilloides. Behavioral Ecology,* **18**, 196–203.

Munday, P.L., Buston, P.M. & Warner, R.R. (2006) Diversity and flexibility of sex-change strategies in animals. *Trends in Ecology & Evolution,* **21**, 89–95.

Mundy, N.I., Badcock, N.S., Hart, T., Scribner, K., Janssen, K. & Nadeau, N. (2004) Conserved genetic basis of a quantitative plumage trait involved in mate choice. *Science,* **303**, 1870–1873.

Murray, M.G. (1987) The closed environment of the fig receptacle and its influence on male conflict in the Old World fig wasp *Philotrypesis pilosa. Animal Behavior,* **35**, 488–506.

Nachman, M.W., Hoekstra, H.E. & D'Agostino, S.L. (2003) The genetic basis of adaptive melanism in pocket mice. *Proceedings of the National Academy of Sciences USA,* **100**, 5268–5273.

Nakagawa, S., Ockendon, N., Gillespie, D.O.S., Hatchwell, B.J. & Burke, T. (2007) Assesssing the function of house sparrows' bib size using a flexible meta-analysis method. *Behavioral Ecology* **18**, 831–840.

Nakamura, M. (1990) Cloacal protuberance and copulatory behaviour of the alpine accentor (*Prunella collaris*). *Auk,* **107**, 284–295.

Nakatsuru, K. & Kramer, D.L. (1982) Is sperm cheap? Limited male fertility and female choice in the lemon tetra (Pisces: Characidae). *Science,* **216**, 753–755.

Neff, B.D. (2003) Decisions about parental care in response to perceived paternity. *Nature,* **422**, 716–719.

Neill, S.R.St.J. & Cullen, J.M. (1974) Experiments on whether schooling by their prey affects the hunting behaviour of cephalods and fish predators. *Journal of Zoology,* **172**, 549–569.

Nelson, X.J. & Jackson, R.R. (2006) Compound mimicry and trading predators by the males of sexually dimorphic Batesian mimics. *Proceedings of the Royal Society of London, Series B,* **273**, 367–372.

Nesse, R.M. & Williams, G.C. (1996) *Why we get sick.* Random House, London.

Nettle, D. (2009) *Evolution and Genetics for Psychology.* Oxford University Press, Oxford.

Neuman, C.R., Safran, R.J. & Lovette, I.J. (2007) Male tail streamer length does not predict apparent or genetic reproductive success in North American barn swallows *Hirundo rustica erythrogaster. Journal of Avian Biology,* **38**, 28–36.

Nichols, H.J., Amos, W., Cant, M.A., Bell, M.B.V. & Hodge, S.J. (2010) Top males gain high reproductive success by guarding more successful females in a cooperatively breeding mongoose. *Animal Behavior,* **80**, 649–657.

Nonacs, P. (1986) Ant reproductive strategies and sex allocation theory. *Quarterly Review of Biology,* **61**, 1–21.

Norberg, A.K. (1994) Swallow tail streamer is a mechanical device for self deflection of tail leading edge, enhancing aerodynamic efficiency and flight manoeuvrability. *Proceedings of the Royal Society of London, Series B,* **257**, 227–233.

Norris, K. & Evans, M.R. (2000) Ecological immunity: life history trade-offs and immune defence in birds. *Behavioral Ecology,* **11**, 19–26.

Nowak, M.A. (2006) Five rules for the evolution of cooperation. *Science,* **314**, 1560–1563.

Nowak, M.A. & Sigmund, K. (2002) Bacterial game dynamics. *Nature,* **418**, 138–139.

Nowak, M.A., Tarnita, C.E. & Wilson, E.O. (2010) The evolution of eusociality. *Nature,* **466**, 1057–1062.

Nunn, C.L. (1999) The evolution of exaggerated sexual swellings in primates and the graded-signal hypothesis. *Animal Behavior,* **58**, 229–246.

Nur, U. (1970) Evolutionary rates of models and mimics in Batesian mimicry. *American Naturalist,* **104**, 477–486.

Nussey, D.H., Postma, E., Gienapp, P. & Visser, M.E. (2005) Selection on heritable phenotypic plasticity in a wild bird population. *Science*, **310**, 304–306.

Ohguchi, O. (1978) Experiments on the selection against colour oddity of water fleas by three-spined sticklebacks. *Zeitschrift für Tierpsychologie*, **47**, 254–267.

Olsson, M., Madsen, T. & Shine, R. (1997) Is sperm really so cheap? Costs of reproduction in male adders, *Vipera berus*. *Proceedings of the Royal Society of London, Series B*, **264**, 455–459.

Orians, G.H. (1969) On the evolution of mating systems in birds and mammals. *American Naturalist*, **103**, 589–603.

Oring, L.W. (1982) Avian mating systems. In: *Avian Biology* (eds D.S. Farner & J.R. King), Vol. 6. Pp. 1–92. Academic Press, London.

Osborne, K.A., Robichon, A., Burgess, E., *et al.* (1997) Natural behaviour polymorphism due to a cGMP-dependent protein kinase of *Drosophila*. *Science*, **277**, 834–836.

Owen, D.F. (1954) The winter weights of titmice. *Ibis*, **96**, 299–309.

Owens, I.P.F. & Bennett, P.M. (1997) Variation in mating systems among birds: ecological basis revealed by hierarchical comparative analysis. *Proceedings of the Royal Society of London, Series B*, **264**, 1103–1110.

Owens, I.P.F. & Hartley, I.R. (1998) Sexual dimorphism in birds: why are there so many different forms of dimorphism? *Proceedings of the Royal Society of London, Series B*, **265**, 397–407.

Owen-Smith, N. (1977) On territoriality in ungulates and an evolutionary model. *Quarterly Review of Biology*, **52**, 1–38.

Packer, C. (1977) Reciprocal altruism in *Papio anubis*. *Nature*, **265**, 441–443.

Packer, C. & Pusey, A.E. (1983a) Male takeovers and female reproductive parameters: a simulation of oestrus synchrony in lions (*Panthera leo*). *Animal Behavior*, **31**, 334–340.

Packer, C. & Pusey, A.E. (1983b) Adaptations of female lions to infanticide by incoming males. *American Naturalist*, **121**, 716–728.

Packer, C., Scheel, D. & Pusey, A.E. (1990) Why lions form groups: food is not enough. *American Naturalist*, **136**, 1–19.

Packer, C., Gilbert, D.A., Pusey, A.E. & O'Brien, S.J. (1991) A molecular genetic analysis of kinship and cooperation in African lions. *Nature*, **351**, 562–565.

Pagel, M. (1993) Honest signalling among gametes. *Nature*, **363**, 539–541.

Pagel, M. (1994) Detecting correlated evolution on phylogenies: a general method for the comparative analysis of discrete characters. *Proceedings of the Royal Society of London, Series B*, **255**, 37–45.

Pagel, M. & Meade, A. (2006) Bayesian analysis of correlated evolution of discrete characters by reversible-jump Markov chain Monte Carlo. *American Naturalist*, **167**, 808–825.

Paley, W. (1802) *Natural Theology*. Wilks & Taylor.

Palmer, A.R. (1999) Detecting publication bias in meta-analysis: a case study of fluctuating asymmetry and sexual selection. *American Naturalist*, **154**, 220–233.

Parker, G.A. (1970a) Sperm competition and its evolutionary effect on copula duration in the fly *Scatophaga stercoraria*. *Journal of Insect Physiology*, **16**, 1301–1328.

Parker, G.A. (1970b) The reproductive behaviour and the nature of sexual selection in *Scatophaga stercoraria* L. (Diptera: Scatophagidae). II. The fertilization rate and the spatial and temporal relationships of each sex around the site of mating and oviposition. *Journal of Animal Ecology*, **39**, 205–228.

Parker, G.A. (1970c) Sperm competition and its evolutionary consequences in the insects. *Biological Reviews*, **45**, 525–567.

Parker, G.A. (1979) Sexual selection and sexual conflict. In: *Sexual Selection and Reproductive Competition in Insects* (eds M.S. Blum & N.A. Blum). pp. 123–166. Academic Press, New York.

Parker, G.A. (1982) Why are there so many tiny sperm? Sperm competition and the maintenance of two sexes. *Journal of Theoretical Biology*, **96**, 281–294.

Parker, G.A. (2006) Sexual conflict over mating and fertilisation: an overview. *Philosophical Transactions of the Royal Society of London. Series B*, **361**, 235–259.

Parker, G.A. & Maynard Smith, J. (1990) Optimality theory in evolutionary biology. *Nature*, **348**, 27–33.

Parker, G.A. & Stuart, R.A. (1976) Animal behaviour as a strategy optimiser: evolution of resource assessment strategies and optimal emigration thresholds. *American Naturalist*, **110**, 1055–1076.

Parker, G.A. & Sutherland, W.J. (1986) Ideal free distributions when individuals differ in competitive ability: phenotype-limited ideal free models. *Animal Behavior*, **34**, 1222–1242.

Parker, G.A., Baker, R.R. & Smith, V.C.F. (1972) The origin and evolution of gamete dimorphism and the male–female phenomenon. *Journal of Theoretical Biology* **36**, 529–553.

Parker, G.A., Royle, N.J. & Hartley, I.R. (2002) Intrafamilial conflict and parental investment: a synthesis. *Philosophical Transactions of the Royal Society of London. Series B*, **357**, 295–307.

Passera, L., Aron, S., Vargo, E.L. & Keller, L. (2001) Queen control of sex ratio in fire ants. *Science*, **293**, 1308–1310.

Pen, I. & Weissing, F. J. (2000) Sex ratio optimization with helpers at the nest. *Proceedings of the Royal Society of London, Series B*, **267**, 539–544.

Perrins, C.M. 1965. Population fluctuations and clutch size in the great tit. *Parus major*. L. *Journal of Animal Ecology*, **34**, 601–647.

Persson, O. & Öhström, P. (1989) A new avian mating system: ambisexual polygamy in the penduline tit *Remiz pendulinus*. *Ornis Scandinavica*, **20**, 105–111.

Peterson, C.C., Nagy, K.A. & Diamond, J. (1990). Sustained metabolic scope. *Proceedings of the National Academy of Sciences USA*, **87**, 2324–2328.

Petrie, M. (1994) Improved growth and survival of offspring of peacocks with more elaborate trains. *Nature*, **371**, 598–599.

Petrie, M. & Williams, A. (1993) Peahens lay more eggs for peacocks with larger trains. *Proceedings of the Royal Society of London, Series B*, **251**, 127–131.

Petrie, M., Halliday, T. & Sanders, C. (1991) Peahens prefer peacocks with elaborate trains. *Animal Behavior*, **41**, 323–331.

Pettifor, R.A., Perrins, C.M. & McCleery, R.H. (1988) Individual optimization of clutch size in great tits. *Nature*, **336**, 160–162.

Pfennig, D.W. & Collins, J.P. (1993) Kinship affects morphogenesis in cannibalistic salamanders. *Nature*, **362**, 836–838.

Pfennig, D.W., Collins, J.P. & Ziemba, R.E. (1999) A test of alternative hypotheses for kin recognition in cannibalistic tiger salamanders. *Behavioral Ecology*, **10**, 436–443.

Pfennig, D.W., Harcombe, W.R. & Pfennig, K.S. (2001) Frequency-dependent Batesian mimicry: predators avoid look-alikes of venemous snakes only when the real thing is around. *Nature*, **410**, 323.

Pianka, E.R. & Parker, W.S. (1975) Age-specific reproductive tactics. *American Naturalist*, **109**, 453–464.

Pietrewicz, A.T. & Kamil, A.C. (1979) Search image formation in the blue jay (*Cyanocitta cristata*). *Science*, **204**, 1332–1333.

Pietrewicz, A.T. & Kamil, A.C. (1981) Search images and the detection of cryptic prey: an operant approach. In: *Foraging Behavior: Ecological, Ethological and Psychological Approaches* (eds A.C. Kamil & T.D. Sargent). pp. 311–332. Garland STPM Press, New York.

Pizzari, T. & Birkhead, T.R. (2000) Female feral fowl eject sperm of subdominant males. *Nature*, **405**, 787–789.

Pizzari, T. & Foster, K.R. (2008) Sperm sociality: cooperation, altruism and spite. *PLoS Biology*, **6**, e130.

Pizzari, T. Cornwallis, C.K., Løvlie, H., Jakobsson, S. & Birkhead, T.R. (2003) Sophisticated sperm allocation in male fowl. *Nature*, **426**, 70–74.

Plowright, R.C., Fuller, G.A. & Paloheimo, J.E. (1989) Shell-dropping by Northwestern crows: a re-examination of an optimal foraging study. *Canadian Journal of Zoology*, **67**, 770–771.

van de Pol, M., Heg, D., Bruinzeel, L.W., Kuijper, B. & Verhulst, S. (2006) Experimental evidence for a causal effect of pair-bond duration on reproductive performance in oystercatchers (*Haematopus ostralegus*). *Behavioral Ecology*, **17**, 982–991.

Pomiankowski, A., Iwasa, Y. & Nee, S. (1991) The evolution of costly mate preferences. I. Fisher and biased mutation. *Evolution*, **45**, 1422–1430.

Pompilio, L., Kacelnik, A. & Behmer, S.T. (2006) State-dependent learned valuation drives choice in an invertebrate. *Science*, **311**, 1613–1615.

Powell, S. & Franks, N.R. (2007) How a few help all: living pothole plugs speed prey delivery in the army ant *Eciton burchellii*. *Animal Behaviour*, **73**, 1067–1076.

Prager, M. & Andersson, S. (2009) Phylogeny and evolution of sexually selected tail ornamentation in widowbirds and bishops (*Euplectes* spp.). *Journal of Evolutionary Biology*, **22**, 2068–2076.

Pratt, S.C. (2005) Quorum sensing by encounter rates in the ant *Temnothorax albipennis*. *Behavioral Ecology*, **16**, 488–496.

Pratt, S.C., Mallon, E.B., Sumpter, D.J.T. & Franks, N.R. (2002) Quorum sensing, recruitment, and collective decision-making during colony emigration by the ant *Leptothorax albipennis*. *Behavioral Ecology and Sociobiology*, **52**, 117–127.

Provine, W.B. (1971) *The Origins of Theoretical Population Genetics*. Chicago University Press, Chicago, IL.

Pravosudov, V.V. (1985) Search for and storage of food by *Parus cinctus lapponicus* and *P. montanus borealis* (Paridae) *Zoologicheskii Zhurnal*, **64**, 1036–1043.

Pravosudov, V.V. & Lucas, J.R. (2001) A dynamic model of short-term energy management in small food-caching and non-caching birds. *Behavioral Ecology*, **12**, 207–218.

Pravosudov, V.V. & Smulders, T.V. (2010) Integrating ecology, psychology and neurobiology within a food-hoarding paradigm. *Philosophical Transactions of the Royal Society of London. Series B*, **365**, 859–867.

Pravosudov, V.V., Kitasysky, A.S. & Ormanska, A. (2006) The relationship between migratory behaviour, memory and the hippocampus. *Proceedings of the Royal Society of London, Series B*, **273**, 2641–2649.

Pribil, S. (2000) Experimental evidence for the cost of polygyny in the red-winged blackbird *Agelaius phoeniceus*. *Behaviour* **137**, 1153–1173.

Pribil, S. & Picman, J. (1996) Polygyny in the red-winged blackbird: do females prefer monogamy or polygamy? *Behavioral Ecology and Sociobiology*, **38**, 183–190.

Pribil, S. & Searcy, W.A. (2001) Experimental confirmation of the polygyny threshold model for red-winged blackbirds. *Proceedings of the Royal Society of London, Series B*, **268**, 1643–1646.

Price, T. (2008) *Speciation in Birds*. Roberts and Co., Greenwood Village, CO.

Prins, H.H.T. (1996) *Ecology and Behaviour of the African Buffalo*. Chapman & Hall.

Prum, R.O. (1997) Phylogenetic tests of alternative intersexual selection mechanisms: trait macroevolution in a polygynous clade (Aves: Pipridae). *American Naturalist*, **149**, 668–692.

Pryke, S.R., Andersson, S. & Lawes, M.J. (2001) Sexual selection of multiple handicaps in red-collared widowbirds: female choice of tail length but not carotenoid display. *Evolution* **55**, 1452–1463.

Pryke, S.R., Andersson, S., Lawes, M.J. & Piper, S.E. (2002) Carotenoid status signalling in captive and wild red-collared widowbirds: independent effects of badge size and colour. *Behavioral Ecology*, **13**, 622–631.

Pulido, F. (2007) The genetics and evolution of avian migration. *BioScience*, **57**, 165–174.

Pulido, F., Berthold, P. & van Noordwijk, A.J. (1996) Frequency of migrants and migratory activity are genetically correlated in a bird population: evolutionary implications. *Proceedings of the National Academy of Sciences USA*, **93**, 14642–14647.

Pulliam, H.R. 1973. On the advantages of flocking. *Journal of Theoretical Biology* **38**, 419–422.

Purvis, A. (1995) A composite estimate of primate phylogeny. *Philosophical Transactions of the Royal Society of London. Series B*, **348**, 405–421.

Queller, D.C. (1989) The evolution of eusociality: reproductive head start of workers. *Proceedings of the National Academy of Sciences USA*, **86**, 3224–3226.

Queller, D.C. (1992) Quantitative genetics, inclusive fitness, and group selection. *American Naturalist*, **139**, 540–558.

Queller, D.C. (1994) Extended parental care and the origin of eusociality. *Proceedings of the Royal Society of London, Series B*, **256**, 105–111.

Queller, D.C. (1994) Genetic relatedness in viscous populations. *Evolutionary Ecology*, **8**, 70–73.

Queller, D.C. (1995) The spaniels of st. marx and the panglossian paradox: a critique of a rhetorical programme. *Quarterly Review of Biology*, **70**, 485–489.

Queller, D.C. (1996) The measurement and meaning of inclusive fitness. *Animal Behaviour*, **51**, 229–232.

Queller, D.C. (1997) Why do females care more than males? *Proceedings of the Royal Society of London, Series B*, **264**, 1555–1557.

Queller, D.C. (2000) Relatedness and the fraternal major transitions. *Philosophical Transactions of the Royal Society of London. Series B*, **355**, 1647–1655.

Queller, D.C. & Goodnight, K.F. (1989) Estimating relatedness using genetic markers. *Evolution*, **43**, 258–275.

Queller, D.C. & Strassmann, J.E. (1998) Kin selection and social insects. *BiocScience*, **48**, 165–175.

Queller, D.C., Ponte, E., Bozzaro, S. & Strassmann, J.E. (2003) Single-gene greenbeard effects in the social amoeba *Dictostelium discoideum*. *Science*, **299**, 105–106.

Queller, D.C., Zacchi, F., Cervo, R., *et al*. (2000) Unrelated helpers in a social insect. *Nature*, **405**, 784–787.

Raby, C.R., Alexis, D.M, Dickinson, A. & Clayton, N.S. (2007) Planning for the future by western scrub jays. *Nature*, **445**, 919–921.

Radford, A.N. & Ridley, A.R. (2008) Close calling regulates spacing between foraging competitors in the group-living pied babbler. *Animal Behavior*, **75**, 519–527.

Radford, A.N., Bell, M.B.V., Hollen, L.I. & Ridley, A.R. (2011) Singing for your supper: sentinel calling by kleptoparasites can mitigate the cost to victims. *Evolution*, **65**, 900–6.

Raihani, N.J., Grutter, A.S. & Bshary, R. (2010) Punishers benefit from third-party punishment in fish. *Science*, **327**, 171.

Rands, S.A., Cowlishaw, G., Pettifor, R.A., Rowcliffe, J.M. & Johnstone, R.A. (2003) Spontaneous emergence of leaders and followers in foraging pairs. *Nature*, **423**, 432–434.

Ratnieks, F.L.W. (1988) Reproductive harmony via mutual policing by workers in eusocial Hymenoptera. *American Naturalist*, **132**, 217–236.

Ratnieks, F.L.W. & Visscher, P.W. (1989) Worker policing in the honeybee. *Nature*, **342**, 796–797.

Ratnieks, F.L.W., Foster, K.R. & Wenseleers, T. (2006) Conflict resolution in insect societies. *Annual Review of Entomology*, **51**, 581–608.

Raymond, M., Pontier, D., Dufour, A-B. & Møller, A.P. (1996) Frequency-dependent maintenance of left-handedness in humans. *Proceedings of the Royal Society of London, Series B*, **263**, 1627–1633.

Read, A.F., Anwar, M., Shutler, D. & Nee, S. (1995) Sex allocation and population structure in malaria and related parasitic protozoa. *Proceedings of the Royal Society of London, Series B*, **260**, 359–363.

Read, A.F., Mackinnon, M.J., Anwar, M.A. & Taylor, L.H. (2002) Kin selection models as evolutionary explanations of malaria. In: *Adaptive Dynamics of Infectious Diseases: In Pursuit of Virulence Management* (eds U. Dieckmann, J.A.J. Metz, M.W. Sabelis & K. Sigmund), pp. 165–178 Cambridge University Press, Cambridge.

Réale, D., Reader, S.M., Sol, D., McDougall, P.T. & Dingemanse, N.J. (2007) Integrating animal temperament within ecology and evolution. *Biological Reviews*, **82**, 291–318.

Reboreda, J.C., Clayton, N.S. & Kacelnik, A. (1996) Species and sex differences in hippocampus size in parasitic and non-parasitic cowbirds. *Neuroreport*, **7**, 505–508.

Reby, D. & McComb, K. (2003) Anatomical constraints generate honesty: acoustic cues to age and weight in the roars of red deer stags. *Animal Behaviour,* **65**, 519–530.

Reby, D., McComb K., Cargnelutti B., Darwin C, Fitch W.T. & Clutton-Brock T.H. (2005) Red deer stags use formants as assessment cues during intrasexual agonistic interactions. *Proceedings of the Royal Society of London, Series B,* **272**, 941–947.

Reece, S.E., Drew, D.R. & Gardner, A. (2008) Sex ratio adjustment and kin discrimination in malaria parasites. *Nature,* **453**, 609–614.

Reeve, H.K. (1989) The evolution of conspecific acceptance thresholds. *American Naturalist,* **133**, 407–435.

Reeve, H.K. (2000) A transactional theory of within-groups conflict. *American Naturalist,* **155**, 365–382.

Reeve, H.K. & Keller, L. (1995) Partitioning of reproduction in mother-daughter versus sibling associations: a test of optimal skew theory. *American Naturalist,* **145**, 119–132.

Reeve, H.K. & Sherman, P.W. (1993) Adaptation and the goals of evolutionary research. *Quarterly Review of Biology,* **68**, 1–32.

Rendall, D., Owren, M.J. & Ryan, M.J. (2009) What do animal signals mean? *Animal Behaviour,* **78**, 233–240.

Reusch, T.B.H., Häberli, M.A., Aeschlimann, P.B. & Milinski, M. (2001) Female sticklebacks count alleles in a strategy of sexual selection explaining MHC polymorphism. *Nature,* **414**, 300–302.

Reynolds, J.D. (1987) Mating system and nesting biology of the red-necked phalarope *Phalaropus lobatus:* what constrains polyandry? *Ibis,* **129**, 225–242.

Reynolds, J.D., Goodwin, N.B. & Freckleton, R.P. (2002) Evolutionary transitions in parental care and live bearing in vertebrates. *Philosophical Transactions of the Royal Society of London. Series B,* **357**, 269–281.

van Rhijn, J.G. (1973) Behavioural dimorphism in male ruffs *Philomachus pugnax* (L.). *Behaviour,* **47**, 153–229.

Rice, W.R. (1996) Sexually antagonistic male adaptation triggered by experimental arrest of female evolution. *Nature,* **381**, 232–234.

Richardson, D.S., Burke, T. & Komdeur, J. (2002) Direct benefits and the evolution of female-biased cooperative breeding in Seychelles warblers. *Evolution,* **56**, 2313–2321.

Richardson, D.S., Burke, T. & Komdeur, J. (2003) Sex-specific associative learning cues and inclusive fitness benefits in the Seychelles warbler. *Journal of Evolutionary Biology,* **16**, 854–861.

Richardson, T.O., Sleeman, P.A., McNamara, J.M., Houston, A.I. & Franks, N.R. (2007) Teaching with evaluation in ants. *Current Biology,* **17**, 1520–1526.

Richner, H. & Heeb, P. (1996) Communal life: honest signalling and the recruitment centre hypothesis. *Behavioral Ecology,* **7**, 115–119.

Ridley, M. (1989) The cladistic solution to the species problems. *Biology and Philosophy,* **4**, 1–16.

Riechert, S.E. (1978) Games spiders play: behavioral variability in teritorial disputes. *Behavioral Ecology and Sociobiology,* **3**, 135–162.

Riechert, S.E. (1984) Games spiders play. III. Cues underlying context-associated changes in agonistic behavior. *Animal Behaviour,* **32**, 1–15.

Riipi, M., Alatalo, R.V., Lindström, L. & Mappes, J. (2001) Multiple benefits of gregariousness cover detectability costs in aposematic aggregations. *Nature,* **413**, 512–514.

Rippin, A.B. & Boag, D.A. (1974) Spatial organization among male sharp-tailed grouse on arenas. *Canadian Journal of Zoology,* **52**, 591–597.

Roberts, G. (1996) Why vigilance declines as group size increases. *Animal Behavior,* **51**, 1077–1086.

Robertson, B.C., Elliot, G.P., Eason, D.K., Clout, M.N. & Gemmell, N.J. (2006) Sex allocation theory aids species conservation *Biology Letters,* **2**, 229–231.

Robertson, K.A. & Monteiro, A. (2005) Female *Bicyclus anynana* butterflies choose males on the basis of their dorsal U-V reflective eyespot pupils. *Proceedings of the Royal Society of London, Series B,* **272**, 1541–1546.

Robinson, G.E., Fernald, R.D. & Clayton, D.F. (2008) Genes and social behaviour. *Science*, **322**, 896–900.

Robinson, S.K. (1986) The evolution of social behaviour and mating systems in the blackbirds (Icterinae). In: *Ecological Aspects of Social Evolution* (eds D.I. Rubenstein & R.W. Wrangham). pp. 175–200. Princeton University Press, Princeton, NJ.

Rodriguez-Girones, M.A. & Lotem, A. (1999) How to detect a cuckoo egg: a signal detection theory model for recognition and learning. *American Naturalist*, **153**, 633–648.

Rohwer, S. (1978) Parent cannibalism of offspring and egg raiding as a courtship strategy. *American Naturalist*, **112**, 429–440.

Rohwer, S. & Rohwer, F.C. (1978) Status signalling in Harris sparrows: experimental deceptions achieved. *Animal Behavior*, **26** 1012–1022.

Roper, T.J. & Redston, S. (1987) Conspicuousness of distasteful prey affects the strength and durability of one-trial avoidance learning. *Animal Behavior*, **35**, 739–747.

Rosenqvist, G. (1990) Male mate choice and female-female competition for mates in the pipefish *Nerophis ophidion*. *Animal Behavior*, **39**, 1110–1116.

Roth, T.C., II, Brodin. A., Smulders, T.V., LaDage, L.D. & Pravosudov, V.V. (2010) Is bigger always better? A critical appraisal of the use of volumetric analysis in the study of the hippocampus. *Philosophical Transactions of the Royal Society of London. Series B*, **365**, 915–931.

Roth, T.C., II., LaDage, L.D. & Pravosudov, V.V. (2011) Variation in hippocampal morphology along an environmental gradient: controlling for the effects of daylength. *Proceedings of the Royal Society of London, Series B*, **278**, 2662–2667.

Rothstein, S.I. (1982) Mechanisms of avian egg recognition: which egg parameters elicit responses by rejector species? *Behavioral Ecology and Sociobiology*, **11**, 229–39.

Rothstein, S.I. (1986) A test of optimality: egg recognition in the eastern phoebe. *Animal Behaviour*, **34**, 1109–1119.

Rothstein, S.I. (2001) Relic behaviours, coevolution and the retention versus loss of host defences after episodes of brood parasitism. *Animal Behavior*, **61**, 95–107.

Roughgarden, J. (2004) *Evolution's Rainbow: Diversity, Gender, and Sexuality in Nature and People*. University of California Press, Berkeley, CA.

Rowe, L. & Houle, D. (1996) The lek paradox and the capture of genetic variance by condition dependent traits. *Proceedings of the Royal Society of London, Series B*, **263**, 1415–1421.

Rowe, L.V., Evans, M.R. & Buchanan, K.L. (2001) The function and evolution of the tail streamer in hirundines. *Behavioral Ecology* **12**, 157–163.

Rowland, H.M., Cuthill, I.C., Harvey, I.F., Speed, M.P. & Ruxton, G.D. (2008) Can't tell the caterpillars from the trees: countershading enhances survival in a woodland. *Proceedings of the Royal Society of London, Series B*, **275**, 2539–2546.

Rowland, H.M., Ihalainen, E., Lindström, L., Mappes, J. & Speed, M.P. (2007) Co-mimics have a mutualistic relationship despite unequal defences. *Nature*, **448**, 64–67.

Royama, T. (1970) Factors governing the hunting behaviour and selection of food by the great tit, *Parus major*. *Journal of Animal Ecology*, **39**, 619–668.

Royle, N.J., Hartley, I.R. & Parker, G.A. (2002) Begging for control: when are offspring solicitation behaviours honest? *Trends in Ecology & Evolution*, **17**, 434–440.

Rubenstein, D.I. (1986) Ecology and sociality in horses and zebras. In *Ecological Aspects of Social Evolution* (eds D.I. Rubenstein & R.W. Wrangham). pp. 282–302. Princeton University Press, Princeton, NJ.

Rubenstein, D.R. & Lovette, I.J. (2007) Temporal variability drives the evolution of cooperative breeding in birds. *Current Biology*, **17**, 1414–1419.

Rubenstein, D.R. & Lovette, I.J. (2009) Reproductive skew and selection on female ornamentation in social species. *Nature*, **462**, 786–789.

Rumbaugh, K.P., Diggle, S.P., Watters, C.M., Ross-Gillespie, A., Griffin, A.S. & West, A.W. (2009) Quorum sensing and the social evolution of bacterial virulence. *Current Biology*, **19**, 341–345.

Russell, A.F. & Hatchwell, B.J. (2001) Experimental evidence for kin-biased helping in a cooperatively breeding vertebrate. *Proceedings of the Royal Society of London, Series B*, **268**, 2169–2174.

Russell, A. F. & Wright, J. (2009) Avian mobbing: byproduct mutualism not reciprocal altruism. *Trends in Ecology & Evolution*, **24**, 3–5.

Russell, A.F., Langmore, N.E., Cockburn, A., Astheimer, L.B. & Kilner, R.M. (2007) Reduced egg investment can conceal helper effects in cooperatively breeding birds. *Science*, **317**, 941–944.

Rutberg, A.T. (1983) The evolution of monogamy in primates. *Journal of Theoretical Biology*, **104**, 93–112.

Ruxton, G.D., Sherratt, T.N. & Speed, M.P. (2004) *Avoiding Atttack: The Evolutionary Ecology of Crypsis, Warning Signals and Mimicry*. Oxford University Press, Oxford.

Ryan, M.J., Tuttle, M.D. & Taft, L.K. (1981) The costs and benefits of frog chorusing behavior. *Behavioral Ecology and Sociobiology*, **8**, 273–278.

Ryan, M.J., Fox, J.H., Wikzynski, W. & Rand, A.S. (1990) Sexual selection for sensory exploitation in the frog. *Physalaemus pustulosus*. *Nature*, **343**, 66–68.

Sachs, J.L. & Wilcox, T.P. (2006) A shift to parasitism in the jellyfish symbiont *Symbiodinium microadriaticum*. *Proceedings of the Royal Society of London, Series B*, **273**, 425–429.

Sachs, J.L., Mueller, U.G., Wilcox, T.P. & Bull, J.J. (2004) The evolution of cooperation. *Quarterly Review of Biology*, **79**, 135–160.

Saether, S.A., Fiske, P. & Kalas, J.A. (2001) Male mate choice, sexual conflict and strategic allocation of copulations in a lekking bird. *Proceedings of the Royal Society of London, Series B*, **268**, 2097–2102.

Saino, N., Primmer, C., Ellegren, H. & Møller, A.P. (1997) An experimental study of paternity and tail ornamentation in the barn swallow (*Hirundo rustica*). *Evolution*, **51**, 562–570.

Santorelli, L.A., Thompson, C.R.L., Villegas, E., *et al.* (2008) Facultative cheater mutants reveal the genetic complexity of cooperation in social amoebae. *Nature*, **451**, 1107–1110.

Santos, J.C., Coloma, L.A. & Cannatella, D.C. (2003) Multiple, recurring origins of aposematism and diet specialization in poison frogs. *Proceedings of the National Academy of Sciences USA*, **100**, 12792–12797.

Sato, N.J., Tokue, K., Noske, R.A., Mikami, O.K. & Ueda, K. (2010) Evicting cuckoo nestlings from the nest: a new anti-parasitism behaviour. *Biology Letters*, **6**, 67–69.

Scarantino, A. (2010) Animal communication between information and influence. *Animal Behaviour*, **79**, e1-e5.

Schaller, G.B. (1972) *The Serengeti Lion*. University of Chicago Press, Chicago, IL.

Schlenoff, D.H. (1985) The startle responses of blue jays to *Catocala* (Lepidoptera: Noctuidae) prey models. *Animal Behavior*, **33**, 1057–1067.

Schluter, D., Price, T., Mooers, A.O. & Ludwig, D. (1997) Likelihood of ancestor states in adaptive radiation. *Evolution*, **51**, 1699–1711.

Schmid-Hempel, P. (1986) Do honeybees get tired? The effect of load weight on patch departure. *Animal Behavior*, **34**, 1243–1250.

Schmid-Hempel, P. & Wolf, T.J. (1988) Foraging effort and life span in a social insect. *Journal of Animal Ecology*, **57**, 509–522.

Schmid-Hempel, P., Kacelnik, A. & Houston, A.I. (1985) Honeybees maximise efficiency by not filling their crop. *Behavioral Ecology and Sociobiology*, **17**, 61–66.

Schoener, T.W. (1983) Simple models of optimal feeding-territory size: a reconciliation. *American Naturalist*, **121**, 608–629.

Schwabl, H. (1996) Maternal testosterone in the avian egg enhances postnatal growth. *Comparative Biochemistry and Physiology*, **114A**, 271–276.

Schwarz, M.P., Richards, M.H. & Danforth, B.N. (2007) Changing paradigms in insect social evolution: insights from halictine and allodapine bees. *Annual Review of Entomology*, **52**, 127–150.

Scott-Phillips, T.C. (2008) Defining biological communication. *Journal of Evolutionary Biology*, **21**, 387–395.

Scott-Phillips, T.C. (2010) Animal communication: insights from linguistic pragmatics. *Animal Behaviour,* **79**, e1-e4.

Scott-Phillips, T.C. & Kirby, S. (2010) Language evolution in the laboratory. *Trends in Cognitive Sciences,* **14**, 411–417.

Scott-Phillips, T.C., Dickins, T.E. & West, S.A. (2011) Evolutionary theory and the ultimate/proximate distinction in the human behavioural Sciences. *Perspectives on Psychological Science,* **6**, 38–47.

Searcy, W.A. (1988) Do female red-winged blackbirds limit their own breeding densities? *Ecology,* **69**, 85–95.

Searcy, W.A. & Nowicki, S. (2005) *The Evolution of Animal Communication.* Princeton University Press, Princeton, NJ.

Searcy, W.A. & Yasukawa, K. (1989) Alternative models of territorial polygyny in birds. *American Naturalist,* **134**, 323–343.

Seddon, N., Merrill, R.M. & Tobias, J.A. (2008) Sexually selected traits predict patterns of species richness in a diverse clade of suboscine birds. *American Naturalist,* **171**, 620–631.

Seeley, T.D. (1995) *The Wisdom of the Hive.* Harvard University Press, Cambridge, MA.

Seeley, T.D. (2003) Consensus building during nest-site selection in honeybee swarms: the expiration of dissent. *Behavioral Ecology and Sociobiology,* **53**, 417–424.

Seeley, T.D. & Buhrman, S.C. (2001) Nest-site selection in honeybees: how well do swarms implement the 'best of N' decision rule? *Behavioral Ecology and Sociobiology,* **49**, 416–427.

Seger, J. (1983) Partial bivoltinism may cause alternating sex-ratio biasses that favour eusociality. *Nature,* **301**, 59–62.

Selander, R.K. (1972) Sexual selection and dimorphism in birds. In *Sexual Selection and the Descent of Man* (ed. B. Campbell). pp. 180–230. Aldine, Chicago, IL.

Setchell, J.M. & Kappeler, P.M. (2003) Selection in relation to sex in primates. *Advances in the Study of Behavior,* **33**, 87–173.

Seyfarth, R.M. & Cheney, D.L. (1990) The assessment by vervet monkeys of their own and another species' alarm calls. *Animal Behavior,* **40**, 754–764.

Seyfarth, R.M., Cheney, D.L. & Marler, P. (1980) Monkey responses to three different alarm calls: evidence of predator classification and semantic communication. *Science,* **210**, 801–803.

Seyfarth, R.M., Cheney, D.L., Bergman, T., Fischer, J., Zuberbuhler, K. & Hammerschmidt, K. (2010) The central importance of information in studies of animal communication. *Animal Behaviour,* **80**, 3–8.

Shafir, S., Reich, T., Tsur, E., Erev, I. & Lotem, A. (2008) Perceptual accuracy and conflicting effects of certainty on risk-taking behaviour. *Nature,* **453**, 917–920.

Shapiro, J.A. (1998) Thinking about bacterial populations as multicellular organisms. *Annual Review of Microbiology,* **52**, 81–104.

Sharp, S.P., McGowan, A., Wood, M.J. & Hatchwell, B.J. (2005) Learned kin recognition cues in a social bird. *Nature,* **434**, 1127–1130.

Sheldon, B.C. (2000) Differential allocation: tests, mechanisms and implications. *Trends in Ecology & Evolution,* **15**, 397–402.

Sheldon, B.C. & Verhulst, S. (1996) Ecological immunity: costly parasite defences and trade-offs in evolutionary ecology. *Trends in Ecology & Evolution,* **11**, 317–321.

Sheldon, B.C. & West, S.A. (2004) Maternal dominance, maternal condition, and offspring sex ratio in ungulate mammals. *American Naturalist,* **163**, 40–54.

Sheldon, B.C., Merilä, J., Qvarnström, A., Gustafsson, L. & Ellegren, H. (1997) Paternal genetic contribution to offspring condition predicted by size of male secondary sexual character. *Proceedings of the Royal Society of London, Series B,* **264**, 297–302.

Sheldon, B.C., Andersson, S., Griffith, S.C., Ornborg, J. & Sendecka, J. (1999) Ultraviolet colour variation influences blue tit sex ratios. *Nature,* **402**, 874–877.

Shelly, T.E. (2001) Lek size and female visitation in two species of tephritid fruit flies. *Animal Behavior,* **62**, 33–40.

Sheppard, P.M. 1959. The evolution of mimicry: a problem in ecology and genetics. *Cold Spring Harbor Symposia in Quantitative Biology* **24**, 131–140.

Sherman, P.W. (1977) Nepotism and the evolution of alarm calls. *Science*, **197**, 1246–12453.

Sherman, P.W. (1981a) Reproductive competition and infanticide in Belding's ground squirrels and other animals. In *Natural Selection and Social Behaviour: Recent Research and New Theory* (eds R.D. Alexander & D.W. Tinkle). pp. 311–331. Chiron Press, New York.

Sherman, P.W. (1981b) Kinship, demography and Belding's ground squirrel nepotism. *Behavioral Ecology and Sociobiology*, **8**, 251–259.

Sherry, D.F. & Hoshooley, J.S. (2010) Seasonal hippocampal plasticity in food-storing birds. *Philosophical Transactions of the Royal Society of London. Series B*, **365**, 933–943.

Sherry, D.F. & Vaccarino, A.L. (1989) Hippocampus and memory for food caches in the black-capped chickadee. *Behavioral Neuroscience*, **103**, 308–318.

Sherry, D.F., Krebs, J.R. & Cowie, R.J. (1981) Memory for the location of stored food in marsh tits. *Animal Behavior*, **29**, 1260–1266.

Shettleworth, S.J. (2010a) *Cognition, Evolution and Behavior*, 2nd edn. Oxford University Press, New York.

Shettleworth, S.J. (2010b) Clever animals and killjoy explanations in comparative psychology. *Trends in Cognitive NeuroSciences*, **14**, 477–481.

Shine, R. (1999) Why is sex determined by nest temperature in many reptiles? *Trends in Ecology & Evolution*, **14**, 186–189.

Shuker, D.M. & West, S.A. (2004) Information constraints and the precision of adaptation: sex ratio manipulation in wasps. *Proceedings of the National Academy of Sciences USA*, **101**, 10363–10367.

Shuster, S.M. (1989) Male alternative reproductive strategies in a marine isopod crustacean (*Paracerceis sculpta*): the use of genetic markers to measure differences in fertilization success among alpha, beta, and gamma- males. *Evolution*, **43**, 1683–1698.

Shuster, S.M. & Sassamann, C. (1997) Genetic interaction between male mating strategy and sex ratio in a marine isopod. *Nature*, **388**, 373–377.

Shuster, S.M. & Wade, M.J. (1991) Equal mating success among male reproductive strategies in a marine isopod. *Nature*, **350**, 608–610.

Shuster, S.M. & Wade, M.J. (2003) *Mating Systems and Strategies*. Princeton University Press, Princeton, NJ.

Sibly, R.M. (1983) Optimal group size is unstable. *Animal Behavior*, **31**, 947–948.

Sih, A., Bell, A. & Johnson, J.C. (2004) Behavioural syndromes: an ecological and evolutionary overview. *Trends in Ecology & Evolution*, **19**, 372–378.

Silk, J.B. (2009) Nepotistic cooperation in non-human primate groups. *Philosophical Transactions of the Royal Society of London. Series B*, **364**, 3243–3254.

Silk, J.B. & Brown, G.R. (2008) Local resource competition and local resource enhancement shape primate birth sex ratios. *Proceedings of the Royal Society of London, Series B* **275**, 1761–1765.

Sillén-Tullberg, B. (1985) Higher survival of an aposematic than of a cryptic form of a distasteful bug. *Oecologia*, **67**, 411–415.

Sillén-Tullberg, B. (1988) Evolution of gregariousness in aposematic butterfly larvae: a phylogenetic analysis. *Evolution*, **42**, 293–305.

Simmons, L.W. (2001) *Sperm Competition and its Evolutionary Consequences in Insects*. Princeton University Press, Princeton, NJ.

Simmons, L.W. & Emlen, D.J. (2006) Evolutionary trade-off between weapons and testes. *Proceedings of the National Academy of Sciences USA*, **103**, 16346–16351.

Simmons, L.W., Parker, G.A. & Stockley, P. (1999a) Sperm displacement in the yellow dungfly, *Scatophaga stercoraria*: an investigation of male and female processes. *American Naturalist*, **153**, 302–314.

Simmons, L.W., Tomkins, J.L., Kotiaho, J.S. & Hunt. J. (1999b) Fluctuating paradigm. *Proceedings of the Royal Society of London, Series B*, **266**, 593–595.

Sinervo, B. & Lively, C.M. (1996) The rock-paper-scissors game and the evolution of alternative male strategies. *Nature*, **380**, 240–243.

Siva-Jothy, M.T. (1984) Sperm competition in the family Libellulidae (Anisoptera) with special reference to *Crocothemis erythraea* (Brulle) and *Orthetrum cancellatum* (L.). *Advances in Odonatology*, **2**, 195–207.

Skelhorn, J., Rowland, H.M. Speed, M.P. & Ruxton, G.D. (2010) Masquerade: camouflage without crypsis. *Science*, **327**, 51.

Smith, B.R. & Blumstein, D.T. (2008) Fitness consequences of personality: a meta-analysis. *Behavioral Ecology*, **19**, 448–455.

Smith, C. & Grieg, D. (2010) The cost of sexual signaling in yeast. *Evolution*, **64**, 3114–3122.

Smith, J.N.M., Yom-Tov, Y. & Moses, R. (1982) Polygyny, male parental care and sex ratios in song sparrows: an experimental study. *Auk*, **99**, 555–564.

Smith, S.M. (1977) Coral snake pattern rejection and stimulus generalisation by naïve great kiskadees (Aves: Tyrannidae). *Nature*, **265**, 535–536.

Smukella, S., M. Caldara, N. Pochet, A. Beauvais, S. Guadagnini, C. Yan, M.D. Vinces, A. Jansen, M.C. Prevost, J.-P. Latge G.R. Fink, K.R. Foster, and K.J. Verstrepen. (2008) *FLO1* is a variable green beard gene that drives biofilm-like cooperation in budding yeast. *Cell*, **135**, 726–737.

Sober, E. & Wilson, D.S. (1998) *Unto Others: The Evolution and Psychology of Unselfish Behavior*. Harvard University Press, Cambridge, MA.

Sokolowski, M.B., Pereira, H.S. & Hughes, K. (1997) Evolution of foraging behaviour in *Drosophila* by density-dependent selection. *Proceedings of the National Academy of Sciences USA*, **94**, 7373–7377.

Sorenson, M.D. & Payne, R.B. (2005) A molecular genetic analysis of cuckoo phylogeny. In: *The Cuckoos* (ed. R.B. Payne). pp. 68–94. Oxford University Press, Oxford.

Speed, M.P. & Ruxton, G.D. (2007) How bright and how nasty: explaining diversity in warning signal strength. *Evolution*, **61**, 623–635.

Spottiswoode, C.N. & Koorevaar, J. (2012) A stab in the dark: chick killing by brood parasitic honeyguides. *Biology Letters*, in press.

Spottiswoode, C.N. & Stevens, M. (2010) Visual modelling shows that avian host parents use multiple visual cues in rejecting parasitic eggs. *Proceedings of the National Academy of Sciences USA*, **107**, 8672–8676.

Spottiswoode, C.N. & Stevens, M. (2012) Host-parasite arms races and changes in bird egg appearance. *American Naturalist*, in press.

Squire, L.R. (2004) Memory systems of the brain: a brief history and current perspective. *Neurobiology of Learning and Memory*, **82**, 171–177.

Stander, P.E. (1992) Cooperative hunting in lions: the role of the individual. *Behavioral Ecology and Sociobiology*, **29**, 445–454.

Stearns, S.C. & Koella, J.C. (2007) *Evolution in health and disease*. Oxford University Press, Oxford.

Steger, R. & Caldwell, R.L. (1983) Intraspecific deception by bluffing: a defence strategy of newly molted stomatopods (Arthropoda: Crustacea). *Science*, **221**, 558–560.

Stenmark, G., Slagsvold, T. & Lifjeld, J.T. (1988) Polygyny in the pied flycatcher *Ficedula hypoleuca:* a test of the deception hypothesis. *Animal Behavior*, **36**, 1646–1657.

Stern, D.L. & Foster, W.A. (1996) The evolution of soldiers in aphids. *Biological Reviews*, **71**, 27–79.

Stern, K. & McClintock, M.K. (1998) Regulation of ovulation by human pheromones. *Nature*, **392**, 177–179.

Stevens, J. R. & Hauser, M. D. (2004) Why be nice? Psychological constraints on the evolution of cooperation. *Trends in Cognitive Sciences*, **8**, 60–65.

Stevens, M. & Merilaita, S. (eds) (2009) Animal camouflage: current issues and new perspectives. *Philosophical Transactions of the Royal Society of London. Series B*, **364**, 421–557.

Stevens, M., Cuthill, I.C., Windsor, A.M.M. & Walker, H.J. (2006) Disruptive contrast in animal camouflage. *Proceedings of the Royal Society of London, Series B*, **273**, 2433–2438.

Stevens, M., Hopkins, E., Hinde, W., *et al.* (2007) Field experiments on the effectiveness of 'eyespots' as predator deterrents. *Animal Behavior*, **74**, 1215–1227.

Stevens, M., Hardman, C.J. & Stubbins, C.L. (2008) Conspicuousness, not eye mimicry, makes 'eyespots' effective antipredator signals. *Behavioral Ecology*, **19**, 525–531.

Stewart, K.J. & Harcourt, A.H. (1994) Gorillas vocalizations during rest periods – signals of impending departure. *Behaviour*, **130**, 29–40.

Stoddard, M.C. & Stevens, M. (2011) Avian vision and the evolution of egg color mimicry in the common cuckoo. *Evolution*, **65**, 2004–2013.

Stokke, B.G. Moksnes, A. & Røskaft, E. (2002) Obligate brood parasites as selective agents for evolution of egg appearance in passerine birds. *Evolution*, **56**, 199–205.

Strassmann, J.E. & Queller, D.C. (2007) Insect societies as divided organisms: the complexities of purpose and cross-purpose. *Proceedings of the National Academy of Sciences USA*, **104**, 8619–8626.

Strassmann, J.E., Zhu, Y. & Queller, D.C. (2000) Altruism and social cheating in the social amoeba *Dictyostelium discoideum*. *Nature*, **408**, 965–967.

Stuart-Fox. D., Moussalli, A. & Whiting, M.J. (2008) Predator-specific camouflage in chameleons. *Biology Letters*, **4**, 326–329.

Summers, K., McKeon, C.S. & Heying, H. (2006) The evolution of parental care and egg size: a comparative analysis in frogs. *Proceedings of the Royal Society of London, Series B*, **273**, 687–692.

Sundström, L. (1994) Sex ratio bias, relatedness asymmetry and queen mating frequency in ants. *Nature*, **367**, 266–268.

Sundström, L. & Boomsma, J.J. (2000) Reproductive alliances and posthumous fitness enhancement in male ants. *Proceedings of the Royal Society of London, Series B*, **267**, 1439–1444.

Sundström, L., Chapuisat, M. & Keller, L. (1996) Conditional manipulation of sex ratios by ant workers: a test of kin selection theory. *Science*, **274**, 993–995.

Sutherland, W.J. (1985) Chance can produce a sex difference in variance in mating success and explain Bateman's data. *Animal Behaviour*, **33**, 1349–1352.

Sweeney, B.W. & Vannote, R.L. (1982) Population synchrony in mayflies: a predator satiation hypothesis. *Evolution*, **36**, 810–821.

Sword, G.A., Simpson, S.J., El Hadi, O.T.M. & Wilps, H. (2000) Density-dependent aposematism in the desert locust. *Proceedings of the Royal Society of London, Series B*, **267**, 63–68.

Szekely, T., Catchpole, C.K., De Voogd, A., Marchl, Z. & De Voogd, T.J. (1996) Evolutionary changes in a song control area of the brain (HVC) are associated with evolutionary changes in song repertoire among European warblers (*Sylviidae*). *Proceedings of the Royal Society of London, Series B*, **263**, 607–610.

Szentirmai, I., Szekely, T. & Komdeur, J. (2007) Sexual conflict over care: antagonistic effects of clutch desertion on reproductive success of male and female penduline tits. *Journal of Evolutionary Biology*, **20**, 1739–1744.

Taborsky, M. (1994) Sneakers, satellites and helpers: parasitic and cooperative behaviour in fish reproduction. *Advances in the Study of Behavior*, **23**, 1–100.

Takahashi, M., Arita, H., Hiraiwa-Hasegawa, M. & Hasegawa, T. (2008) Peahens do not prefer peacocks with more elaborate trains. *Animal Behavior*, **75**, 1209–1219.

Tallamy, D.W. (2000) Sexual selection and the evolution of exclusive paternal care in arthropods. *Animal Behavior*, **60**, 559–567.

Tanaka, K.D. & Ueda, K. (2005) Horsfield's hawk-cuckoo nestlings simulate multiple gapes for begging. *Science*, **308**, 653.

Taylor, A.H., Hunt, G.R., Medina, F.S. & Gray, R.D. (2009) Do New Caledonian crows solve physical problems through causal reasoning? *Proceedings of the Royal Society, Biological Sciences*, **276**, 247–254.

Taylor, A.H., Elliffe, D., Hunt, G.R. & Gray, R.D. (2010) Complex cognition and behavioural innovation in New Caledonian crows. *Proceedings of the Royal Society, Biological Sciences*, **277**, 2637–2643.

Taylor, P.D. (1981) Intra-sex and inter-sex sibling interactions as sex determinants. *Nature*, **291**, 64–66.

Taylor, P.D. (1992) Altruism in viscous populations – an inclusive fitness model. *Evolutionary Ecology*, **6**, 352–356.

Tebbich, S. & Bshary, R. (2004) Cognitive abilities related to tool use in the woodpecker finch, *Cactospiza pallida*. *Animal Behavior*, **67**, 689–697.

Thayer, G.H. 1909. *Concealing-Coloration in the Animal Kingdom: an exposition of the laws of disguise through color and pattern: being a summary of Abbott H. Thayer's discoveries*. Macmillan, New York.

Théry, M. & Casas, J. (2009) The multiple disguises of spiders: web colour and decorations, body colour and movement. *Philosophical Transactions of the Royal Society of London. Series B*, **364**, 471–480.

Thomas, A.L.R. & Rowe, L. (1997) Experimental tests on tail elongation and sexual selection in swallows (*Hirundo rustica*) do not affect the tail streamer and cannot test its function. *Behavioral Ecology*, **8**, 580–581.

Thomas, J.A. & Settele, J. (2004) Butterfly mimics of ants. *Nature*, **432**, 283–284.

Thomas, J.A., Simcox, D.J. & Clarke, R.T. (2009) Successful conservation of a threatened *Maculina* butterfly. *Science*, **325**, 80–83.

Thorne, B.L. (1997) Evolution of eusociality in termites. *Annual Review of Ecology and Systematics*, **28**, 27–54.

Thornhill, R. (1976) Sexual selection and nuptial feeding behaviour in *Bittacus apicalis* (Insecta: Mecoptera). *American Naturalist*, **110**, 529–548.

Thornhill, R. (1980) Rape in *Panorpa* scorpionflies and a general rape hypothesis. *Animal Behavior*, **28**, 52–59.

Thornhill, R. & Alcock, J. (1983) *The Evolution of Insect Mating Systems*. Harvard University Press, Cambridge, MA.

Thornton, A. & Malapert, A. (2009) Experimental evidence for social transmission of food aquisition techniques in wild meerkats. *Animal Behavior*, **78**, 255–264.

Thornton, A. & McAuliffe, K. (2006) Teaching in wild meerkats. *Science*, **313**, 227–229.

Thorogood, R., Ewen, J.G. & Kilner, R.M. (2011) Sense and sensitivity: responsiveness to offspring signals varies with the parents' potential to breed again. *Proceedings of the Royal Society of London, Series B*, **278**, 2638–2645.

Tibbetts, E.A. & Dale, J. (2004) A socially enforced signal of quality in a paper wasp. *Nature*, **432**, 218–222.

Tibbetts, E.A. & Lindsay, R. (2008) Visual signals of status and rival assessment in *Polistes dominulus* paper wasps. *Biology Letters*, **4**, 237–239.

Tinbergen, J.M. & Both, C. (1999) Is clutch size individually optimized? *Behavioral Ecology*, **10**, 504–509.

Tinbergen, J.M. & Daan, S. (1990) Family planning in the great tit (*Parus major*): optimal clutch size as integration of parent and offspring fitness. *Behaviour*, **114**, 161–190.

Tinbergen, L. (1960) The natural control of insects in pinewoods. I. Factors influencing the intensity of predation by song birds. *Archs. Neerl. Zool* **13**, 265–343.

Tinbergen, N. (1963) On aims and methods of ethology. *Zeitschrift für Tierpsychologie*, **20**, 410–433.

Tinbergen, N. (1974) *Curious Naturalists*. Penguin Education, Harmondsworth, UK.

Tinbergen, N., Broekhuysen, G.J., Feekes, F., Houghton, J.C.W., Kruuk, H. & Szulc, E. (1963) Egg shell removal by the black-headed gull, *Larus ridibundus* L.: a behaviour component of camouflage. *Behaviour*, **19**, 74–117.

Tinbergen, N., Impekoven, M. & Franck, D. (1967) An experiment on spacing out as a defence against predators. *Behaviour*, **28**, 307–21.

Tobias, J.A. & Seddon, N. (2009) Signal jamming mediates sexual conflict in a duetting bird. *Current Biology*, **19**, 1–6.

Todrank, J. & Heth, G. (2006) Crossed assumptions foster misinterpretations about kin recognition mechanisms. *Animal Behaviour*, **72**, e1-e3.

Tomkins, J.L. & Brown, G.S. (2004) Population density drives the local evolution of a threshold dimorphism. *Nature*, **431**, 1099–1103.

Tomkins, J.L. & Hazel, W. (2007) The status of the conditional evolutionarily stable strategy. *Trends in Ecology & Evolution*, **22**, 522–528.

Tóth, E. & Duffy, J. E. (2004) Coordinated group response to nest intruders in social shrimp. *Biology Letters*, **1**, 49–52.

Traulsen, A. & Nowak, M.A. (2006) Evolution of cooperation by multilevel selection. *Proceedings of the National Academy of Sciences USA*, **103**, 10952–10955.

Tregenza, T. & Wedell, N. (2002) Polyandrous females avoid costs of inbreeding. *Nature*, **415**, 71–73.

Tregenza, T. (1995) Building on the ideal free distribution. *Advances in Ecological Research*, **26**, 253–307.

Treherne, J.E. & Foster, W.A. (1980) The effects of group size on predator avoidance in a marine insect. *Animal Behavior*, **28**, 1119–1122.

Treherne, J.E. & Foster, W.A. (1981) Group transmission of predator avoidance behaviour in a marine insect: the Trafalgar effect. *Animal Behavior*, **29**, 911–917.

Trillmich, F. & Wolf, J.B.W. (2008) Parent-offspring and sibling conflict in Galapagos fur seals and sea lions. *Behavioral Ecology and Sociobiology*, **62**, 363–375.

Trivers, R.L. 1971. The evolution of reciprocal altruism. *Quarterly Review of Biology*, **46**, 35–57.

Trivers, R.L. (1972) Parental investment and sexual selection. In: *Sexual Selection and the Descent of Man* (ed. B. Campbell). pp. 139–179. Aldine, Chicago, IL.

Trivers, R.L. (1974) Parent–offspring conflict. *American Zoologist*, **14**, 249–264.

Trivers, R. (2000) The elements of a scientific theory of self-deception. *Annals of the New York Academy of Sciences*, **907**, 114–131.

Trivers, R. (2011) *Deceit and Self-Deception: Fooling yourself the better to fool others*. Allen Lane, Penguin.

Trivers, R.L. & Hare, H. (1976) Haplodiploidy and the evolution of social insects. *Science*, **191**, 249–263.

Trivers, R.L. & Willard, D.E. (1973) Natural selection of parental ability to vary the sex ratio of offspring. *Science*, **179**, 90–92.

Tullberg, B.S., Leimar, O. & Gamberale-Stille, G. (2000) Did aggregation favour the initial evolution of warning coloration? A novel world revisited. *Animal Behavior*, **59**, 281–287.

Úbeda, F. (2008) Evolution of genomic imprinting with biparental care?: implications for Prader-Willi and Angelman syndromes. *PLOS Biology*, **6**, 1678–1692.

Vahed, K., Parker, D.J. & Gilbert, J.D.J. (2011) Larger testes are associated with a higher level of polyandry, but a smaller ejaculate volume, across bushcricket species (*Tettigoniidae*). *Biology Letters*, in press.

van Valen, L. (1973) A new evolutionary law. *Evolutionary Theory*, **1**, 1–30.

Vallin, A., Jakobsson, S., Lind, J. & Wiklund, C. (2005) Prey survival by predator intimidation: an experimental study of peacock butterfly defence against blue tits. *Proceedings of the Royal Society of London, Series B*, **272**, 1203–1207.

VanderWall, S.B. (1990) *Food hoarding in animals*. Chicago University Press, Chicago, IL.

Vane-Wright, R.I., Raheem, D.C., Cieslak, A. & Vogler, A.P. (1999) Evolution of the mimetic African swallowtail butterfly *Papilio dardanus*: molecular data confirm relationships with *P. phorcas* and *P. constantinus*. *Biological Journal of the Linnean Society*, **66**, 215–229.

Vehrencamp, S.L. (1983) A model for the evolution of despotic versus egalitarian societies. *Animal Behavior*, **31**, 667–82.

Velicer, G.J., Kroos, L. & Lenski, R.E. (2000) Developmental cheating in the social bacterium *Myxococcus xanthus*. *Nature*, **404**, 598–601.

Verner, J. & Willson, M.F. 1966. The influence of habitats on mating systems of North American passerine birds. *Ecology*, **47**, 143–147.

Visscher, P.K. & Camazine, S. (1999) Collective decisions and cognition in bees. *Nature*, **397**, 400.

Visser, M.E. & Lessells, C.M. (2001) The costs of egg production and incubation in great tits (*Parus major*). *Proceedings of the Royal Society of London, Series B*, **268**, 1271–1277.

Visser, M.E., van Noordwijk, A.J., Tinbergen, J.M. & Lessells, C.M. (1998) Warmer springs lead to mistimed reproduction in great tits (*Parus major*). *Proceedings of the Royal Society of London, Series B*, **265**, 1867–1870.

Vogel, S., Ellington, C.P. & Kilgore, D.L. 1973. Wind-induced ventilation of the burrows of the prairie dog *Cynomys ludovicianus*. *Journal of Comparative Physiology*, **85**, 1–14.

Waage, J.K. (1979) Dual function of the damselfly penis: sperm removal and transfer. *Science*, **203**, 916–918.

Wade, M.J. (1979) Sexual selection and variance in reproductive success. *American Naturalist*, **114**, 742–7.

Wade, M.J. & Shuster, S.M. (2002) The evolution of parental care in the context of sexual selection: a critical reassessment of parental investment theory. *American Naturalist*, **160**, 285–292.

Ward, P. & Zahavi, A. (1973) The importance of certain assemblages of birds as 'information-centres' for food finding. *Ibis*, **115**, 517–534.

Ward, R.J.S., Cotter, S.C. & Kilner, R.M. (2009) Current brood size and residual reproductive value predict offspring desertion in the burying beetle *Nicrophorus vespilloides*. *Behavioral Ecology*, **20**, 1274–1281.

Warner, D.A. & Shine, R. (2008) The adaptive significance of temperature-dependent sex determination in a reptile. *Nature*, **451**, 566–568.

Warner, R.R. (1987) Female choice of sites versus mates in a coral reef fish *Thalassoma bifasciatum*. *Animal Behavior*, **35**, 1470–1478.

Warner, R.R. (1988) Traditionality of mating-site preferences in a coral reef fish. *Nature*, **335**, 719–721.

Warner, R.R. (1990) Male versus female influences on mating site determination in a coral reef fish. *Animal Behavior*, **39**, 540–548.

Warner, R.R., Robertson, D.R. & Leigh, E.G.J. (1975) Sex change and sexual selection. *Science*, **190**, 633–638.

Warner, R.R., Shapiro, D.Y., Marcanato, A. & Petersen, C.W. (1995) Sexual conflict: males with highest mating success convey the lowest fertilization benefits to females. *Proceedings of the Royal Society of London, Series B*, **262**, 135–139.

Watson, A. 1967. Territory and population regulation in the red grouse. *Nature*, **215**, 1274–1275.

Weatherhead, P.J. & Robertson, R.J. (1979) Offspring quality and the polygyny threshold: the 'sexy son hypothesis'. *American Naturalist*, **113**, 201–208.

Wedell, N., Gage, M.J.G. & Parker, G.A. (2002) Sperm competition, male prudence and sperm-limited females. *Trends in Ecology & Evolution*, **17**, 313–320.

Wehner, R. (1987) 'Matched filters' - neural models of the external world. *Journal of Comparative Physiology A*. **161**, 511–531.

Wells, K.D. (1977) The social behaviour of anuran amphibians. *Animal Behavior*, **25**, 666–693.

Wenseleers, T. & Ratnieks, F.L.W. (2006a) Comparative analysis of worker reproduction and policing in eusocial hymenoptera supports relatedness theory. *American Naturalist*, **168**, E163-E179.

Wenseleers, T. & Ratnieks, F. L. W. (2006b) Enforced altruism in insect societies. *Nature*, **444**, 50.

Werner, E.E., Gilliam, J.F., Hall, D.J. & Mittelbach, G.E. (1983) An experimental test of the effects of predation risk on habitat use in fish. *Ecology*, **64**, 1540–1548.

Werren, J.H. (1983) Sex ratio evolution under local mate competition in a parasitic wasp. *Evolution*, **37**, 116–124.

West, S.A. (2009) *Sex Allocation*. Princeton University Press, Princeton, NJ.

West, S.A. & Gardner, A. (2010) Altruism, spite and greenbeards. *Science*, **327**, 1341–1344.

West, S.A., Murray, M.G., Machado, C.A., Griffin, A.S. & Herre, E.A. (2001) Testing Hamilton's rule with competition between relatives. *Nature*, **409**, 510–513.

West, S.A., Griffin, A.S., Gardner, A. & Diggle, S.P. (2006) Social evolution theory for microbes. *Nature Reviews Microbiology*, **4**, 597–607.

West, S. A., Griffin, A. S. & Gardner, A. (2007a) Social semantics: altruism, cooperation, mutualism, strong reciprocity and group selection. *Journal of Evolutionary Biology*, **20**, 415–432.

West, S. A., Griffin, A. S. & Gardner, A. (2007b) Evolutionary explanations for cooperation. *Current Biology*, **17**, R661-R672.

West, S.A., Griffin, A.S. & Gardner, A. (2008) Social semantics: how useful has group selection been? *Journal of Evolutionary Biology*, **21**, 374–385.

Westneat, D.F. & Stewart, I.R.K. (2003) Extra-pair paternity in birds: causes, correlates and conflict. *Annual Review of Ecology and Systematics*, **34**, 365–396.

Wheeler, D.A., Kyriacou, C.P., Greenacre, M.L., *et al.* (1991) Molecular transfer of a species-specific behavior from *Drosophila simulans* to *Drosophila melanogaster*. *Science*, **251**, 1082–1085.

Wheatcroft, D. J. & Krams, I. (2009) Response to Russell and Wright: avian mobbing. *Trends in Ecology & Evolution*, **24**, 5–6.

Whiten, A., Goodall, J., McGrew, W.C., *et al.* (1999) Cultures in chimpanzees. *Nature*, **399**, 682–685.

Whiten, A., Horner, V. & de Waal, F.B.M. (2005) Conformity to cultural norms of tool use in chimpanzees. *Nature*, **437**, 737–740.

Whitfield, D.P. (1990) Individual feeding specializations of wintering turnstone *Arenaria interpres*. *Journal of Animal Ecology*, **59**, 193–211.

Whitham, T.G. (1978) Habitat selection by *Pemphigus* aphids in response to resource limitation and competition. *Ecology*, **59**, 1164–1176.

Whitham, T.G. (1979) Territorial behaviour of *Pemphigus* gall aphids. *Nature*, **279**, 324–325.

Whitham, T.G. (1980) The theory of habitat selection examined and extended using *Pemphigus* aphids. *American Naturalist*, **115**, 449–466.

Wickler, W. (1985) Coordination of vigilance in bird groups: the 'watchman's song' hypothesis. *Zeitschrift für Tierpsychologie*, **69**, 250–253.

Wiens, J.J. (2001) Widespread loss of sexually selected traits: how the peacock lost its spots. *Trends in Ecology & Evolution*, **16**, 517–523.

Wiley, R.H. 1973. Territoriality and non-random mating in the sage grouse. *Centrocercus urophasianus. Animal Behavior* (Monograph), **6**, 87–169.

Wilkinson, G.S. (1984) Reciprocal food sharing in the vampire bat. *Nature*, **308**, 181–184.

Wilkinson, G.S. & Reillo, P.R. (1994) Female choice response to artificial selection on an exaggerated male trait in a stalk-eyed fly. *Proceedings of the Royal Society of London, Series B*, **255**, 1–6.

Williams, G.C. (1966a) *Adaptation and Natural Selection*. Princeton University Press, Princeton, NJ.

Williams, G.C. (1966b) Natural selection, the costs of reproduction, and a refinement of Lack's principle. *American Naturalist*, **100**, 687–690.

Williams, G.C. (1975) *Sex and Evolution*. Princeton University Press, Princeton, NJ.

Williams, G.C. & Nesse, R.M. (1991) The dawn of darwinian medicine. *Quarterly Review of Biology*, **66**, 1–22.

Williams, P., Winzer, K., Chan, W. & Cámara, M. (2007) Look who's talking: communication and quorum sensing in the bacterial world. *Philosophical Transactions of the Royal Society of London. Series B*, **362**, 1119–1134.

Wilson, D.S. (2008) Social semantics: towards a genuine pluralism in the study of social behaviour. *Journal of Evolutionary Biology*, **21**, 368–373.

Wilson, D.S. & Wilson, E.O. (2007) Rethinking the theoretical foundation of sociobiology. *Quarterly Review of Biology*, **82**, 327–348.

Wilson, E.O. (1971) *The Insect Societies*. Belknap Press, Cambridge, MA.

Wilson, E.O. (1975) *Sociobiology*. Harvard University Press, Cambridge, MA.

Wilson, E.O. & Hölldobler, B. (2005) Eusociality: origin and consequences. *Proceedings of the National Academy of Sciences USA*, **102**, 13367–13371.

Wilson, E.O. & Hölldobler, B. (2009) *The Superorganism*. W.W. Norton, London.

Wilson, K. (1994) Evolution of clutch size in insects: II A test of static optimality models using the beetle *Callosobruchus maculatus* (Coleoptera: Bruchidae). *Journal of Evolutionary Biology*, **7**, 365–386.

Wolf, L., Ketterson, E.D. & Nolan, V., Jr (1990) Behavioural response of female dark-eyed juncos to experimental removal of their mates: implications for the evolution of male parental care. *Animal Behavior*, **39**, 125–134.

Wolf, M., van Doorn, G.S., Leimar O & Weissing, F.J. (2007) Life-history trade-offs favour the evolution of animal personalities. *Nature*, **447**, 581–584.

Wolf, T.J. & Schmid-Hempel, P. (1989) Extra loads and foraging lifespan in honeybee workers. *Journal of Animal Ecology*, **58**, 943–954.

Wong, M.Y.L., Buston, P., Munday, P.L. & Jones, G.P. (2007) The threat of punishment enforces peaceful cooperation and stabilises queues in a coral-reef fish. *Proceedings of the Royal Society of London, Series B*, **274**, 1093–1099.

Wong, M.Y.L., Munday, P.L., Buston, P.M. & Jones, G.P. (2008) Fasting or feasting in a fish social hierarchy. *Current Biology*, **18**, R372-R373.

Woyciechowski, M. & Lomnicki, A. (1987) Multiple mating of queens and the sterility of workers among eusocial Hymenoptera. *Journal of Theoretical Biology*, **128**, 317–327.

Wright, J., Maklakov, A.A. & Khazin, V. (2001) State-dependent sentinels: an experimental study in the Arabian babbler. *Proceedings of the Royal Society of London, Series B*, **268**, 821–826.

Wright, J., Stone, R.E. & Brown, N. (2003) Communal roosts as structured information centres in the raven, *Corvus corax*. *Journal of Animal Ecology*, **72**, 1003–1014.

Wynne-Edwards, V.C. (1962) *Animal Dispersion in Relation to Social Behaviour*. Oliver & Boyd, Edinburgh.

Wynne-Edwards, V.C. (1986) *Evolution Through Group Selection*. Blackwell Scientific Publications, Oxford.

Yom-Tov, Y. (1980) Intraspecific nest parasitism in birds. *Biol. Rev.* **55**, 93–108.

Young, A.J. & Clutton-Brock, T.H. (2006) Infanticide by subordinates influences reproductive sharing in cooperatively breeding meerkats. *Biology Letters*, **2**, 385–387.

Young, A.J., Carlson, A.A., Monfort, S.L., Russell, A.F., Bennett, N.C. & Clutton-Brock, T. (2006) Stress and the suppression of subordinate reproduction in cooperatively breeding meerkats. *Proceedings of the National Academy of Sciences USA*, **103**, 12005–12010.

Zach, R. (1979) Shell dropping: decision making and optimal foraging in Northwestern crows. *Behaviour*, **68**, 106–117.

Zahavi, A. (1975) Mate selection – a selection for a handicap. *Journal of Theoretical Biology*, **53**, 205–14.

Zahavi, A. (1977) The cost of honesty (further remarks on the handicap principle). *Journal of Theoretical Biology*, **67**, 603–605.

Zamudio, K.R. & Sinervo, B. (2000) Polygyny, mate-guarding and posthumous fertilization as alternative male mating strategies. *Proceedings of the National Academy of Sciences USA*, **97**, 14427–14432.

Zeh, D.W. & Smith, R.L. (1985) Paternal investment by terrestrial arthropods. *American Zoologist*, **25**, 785–805.

Zuberbühler, K. (2009) Survivor signals: the biology and psychology of animal alarm calling. *Advances in the Study of Behavior*, **40**, 277–322.

van Zweden, J.S. Brask, J.B., Christensen, J.H., Boomsma, J.J., Linksvayer, T.A. & d'Ettorre, P. (2010) Blending of heritable recognition cues among ant nestmates creates distinct colony gestalt odours but prevents within-colony nepotism. *Journal of Evolutionary Biology*, **23**, 1498–1508.

Index

Photo © Joseph Tobias

Note: Page reference in *italics* refer to Figures and Tables; those in **bold** refer to Boxes